The Origins of Musicality

The Origins of Musicality

edited by Henkjan Honing

The MIT Press
Cambridge, Massachusetts
London, England

This book was set in Syntax LT Std and Times New Roman by Toppan Best-set Premedia Limited. Printed and bound in the United States of America.

Library of Congress Cataloging-in-Publication Data

Names: Honing, Henkjan.
Title: The origins of musicality / edited by Henkjan Honing ; foreword by W. Tecumseh Fitch.
Description: Cambridge, MA : MIT Press, [2018] | Includes bibliographical references and index.
Identifiers: LCCN 2017033345 | ISBN 9780262037457 (hardcover : alk. paper)
9780262538510 (pb.)
Subjects: LCSH: Music--Origin. | Musical ability.
Classification: LCC ML3800 .O744 2018 | DDC 781.1/1—dc23 LC record available at
https://lccn.loc.gov/2017033345

10 9 8 7 6 5 4 3 2

Contents

Foreword

The past two decades have been an exciting period for those interested in the cognitive and biological underpinnings of human musicality—an excitement well captured by the chapters in this book. The years since 2000 have witnessed increasing interdisciplinary synergy among musicology, ethnomusicology, psychology, neuroscience, and biology, leading to a nascent field of biomusicology that spans all of these disciplines. The chapters in this book, by leading international experts in these fields, offer an up-to-date snapshot of this exciting and broad field of scientific research.

A fundamental insight helping to drive this progress, driven largely by work in the psychology and neuroscience of music, has been the increasing realization that everyone is a skilled and sophisticated musical listener, even those who consider themselves to be "unmusical" or do not themselves produce music. In fact, abundant research now demonstrates that even musically untrained individuals have detailed, implicit knowledge of the musical forms and styles of their culture. In the same way that we all unconsciously know the structure and rules of our native language even if we cannot express them explicitly, we all have an implicit understanding of the melodic, rhythmic, and harmonic regularities of our cultures' musics. Musicality—the underlying capacity to acquire this knowledge—is thus part of our universal birthright as human beings. This realization has fueled a welcome extension of empirical research beyond the confines of traditional Western art music to include not just the art music of many other cultures, but also popular folk and dance music and, indeed, an increasing willingness to include dance itself as one important facet of human musicality.

Progress in understanding human musicality has been made on all fronts. Perhaps the most striking has been the meteoric rise of neuroscientific investigations of music, driven by the phenomenal increase in the power and availability of neuroimaging methods. Although the psychology of music has long been an active and sophisticated subfield, it has recently extended its scope from adults into childhood and infancy to show that even newborn infants are musically sensitive and rapidly develop expertise in their cultures' musics in the first few years of life. Ethnomusicologists have gradually overcome a nearly sixty-year aversion to comparing the musics of different cultures (a pause broken only by Alan Lomax's groundbreaking "cantometrics" research in the 1970s) and resumed the search for characteristics of human music that are universal, or nearly so—an important foundation for any biological characterization of human musicality.

From my viewpoint as a biologist, investigations of animal analogues to human music have been particularly exciting as an increasing number of researchers have embraced an explicitly comparative approach to understanding the biological basis and evolutionary history of musicality. This line of research has revealed detailed commonalities of bird or whale "song" with our own music (in this case, convergently evolved), as well as some specific homologs such as the drumming behavior regularly engaged in by our nearest relatives, the chimpanzees and gorillas. This comparative research shows that although music itself may be specifically human, some of the fundamental mechanisms that underlie human musicality are shared with other species.

This book was born at a week-long international workshop on the cognitive and biological basis of musicality organized by Henkjan Honing, held in Leiden at the Lorentz Center in 2014, a workshop I was privileged to attend. The remarkable open-minded, collegial, and fearlessly interdisciplinary atmosphere that developed at this workshop is something I will never forget. This workshop led to a series of articles published in a special issue of *Philosophical Transactions of the Royal Society* in 2015. Most of the chapters in this book are reworked and revised versions of those articles, updated to reflect the latest research. Readers eager to learn more about human musicality and its biological basis will find no better starting point than this book.

W. Tecumseh Fitch
Vienna, April 2017

Preface

Why do we have music? What is music for, and why does every human culture have it? Is it a uniquely human capability, as language is? Or are some of its core components also present in nonhuman animals? And what biological and cognitive mechanisms are essential for perceiving, appreciating, and making music?

The search for a possible answer to these and other questions forms the backdrop of this book that brings together a collection of papers written by contributors to an interdisciplinary workshop, "What Makes Us Musical Animals? Cognition, Biology and the Origins of Musicality," held at the Lorentz Center, Leiden, the Netherlands, April 7–11, 2014. The workshop was made possible through a generous Distinguished Lorentz Fellowship granted to me by the Netherlands Institute for Advanced Study in the Humanities and Social Sciences (NIAS) and the Lorentz Center for the Sciences. During this workshop, it became clear that reframing the empirical evidence from a variety of fields and proposing a research agenda on musicality were both important and timely. Moreover, the twenty-three contributing experts from a wide range of disciplines (cognitive biology, cognitive neuroscience, neurobiology, animal cognition, ethnomusicology, molecular genetics, anthropology, developmental psychology, and computational cognition) agreed on a list of key questions for a future agenda on musicality, providing the momentum for this book (see table P.1).

Together the chapters set a research agenda for the study of musicality in the years to come, an endeavor that is multidisciplinary, as is the background of the authors. The topics of the fourteen, mostly coauthored chapters resulted from a bottom-up selection process during the Lorentz workshop, prompted by a series of position statements and reviews. These topics formed the basis of working sessions in which the key ingredients of the chapters were formulated.

The immediate outcome of the workshop was published in a theme issue on musicality in *Philosophical Transactions of the Royal Society B*, coedited with Carel ten Cate, Isabelle Peretz, and Sandra Trehub (Honing, ten Cate, Peretz, & Trehub, 2015). This book includes, next to updated and extended versions of the original papers, new theoretical material,

Table P.1
Some key questions for a future research agenda on musicality

1. What is the most promising means of carving musicality into component skills?
2. What kinds of natural behavior in other species might be related to musicality?
3. How can we more clearly differentiate biological and cultural contributions to musicality?
4. What is the neuronal circuitry associated with different aspects of musicality?
5. How do the relevant genes contribute to building a musical brain (i.e., using functional studies to bridge the gap between genes, neurons, circuits, and behavior)?
6. Can we use such genes to trace the evolutionary history of our musical capacities in human ancestors and study parallels in nonhuman animals?
7. Can nonhuman animals detect higher-order patterns in sounds (e.g., auditory grouping), as humans do?
8. Is entrainment or beat induction restricted to species capable of vocal learning?
9. Can nonhuman animals generalize across timbres?
10. Do absolute and relative processing of pitch, duration, and timbre depend on context, stimuli, or species?
11. How can we study the evolution of musicality relative to language?

novel empirical results, as well as a historical chapter that positions these recent developments in the relatively long history of studying the origins of musicality. The chapters are grouped in four parts. Part I presents an outline of a research program on musicality and discusses the pitfalls and prospects of studying the evolution of music. Part II consists of four position statements on the origins of musicality. Part III presents five reviews on the origins of musicality from the perspective of anthropology, (ethno)musicology, paleontology, behavioral biology, cognitive neuroscience, neurobiology, and genetics. The chapters in part IV review the computational modeling of animal song and creativity, relating it to large body of empirical work from the field of biology. This is complemented by a chapter on affect in relation to behavioral biology. The book closes with a case study on the history of musicality research.

The Lorentz workshop on musicality—and hence this book—would not have been possible without the help of many. First, special thanks go to Carel ten Cate, W. Tecumseh Fitch, Isabelle Peretz, and Sandra Trehub, members of the scientific organizing committee of the Lorentz workshop, for their expertise, guidance, and enthusiasm that resulted in an inspiring and exceptionally productive meeting. I am very grateful to NIAS and the Lorentz Center for the Sciences that—through a Distinguished Lorentz Fellowship—facilitated several research activities on musicality in the academic year 2013–14. Thanks to all staff at both the Lorentz Center and the NIAS for their support and contributions to the fellowship and the associated workshop, particularly Aafke Hulk, Arjen Doelman, Mieke Schutte, and Petry Kievit. I also thank several students who served in different roles during the workshop, especially Merwin Olthof and Fleur Bouwer. Very special thanks go

Figure P.1
Participants of the 2014 Lorentz Workshop "What Makes Us Musical Animals? Cognition, Biology and the Origins of Musicality." Top row, from left to right: Carel ten Cate, Willem Zuidema, Hugo Merchant, Marisa Hoeschele. Middle row: Simon E. Fisher, Yukiko Kikuchi, David Huron, Laurel J. Trainor, Martin Rohrmeier, Judith Becker, Jessica Grahn, Yuko Hattori, Bruno Gingras, Geraint A. Wiggins. Bottom row: Isabelle Peretz, W. Tecumseh Fitch, Ani Patel, Björn Merker, Henkjan Honing, Iain Morley, Sandra E. Trehub. Not in Picture: Peter Tyack, Constance Scharff, and Julia Kursell. Photographer: Merwin Olthof.

to Helen Eaton, commissioning editor of *Philosophical Transactions of the Royal Society B*, for the kind invitation to edit a theme issue on musicality (Honing et al., 2015) on which this book is based.

With regard to this book, I am very grateful to all referees for their time and effort in critically evaluating the draft chapters, particularly Simon E. Fisher, Björn Merker, Mark Steedman, W. Tecumseh Fitch, and Andrea Ravignani. I thank the Netherlands Organization for Scientific Research for its support through a Horizon grant (317-70-010). Iza Korsmit is warmly thanked for her editorial help in preparing the manuscript. Thanks also to Bob Prior of the MIT Press for his continued support during the preparation of this book. Finally, I thank all of the chapter authors, including W. Tecumseh Fitch for his foreword.

I hope this book, and the chapters that are part of it, will contribute to formation of a field bound for future growth: the interdisciplinary study of musicality.

Henkjan Honing
Amsterdam, March 2017

Reference

Honing, H., ten Cate, C., Peretz, I., & Trehub, S. E. (2015). Without it no music: Cognition, biology and evolution of musicality. *Philosophical Transactions of the Royal Society of London, Series B: Biological Sciences, 370*(1664). doi:10.1098/rstb.2014.0088

I

INTRODUCTION

1

Musicality as an Upbeat to Music: Introduction and Research Agenda

Henkjan Honing

Over the years, it has become clear that all humans share a predisposition for music, just like we have for language. We all can perceive and enjoy music. This view is supported by a growing body of research from developmental psychology (Trainor, 2008; Trehub, 2003), cognitive biology (Fitch, 2006; Wilson & Cook, 2016), neuroscience (Peretz & Coltheart, 2003; Peretz & Zatorre, 2005; Zatorre, 2005), and the many contributions from the field of music cognition (Deutsch, 2013; Hallam, Cross, & Thaut, 2009). These studies indicate that our capacity for music has an intimate relationship with our cognition and underlying biology, which is particularly clear when the focus is on perception rather than production (Honing, 2013).

Until relatively recently, most scholars were wary of the notion that music cognition could have a biological basis. Instead, music was viewed as a cultural product with no evolutionary history and no biological constraints on its manifestation (e.g., Repp, 1991). Such a view is indicative of a Western perspective on music, in which music is viewed as the preserve of professional musicians who have honed their skills through years of practice (Blacking, 1973). Obviously such notions do not explain the presence of music in all cultures and time periods, let alone other species. There is increasing evidence that all humans, not just highly trained individuals, share a predisposition for music in the form of musicality—defined as a spontaneous developing set of traits based on and constrained by our cognitive abilities and their underlying biology. To recognize a melody and perceive the beat of music are trivial skills for most humans, but they are in fact fundamental features of our musicality (Honing, 2012; Trehub, 2003). Even infants are sensitive to such features, which are common across cultures (Savage, Brown, Sakai, & Currie, 2015; Trehub, 2015). Though we are learning more and more about our own musicality (Deutsch, 2013; Fitch, 2006; Honing, 2013), the cognitive and biological mechanisms underlying it remain unclear.

The aim of this book is to identify the cognitive, biological, and mechanistic underpinnings for melodic and rhythmic cognition as key ingredients of musicality, assess to what extent these are unique to humans, and by doing so provide insight in their biological origins. Our aspiration as editor and chapter authors is that this book will lay a new interdisciplinary and comparative foundation for the study of musicality.

The key ingredients for this endeavor were generated at the Lorentz workshop, "What Makes Us Musical Animals?" that was held in April 2014 in Leiden, the Netherlands (see the preface to this volume for more details). This workshop aimed to assemble the key ingredients for a research agenda on musicality and thus provided the impetus for this book as a whole.

Before outlining this research agenda and relating its key ingredients to the chapters that follow, I present some preliminaries on the interdisciplinary field of musicality.

Music versus Musicality

Within any given culture, people agree, more or less, on what constitutes music. However, there is considerably less agreement across cultures (see chapter 6, this volume). Venturing across species caused even more debate. Though some people agree that the songs of birds, humpback whales, a Thai elephant orchestra, or the interlocking duets of gibbons are instances of music (Wallin, Merker, & Brown, 2000), others would argue instead that human listeners use a musical frame of reference that makes many things seem musical (Honing, 2013).

Addressing these issues productively requires a distinction between the notions of music and musicality (Honing & Ploeger, 2012; Morley, 2013; Trehub, 2003). *Musicality* in all its complexity can be defined as a natural, spontaneously developing set of traits based on and constrained by our cognitive and biological system. *Music* in all its variety can be defined as a social and cultural construct based on that very musicality. As such, the study of the origins of music is conditional on the study of musicality.

One approach in the investigation of the origins of music is to study the structure of music, seeking key similarities and differences of musical form and activity in a variety of human cultures (Brown & Jordania, 2011; Brown et al., 2014; Lomax & Berkowitz, 1972; Nettl, 2006; Savage et al., 2015). Although there is no widely shared definition of music, the presence of several cross-cultural similarities supports the notion of musicality as a prominent characteristic of humankind. These similarities are suggestive of underlying cognitive and biological mechanisms that may constrain and shape musical behaviors across cultures (Savage et al., 2015; Trehub, 2015).

An alternative approach is to study the structure of musicality by attempting to identify the basic underlying cognitive and biological mechanisms, their function and developmental course, and effective ways to study these mechanisms in human and nonhuman animals (Honing, ten Cate, Peretz, & Trehub, 2015). The major challenge of this approach, and hence of this book, is to delineate the traits that constitute the musicality phenotype.

A Multicomponent Approach to Musicality

Certain cognitive capacities such as language and music are viewed as typically human. However, we still know very little about whether other species share one or more of the basic mechanisms that constitute musicality. Revealing such common mechanisms requires employing a bottom-up perspective focusing on the constituent capacities underlying musicality. Instead of asking which species are musical, we should ask how musicality works, what the necessary components of musicality are, which ones are shared with other species, and how they evolved. Such an approach has resulted in important insights into the domains of animal cognition (Call & Tomasello, 2008; de Waal & Ferrari, 2010, 2012) and the evolution of language (Fitch, 2010). Interestingly, all participants of the 2014 workshop on musicality quickly agreed on the importance of a multicomponent approach (see also chapter 2, this volume).

A multicomponent perspective aims to combine functional, developmental, phylogenetic, and mechanistic approaches in order to generate an integrated theory of musicality while focusing on the constituent capacities underlying the musicality phenotype. This can be addressed by using the four explanatory levels that Tinbergen (1963) posited, describing the mechanisms, functions, and developmental course of musicality in a variety of animals and cultures, with input from anthropological, neuroscientific and genetic sources, and consequently examining in greater depth how music evolved.

This multicomponent approach is based on the neo-Darwinian assumption that if closely related species (e.g., humans and apes or walruses and sea lions) exhibit similar solutions to similar problems, they are probably engaging similar mechanisms (see figure 1.1). When two such species share a particular musical trait, one can infer that their common ancestor also had that trait. By examining these homologous traits in a natural group of species (i.e., clade), one can date the origin of that particular trait. Species that are closely related to humans can be assumed to share some cognitive abilities and may therefore be good models for experiments to tease apart various neurological, genetic, or epigenetic contributions to a certain musical trait. This is the principal motivation for studying musicality in closely related species (see chapters 7 and 8, this volume). The most promising group for such comparative research are monkeys (e.g., macaques or marmosets), a well-known animal model that, because of its homology, has been instrumental in understanding human brain function in domains ranging from audition and vision (Mendoza & Merchant, 2014) to language and music (see chapter 8, this volume; Rilling et al., 2008).

The study of more distant or unrelated species that share a similar trait (i.e., not homologous) can also contribute to an understanding of underlying mechanisms. The convergent evolution of particular traits in distant species (analogous trait; a form of homoplasy) will therefore allow identification of the biological constraints or mechanisms required for that trait, as well as yielding hypotheses for the selection pressures giving rise to it (see figure 1.2). Birds are arguably the most promising group for such comparative research. Many

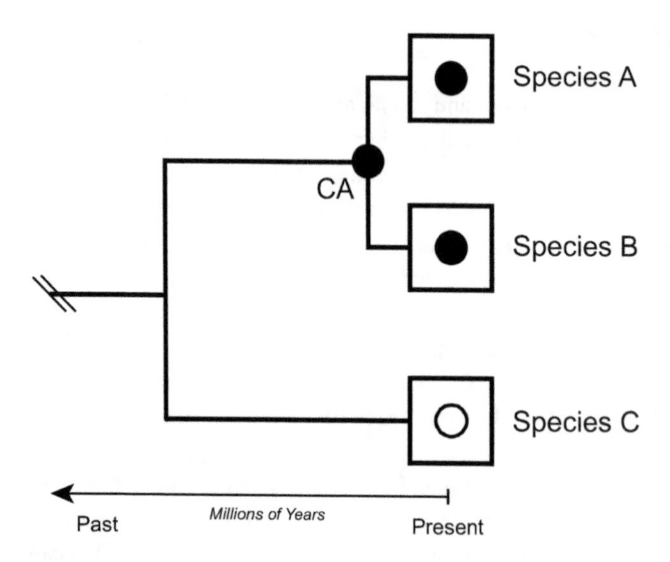

Figure 1.1
Neo-Darwinian perspective on the evolution of musicality. Diagrammatic representation of a hypothetical phylogenetic tree illustrating the Darwinian assumption that closely related species share similar traits. When two species (A and B) share a certain musical trait, one can infer that their common ancestor (CA) also had that trait (referred to as a homologous trait). Filled circles represent a trait; open circles indicate the absence of that trait.

bird species produce vocalizations characterized by brief elements ("notes") of varying complexity and frequency and often structured with rhythmic regularity. Various bird species (e.g., starlings, zebra finches, budgerigars) are extensively studied for their auditory perceptual abilities, and studying their perception of musically inspired stimuli (Spierings & ten Cate, 2014; ten Cate, Spierings, Hubert, & Honing, 2016) has contributed a lot to current ideas about the apparent uniqueness of human musicality.

By combining these two approaches—comparing humans with genetically close and genetically distant species—we can study musicality as a composite of several traits, each with its own underlying neurocognitive mechanisms and evolutionary history (see figure 1.3). We can study this by addressing some critiques regarding the study of the evolution of music cognition later in this chapter.

Core Components of Musicality

Potential candidates for the basic components of musicality that have been proposed in the recent literature are relative pitch (e.g., contour and interval analysis; Justus & Hutsler, 2005; Trehub, 2003), regularity and beat perception (Honing, 2012; ten Cate et al., 2016), tonal encoding of pitch (Hoeschele, Cook, Guillette, Hahn, & Sturdy, 2014; Peretz & Coltheart, 2003), and metrical encoding of rhythm (Fitch, 2013b; Winkler, Háden, Ladinig, Sziller, & Honing, 2009). Some of these musical traits may be common to humans

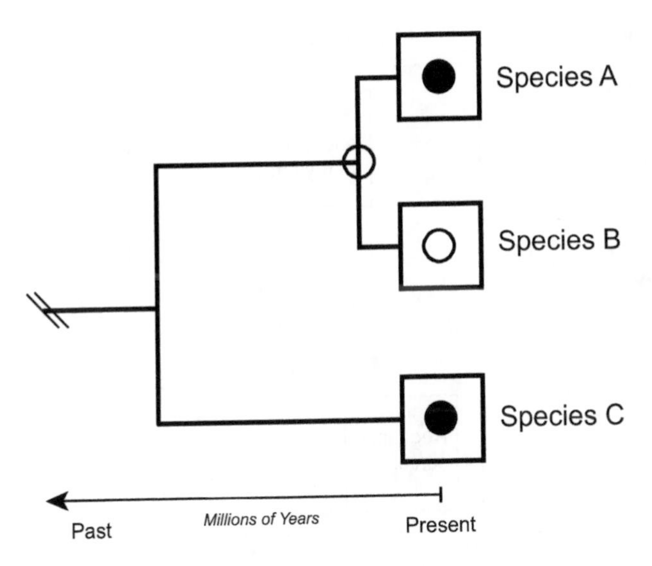

Figure 1.2
Convergent evolution of musicality. Diagrammatic representation of a hypothetical phylogenetic tree illustrating an analogous trait (homoplasy) in which a distant species (C compared to A) developed a musical trait that is lacking in a more closely related species (B compared to A). Filled circles represent a trait; open circles indicate the absence of that trait.

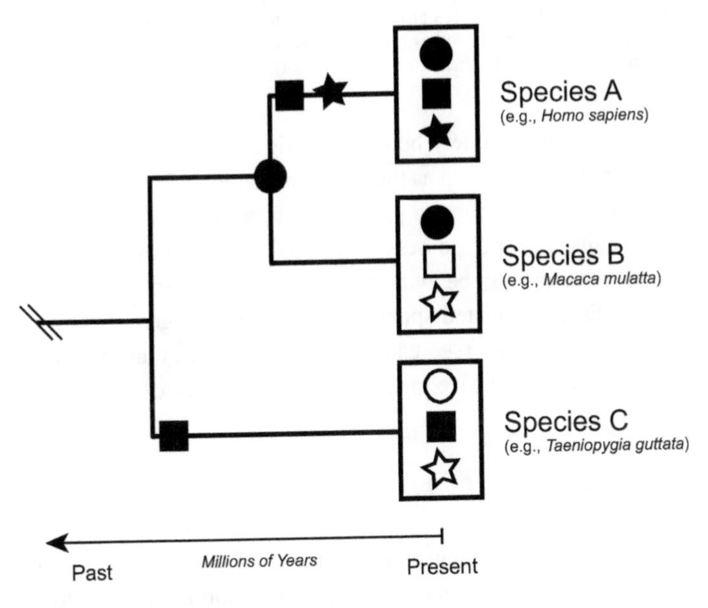

Figure 1.3
Multicomponent perspective on musicality. Diagrammatic representation of the structure and evolution of musicality, illustrating the hypothesized contributions of several traits to musicality as a complex or multicomponent phenotype. Shaded shapes represent a single trait; open shapes indicate the absence of that same trait. The positions of shapes on the tree stand for the hypothesized dates of origin of those traits.

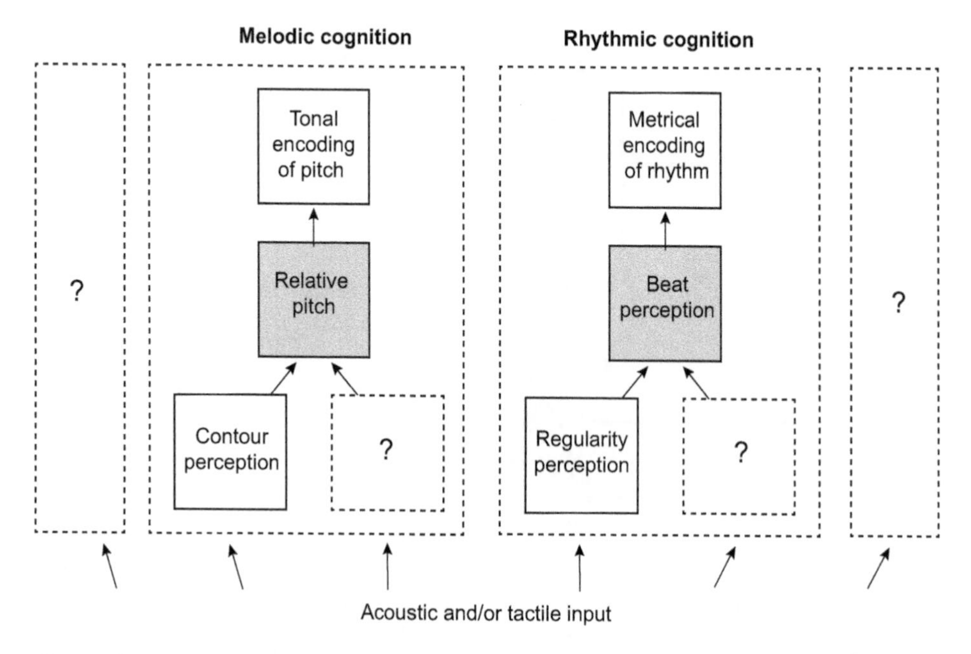

Figure 1.4
Neurocognitive model of musicality. Diagrammatic sketch of a neurocognitive model of musicality, indicating the core components hypothesized in the current literature. Each box represents a processing component, with the arrows representing pathways of communication between processing components. Boxes with a question mark refer to other, as yet undiscovered, components. Hypothesized components adapted from Peretz and Coltheart (2003), Honing et al. (under review), and Merchant and Honing (2014).

and other species, and others might be uniquely human (see figure 1.4). A comparative approach using well-known animal models, such as the rhesus macaque (*Macaca mulatta*) and the zebra finch (*Taeniopygia guttata*), offers a promising strategy for studying these components. Although only humans appear to possess the full suite of abilities that constitutes musicality, some core components may be evident in closely related species, implying biological precursors or, in distantly related species, implying convergent evolution. This will inform a phenomics of musicality (see chapter 10, this volume) and facilitate the development of an integrated neurocognitive model of musicality (see chapter 9, this volume).

Is Musicality Grounded in Our Biology?

In the past decade, it has become popular to consider the origins of music from an evolutionary perspective (Cross, 2007; Justus & Hutsler, 2005; McDermott & Hauser, 2005; Vitouch & Ladinig, 2009; Wallin et al., 2000). Yet disagreement remains about the extent to which music is grounded in our biology, if it played a role in our survival as a species, and, if so, whether musicality resulted from natural or sexual selection (or even group selection; Brown, 2000).

At least three adaptationist explanations of music have been proposed. Charles Darwin (1871) was the first to suggest that sexual selection played a role in the origins of music, a view that has been revived and reevaluated in recent years (Merker, 2000; Miller, 2000; but see, e.g., Mosing et al., 2015). For Darwin, music had no survival benefits but offered a means of impressing potential partners, thereby contributing to reproductive success. Like other subsequent scholars (Brown, 2000; Mithen, 2009), Darwin argued that musical vocalizations preceded language (Fitch, 2013a). The latter view supports the idea that musicality might actually be a precursor of both music and language, the latter recycling parts of the neural structures that evolved for musicality (see chapter 9, this volume).

Another view traces the origins of music to caregivers' music-like vocalizations to infants, which are thought to enhance parent-infant bonds, ease the burdens of caregiving, and promote infant well-being and survival (Dissanayake, 2008; Trehub, 2003). Proponents of this view see such vocalizations as having paved the way not only for language but also for music (Brown, 2000).

A third view stresses the role of music in promoting and maintaining group cohesion. Music is thought to be the social glue that enhances cooperation and strengthens feelings of unity (Cross, 2009; Merker, Madison, & Eckerdal, 2009). According to Dunbar (2012), group singing and dancing among our hominin ancestors replaced social grooming (i.e., grooming of others involving touch) as a means of maintaining social connections as groups expanded in size. Song and dance have similar neurochemical effects to social grooming, such as endorphin release (Dunbar, 2010), which has important social consequences.

A prominent nonadaptationist view considers music to be a technology or transformative invention that makes use of existing skills and has important consequences for culture and biology (Patel, 2010). In this view, music is an exaptation, spandrel, or evolutionary by-product of other skills (cf. Pinker, 1997). This notion has parallels with the transformative control of fire by early humans that made it possible for them to cook food and obtain warmth, which also had an important cultural and biological impact (Goudsblom, 1995; Wrangham & Conklin-Brittain, 2003). Two more recent viewpoints, combining adaptive and nonadaptive perspectives, are discussed in more detail in chapters 4 and 5, this volume.

Can the Evolution of Music Cognition Be Studied?

Despite the many theories on the biological origins of music, most scholars until relatively recently were wary of the notion that music cognition could have a biological basis:

> There is no reason to believe there is a universally shared, innate basis for music perception. Although the possible survival value of music has often been speculated about, music has not been around long enough to have shaped perceptual mechanisms over thousands of generations. Clearly, music is a cultural artifact, and knowledge about it must be acquired. Moreover, in contrast to speech, this knowledge is acquired relatively slowly and not equally by all individuals of a given nature. (Repp, 1991, p. 260)

This position is typical of scholarly thought over the past fifty years, with music viewed as a cultural product with no evolutionary history and no biological constraints on its manifestation.

The available unambiguous fossil record—beautiful bone flutes—dates musical activity to at least 45,000 years ago (Conard, Malina, & Münzel, 2009; Morley, 2013), a modest time frame in evolutionary terms. For comparison, most researchers date the presence of modern human language to at least 100,000 years ago (Fitch, 2017). It is impossible, however, to conclude from this that music has not been around long enough to shape perception or cognition. Vocal music and percussive use of the body leave no physical traces, so the archaeological record can only provide evidence of musical instruments, and only of instruments made of durable material such as bone. Opposing claims that "we may safely infer ... that music is among the most ancient of human cognitive traits" (Zatorre & Salimpoor, 2013, p. 10430) are equally speculative. For the moment, at least, definitive conclusions about the prehistory and origins of music cannot be formulated.

Furthermore, there is much skepticism about whether it is even possible to gain insight into the evolution of cognition in general (Bolhuis & Wynne, 2009; Honing & Ploeger, 2012; Lewontin, 1998) and, by extension, into musicality. According to Lewontin (1998), evolutionary theory rests on three principles—variation, heredity, and natural selection—that limit scientific inquiry into cognition. To understand the evolution of cognition, it is necessary to understand the variation in cognitive traits in ancestral times. Because cognition does not fossilize, however, we cannot acquire the requisite evidence about variability (Lewontin, 1998). On the issue of the heritability of musicality, many studies provide evidence to support this (see chapter 10, this volume), but it is difficult to specify the genes because cognitive traits are polygenic. It would also be necessary to gather evidence about the possibility that cognitive traits were the target of natural selection. Without reconstructing the minds of our hunter-gatherer predecessors, for example, we can only guess at the selection pressures they faced (Bolhuis & Wynne, 2009). But these skeptical comments apply to virtually all aspects of human cognition, not just music.

It is worth noting, however, that the possible adaptive function of music is only one of several indispensable levels at which the cognitive and biological phenomena that may underlie musicality can be analyzed. In addition to the possible survival or reproductive value of music (adaptation), one can examine the neurobiological substrates (mechanisms), developmental trajectory (ontogeny), and their evolutionary history (phylogeny; see Tinbergen, 1963). For traits that are shared with other animals, the comparative approach provides a rigorous way to analyze both phylogeny (via homology) and adaptation (via analogy). Thus, multiple divergent perspectives are necessary for understanding the full complexity of musicality, making its study a truly interdisciplinary endeavor.

A Research Agenda on Musicality

Given these considerations and as an outcome of the discussions held at the Lorentz workshop in 2014, three groups of interrelated questions arise. I present them below in the form of aims for future research in musicality.

One important question concerns the identification of the constituent components of musicality in humans. What is the most promising means of carving musicality into component skills? How can these constituent components be effectively probed in humans? What is the neural circuitry associated with the core components of musicality? How can we more clearly differentiate biological and cultural contributions to musicality?

Second, and in parallel to studying musicality in humans, the foreseen research agenda aims to investigate the core components of musicality in closely related animal species (e.g., monkeys) and distantly related animal species (e.g., parrots, songbirds, seals). The central questions in these comparative studies evolve around rhythmic and melodic cognition: What are the contributions of regularity perception and beat perception in the perception of rhythm, and to what extent are these species-dependent? What are the contributions of relative pitch and timbre on the perception of melody, and in how far are these species dependent? Can animals detect higher-order (e.g., hierarchical) patterns in sound, as humans do?

Once the main core components of musicality are identified, the stage is set to work on a phenomics of musicality, addressing questions like: How can evaluation and measurement of individual musical phenotypes be operationalized? Which genes can be associated with the constituent components of musicality?

Finally, given the acquired evidence on the components of musicality in humans, monkeys, and birds and the relationship between musicality and genetics, the aim is to combine these into an integrated model of musicality that will further constrain evolutionary theories of music and musicality.

I next elaborate on this research agenda and its aims, and make further links to the chapters that discuss these issues.

Decomposing Musicality into Constituent Components

The primary task of the proposed research agenda is to carve musicality into its constituent components using a divide-and-conquer strategy (see figure 1.1). However, our current knowledge about such musical traits is fragmentary for animals, and there are many open questions even for humans concerning the cognitive abilities involved. Therefore, what is needed are carefully designed experiments with artificially generated stimuli that examine and compare the presence of specific traits in both humans and nonhuman animals.

While musicality is likely made up of many components, it appears to be good strategy to start with a focus on core aspects like melody and rhythmic cognition (see figure 1.4). Both domains are well suited for comparative studies, both cross-cultural and cross-species, and the nature and extent of their presence in nonhuman animals have attracted considerable debate in the recent literature. These recent discussions, combined with the availability of suitable experimental techniques for tracking these phenomena in human and nonhuman animals, make this a timely and feasible enterprise (see chapters 7, 8, and 9, this volume).

Of course, we need to remain cautious about making claims on music-specific modes of processing until more general accounts have been ruled out. It still has to be demonstrated that the constituent components of musicality, when identified, are indeed domain specific. In contrast, the argument that music is a human invention (see chapter 5, this volume) depends on the demonstration that the components of musicality are not domain specific, but each cognitively linked to some nonmusical mental ability. So while there might be quite some evidence that components of musicality overlap with nonmusical cognitive features (see chapter 5, this volume), this is in itself no evidence against musicality as an evolved biological trait or set of traits. As in language, musicality could have evolved from existing elements that are brought together in unique ways, and that system may still have emerged as a biological product through evolutionary processes, such as natural or sexual selection. As such there is no need for musicality to be modular or show a modular structure (S. Fisher, personal communication; cf. Fodor, 1983). Alternatively, based on the converging evidence for music-specific responses along specific neural pathways, it could be that brain networks that support musicality are partly recycled for language, thus predicting more overlap than segregation of cognitive functions (see chapter 9, this volume). In fact, this is one possible route to test the Darwin-inspired conjecture that musicality precedes music and language.

Probing Melodic Cognition

Many animals—insects, fish, frogs, birds, and mammals—sing sequences of pitched melodic tones to communicate with conspecifics (Doolittle, Gingras, Endres, & Fitch, 2014; Richner, 2016). One promising way of studying the biological basis of melodic cognition is to study it in humans and birds. Humans and songbirds share many interesting similarities with regard to their auditory processing capabilities. For example, humans and European starlings (*Sturnus vulgaris*) have similar frequency sensitivity, perceive the pitch of the missing fundamental, and parse multiple pure-tone sequences into separate auditory streams (Bregman, Patel, & Gentner, 2016). At higher levels, the musical nature of birdsong has long been appreciated by humans, and some songbirds can readily learn to discriminate and even imitate human melodic sequences (Nicolai, Gundacker, Teeselink, & Güttinger, 2014).

Given these similarities, it is surprising to find major differences in how humans and songbirds perceive sequences of tones. Humans readily recognize tone sequences that are shifted up or down in log frequency because the pattern of relative pitches is maintained (referred to as interval perception or relative pitch; see figure 1.4). In contrast, songbirds were assumed to have a strong bias to rely on absolute pitch for the recognition of tone sequences (with a pitch-shifted melody to be perceived as an altogether different melody). However, a recent study (Bregman et al., 2016) suggests that songbirds attend more to the acoustic spectral contour (the distribution of energy over different frequency bands, changing over time) than to the absolute pitch. Stimuli that preserve this shape, even when pitch cues are removed (noise-band vocoded signal), seem to allow for perceptual generalization of learned acoustic patterns. Hence it could well be that a sensitivity to spectral contour is present in both human and avian cognition of musical signals, but that relative pitch is the preferred mode of listening for humans.

Probing Rhythmic Cognition

Most animals show at least some sort of rhythmic behavior, like walking, flying, crawling, or swimming.[1] It is hence not unnatural to think that the perception (and enjoyment) of rhythm might well be shared by most animals, as Darwin (1871) argued. While recent experimental research has been finding some support for this claim (Patel, Iversen, Bregman, & Schulz, 2009; Ravignani & Cook, 2016; Schachner, Brady, Pepperberg, & Hauser, 2009; Wilson & Cook, 2016), there are also aspects of rhythmic cognition that appear to be species specific (Fitch, 2013b), such as the ability to perceive a regular pulse in a varying rhythm (i.e., beat perception) and consequently being able to synchronize to it (see chapter 8, this volume). However, if the production of synchronized movement to sound or music is not observed in certain species, this is no evidence for the absence of beat perception. It could well be that while certain species are not able to synchronize their movements to a regular beat, they are capable of beat perception. Hence, the development of probing techniques that are not dependent on overt behavior is crucial. One possibility is the application of electrophysiological techniques (e.g., scalp-recorded evoked potentials) that were shown to allow for a direct comparison between human and nonhuman primates (Fukushima et al., 2010; Honing, Háden, Prado, Bartolo, & Merchant, 2012; Ueno et al., 2010).

With regard to rhythmic cognition in primates, the empirical evidence suggests beat-based timing to be gradually developed, peaking in humans but present in only limited form in other nonhuman primates, while humans share interval-based timing with all nonhuman primates and related species (see figure 1.5).

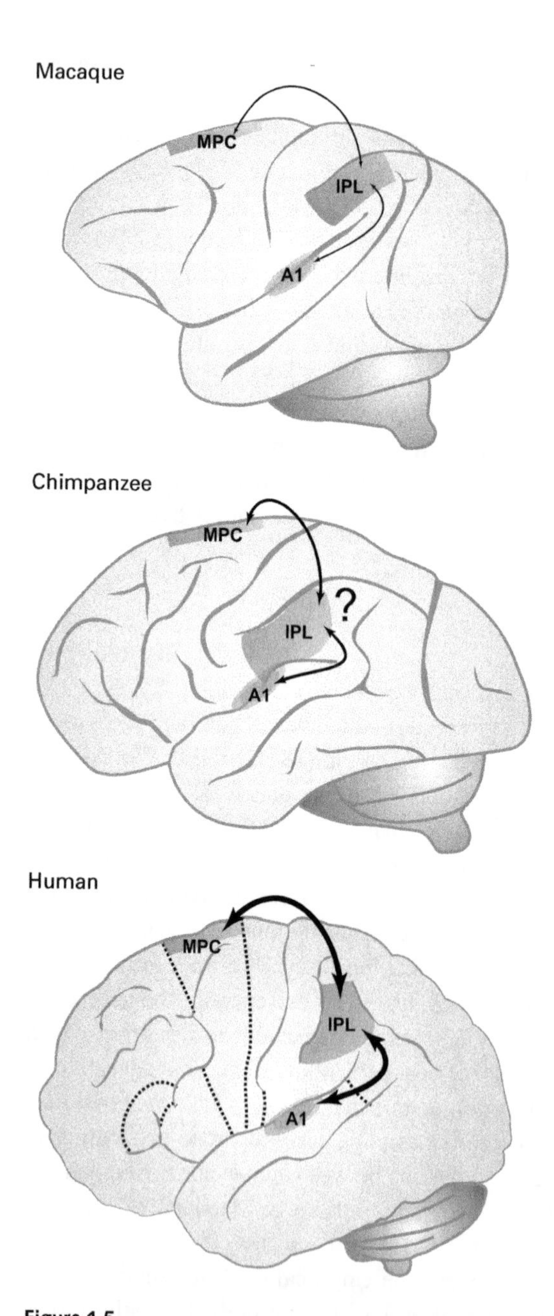

Figure 1.5
The gradual audiomotor evolution (GAE) hypothesis. This hypothesis suggests connections between medial premotor cortex (MPC), inferior parietal lobe (IPL), and primary auditory area (A1) to be stronger in humans than in other primates (marked with solid lines), suggesting beat-based timing to have gradually evolved. Line thickness indicates the hypothesized connection strength, the question mark indicates absence of evidence. (Adapted from Honing et al., under review; Merchant and Honing, 2014.)

With regard to rhythmic cognition in birds, the literature remains partial and divided. Initially a causal link between vocal learning and beat perception and synchronization was proposed (Patel, 2006; Patel et al., 2009), associated with evolutionary modifications to the basal ganglia that play a key role in mediating a link between auditory input and motor output during learning (Petkov & Jarvis, 2014); however, other studies have questioned such an explicit link (Honing & Merchant, 2014; Wilson & Cook, 2016) or suggested at least a graded scale in avian species (ten Cate et al., 2016). Some species attend more strongly to specific local features of the individual stimuli (e.g., the exact duration of time intervals) rather than the overall regularity of the stimuli, a main feature human listeners attend to; van der Aa, Honing, & ten Cate, 2015). These findings call for a reexamination of the nature and mechanisms underlying rhythmic cognition and its core components such as regularity (i.e., isochrony) and beat perception (see figure 1.4).

Operationalizing the Musical Phenotype

A comparative approach to musicality faces numerous challenges, including the identification of candidate skills and credible means of distinguishing biological from cultural contributions to human musicality. Nevertheless, consensus is growing that musicality has deep biological foundations, based on accumulating evidence for the involvement of genetic variation (Liu et al., 2016; Oikkonen, Onkamo, Järvelä, & Kanduri, 2016; Peretz, 2016). Recent advances in molecular technologies provide an effective way of exploring these biological foundations. Next to examining clustering in families and co-occurrence in twins of extreme levels of musical ability, genome-wide association studies offer a promising route to capture the polymorphic content of a large phenotyped population sample. The success of genetic studies of musical ability is, however, critically dependent on a robust, objective, and reliable measure of the musicality phenotype, the primary aim of this research agenda. But once a small set of reliably measurable core components is identified, opportunities to administer standardized aptitude tests online (using web-based or gaming techniques, for example) provide an important step toward high-powered genome-wide screens to be able to analyze musical phenotypes (see chapter 10, this volume).

Constraining Evolutionary Theories of Music and Musicality

While I have argued that the comparative method is an effective strategy in avoiding "just-so stories" on the evolution of music, the ultimate aim is to arrive at a neurocognitive model of musicality that will further constrain theories of music and musicality (see chapters 11 through 13, this volume). Fitch (in chapter 2, this volume) and Merker, Morley, and Zuidema (in chapter 3, this volume) make an important first step in identifying these constraints. In addition to studying the cognitive and biological aspects of musicality, scholars of the history of science are in a position to evaluate and critically situate recent developments in the longer tradition of the study of the origins of music, spanning the eighteenth century (e.g., Rousseau, Rameau), nineteenth century (e.g., Darwin, Spencer), and twentieth century (e.g., Stumpf, Merker, Brown). (A fine example of this is in chapter 14, this volume.)

Summary

The aim of the research agenda outlined in this chapter is threefold. The primary aim is to identify the basic neurocognitive mechanisms that constitute musicality, as well as effective ways of studying these in human and nonhuman animals. The second aim is to develop a comprehensive operational tool for the analysis of musical phenotypes that will be able to identify and define the biological basis of musicality. Based on the outcome of the first two aims, the third aim is to develop an integrated neurocognitive model of musicality that will further constrain existing evolutionary theories of music and musicality.

Acknowledgments

This chapter is based on material that served as a starting point for Honing et al. (2015) and benefited from the feedback of Carel ten Cate, Isabelle Peretz, and Sandra Trehub. I thank W. Tecumseh Fitch for his comments on an earlier version of this chapter. The preparation of this chapter (and the book as a whole) was supported by a Distinguished Lorentz Fellowship granted to H.H. by the Lorentz Center for the Sciences and the Netherlands Institute for Advanced Study in the Humanities and Social Sciences and a Horizon grant (317-70-010) of the Netherlands Organization for Scientific Research.

Note

1. The specific question of whether animals can detect regularity in a stimulus and synchronize their own behavior to arbitrary rhythmic patterns got sudden attention with the discovery of Snowball, a sulfur-crested cockatoo that could entrain head and body movements with the beat in several popular songs (Patel, Iversen, Bregman, & Schulz, 2009). Parrots such as Snowball are vocal learners, and vocal learning is associated with evolutionary modifications to the basal ganglia, which play a key role in mediating a link between auditory input and motor output during learning (Petkov & Jarvis, 2014). Because such linkage between auditory and motor areas in the brain is also required for beat entrainment, Patel (2006) suggested that only vocal learning species might be able to show beat perception (see chapters 7 and 8, this volume).

References

Blacking, J. (1973). *How musical is man?* The Jessie and John Danz Lectures. Seattle: University of Washington Press.

Bolhuis, J. J., & Wynne, C. D. L. (2009). Can evolution explain how minds work? *Nature, 458*(7240), 832–833. doi:10.1038/458832a

Bregman, M. R., Patel, A. D., & Gentner, T. Q. (2016). Songbirds use spectral shape, not pitch, for sound pattern recognition. *Proceedings of the National Academy of Sciences of the United States of America, 113*(6), 1666–1671. doi:10.1073/pnas.1515380113

Brown, S. (2000). The "musilanguage" model of music evolution. In N. Wallin, S. Brown, & B. Merker (Eds.), *The origins of music* (pp. 271–300). Cambridge, MA: MIT Press.

Brown, S., & Jordania, J. (2011). Universals in the world's musics. *Psychology of Music, 41*(2), 229–248. doi:10.1177/0305735611425896

Brown, S., Savage, P. E., Ko, A. M.-S., Stoneking, M., Ko, Y.-C., Loo, J.-H.,& Trejaut, J. A. (2014). Correlations in the population structure of music, genes and language. *Proceedings of the Royal Society of London B: Biological Sciences, 281*(1774): 20132072. http://doi.org/http://dx.doi.org/10.1098/rspb.2013.2072

Call, J., & Tomasello, M. (2008). Does the chimpanzee have a theory of mind? Thirty years later. *Trends in Cognitive Sciences, 12*(5), 187–192. doi:10.1016/j.tics.2008.02.010

Conard, N. J., Malina, M., & Münzel, S. C. (2009). New flutes document the earliest musical tradition in southwestern Germany. *Nature, 460*(7256), 737–740. doi:10.1038/nature08169

Cross, I. (2007). Music and cognitive evolution. In R. I. M. Dunbar & L. Barrett (Eds.), *Oxford handbook of evolutionary psychology* (pp. 649–667). Oxford: Oxford University Press.

Cross, I. (2009). The nature of music and its evolution. In S. Hallam, I. Cross, & M. Thaut (Eds.), *Oxford handbook of music psychology* (pp. 3–13). Oxford: Oxford University Press.

Darwin, C. (1871). *The descent of man, and selection in relation to sex.* London: John Murray.

de Waal, F. B. M., & Ferrari, P. F. (2010). Towards a bottom-up perspective on animal and human cognition. *Trends in Cognitive Sciences, 14*(5), 201–207. doi:10.1016/j.tics.2010.03.003

de Waal, F. B. M., & Ferrari, P. F. (2012). *The primate mind.* Cambridge, MA: Harvard University Press.

Deutsch, D. (2013). *The psychology of music* (3rd ed.). London: Academic Press. doi:10.1016/B978-0-12-381460-9.00007-9

Dissanayake, E. (2008). If music is the food of love, what about survival and reproductive success? *Musicae Scientiae* [Special issue], *12*(1), 169–195.

Doolittle, E. L., Gingras, B., Endres, D. M., & Fitch, W. T. (2014). Overtone-based pitch selection in hermit thrush song: Unexpected convergence with scale construction in human music. *Proceedings of the National Academy of Sciences, 11*(46), 1–6. http://doi.org/ 10.1073/pnas.1406023111

Dunbar, R. I. M. (2010). The social role of touch in humans and primates: Behavioural function and neurobiological mechanisms. *Neuroscience and Biobehavioral Reviews, 34*(2), 260–268. doi:10.1016/j.neubiorev.2008.07.001

Dunbar, R. I. M. (2012). On the evolutionary function of song and dance. In N. Bannan (Ed.), *Music, language, and human evolution* (pp. 201–214). Oxford: Oxford University Press.

Fitch, W. T. (2006). The biology and evolution of music: A comparative perspective. *Cognition, 100*(1), 173–215. doi:10.1016/j.cognition.2005.11.009

Fitch, W. T. (2010). *The evolution of language*. Cambridge: Cambridge University Press.

Fitch, W. T. (2013a). Musical protolanguage: Darwin's theory of language evolution revisited. In J. J. Bolhuis & M. Everaert (Eds.), *Birdsong, speech, and language: Exploring the evolution of mind and brain* (pp. 489–503). Cambridge, MA: MIT Press.

Fitch, W. T. (2013b). Rhythmic cognition in humans and animals: Distinguishing meter and pulse perception. *Frontiers in Systems Neuroscience, 7*, 1–16. doi:10.3389/fnsys.2013.00068

Fitch, W. T. (2017). Empirical approaches to the study of language evolution. *Psychonomic Bulletin and Review, 24*, 3–33. doi:10.3758/s13423-017-1236-5

Fodor, J. A. (1983). *The modularity of mind.* Cambridge, MA: MIT Press; 10.2307/2184717

Fukushima, H., Hirata, S., Ueno, A., Matsuda, G., Fuwa, K. K., Sugama, K., et al. (2010). Neural correlates of face and object perception in an awake chimpanzee (*Pan troglodytes*) examined by scalp-surface event-related potentials. *PLoS One, 5*(10), e13366. doi:10.1371/journal.pone.0013366

Goudsblom, J. (1995). *Fire and civilization.* London: Penguin Books.

Hallam, S., Cross, I., & Thaut, M. (2009). *Oxford handbook of music psychology*. Oxford: Oxford University Press.

Hoeschele, M., Cook, R. G., Guillette, L. M., Hahn, A. H., & Sturdy, C. B. (2014). Timbre influences chord discrimination in black-capped chickadees (*Poecile atricapillus*) but not humans (*Homo sapiens*). *Journal of Comparative Psychology, 128*(4), 387–401. doi:10.1037/a0037159

Honing, H. (2012). Without it no music: Beat induction as a fundamental musical trait. *Annals of the New York Academy of Sciences, 1252*(1), 85–91. doi:10.1111/j.1749-6632.2011.06402.x

Honing, H. (2013). *Musical cognition: A science of listening.* New Brunswick, NJ: Transaction Publishers.

Honing, H., Bouwer, F. L., Prado, L., & Merchant, H. (under review). Rhesus monkeys (*Macaca mulatta*) sense isochrony in rhythm, but not the beat. *Frontiers in Neuroscience.*

Honing, H., Háden, G. P., Prado, L., Bartolo, R., & Merchant, H. (2012). Rhesus monkeys (*Macaca mulatta*) detect rhythmic groups in music, but not the beat. *PLoS One, 7*(12), 1–10. doi:10.1371/journal.pone.0051369

Honing, H., & Merchant, H. (2014). Differences in auditory timing between human and non-human primates. *Behavioral and Brain Sciences, 27*(6), 557–558.

Honing, H., & Ploeger, A. (2012). Cognition and the evolution of music: Pitfalls and prospects. *Topics in Cognitive Science, 4*(4), 513–524. doi:10.1111/j.1756-8765.2012.01210.x

Honing, H., ten Cate, C., Peretz, I., & Trehub, S. E. (2015). Without it no music: Cognition, biology and evolution of musicality. *Philosophical Transactions of the Royal Society of London B: Biological Sciences, 370*(1664), 20140088. doi:10.1098/rstb.2014.0088

Justus, T., & Hutsler, J. (2005). Fundamental issues in the evolutionary psychology of music: Assessing innateness and domain specificity. *Music Perception, 23*(1), 1–27.

Lewontin, R. (1998). The evolution of cognition: Questions we will never answer. In D. Scarborough & S. Sternberg (Eds.), *Methods, models, and conceptual issues: An invitation to cognitive science* (pp. 107–132). Cambridge, MA: MIT Press.

Liu, X., Kanduri, C., Oikkonen, J., Karma, K., Raijas, P., Ukkola-Vuoti, L., et al. (2016, February). Detecting signatures of positive selection associated with musical aptitude in the human genome. *Scientific Reports, 6,* 21198. doi:10.1038/srep21198

Lomax, A., & Berkowitz, N. (1972). The evolutionary taxonomy of culture. *Science, 177*(4045), 228–239. doi:10.1126/science.177.4045.228

McDermott, J. H., & Hauser, M. D. (2005). The origins of music: Innateness, uniqueness, and evolution. *Music Perception, 23*(1), 29–59.

Mendoza, G., & Merchant, H. (2014). Motor system evolution and the emergence of high cognitive functions. *Progress in Neurobiology, 122,* 73–93. doi:10.1016/j.pneurobio.2014.09.001

Merchant, H., & Honing, H. (2014). Are non-human primates capable of rhythmic entrainment? Evidence for the gradual audiomotor evolution hypothesis. *Frontiers in Auditory Cognitive Neuroscience, 7*(274), 1–8. doi:10.3389/fnins.2013.00274

Merker, B. (2000). Synchronous chorusing and human origins. In N. L. Wallin, B. Merker, & S. Brown (Eds.), *The origins of music* (pp. 315–327). Cambridge, MA: MIT Press.

Merker, B., Madison, G. S., & Eckerdal, P. (2009). On the role and origin of isochrony in human rhythmic entrainment. *Cortex, 45*(1), 4–17. doi:10.1016/j.cortex.2008.06.011

Miller, G. F. (2000). Evolution of human music through sexual selection. In N. L. Wallin, B. Merker, & S. Brown (Eds.), *The origins of music* (pp. 329–360). Cambridge, MA: MIT Press.

Mithen, S. J. (2009). The music instinct: The evolutionary basis of musicality. *Annals of the New York Academy of Sciences, 1169,* 3–12.

Morley, I. (2013). *The prehistory of music: Human evolution, archaeology, and the origins of musicality.* Oxford: Oxford University Press.

Mosing, M. A., Verweij, K. J. H., Madison, G., Pedersen, N. L., Zietsch, B. P., & Ullén, F. (2015). Did sexual selection shape human music? Testing predictions from the sexual selection hypothesis of music evolution using a large genetically informative sample of over 10,000 twins. *Evolution and Human Behavior, 36*(5), 359–366. doi:10.1016/j.evolhumbehav.2015.02.004

Nettl, B. (2006). Response to Victor Grauer: On the concept of evolution in the history of ethnomusicology. *World of Music, 48*(2), 59–72.

Nicolai, J., Gundacker, C., Teeselink, K., & Güttinger, H. R. (2014). Human melody singing by bullfinches (*Pyrrhula pyrrula*) gives hints about a cognitive note sequence processing. *Animal Cognition, 17*(1), 143–155. doi:10.1007/s10071-013-0647-6

Oikkonen, J., Onkamo, P., Järvelä, I., & Kanduri, C. (2016). Convergent evidence for the molecular basis of musical traits. *Scientific Reports, 6*, 1–10. doi:10.1038/srep39707

Patel, A. D. (2006). Musical rhythm, linguistic rhythm, and human evolution. *Music Perception, 24*(1), 99–104. doi:10.1525/mp.2006.24.1.99

Patel, A. D. (2010). Music, biological evolution, and the brain. In M. Bailar (Ed.), *Emerging disciplines* (pp. 1–37). Houston, TX: Rice University Press.

Patel, A. D., Iversen, J. R., Bregman, M. R., & Schulz, I. (2009). Experimental evidence for synchronization to a musical beat in a nonhuman animal. *Current Biology, 19*(10), 827–830. doi:10.1016/j.cub.2009.03.038

Peretz, I. (2016). Neurobiology of congenital amusia. *Trends in Cognitive Sciences, 20*(11), 857–867. doi:10.1016/j.tics.2016.09.002

Peretz, I., & Coltheart, M. (2003). Modularity of music processing. *Nature Neuroscience, 6*(7), 688–691. doi:10.1038/nn1083

Peretz, I., & Zatorre, R. J. (2005). Brain organization for music processing. *Annual Review of Psychology, 56*, 89–114. doi:10.1146/annurev.psych.56.091103.070225

Petkov, C. I., & Jarvis, E. D. (2014). The basal ganglia within a cognitive system in birds and mammals. *Behavioral and Brain Sciences, 37*(6), 568–604. doi:10.1017/S0140525X13004160

Pinker, S. (1997). *How the mind works*. New York: Norton

Ravignani, A., & Cook, P. F. (2016). The evolutionary biology of dance without frills. *Current Biology, 26*(19), R878–R879. doi:10.1016/j.cub.2016.07.076

Repp, B. H. (1991). Some cognitive and perceptual aspects of speech and music. In J. Sundberg, L. Nord, & R. Carlson (Eds.), *Music, language, speech and brain* (pp. 257–268). Stockholm: Macmillan.

Richner, H. (2016). Interval singing links to phenotypic quality in a songbird. *Proceedings of the National Academy of Sciences of the United States of America, 113*(45), 12763–12767. doi:10.1073/pnas.1610062113

Rilling, J. K., Glasser, M. F., Preuss, T. M., Ma, X., Zhao, T., Hu, X., et al. (2008). The evolution of the arcuate fasciculus revealed with comparative DTI. *Nature Neuroscience, 11*(4), 426–428. doi:10.1038/nn2072

Savage, P. E., Brown, S., Sakai, E., & Currie, T. E. (2015). Statistical universals reveal the structures and functions of human music. *Proceedings of the National Academy of Sciences of the United States of America, 112*(29), 8987–8992. doi:10.1073/pnas.1414495112

Schachner, A., Brady, T. F., Pepperberg, I. M., & Hauser, M. D. (2009). Spontaneous motor entrainment to music in multiple vocal mimicking species. *Current Biology, 19*(10), 831–836. doi:10.1016/j.cub.2009.03.061

Spierings, M. J., & ten Cate, C. (2014). Zebra finches are sensitive to prosodic features of human speech. *Proceedings of the Royal Society B: Biological Sciences 281*(1787). doi:10.1098/rspb.2014.0480

ten Cate, C., Spierings, M., Hubert, J., & Honing, H. (2016, May). Can birds perceive rhythmic patterns? A review and experiments on a songbird and a parrot species. *Frontiers in Psychology, 7*, 1–14. doi:10.3389/fpsyg.2016.00730

Tinbergen, N. (1963). On aims and methods of ethology. *Zeitschrift für Tierpsychologie, 20*, 410–433.

Trainor, L. J. (2008). Science and music: The neural roots of music. *Nature, 453*, 598–599. doi:10.1038/453598a

Trehub, S. E. (2003). The developmental origins of musicality. *Nature Neuroscience, 6*, 669–673.

Trehub, S. E. (2015). Cross-cultural convergence of musical features. *Proceedings of the National Academy of Sciences of the United States of America, 112*(29), 8809–8810. doi:10.1073/pnas.1510724112

Ueno, A., Hirata, S., Fuwa, K. K., Sugama, K., Kusunoki, K., Matsuda, G., et al. (2010). Brain activity in an awake chimpanzee in response to the sound of her own name. *Biology Letters, 6*(3), 311–313. doi:10.1098/rsbl.2009.0864

van der Aa, J., Honing, H., & ten Cate, C. (2015). The perception of regularity in an isochronous stimulus in zebra finches (*Taeniopygia guttata*) and humans. *Behavioural Processes, 115*, 37–45. doi:10.1016/j.beproc.2015.02.018

Vitouch, O., & Ladinig, O. (2009). Preface to Special Issue: Music and evolution. *Musicae Scientiae, 13*(2 Suppl.), 7–11. doi:10.1177/1029864909013002021

Wallin, N., Merker, B., & Brown, S. (Eds.). (2000). *The origins of music*. Cambridge, MA: MIT Press.

Wilson, M., & Cook, P. F. (2016). Rhythmic entrainment: Why humans want to, fireflies can't help it, pet birds try, and sea lions have to be bribed. *Psychonomic Bulletin and Review*. doi:10.3758/s13423-016-1013-x

Winkler, I., Háden, G. P., Ladinig, O., Sziller, I., & Honing, H. (2009). Newborn infants detect the beat in music. *Proceedings of the National Academy of Sciences of the United States of America, 106*(7), 2468–2471. doi:10.1073/pnas.0809035106

Wrangham, R., & Conklin-Brittain, N. (2003). Cooking as a biological trait. *Comparative Biochemistry and Physiology A: Molecular and Integrative Physiology, 136*, 35–46. doi:10.1016/S1095-6433(03)00020-5

Zatorre, R. J. (2005). Music the food of neuroscience? *Nature, 434*, 312–315.

Zatorre, R. J., & Salimpoor, V. N. (2013). From perception to pleasure: Music and its neural substrates. *Proceedings of the National Academy of Sciences of the United States of America, 110*(Suppl.), 10430–10437. doi:10.1073/pnas.1301228110

II

ORIGINS, PRINCIPLES, AND CONSTRAINTS

2

Four Principles of Biomusicology

W. Tecumseh Fitch

In April 2014, the organizers invited me to present a short position statement on the first day of the Lorentz Conference on Musicality (see chapter 1, this volume). My goal was to present several principles that I believe are necessary foundations for a future discipline of biomusicology but that I also suspected might be controversial and spark discussion and debate. To my surprise, however, with few exceptions, these proposed principles were readily accepted by the very diverse set of scholars assembled at that conference: psychologists, biologists, neuroscientists, anthropologists, musicologists, and computer scientists. This speaks well for an interdisciplinary endeavor like biomusicology because it is often differences in fundamental principles and approach that stand in the way of productive interdisciplinary exchange. In this case, I was pleased to find a substantial amount of common ground among researchers as to both the conceptualization of the core problems and, in many cases, the strategies for solving them.

In the first half of this chapter, I present updated versions of these principles and briefly explore some of their implications for current and future biomusicological research. I do not hope in a short piece like this to outline all of the interesting problems and approaches in biomusicology, but rather to propose some foundational principles for approaching any problem in this domain. In the second half, I put some flesh on the bones of one these principles—the multicomponent approach—by offering my opinion about four core components of human musicality. But first, I should clarify my terms.

Defining the Object of Study: "Musicality" versus Music

Biomusicology is the biological study of musicality in all its forms. Human *musicality* refers to the set of capacities and proclivities that allows our species to generate and enjoy music in all of its diverse forms. A core tenet of biomusicology is that musicality is deeply rooted in human biology, in a form that is typical of our species and broadly shared by members of all human cultures. While music, the product of human musicality, is extremely diverse, musicality itself is a stable aspect of our biology and thus can be studied productively from comparative, neural, developmental, and cognitive perspectives. The biomusicological approach is comparative in at least two senses. First, it takes as its domain all of human music making across cultures (not privileging any one culture or "art music" created by music professionals). Second, this approach is comparative in that it seeks insight into the biology of human musicality, wherever possible, by looking at related traits in other animals.

Note that there is no contradiction in seeing musicality as a universal aspect of human biology while accepting the vast diversity of music itself, across cultures or over historical time within a culture (Brown & Jordania, 2013; Mauch, MacCallum, Levy, & Leroi, 2015). While the number of possible songs is unlimited, singing as an activity can be insightfully analyzed using a relatively small number of parameters (Is singing done in groups or alone? With or without instrumental accompaniment? Is it rhythmically regular or not?). As Alan Lomax showed in his groundbreaking cantometrics research program, such a classification can provide insight into both the unity and diversity of music, as instantiated in human cultures across the globe (Lomax, 1977, 1980; Lomax & Berkowitz, 1972). Furthermore, the form and function of the vocal apparatus that produces song are shared by all normal humans, from a newborn to Pavarotti (Titze, 1994), and indeed the overall form and function of our vocal apparatus is shared with many other mammal species from mice to elephants (Fitch, 2000, 2016; Herbst et al., 2012).

While ethnomusicology traditionally focuses on the form and social function of songs (and other products of musicality), biomusicology seeks an understanding of the more basic and widely shared capabilities underlying our capacity to make music, such as singing. There is no conflict between these endeavors, and indeed there is great potential for synergy among them since each can feed the other with data, hypotheses, and potential generalizations.

Four Foundational Principles of Biomusicology

I propose four core principles that provide a firm initial foundation for effective, productive scientific inquiry into musicality.

1 The Multicomponent Principle: Musicality Encompasses Multiple Components

The first principle is uncontroversial among musicologists (if not always clearly recognized by biologists): productive research into musicality requires that we identify and study its multiple interacting components. This basic notion is familiar from music theory, where Western music is commonly dissected into the separate components of rhythm, melody, and harmony, each considered to be important aspect of a typical piece of music. But we cannot assume that this particular traditional theoretical breakdown is the appropriate one from a biological perspective or that rhythm or harmony are themselves monolithic capacities. Rather, we should be ready to explore multiple componential frameworks open-mindedly and allow the data to steer us to the insightful subdivisions. We should also accept that different componential breakdowns might be appropriate for different purposes. For example, from a biological comparative perspective, it is useful to seek aspects of human musicality that have parallels in other species (I explore this approach in the second half of this chapter, concluding that singing, drumming, and dancing all find meaningful homologues or analogues in nonhuman animals). But a developmental researcher investigating the time course of musical development might find a different taxonomy appropriate and a neuroscientist yet another. There is no one "true" or "correct" breakdown.

The multicomponent perspective is crucial for the biological study of musicality, for although it seems true that no nonhuman species possesses "music" in its full human forms, it is also true that many animal species share some of the capacities underlying human musicality, spanning broadly shared capabilities like pitch and time perception, to less common abilities like synchronization or vocal learning. Indeed, based on current data, it seems likely that most of the basic capacities comprising human musicality are shared with at least some other animal species. What is unusual about humans may simply be that we combine all of these abilities. I discuss this hypothesis in a later section, as well as the question of meaningful possibilities for subdivision. Principle 1 does not entail accepting any particular taxonomy of components, but rather the general need for some such multicomponent viewpoint. Thus, in a nutshell, principle 1 exhorts us to "divide and conquer."

2 The Principle of Explanatory Pluralism: Consider All of Tinbergen's Explanatory Levels

The second principle is familiar to biologists but less so to psychologists or musicologists. The essential insight for this second principle was provided over fifty years ago by Nobel Prize–winning ethologist Niko Tinbergen (1963): that any biological phenomenon can be understood, and its causation explained, at multiple different levels. These levels can be divided into two broad families: proximate and ultimate explanations. *Proximate factors* are all those that help explain why some particular organism does something and include *mechanistic* explanations ("How does it work?") and *ontogenetic* or developmental explanations ("How did it develop in this particular organism's lifetime?"). These are the domains of (neuro) physiology and developmental biology, respectively.

But thanks to Darwin, biologists are not fully satisfied by just these two levels of explanation: we also strive to understand life from the viewpoint of the longer timescale of evolution and to understand how and why some particular capability arose in a species (or group of species). This is the domain of ultimate factors, traditionally divided into questions about *phylogeny* (the evolutionary history of acquisition and modification of a trait) and questions concerning the ultimate *function*, or "survival value," of the trait ("How does it help those that possess the trait to survive and reproduce more effectively than others in a population?"). Both of these levels are core components of modern evolutionary biology.

Tinbergen's four levels of explanation, sometimes called his "four whys," were extremely important when he proposed them because they provided a resolution to a long-running and unproductive debate between (mostly) English-speaking scientists like Theodore Schneirla and Daniel Lehrman, who focused on mechanistic and ontogenetic explanations (Lehrman, 1953) and the (mostly) continental European scientists like Konrad Lorenz and Tinbergen, who were comparative biologists interested in ultimate explanations. Tinbergen pointed out that there is actually no conflict between these different types of explanation and that full understanding of any biological trait requires answers at all four levels of causation. Thus, we know that male songbirds sing in spring because their testosterone levels are high (a mechanistic explanation), but we also know that an important function of song is to defend a territory and attract mates (an ultimate functional explanation). In this well-understood case, we know that both explanations are correct and important, and it would be a waste of time to argue that one of these factors and not the other provides the "true" explanation. Tinbergen's rule—"Attend to all levels of biological explanation!"—provides a widely accepted antidote to such unproductive debate, generally taught to students of biology early in their training.

Applying Tinbergen's approach to musicality yields several important insights. Mechanistic questions in the domain of musicality include issues such as, "What are the neural bases for rhythm perception?" (see chapter 8, this volume) or, "What physiological and cognitive factors underlie a skilled singer's abilities?" Ontogenetic issues include, "At what age do infants perceive relative pitch relationships?" or, "Does early exposure to musical performance enhance pitch perception?" (Trainor, 2005; Trainor & Trehub, 1992; Trehub, 2003). Of course, there is no hard-and-fast line dividing these two types of explanations, and for many (perhaps most) traits, they are tightly intertwined. For example, it now seems clear that early and intensive exposure to music during ontogeny causes measurable changes in neural mechanisms later in life (e.g., Elbert, Pantev, Wienbruch, Rockstroh, & Taub, 1995; Gaser & Schlaug, 2003; Schlaug, 2001). Of Tinbergen's four main questions, these two proximate foci are currently active research areas and represent core empirical domains of biomusicology.

Regarding ultimate questions, it is often thought that the core evolutionary question in biomusicology concerns whether music is an adaptation and, if so, for what. Thus, for example, Steven Pinker (1997) provocatively suggested that music is simply a by-product of other cognitive abilities (a form of "auditory cheesecake"), and not itself an adaptation. Many subsequent scholars have challenged this hypothesis with specific proposals that music is an adaptation for particular functions (Ball, 2010; Brown, 2000; Hagen & Bryant, 2003; Hagen & Hammerstein, 2009; Huron, 2001; McDermott & Hauser, 2005; Merker, 2000; Roederer, 1984). This debate is reviewed in Ball (2010), Fitch (2006), and Morley (2014); I do not find it particularly productive, so I will not discuss it further here. But note that Tinbergen stressed that the "function" question must be construed more broadly than the related question of whether a trait is an adaptation per se (a trait shaped by natural selection to its current function). A trait can be useful, and increase survival and reproduction, without being an adaptation: an aversion to birth control might increase an individual's reproductive output but is obviously not an adaptation per se. Thus, in following Tinbergen's rule, we should clearly separate questions about what music is good for (e.g., seduction, social bonding, making a living) from the much harder questions about whether it is an adaptation for that or those purposes. Furthermore, questions of phylogeny (When did some trait evolve?) are just as important as the "why" question of function (see below).

Although Tinbergen's four questions provide excellent coverage for many biological traits, he apparently overlooked one domain of causation: the domain of *cultural change* over historical time. This is a class of causal explanations spanning, in temporal terms, between the domain of individual ontogeny and species phylogeny (and is sometimes confusingly referred to as evolution, as in "the evolution of English" or "the evolution of rap music"). This level of explanation is linked to, but independent of, both ontogeny and phylogeny. The issue is clearly exemplified by historical change in human language: there are many interesting questions concerning language where neither ontogenetic nor phylogenetic answers would be fully satisfying. For example, we might ask why an English-speaking child tends to place the verb second in declarative sentences, after the subject and before the object (so-called SVO basic word order). An ontogenetic answer would be, "because that's what her parents do," and an ultimate answer, "because her ancestors evolved the capacity to learn language." Although neither is incorrect, these answers leave out a crucial intervening level of explanation, concerning English as a language. English, like all other languages, changes gradually over multiple generations by virtue of being learned anew, with minor variations, by each child. This iterated process of learning leads to a novel cultural level of explanation, sometimes termed "glossogeny" (Fitch, 2008; Hurford, 1990), that can be studied productively in computational models and laboratory experiments (Kirby, 2017; Scott-Phillips & Kirby, 2010; Smith & Kirby, 2008). The glossogenetic answer to the SVO question is complex and part of the general domain of historical linguistics; it involves such factors as basic word order in proto-Germanic and the overlay of French after the Norman conquest (Baugh & Cable, 1978).

We know much less about the cultural evolution of most musical genres and idioms over time than we do about historical change in language. Nonetheless, it seems safe to assume that many interesting musical phenomena will find insightful explanations at this level (see chapter 3, this volume). One example concerns the dual origins of much contemporary popular music in the fusion of the harmonic and melodic traditions of Western Europe with the syncopated, polyrhythmic traditions of West Africa, brought together historically due to slavery in the Americas (Bolcom & Harrison, 1986; Harrison, 1986; Temperley, 2004).

Summarizing, Tinbergen's principle exhorts us to investigate each meaningful level of biological causation and not to prioritize any single level over the others. Ultimately biomusicology will seek an understanding of musicality from mechanistic, ontogenetic, phylogenetic, functional, and cultural viewpoints. Even if a particular researcher chooses to focus, for reasons of personal interest or empirical expedience, on some subset of these questions, the field as a whole should seek answers to all of them.

3 The Comparative Principle: Adopt a Comparative Approach, Embracing Both Homology and Analogy

The first two principles urge us to isolate and analyze subcomponents of musicality and to approach their biology and evolution from a multifaceted Tinbergian viewpoint. The third and fourth principles concern our sources of data in this endeavor.

The third principle—"be broadly comparative!"—urges a biologically comparative approach, involving the study of behavioral capacities resembling or related to components of human musicality in a wide range of nonhuman animal species. This principle is of course familiar to most biologists but remains contentious in musicology or psychology. "Broad" in this context means that we should not limit our biological investigations to close relatives of humans (e.g., nonhuman primates) but should rather investigate any species exhibiting traits relevant to human musicality.

The capacity for complex vocal learning nicely illustrates the need for broad comparison. This capacity underlies our ability to learn and share new sung melodies and is shared with a diverse set of bird and mammal species (the current species count includes songbirds, parrots, cetaceans, hummingbirds, seals, bats, and elephants) but is not found in any nonhuman primate. In contrast, the human propensity to generate percussive sounds using limb movements ("drumming") is shared with both our nearest primate relatives (gorillas and chimpanzees) and also with woodpeckers, kangaroo rats, and palm cockatoos (Fitch, 2006). Similarly, chorusing and turn-taking among two or more individuals, a "design feature" of human musicality, is seen in various forms in duetting primate and bird pairs and in a wide diversity of frog and insect species (Bowling & Fitch, 2014; Greenfield, 2005; Haimoff, 1986; Ravignani et al., 2014; Thorpe, 1972). Thus, depending on the specific component under investigation, the set of animal species that are relevant may be quite different, and quite diverse.

Similar traits can be found in different species for several different reasons, and these are given specific names by biologists. In one type, termed *homology*, a shared trait is present in related species because a common ancestor of those species possessed the trait. Thus, all birds have feathers because the last common ancestor (LCA) of all living birds had feathers. All living mammal species produce milk to suckle their young because their LCA produced milk. These are canonical examples of homology. A second class of shared traits are those that evolved independently, or "convergently," in two different clades; such traits can be termed *analogies* (the more technical biological term, *homoplasy*, refers to all shared traits that are not homologies and includes analogy as a special case). Canonical examples of analogy include the independent evolution of wing from forelimbs in birds and bats, or the evolution of bipedalism (walking on two feet) in humans and birds. Neither wings nor bipedalism were present in the quadrupedal reptilian LCA of mammals and birds, but instead evolved convergently in each of these clades.

Analogous and homologous traits play different roles in biology, but both are important. Homologous traits are the most useful in classification and taxonomy (for this purpose, analogous traits are just a nuisance variable). More relevant to biomusicology, homologies often allow us to make inferences about traits that were present in an ancestral species because a set of homologous traits in a particular clade is by definition inherited by descent from a common ancestor of that clade. Often, particularly for behavioral or cognitive capacities, homology-based phylogenetic inference is the only means we have of reconstructing these extinct ancestors, because behavioral traits typically leave no fossils (fossil footprints providing one exception). For example, although we will probably never find a fossilized Cretaceous stem mammal in the act of suckling her young, we can nonetheless infer, with great confidence, that the ancestral mammal did so from the fact that all living descendants of this species still do. Thus, a careful analysis of living species, combined with comparative inference, provides a sort of evolutionary time machine to reconstruct the behavior and physiology of long-extinct species.

Analogous traits serve a different and complementary purpose: they provide a means for testing hypotheses using multiple independent data points. Although all of the more than 5,000 existing species of mammals suckle their young, this ability derives from their singular evolutionary origin at the base of the clade, and thus statistically constitutes a single evolutionary data point (not 5,000). In contrast, convergently evolved traits are by definition independent evolutionary events, and each clade independently possessing a trait therefore represents an independent data point. Only a set of convergently evolved traits provides an adequate database for statistically valid tests of evolutionary hypotheses. This point is often ignored, even by biologists discussing music evolution (e.g., McDermott & Hauser, 2005). Fortunately, for many cases of convergent evolution, such as bipedalism or vocal learning, a trait has evolved independently enough times to provide a rich source of evidence to test hypotheses concerning both evolution and mechanistic function. Thus, for example, we can test mechanistic hypotheses about the requirements of vocal learning by examining its neural correlates in the many species that have evolved this ability convergently (cf. Fitch & Jarvis, 2013; Fitch 2017). Similarly, we can test functional hypotheses about why the capacity for vocal synchrony or antiphony is adaptive by examining the many bird, mammal, frog, and insect species that have convergently evolved this ability (Ravignani et al., 2014).

While the conceptual distinction between homology and analogy is clear, recent discoveries in genetics and neuroscience suggest that in some cases, a trait can be both homologous and analogous, depending on the level of explanation. For example, while eye and wings have evolved independently in insects and vertebrates, it turns out that they rely in both cases on an identical set of genes and developmental pathways. This situation of convergent evolution taking the same developmental and genetic path twice has been termed *deep homology* (Shubin, Tabin, & Carroll, 1997, 2009). Deep homology also appears to characterize the capacity for complex vocal learning. Despite vocal production learning evolving convergently and independently many times (reviewed in Fitch & Jarvis, 2013), comparisons of birds and humans reveal that the same genes (e.g., *FOX-P2*) play a role in vocal learning in both groups (Haesler et al., 2007) and also that homologous neural mechanisms have been independently harnessed into vocal learning systems in birds and humans (Feenders et al., 2008). In both cases, there appears to be a deep mechanistic homology between birdsong and human vocal learning, despite their independent evolutionary origins (cf. Fitch, 2009; Fitch & Mietchen, 2013; Scharff & Petri, 2011).

In summary, principle 3 exhorts biomusicologists to adopt a broad comparative approach to any specific capability proposed as relevant to musicality. While it is important to distinguish homologous traits from those that convergently evolved, there is no justification for ignoring the latter (e.g., McDermott & Hauser, 2005), because both serve useful roles in comparative biology.

4 The Ecological Principle: Seek Broad Ecological Validity Including Popular Styles, Eschewing Elitism

Like the previous one, this principle is also broadly comparative but this time involves comparisons within our own species. According to this populist "ecological" principle, biomusicologists should seek to understand all manifestations of human musicality, from simple nursery tunes or singing in the shower, to expert bowmanship on a Stradivarius or the complex polyrhythmic improvisations of a Ghanaian master drummer. This principle is familiar to ethnomusicologists but not as widely appreciated by researchers in music cognition or neuroscience, where a focus on the Western "high art" canon remains evident. Although it is of course important to understand highly developed musical forms, performed by elite musicians, this should not lead us to neglect more basic and widespread expressions of musicality. Indeed, such everyday occurrences of music made by normal individuals with little or no musical training may in fact provide a clearer view of the underlying biology of musicality than the highly codified practices of sophisticated art music.

The ecological principle is particularly important when addressing questions about the functional, adaptive relevance of music in our species (see Morley, 2013). It makes little sense to ask about the evolutionary "survival value" of writing or performing a modern orchestral piece, but it is not unreasonable to ask about the potential adaptive value of a mother singing to her child or of a tribal group singing and dancing together. Much of traditional musicology adopts an implicitly elitist attitude, where the proper object of study is high art, composed and performed by a musical elite. Sometimes such elitism is explicit: a textbook intended to introduce students to music and art appreciation states that art "which aims merely to amuse and to provide a pleasant diversion … has little or no lasting quality." In particular, the authors state that, "art which caters to the masses ... is of little aesthetic value and will not be considered" (Wold & Cykler, 1976, p. 1). But if we ever hope to understand the shared biological basis of music, it is precisely popular music styles (e.g., dance music) that will be most relevant, along with behaviors such as a mother singing lullabies in order to soothe her infant–one of the functions of song for which the empirical data are most convincing (Trehub & Nakata, 2001; Trehub & Trainor, 1998). An elitist attitude can thus lead us to overlook aspects of musicality that are centrally relevant biologically.

Equally important are the cognitive abilities of self-avowed "nonmusicians." One of the most fundamental findings in the previous two decades of music cognition research is that untrained listeners, including those who claim they know nothing about music, exhibit sophisticated perceptual and cognitive abilities implying rich implicit understanding of musical principles (cf. Bigand & Poulin-Charronnat, 2006; Honing, 2009; Madison & Merker, 2004). In many cases, such capabilities are already present in infants and children as well (Schellenberg & Trehub, 1999; Trainor & Trehub, 1992; Trehub, 2003). Any scientific exploration of the biological basis of human musicality should therefore take a broad view of musicality across ages and over multiple levels of skill or training. This is not to say that musical expertise should be ignored as an explanatory factor; contrasts between highly skilled musicians and untrained listeners can provide a valuable source of data to help address mechanistic and developmental questions. But a focus only on the musical elite may often prove fundamentally misleading.

A third important facet of this principle concerns the diverse functions of music in human societies, with different functions shaping the expression of musicality in fundamental ways. For example, music created for dancers typically has a clear and steady rhythm, as does most music intended for simultaneous performance by multiple individuals (Savage, Brown, Sakai, & Currie, 2015; Temperley, 2004). In both cases, a steady and explicit rhythmic framework is a crucial asset in group synchronization. In contrast, music for solo performance that is intended to express sorrow develops under very different constraints and may show no clear isochronic beat at all (Clayton, 1996; Frigyesi, 1993; Tolbert, 1990). Only by studying the multiple contexts in which human musicality is expressed can we begin to make meaningful generalizations about the overall functions of music (cf. Huron, 2001).

Principle 4 thus states that in order to obtain an ecologically valid overview of human musicality, we need to take a broad, populist, and nonelitist viewpoint about what "counts" as music. While high art music of many cultures is certainly relevant in this endeavor (including Western orchestral symphonies, Ghanaian agbekor improvisations, North Indian ragas, and Balinese gamelan), so are folk music, nursery tunes, working chants, whistling while you work, and singing in the shower. Dance music in particular should be embraced as one of the core universal behavioral contexts for human music, and dance itself accepted as a core component of human musicality (cf. Fitch, 2016; Laland, Wilkins, & Clayton, 2016).

Four Core Components of Musicality

To illustrate how the four principles interact, let us return to the question raised by the multicomponent principle: "What are the biologically relevant components underlying human musicality?" One first attempt at answering this question might combine the comparative and ecological principles to ask what functions music performs in human societies and to what extent we can identify mechanisms underlying those functions in nonhuman animals. This approach leads to at least four proposed subcomponents of music.

1 Song: Complex, Learned Vocalizations

Let us start with song, one of the few aspects of human musicality that virtually all commentators agree is universally found in all human cultures (Brown & Jordania, 2013; Lomax, 1977; Nettl, 2000; Trehub, 2000). Perhaps the most obvious fact about human song is that it varies considerably among cultures and much less so within cultures (e.g., Lomax, 1980). That is, each culture has both a shared, open-ended repertoire of specific songs and culturally specific styles or idioms that encompass multiple songs. This situation is possible only when songs can be learned—so a child or newcomer can absorb the song repertoire of its community—and new songs can be generated within the style. This aspect of human song therefore entails the capacity for complex vocal learning, where novel sounds can be internalized and reproduced (cf. chapter 3, this volume). Having identified this particular design feature of human singing, we can now ask which nonhuman species share this feature (cf. Fitch, 2006). As already noted, many different species have independently evolved the capacity for complex vocal learning, providing a rich comparative database for understanding singing from the multiple perspectives of Tinbergen's rule.

The criterion of vocal learning also provides a nonarbitrary way in which we can decide whether an animal species has "song." Other commentators have typically used implicit, intuitive criteria to decide this issue. For example, Hauser and McDermott (2003) suggest that three animal groups have "animal song": songbirds, humpbacked whales, and gibbons. In contrast, Geissmann's (2000) review of gibbon song suggests that song exists in four primate groups: gibbons, tarsiers, indri, and langurs, a list that has been propagated uncritically in the literature (e.g., Morley, 2014). These authors provide no definition of animal song or any justification for their different lists. In contrast, Haimoff (1986) does offer a definition of *song*—animal sounds that "are for the most part pure in tone and musical in nature" (p. 53) —and then nominates the same four primate clades listed by Geissmann as *duet* singers. But lacking wide agreement about what "musical in nature" means, this definition is not very helpful. It remains entirely unclear why none of these authors consider the complex, multinote pant-hoot displays of chimpanzees, with their marked crescendi and drummed finale (Arcadi, 1996), or the tonal "combination long calls" of cotton-top tamarins (Weiss, Garibaldi, & Hauser, 2001), or a host of other primate vocalizations to be "song." Explicitly stating without justification that chimpanzees do not have song, Hauser and McDermott (2003) go on to conclude that "animal song thus likely has little to do with human music" (p. 667). But here the attempt at a comparative analysis has misfired at the first step: without any objective and noncircular criteria to define *song*, we cannot even objectively state what species have, or lack, song—much less evaluate its potential relevance to human music.

In contrast, if we identify vocal learning as a core defining feature of human, bird, and whale "singing," we obtain a clear and unambiguous criterion that allows us to adopt a meaningful comparative perspective (Fitch, 2006). This is why I have previously argued that a musically relevant definition of *song* is "complex, learned vocalization," regardless of tonality or any aesthetic qualities these complex vocal displays might possess to our ears. While the aesthetic virtues of the rough and sputtering underwater vocal displays of a harbor seal remain a matter of taste (Hanggi & Schusterman, 1992; Thomson & Richardson, 1995), it is clear that this species does have a capacity for vocal learning (Ralls, Fiorelli, & Gish, 1985). Furthermore, dialectal variations among populations of harbor seals and some other pinniped species suggest that this ability allows seals to learn locale-specific vocal displays (Hanggi & Schusterman, 1994; Janik & Slater, 1997; Schusterman, Balliet, & St. John, 1970). By my definition, the displays of songbirds, parrots, whales, or seals can be termed animal song and considered analogous to human singing, but the displays of chimpanzees, gibbons, indri, and other nonhuman primates cannot because these primate displays, though complex and beautiful, are not learned. I do not object if scientists studying the haunting choruses of the indri or the territorial displays of gibbon pairs continue to use the traditional term *songs* for these unlearned vocalizations. For that matter, people can freely apply the term to frog, cricket, or fish "songs," or even "the song of the forest." But in the scientific context of comparisons with music, I think that such colloquial use, without any clear and nonarbitrary guidelines or objective justification, is deeply misleading.

2 Instrumental Music: Percussion and Drumming

Of course, we humans do not express our musicality solely by singing; virtually all human cultures also have instrumental musical traditions. By *instrumental music* I mean the creation of communicative acoustic signals through nonvocal means. This broad definition includes the highly developed harmonic string and wind ensembles typical across Eurasia, the timbrally complex and more percussive gamelan tradition of Southeast Asia, and the complex polyrhythmic drum ensembles of sub-Saharan Africa. The earliest unequivocal archaeological evidence for musicality in our species is represented by instruments: numerous bone flutes have been found throughout Eurasia that document sophisticated human music making at least forty thousand years ago (Buisson, 1990; Conard, Malina,

& Münzel, 2009; Hahn & Münzel, 1995; Kunej & Turk, 2000) and other putative musical instruments are also known (cf. Morley, 2013). However, while "aereophones" are certainly common in human music across the world, they are not universal. The one form of instrumental music that is very nearly universal is the use of percussive instruments: ideophones and drums (Brown & Jordania, 2013; Nettl, 2000; Savage et al., 2015). I thus focus on percussive drumming here as a second core component of human musicality.

From a biological comparative viewpoint, there are many interesting parallels with human drumming in nature. It is much harder to find parallels with other instrument types, but spiders plucking and vibrating their webs might be considered a distant analogue of stringed instruments (Riechert, 1978). Defining percussive drumming as the production of structured communicative sounds by striking objects with limbs, other body parts, or other objects, we find several instances in other species. Starting with analogues, woodpeckers (bird family Picidae) produce displays by striking hollow trees with the bill (Dodenhoff, Stark, & Johnson, 2001; Winkler & Short, 1978), and multiple species of desert rodents produce audible and far-carrying seismic signals by pounding the ground with their feet (Randall, 2001). Both examples help to clarify the distinction between structured communicative sounds and sounds that are an incidental by-product of other behaviors. Any organism generates footfall sounds when it locomotes, but rodents' communicative drumming displays are produced without locomoting, in particular locations (often within their burrow), and in specific contexts (territorial displays and predator alarms; Randall, 2001). Similarly, woodpeckers make incidental sounds when foraging for wood-boring larvae, but during their drumming displays, they seek out particularly resonant trees (or, in urban environments, other resonant objects such as hollow metal containers on poles). Again, these displays are made in particular contexts, including territorial defense and advertisement, and often they are both identifiable as to species and bear individual-specific signatures (Dodenhoff et al., 2001; Stark, Dodenhoff, & Johnson, 1998). Thus, these displays show every sign of having evolved for the purpose of influencing others and thus constitute animal signals by most definitions (e.g., Maynard Smith & Harper, 2003; Searcy & Nowicki, 2005).

Among primates, many ape and monkey species generate nonvocal sounds as part of communicative displays (e.g., branch shaking or, in captivity, cage rattling; Remedios, Logothetis, & Kayser, 2009). Orangutans have been reported to modify the frequency content of their vocal displays using leaves placed in front of the mouth, an example of tool use that blurs the line between vocal and instrumental displays (Hardus, Lameira, van Schaik, & Wich, 2009). The most striking example of instrumental behaviors in primates comes from the drumming behavior of our nearest living relatives, the African great apes (gorillas, chimpanzees, and bonobos). While still little studied, these behaviors include drumming on resonant objects with the feet or hands, typical of chimpanzees, and drumming with the hands on the chest or other body parts, by gorillas (Arcadi, Robert, & Mugurusi, 2004; Fitch, 2006; Redshaw & Locke, 1976; Schaller, 1963). Clapping by striking the hands together is also commonly seen in all three species in captivity and has been observed in the wild in chimpanzees and gorillas (Babiszewska, Schel, Wilke, & Slocombe, 2015; Kalan & Rainey, 2009; Koops & Matsuzawa, 2006). There is strong evidence that such percussive drumming is part of the evolved behavioral repertoire of African great apes; it is consistently observed in both wild and captive animals, exhibited in particular contexts (displays and play), and when it involves objects, they are often particularly resonant objects apparently sought out for their acoustic properties (Arcadi et al., 2004). Drumming thus represents not just a universal human behavior but also one that we share with our nearest living relatives. Drumming is thus a clear candidate for a homologous behavioral component of the entire African great ape clade, of which humans are one member. Applying the phylogenetic logic of the comparative principle, this suggests that drumming evolved in the LCA of gorillas, chimpanzees, and humans, who lived roughly 7 or 8 million years ago in the forests of Africa (Glazko & Nei, 2003).

Even a brief survey of animal instrumental music would be incomplete without mentioning the palm cockatoo *Probosciger aterrimus*, a large parrot species living in Australia and New Guinea. Male palm cockatoos use a detached stick, held in the foot, to strike on resonant hollow branches as part of their courtship displays (Heinsohn, Zdenek, Cunningham, Endler, & Langmore, 2017; Wood, 1984, 1988). They are also occasionally seen to drum with the clenched foot alone but much more quietly, suggesting that this sole animal example of tool-assisted drumming may have evolved from a limb-based drumming comparable to that seen in chimpanzees. This provides an interesting parallel to human drumming, where the hand drumming that we share with other apes is often augmented by drumming with tools like sticks or mallets.

In summary, drumming appears to constitute another core component of human musicality with clear animal analogues. In the case of the African great apes, percussive drumming appears to constitute a homologous trait, suggesting that this component of human musicality evolved in the LCA of humans, gorillas, and chimpanzees more than 7 million years ago.

3 Social Synchronization: Entrainment, Duets, and Choruses

A third core component of human musicality is our capacity to synchronize our musical behaviors with others. This may be by performing the same action at the same time (e.g., clapping or chanting in unison—synchronization *sensu strictu*) or various more complex forms of entrainment such as antiphony or the complex interlocking patterns of an agbekor drum ensemble. Although solo music, performed by a single individual, is not uncommon, music performed in groups is a far more typical expression of human musicality. This is again a universal behavior seen in at least some of the music of all human cultures (Brown & Jordania, 2013; Savage et al., 2015), and such coordinated group displays also find important parallels in the animal world.

Social synchronization requires the individual capacity for synchronization to some external time giver. The most sophisticated form of synchronization involves beat-based predictive timing, where an internal beat is tuned to the frequency and phase of an isochronous time-giver, allowing perfect 0-degree phase alignment. This capacity to extract an isochronic beat and synchronize to it is termed beat perception and synchronization (BPS; Patel, 2006). Although the majority of research in both human and animal studies BPS to either a metronome or recorded musical stimuli (Repp, 2005; Repp & Su, 2013), human rhythmic abilities obviously did not arise to allow people to synchronize to metronomes but rather to the actions of other humans in groups. Thus, by the ecological principle, the concept of mutual entrainment among two or more individuals should be the ability of central interest rather than BPS to a mechanical timekeeper.

Despite a long tradition of suggesting that BPS is uniquely human, recent findings clearly document this ability in several species, including many parrot species (Hasegawa, Okanoya, Hasegawa, & Seki, 2011; Patel, Iversen, Bregman, & Schulz, 2009; Schachner, 2010) and more recently a California sea lion *Zalophus californianus* (Cook, Rouse, Wilson, & Reichmuth, 2013). In contrast, the evidence for BPS in nonhuman primates remains weak, with partial BPS by a single chimpanzee and not others (Hattori, Tomonaga, & Matsuzawa, 2013). The literature suggests a lack of BPS abilities in other nonhuman primates (see Honing, Merchant, Háden, Prado, & Bartolo, 2012; Merchant & Honing, 2013; Merchant, Zarco, Pérez, Prado, & Bartolo, 2011; chapter 6, this volume). Thus, while human BPS clearly finds analogues in the animal kingdom, it is too early to say whether homologous behaviors exist in our primate relatives. But again this aspect of human musicality provides ample scope for further comparative investigation (cf. Fitch, 2013).

Synchronization in larger groups—"chorusing"—is also very broadly observed in a wide variety of nonhuman species, including frogs and crickets in the acoustic domain and fireflies and fiddler crabs in the visual domain (for reviews, see Fitch, 2015; Greenfield, 2005; Ravignani et al., 2014). In some cases, choruses involve BPS. For example, in certain firefly species, all individuals in a tree synchronize their flashing to produce one of the most impressive visual displays in the animal kingdom (Buck, 1938, 1988; Moiseff & Copeland, 2010). These cases all represent convergently evolved analogues of human BPS and thus provide ideal data for testing evolutionary hypotheses about why such synchronization capacities might evolve, along with mechanistic hypotheses about the minimal neural requirements supporting these capacities. Although frog, cricket, and firefly examples are often neglected in discussions of music evolution, presumably because they are limited to a particular signaling dimension and a narrow range of frequencies, some species show a flexibility and range of behaviors that is musically interesting. For example, the chirps of tropical *Mecapoda* katydids are typically synchronized (predictively entrained at 0-degree phase) but under certain circumstances can also alternate (180-degree phase) or show more complex entrainment patterns, and over a broad range of tempos (chirp periods from 1.5 to 3.0 seconds; Sismondo, 1990). Thus, even very small brains are capable of generating an interesting variety of ensemble behaviors in chorusing animals—raising the fascinating question of why such behaviors are rare in so-called higher vertebrates like birds and mammals.

Other less demanding forms of temporal coordination also exist, but these forms of multi-individual coordination have been less researched and discussed (even in humans). These include turn taking and call-and-response patterns and can be accomplished using reactive rather than predictive mechanisms (e.g., "don't call until your partner has finished"). Again, such abilities find many parallels in the animal world. The most widespread examples are found in duetting birds or primates, typically between the male and female of a mated pair. Over 90 percent of bird species form (socially) monogamous pairs, exhibiting joint parental care and often joint territory defense. It is thus unsurprising that coordinated duetting is common, and better studied, in birds than in most other groups (Brenowitz, Arnold, & Levin, 1985; Farabaugh, 1982; Thorpe, 1963, 1972; Thorpe & North, 1965; Wickler, 1980). Avian duetting, like female song more generally, is more common in tropical nonmigratory species than in temperate climates (Langmore, 2000; Riebel, 2003), and the ancestral state of songbirds may have included both male and female song (Odom, Hall, Riebel, Omland, & Langmore, 2014).

Duets have also evolved convergently in at least four monogamous primate species (Haimoff, 1986). Typically, in duets, the male and female parts are temporally coordinated and interlock antiphonally, and this temporal coordination requires some learning by the pair members to become fluent. However, there is no evidence for vocal learning of the calls themselves, which (especially for gibbons) are innately determined (Geissmann, 2000). Gibbon duets probably rely on reaction-based turn taking and do not appear to require predictive BPS mechanisms, but this remains an understudied area.

Although it is rare, some bird species also show a mixture between duetting and chorusing. The plain-tailed wren (*Thryothorus = Pheugopedius euophrys*) is a member of a clade in which all species show duetting (Mann, Dingess, Barker, Graves, & Slater, 2009), but unique to this species, the birds often live in larger mixed-sex groups that sing together. During territorial song displays, the female and male parts interlock antiphonally in the normal way, but multiple females sing the female part in perfect synchrony, while the males also combine their parts synchronously, with remarkably exact timing (Mann, Dingess, & Slater, 2006). In general, duetting and chorusing provide a rich set of analogues to human ensemble behavior, allowing both the evolution and mechanistic basis of such behaviors to be analyzed using the comparative method.

4 Dance: A Neglected Core Component of Musicality

I conclude with a component of human musicality that has been unjustly neglected in most discussions of the cognition and neuroscience of music: our capacity to dance (but see Fitch, 2016; Laland et al., 2016). Although English and many other European languages distinguish "music" from "dance," this distinction is not made in many other languages, where music and dance are considered together to comprise a distinctive mode of human interaction (cf. Merker, 2000; Morley, 2014; Nettl, 2000). A close linkage between music and dance is also evident in most European music outside the concert hall, and although dance may be distinguished from music, it is almost always accompanied by it. Furthermore, so much of human music is created for the express purpose of dancing that in the development of many musical styles (e.g., waltz or swing), dance and music have undoubtedly influenced each other deeply (Crease, 2000; Fitch, 2016). Finally, dancers make use of the synchronization abilities to synchronize with the music and with other dancers. Thus, I nominate dance as another core component of human musicality.

It is not trivial to define dance—although roughly speaking dance involves making movements to music that aren't necessary to produce that music—and it is probably foolhardy to seek a definition that clearly distinguishes it from other aspects of musicality. Again starting from the comparative viewpoint, there is a vast array of visual displays among animals, from claw waving in crabs to begging gestures in apes, which are probably not relevant to human musicality. With such comparisons in mind, I will provisionally define *dance* as "complex, communicative body movements, typically produced as optional accompaniments to a multimodal display that includes sound production." This definition picks out the core of most human dancing without attempting to distinguish it strictly from drumming: by this definition, tap dancing constitutes both dancing and drumming simultaneously. Chimpanzee drumming is typically the culmination of a multimodal display that includes both vocal elements (pant-hoot) and a swaggering and rushing about; I am happy to consider this a form of dancing. By my definition, the expressive movements often made by instrumentalists as they play, over and above those necessary to produce the sounds, would also be classified as dancing, as would head bobbing, foot tapping, or hand movements that listeners make in synchrony with music. While I am aware that pantomime, or some "high art" dance, may be performed silently, I don't find such rare exceptions particularly troublesome (any more than John Cage's famous *4′33″*—a "musical" piece involving no sound—should constitute a central problem in defining music). If we seek comparisons that help fuel scientific, biologically oriented research, we should seek useful generalizations rather than perfect definitions.

When searching for animal analogues of dance, it is important to note that multimodal signaling is a ubiquitous aspect of advertisement displays in animals and probably represents the rule rather than the exception (cf. Coleman, 2009; Partan & Marler, 2005; Starnberger, Preininger, & Hödl, 2014). For example, many frogs have air sacs that are inflated when the frog calls. In some species, these sacs are decorated in various ways and thus serve as simultaneous visual displays; studies with robot frogs demonstrate that both components of these multimodal displays are attended to by other frogs (Narins, Grabul, Soma, Gaucher, & Hödl, 2005). But because vocal sac inflation is a mechanically necessary part of the vocal display rather than an accompaniment to that display, I would not consider this to be "dance." However, a frog that waves its feet while calling would be dancing by my definition (cf. Krishna & Krishna, 2006; Preininger et al., 2013). The clearest potential analogues of human dancing are seen in the elaborate and stereotyped visual or vocal displays seen during courtship in many bird species, such as birds of paradise, ducks, grebes, and cranes. In the case of cranes, for example, courtship is a protracted affair that includes elaborate, synchronized species-typical body and neck movement in addition to the pairs' synchronized calling behavior (Archibald, 1976; Johnsgard, 1983). These are typically, and I think rightly, referred to as "dance." Other multimodal displays exist that seem intuitively to be dancelike, for example, the "stiff walking" seen during aggressive display in red deer, accompanied by roaring, or the "swaggering" gait, with full piloerection, often seen during pant-hoot displays in chimpanzees, are quite difficult to quantify but deserve further study.

Although animal "dancing" behaviors remain relatively unexplored, particularly in the context of biomusicology, I suggest that accepting dance as a core component of human musicality will open the door to further fruitful comparisons, uncovering both analogues and possible homologues in other species. More generally, I suggest that biomusicology will profit greatly by explicitly incorporating dance into discussions of the biology and evolution of human music (Laland et al., 2016). It is time to recognize dance as a full peer of song or drumming in human expressions of musicality.

Conclusion

In closing, I reemphasize that both the principles and components discussed in this chapter are offered as points of departure for future work, not final pronouncements. I fully expect, and hope, that as the field of biomusicology progresses, more principles will be developed and/or the ones presented here augmented and refined. In particular, the four-component breakdown I have given is just one way to slice the pie of musicality, developed specifically for the purposes of fruitful comparisons among species. Two other important multicomponent analyses include the search for musical universal of various types and the attempt to break music into "design features," which allow a matrix of comparisons between music and other human cognitive features (such as language or architecture) and with other animal communication systems, following Charles Hockett (1960). Hockett's list of design features of language provided an important starting point for subsequent research in animal communication, and I have offered a list of musical design features extending his (Fitch, 2005, 2006). My list includes some features that are shared with language (such as generativity and complexity), as well as features that differentiate most music from language (such as the use of discrete pitches or of isochronic rhythms), but shorter lists of musical design features have also been proposed (Bispham, 2009). The design feature approach focuses on characteristics of music rather than on the cognitive abilities making up musicality, but may be preferable in cases where we have empirical access only to surface behaviors. There is thus plenty of room for expansion and exploration of this feature-based approach to analyzing musicality into component parts.

Another important alternate approach to analyzing the components underlying musicality is much older and much more controversial: the search for musical universals. This was a core desideratum of the first wave of comparative musicologists, centered in Germany in the twentieth century between the wars (Sachs, 1926, 1940; von Hornbostel, 1928). Unfortunately, with a few exceptions (Harwood, 1976; Lomax, 1980; Lomax & Berkowitz, 1972; Sachs, 1953, 1962) the search for universal principles or traits of music was abandoned after the breakup of this group of researchers by the Nazis. Indeed, in postwar ethnomusicology, the very notion of musical universals became somewhat taboo, and in line with prevailing anthropological attitudes concerning culture more generally, music was seen as a system free to vary with virtually no constraints (cf. Nettl, 1983, 1995, 2000). But the steady increase in the scientific study of music, particularly music neuroscience and music cognition, has led a few brave scholars to reopen this search (Brown & Jordania, 2013; Nettl, 2000; Savage et al., 2015). This empirical quest to derive broad generalizations about human music is clearly another important component of biomusicology that has been neglected for too long.

Biomusicologists may learn some important lessons from the long-running discussions of language universals in linguistics (cf. Fitch, 2011). The earliest modern attempts to empirically analyze language universals were led by comparative linguist Joseph Greenberg (1963), who clearly distinguished between truly universal traits (e.g., "all languages have both nouns and verbs"), statistical universals ("most languages have trait *x*"), and implicational universals. Implicational universals are the most interesting: they take the form, "if a language has trait *x*, it will also have trait *y*," and again may be truly universal or just strong statistical generalizations. I know of few discussions of this type of universals concerning musicality, but Temperley (2004) has offered a fascinating set of candidate topics for this type of implicational generalization in music. For example, Temperley suggests a trade-off between syncopation and rubato (free expressive variation in tempo) as a musical style evolves, arguing convincingly that syncopation works well only in the context of a relatively strict isochronic beat (otherwise time shifts intended as syncopations become indistinguishable from expressive temporal dynamics).

After Greenberg, the discussion of language universals became more heated when Noam Chomsky (1965) introduced his controversial concept of universal grammar (UG), adopting an old seventeenth-century term to a new purpose (Chomsky, 1965). The debate this concept sparked has often been unproductive, mainly due to the frequent conflation of UG (the capacity to acquire language) with superficial traits found in all human languages (Greenberg's "true universals"). Since true universals are unusual, their rarity has frequently been claimed to disprove the concept of UG itself (e.g., Evans & Levinson, 2009; Vaneechoutte, 2014), despite the fact that Chomsky stressed his focus on "deep-seated regularities"—very general aspects of the capacity to acquire and use language, such as its creative aspect—and not on traits found in all human languages (Chomsky, 1965, pp. 5–7). Biomusicology, and musicology more generally, will do well to learn from this history of linguistic debate over language universals lest we be doomed to repeat it. The key point is that some particular capacity may well be a universal trait of human *musicality* (available as part of the cognitive tool kit of any normal human) without being expressed in all musical styles or observed in all human cultures. For example, humans around the world have a capacity to entrain our movements to musical rhythms, but we do not express this ability with every form of music. Indeed, for some nonisochronic "free" rhythms, this would be both difficult and culturally inappropriate (Clayton, 1996). But there is no conflict in claiming that synchronization to isochronic rhythms is a universal human capacity and simultaneously observing that it is not observed in all musical pieces, styles or cultures (cf. Brown & Jordania, 2013). Similar points could be made, mutatis mutandis, concerning melodic grouping or harmonic "syntax."

Although the principles and components introduced here are preliminary and by no means exhaust the store, I hope to have shown how adopting some breakdown and then proceeding to study each component comparatively opens the door to rich and exciting sources of data to help understand the biology and evolution of music. Asking monolithic questions like, "When did music evolve?" is unlikely to be productive, but questions like, "When did our propensity to drum with our limbs evolve?" can already be tentatively answered ("around 8 million years ago").

Similarly, a question like, "*Why* did music evolve?" must immediately grapple with the broad range of uses to which music is put in human cultures. In contrast, the question, "Why did the human capacity to *entrain* evolve?" is one that we can begin to answer by employing the comparative approach, given the many species that have convergently evolved this ability. Again, the exact breakdown is likely to remain a matter of debate for the foreseeable future and will be dependent on the specific problem being addressed. But I suggest that the need for some breakdown is a core prerequisite for future progress in this fascinating field of research.

Acknowledgments

An earlier version of this chapter was previously published as Fitch (2015). This work was supported by ERC Advanced Grant SOMACCA (230604) and Austrian Science Fund Grant "Cognition and Communication" (FWF W1234-G17). I thank the Lorentz Center for hosting the productive and informative workshop leading to this chapter and all of the participants for the lively and constructively critical discussion. I greatly profited from the comments of Gesche Westphal-Fitch, Henkjan Honing, and two anonymous reviewers on a previous version of this chapter, as well as from discussions with Jessica Grahn, Anirudh Patel, Björn Merker, and Henkjan Honing.

References

Arcadi, A. C. (1996). Phrase structure of wild chimpanzee pant hoots: Patterns of production and interpopulation variability. *American Journal of Primatology, 39*(3), 159–178.

Arcadi, A. C., Robert, D., & Mugurusi, F. (2004). A comparison of buttress drumming by male chimpanzees from two populations. *Primates, 45*(2), 135–139. doi:10.1007/s10329-003-0070-8

Archibald, G. W. (1976). *The unison call of cranes as a useful taxonomic tool* (Unpublished doctoral dissertation). Cornell University, Ithaca, NY.

Babiszewska, M., Schel, A. M., Wilke, C., & Slocombe, K. E. (2015). Social, contextual, and individual factors affecting the occurrence and acoustic structure of drumming bouts in wild chimpanzees (*Pan troglodytes*). *American Journal of Physical Anthropology, 156*, 125–134.

Ball, P. (2010). *The music instinct: How music works and why we can't do without it.* Oxford: Oxford University Press.

Baugh, A. C., & Cable, T. (1978). *A history of the English language*. Englewood Cliffs, NJ: Prentice Hall.

Bigand, E., & Poulin-Charronnat, B. (2006). Are we "experienced listeners"? A review of the musical capacities that do not depend on formal musical training. *Cognition, 100*(1), 100–130. doi:10.1016/j.cognition.2005.11 .007

Bispham, J. C. (2009). Music's "design features": Musical motivation, musical pulse, and musical pitch. *Musicae Scientiae, 13*(2), 41–46. doi:10.1177/1029864909013002041

Bolcom, W., & Harrison, M. (1986). Ragtime. In P. Oliver, M. Harrison, & W. Bolcom (Eds.), *The New Grove: Gospel, blues and jazz with spirituals and ragtime* (pp. 23–35). London: Macmillan.

Brenowitz, E. A., Arnold, A. P., & Levin, R. N. (1985). Neural correlates of female song in tropical duetting birds. *Brain Research, 343*(1), 104–112. doi:10.1016/0006-8993(85)91163-1

Brown, S. (2000). Evolutionary models of music: From sexual selection to group selection. In F. Tonneau & N. S. Thompson (Eds.), *Perspectives in ethology: Evolution, culture and behavior* (Vol. 13, pp. 231–283). New York: Kluwer Academic; 10.1007/978-1-4615-1221-9_9

Brown, S., & Jordania, J. (2013). Universals in the world's musics. *Psychology of Music, 41*(2), 229–248. doi:10.1177/0305735611425896

Buck, J. (1938). Synchronous rhythmic flashing in fireflies. *Quarterly Review of Biology, 13*(3), 301–314. doi:10.1086/394562

Buck, J. (1988). Synchronous rhythmic flashing in fireflies. II. *Quarterly Review of Biology, 63*(3), 265–287. doi:10.1086/415929

Buisson, D. (1990). Les flûtes paléolithiques d'Isturitz (Pyrénées-Atlantiques). *Bulletin de la Société Préhistorique Francaise, 87*(10), 420–433. doi:10.3406/bspf.1990.9925

Chomsky, N. (1965). *Aspects of the theory of syntax.* Cambridge, MA: MIT Press.

Clayton, M. R. L. (1996). Free rhythm: Ethnomusicology and the study of music without metre. *Bulletin of the School of Oriental and African Studies, University of London, 59*(2), 323–332. doi:10.1017/ S0041977X00031608

Coleman, S. W. (2009). Taxonomic and sensory biases in the mate-choice literature: There are far too few studies of chemical and multimodal communication. *Acta Ethologica, 12*(1), 45–48. doi:10.1007/s10211-008 -0050-5

Conard, N. J., Malina, M., & Münzel, S. C. (2009). New flutes document the earliest musical tradition in southwestern Germany. *Nature, 460*, 737–740. doi:10.1038/nature08169

Cook, P., Rouse, A., Wilson, M., & Reichmuth, C. J. (2013). A California sea lion (*Zalophus californianus*) can keep the beat: Motor entrainment to rhythmic auditory stimuli in a non vocal mimic. *Journal of Comparative Psychology, 127*(4), 412–427. doi:10.1037/a0032345

Crease, R. P. (2000). Jazz and dance. In B. Kirchner (Ed.), *The Oxford companion to jazz* (pp. 696–705). New York: Oxford University Press.

Dodenhoff, D. J., Stark, R. D., & Johnson, E. V. (2001). Do woodpecker drums encode information for species recognition? *Condor, 103*(1), 143–150. doi:10.1650/0010-5422(2001)103[0143:DWDEIF]2.0.CO;2

Elbert, T., Pantev, C., Wienbruch, C., Rockstroh, B., & Taub, E. (1995). Increased cortical representation of the fingers of the left hand in string players. *Science, 270*(5234), 305–307. doi:10.1126/science.270.5234.305

Evans, N., & Levinson, S. C. (2009). The myth of language universals: Language diversity and its importance for cognitive science. *Behavioral and Brain Sciences, 32*(5), 429–448. doi:10.1017/s0140525x0999094x

Farabaugh, S. M. (1982). The ecological and social significance of duetting. In D. S. Kroodsma & E. H. Miller (Eds.), *Acoustic communication in birds* (Vol. 2, pp. 85–124). New York: Academic Press; 10.1007/978-1-4684-2901-5_2

Feenders, G., Liedvogel, M., Rivas, M., Zapka, M., Horita, H., Hara, E., et al. (2008). Molecular mapping of movement-associated areas in the avian brain: A motor theory for vocal learning origin. *PLoS One, 3*(3), e1768. doi:10.1371/journal.pone.0001768

Fitch, W. T. (2000). The phonetic potential of nonhuman vocal tracts: Comparative cineradiographic observations of vocalizing animals. *Phonetica, 57*(2–4), 205–218. doi:10.1159/000028474

Fitch, W. T. (2005). The evolution of music in comparative perspective. *Annals of the New York Academy of Sciences, 1060*, 29–49. doi:10.1196/annals.1360.004

Fitch, W. T. (2006). The biology and evolution of music: A comparative perspective. *Cognition, 100*(1), 173–215. doi:10.1016/j.cognition.2005.11.009

Fitch, W. T. (2008). Glossogeny and phylogeny: Cultural evolution meets genetic evolution. *Trends in Genetics, 24*(8), 373–374. doi:10.1016/j.tig.2008.05.003

Fitch, W. T. (2009). The biology and evolution of language: "Deep homology" and the evolution of innovation. In M. S. Gazzaniga (Ed.), *The cognitive neurosciences IV* (pp. 873–883). Cambridge, MA: MIT Press.

Fitch, W. T. (2011). Unity and diversity in human language. *Philosophical Transactions of the Royal Society of London B: Biological Sciences, 366*(1563), 376–388. doi:10.1098/rstb.2010.0223

Fitch, W. T. (2013). Rhythmic cognition in humans and animals: Distinguishing meter and pulse perception. *Frontiers in Systems Neuroscience, 7*(68), 1–16. doi:10.3389/fnsys.2013.00068

Fitch, W. T. (2015). Four principles of bio-musicology. *Philosophical Transactions of the Royal Society of London B: Biological Sciences, 370*(1664), 20140091. doi:10.1098/rstb.2014.0091

Fitch, W. T. (2015). The biology and evolution of musical rhythm: An update. In I. Toivonen, P. Csúri, & E. van der Zee (Eds.), *Structures in the mind: Essays on language, music, and cognition in honor of Ray Jackendoff* (pp. 293–324). Cambridge, MA: MIT Press.

Fitch, W. T. (2016). Dance, music, meter and groove: A forgotten partnership. *Frontiers in Human Neuroscience, 10*(64), 1–7. doi:10.3389/fnhum.2016.00064

Fitch, W. T. (2017). Empirical approaches to the study of language evolution. *Psychonomic Bulletin and Review, 24*(1), 3–33. doi:10.3758/s13423-017-1236-5

Fitch, W. T., & Jarvis, E. D. (2013). Birdsong and other animal models for human speech, song, and vocal learning. In M. A. Arbib (Ed.), *Language, music, and the brain: A mysterious relationship* (pp. 499–539). Cambridge, MA: MIT Press.

Fitch, W. T., & Mietchen, D. (2013). Convergence and deep homology in the evolution of spoken language. In J. J. Bolhuis & M. B. H. Everaert (Eds.), *Birdsong, speech and language: Exploring the evolution of mind and brain* (pp. 45–62). Cambridge, MA: MIT Press.

Fitch, W. T., & Suthers, R. A. (2016). Vertebrate vocal production: An introductory overview. In R. A. Suthers, W. T. Fitch, A. N. Popper, & R. R. Fay (Eds.), *Vertebrate sound production and acoustic communication* (pp. 1–18). New York: Springer.

Frigyesi, J. (1993). Preliminary thoughts toward the study of music without clear beat: The example of "flowing rhythm" in Jewish "nusah." *Asian Music, 24*(2), 59–88. doi:10.2307/834467

Gaser, C., & Schlaug, G. (2003). Brain structures differ between musicians and non-musicians. *Journal of Neuroscience, 23*(27), 9240–9245.

Geissmann, T. (2000). Gibbon song and human music from an evolutionary perspective. In N. L. Wallin, B. Merker, & S. Brown (Eds.), *The origins of music* (pp. 103–123). Cambridge, MA: MIT Press.

Glazko, G. V., & Nei, M. (2003). Estimation of divergence times for major lineages of primate species. *Molecular Biology and Evolution, 20*(3), 424–434. doi:10.1093/molbev/msg050

Greenberg, J. (1963). *Universals of language*. Cambridge, MA: MIT Press.

Greenfield, M. D. (2005). Mechanisms and evolution of communal sexual displays in arthropods and anurans. *Advances in the Study of Behavior, 35*, 1–62. doi:10.1016/s0065-3454(05)35001-7

Haesler, S., Rochefort, C., Geogi, B., Licznerski, P., Osten, P., & Scharff, C. (2007). Incomplete and inaccurate vocal imitation after knockdown of FoxP2 in songbird basal ganglia nucleus Area X. *PLoS Biology, 5*(12), e321. doi:10.1371/journal.pbio.0050321

Hagen, E. H., & Bryant, G. A. (2003). Music and dance as a coalition signaling system. *Human Nature, 14*(1), 21–51. doi:10.1007/s12110-003-1015-z

Hagen, E. H., & Hammerstein, P. (2009). Did Neanderthals and other early humans sing? Seeking the biological roots of music in the territorial advertisements of primates, lions, hyenas, and wolves. *Musicae Scientiae, 13*(2), 291–320. doi:10.1177/1029864909013002131

Hahn, J., & Münzel, S. (1995). Knochenflöten aus dem Aurignacien des Geissenklösterle bei Blaubeuren, Alb-Donau-Kreis. [Bone flutes from the Aurignacian of Geissenklösterle, near Blaubeuren in the Alb-Donau-Kreis (by Fitch)]. *Fundberichte aus Baden-Würtemberg, 20*, 1–12.

Haimoff, E. H. (1986). Convergence in the duetting of monogamous Old World primates. *Journal of Human Evolution, 15*(1), 51–59. doi:10.1016/s0047-2484(86)80065-3

Hanggi, E. B., & Schusterman, R. J. (1992). Underwater acoustical displays by male harbour seals (*Phoca vitulina*): Initial results. In J. A. Thomas, R. A. Kastelein, & A. Y. Supin (Eds.), *Marine mammal sensory systems* (pp. 601–629). New York: Plenum Press; 10.1007/978-1-4615-3406-8_31

Hanggi, E. B., & Schusterman, R. J. (1994). Underwater acoustic displays and individual variation in male harbor seals, *Phoca vitulina. Animal Behaviour, 48*(6), 1275–1283. doi:10.1006/anbe.1994.1363

Hardus, M. E., Lameira, A. R., van Schaik, C. P., & Wich, S. A. (2009). Tool use in wild orangutans modifies sound production: A functionally deceptive innovation? *Proceedings of the Royal Society of London B: Biological Sciences, 276*(1673), 3689–3694. doi:10.1098/rspb.2009.1027

Harrison, M. (1986). Jazz. In P. Oliver, M. Harrison, & W. Bolcom (Eds.), *The New Grove gospel: Blues and jazz with spirituals and ragtime* (pp. 223–356). London: Macmillan.

Harwood, D. L. (1976). Universals in music: A perspective from cognitive psychology. *Ethnomusicology, 20*(3), 521–533. doi:10.2307/851047

Hasegawa, A., Okanoya, K., Hasegawa, T., & Seki, Y. (2011). Rhythmic synchronization tapping to an audio-visual metronome in budgerigars. *Scientific Reports, 1*, 1–8. doi:10.1038/srep00120

Hattori, Y., Tomonaga, M., & Matsuzawa, T. (2013). Spontaneous synchronized tapping to an auditory rhythm in a chimpanzee. *Scientific Reports, 3*, 1–6. doi:10.1038/srep01566

Hauser, M. D., & McDermott, J. (2003). The evolution of the music faculty: A comparative perspective. *Nature Neuroscience, 6*, 663–668. doi:10.1038/nn1080

Heinsohn, R., Zdenek, C. N., Cunningham, R. B., Endler, J. A., & Langmore, N. E. (2017). Tool-assisted rhythmic drumming in palm cockatoos shares key elements of human instrumental music. *Science Advances, 3*(6), e1602399. doi:10.1126/sciadv.1602399

Herbst, C. T., Stoeger, A. S., Frey, R., Lohscheller, J., Titze, I. R., Gumpenberger, M., et al. (2012). How low can you go? Physical production mechanism of elephant infrasonic vocalizations. *Science, 337*(6094), 595–599. doi:10.1126/science.1219712

Hockett, C. F. (1960). Logical considerations in the study of animal communication. In W. E. Lanyon & W. N. Tavolga (Eds.), *Animal sounds and communication* (pp. 392–430). Washington, DC: American Institute of Biological Sciences.

Honing, H. (2009). *Musical cognition: A science of listening*. New Brunswick, NJ: Transaction.

Honing, H., Merchant, H., Háden, G. P., Prado, L., & Bartolo, R. (2012). Rhesus monkeys (*Macaca mulatta*) detect rhythmic groups in music, but not the beat. *PLoS One, 7*(12), e51369. doi:10.1371/journal.pone.0051369

Honing, H., ten Cate, C., Peretz, I., & Trehub, S. E. (2015). Without it no music: Cognition, biology and evolution of musicality. *Philosophical Transactions of the Royal Society of London B: Biological Sciences, 370*(1664), 1–8. doi:10.1098/rstb.2014.0088

Hurford, J. (1990). Nativist and functional explanations in language acquisition. In I. M. Roca (Ed.), *Logical issues in language acquisition* (pp. 85–136). Dordrecht: Foris.

Huron, D. (2001). Is music an evolutionary adaptation? *Annals of the New York Academy of Sciences, 930*, 43–61. doi:10.1111/j.1749-6632.2001.tb05724.x

Janik, V. M., & Slater, P. B. (1997). Vocal learning in mammals. *Advances in the Study of Behavior, 26*, 59–99. doi:10.1016/s0065-3454(08)60377-0

Johnsgard, P. A. (1983). *Cranes of the world.* Bloomington: Indiana University Press.

Kalan, A. K., & Rainey, H. J. (2009). Hand-clapping as a communicative gesture by wild female swamp gorillas. *Primates, 50*(3), 273–275. doi:10.1007/s10329-009-0130-9

Kirby, S. (2017). Culture and biology in the origins of linguistic structure. *Psychonomic Bulletin and Review, 24*(1), 118-137. doi:10.3758/s13423-016-1166-7

Koops, K., & Matsuzawa, T. (2006). Hand clapping by a chimpanzee in the Nimba mountains, Guinea, West Africa. *Pan Africa News, 13*(2), 19–21.

Krishna, S. N., & Krishna, S. B. (2006). Visual and acoustic communication in an endemic stream frog, *Micrixalus saxicolus* in the Western Ghats, India. *Amphibia-Reptilia, 27*(1), 143–147. doi:10.1163/156853806776052056

Kunej, D., & Turk, I. (2000). New perspectives on the beginnings of music: Archaeological and musicological analysis of a middle Paleolithic bone "flute." In N. L. Wallin, B. Merker, & S. Brown (Eds.), *The origins of music* (pp. 235–268). Cambridge, MA: MIT Press.

Laland, K., Wilkins, C., & Clayton, N. (2016). The evolution of dance. *Current Biology, 26*(1), R5–R9. doi:10.1016/j.cub.2015.11.031

Langmore, N. E. (2000). Why female birds sing. In Y. Espmark, T. Amundsen, & G. Rosenqvist (Eds.), *Signalling and signal design in animal communication* (pp. 317–327). Trondheim: Tapir Academic Press.

Lehrman, D. S. (1953). A critique of Konrad Lorenz's theory of instinctive behavior. *Quarterly Review of Biology, 28*(4), 337–363. doi:10.1086/399858

Lomax, A. (1977). Universals in song. *World of Music, 19*(1/2), 117–129.

Lomax, A. (1980). Factors of musical style. In S. Diamond (Ed.), *Theory and practice: Essays presented to Gene Weltfish* (pp. 29–58). The Hague: Mouton; 10.1515/9783110803211.29

Lomax, A., & Berkowitz, N. (1972). The evolutionary taxonomy of culture. *Science, 177*(4045), 228–239. doi:10.1126/science.177.4045.228

Madison, G., & Merker, B. (2004). Human sensorimotor tracking of continuous subliminal deviations from isochrony. *Neuroscience Letters, 370*(1), 69–73. doi:10.1016/j.neulet.2004.07.094

Mann, N. I., Dingess, K. A., Barker, K., Graves, J. A., & Slater, P. J. B. (2009). A comparative study of song form and duetting in neotropical *Thryothorus* wrens. *Behaviour, 146*(1), 1–43. doi:10.1163/156853908x390913

Mann, N. I., Dingess, K. A., & Slater, P. J. B. (2006). Antiphonal four-part synchronized chorusing in a neotropical wren. *Biology Letters, 2*(1), 1–4. doi:10.1098/rsbl.2005.0373

Mauch, M., MacCallum, R. M., Levy, M., & Leroi, A. M. (2015). The evolution of popular music: USA 1960–2010. *Royal Society Open Science, 2*(5), 1–19. doi:10.1098/rsos.150081

Maynard Smith, J., & Harper, D. G. C. (2003). *Animal signals.* Oxford: Oxford University Press.

McDermott, J., & Hauser, M. D. (2005). The origins of music: Innateness, uniqueness, and evolution. *Music Perception, 23*(1), 29–59. doi:10.1525/mp.2005.23.1.29

Merchant, H., & Honing, H. (2013). Are non-human primates capable of rhythmic entrainment? Evidence for the gradual audiomotor evolution hypothesis. *Frontiers in Neuroscience, 7*(274), 1–8. doi:10.3389/fnins.2013.00274

Merchant, H., Zarco, W., Pérez, O., Prado, L., & Bartolo, R. (2011). Measuring time with different neural chronometers during a synchronization-continuation task. *Proceedings of the National Academy of Sciences of the United States of America, 108*(49), 19784–19789. doi:10.1073/pnas.1112933108

Merker, B. (2000). Synchronous chorusing and human origins. In N. L. Wallin, B. Merker, & S. Brown (Eds.), *The origins of music* (pp. 315–327). Cambridge, MA: MIT Press.

Moiseff, A., & Copeland, J. (2010). Firefly synchrony: A behavioral strategy to minimize visual clutter. *Science, 329*(5988), 181. doi:10.1126/science.1190421

Morley, I. (2013). *The prehistory of music: Human evolution, archaeology, and the origins of musicality*. Oxford: Oxford University Press.

Morley, I. (2014). A multi-disciplinary approach to the origins of music: Perspectives from anthropology, archaeology, cognition and behaviour. *Journal of Anthropological Sciences, 92*, 147–177. doi:10.4436/JASS.92008

Narins, P. M., Grabul, D. S., Soma, K. K., Gaucher, P., & Hödl, W. (2005). Cross-modal integration in a dart-poison frog. *Proceedings of the National Academy of Sciences of the United States of America, 102*(7), 2425–2429. doi:10.1073/pnas.0406407102

Nettl, B. (1983). *The study of ethnomusicology: Twenty-nine issues and concepts*. Urbana: University of Illinois Press.

Nettl, B. (1995). *Heartland excursions: Ethnomusicological reflections on schools of music*. Urbana: University of Illinois Press.

Nettl, B. (2000). An ethnomusicologist contemplates universals in musical sound and musical culture. In N. L. Wallin, B. Merker, & S. Brown (Eds.), *The origins of music* (pp. 463–472). Cambridge, MA: MIT Press.

Odom, K. J., Hall, M. L., Riebel, K., Omland, K. E., & Langmore, N. E. (2014). Female song is widespread and ancestral in songbirds. *Nature Communications, 5*(3379), 1–6. doi:10.1038/ncomms4379

Partan, S. R., & Marler, P. (2005). Issues in the classification of multimodal communication signals. *American Naturalist, 166*(2), 231–245. doi:10.1086/431246

Patel, A. D. (2006). Musical rhythm, linguistic rhythm, and human evolution. *Music Perception, 24*(1), 99–104. doi:10.1525/mp.2006.24.1.99

Patel, A. D., Iversen, J. R., Bregman, M. R., & Schulz, I. (2009). Experimental evidence for synchronization to a musical beat in a nonhuman animal. *Current Biology, 19*(10), 827–830. doi:10.1016/j.cub.2009.03.038

Pinker, S. (1997). *How the mind works*. New York: Norton.

Preininger, D., Boeckle, M., Freudmann, A., Starnberger, I., Sztatecsny, M., & Hödl, W. (2013). Multimodal signaling in the small torrent frog (*Micrixalus saxicola*) in a complex acoustic environment. *Behavioral Ecology and Sociobiology, 67*, 1449–1456. doi:10.1007/s00265-013-1489-6

Ralls, K., Fiorelli, P., & Gish, S. (1985). Vocalizations and vocal mimicry in captive harbor seals, *Phoca vitulina. Canadian Journal of Zoology, 63*(5), 1050–1056. doi:10.1139/ z85–157

Randall, J. A. (2001). Evolution and function of drumming as communication in mammals. *American Zoologist, 41*(5), 1143–1156. doi:10.1093/icb/41.5.1143

Ravignani, A., Bowling, D., & Fitch, W. T. (2014). Chorusing, synchrony and the evolutionary functions of rhythm. *Frontiers in Psychology, 5*(1118), 1–15. doi:10.3389/fpsyg.2014.01118

Redshaw, M., & Locke, K. (1976). The development of play and social behaviour in two lowland gorilla infants. *Journal of the Jersey Wildlife Preservation Trust, 13th Annual Report*, 71–86.

Remedios, R., Logothetis, N. K., & Kayser, C. (2009). Monkey drumming reveals common networks for perceiving vocal and nonvocal communication sounds. *Proceedings of the National Academy of Sciences of the United States of America, 106*(42), 18010–18015. doi:10.1073/pnas.0909756106

Repp, B. H. (2005). Sensorimotor synchronization: A review of the tapping literature. *Psychonomic Bulletin and Review, 12*(6), 969–992. doi:10.3758/bf03206433

Repp, B. H., & Su, Y.-H. (2013). Sensorimotor synchronization: A review of recent research (2006–2012). *Psychonomic Bulletin and Review, 20*(3), 403–452. doi:10.3758/s13423-012-0371-2

Riebel, K. (2003). The “mute” sex revisited: Vocal production and perception learning in female songbirds. *Advances in the Study of Behavior, 33*, 49–86. doi:10.1016/s0065-3454(03)33002-5

Riechert, S. E. (1978). Games spiders play: Behavioural variability in territorial disputes. *Behavioral Ecology and Sociobiology, 3*(2), 135–162. doi:10.1007/bf00294986

Roederer, J. G. (1984). The search for a survival value for music. *Music Perception, 1*(3), 350–356. doi:10.2307/40285265

Sachs, C. (1926). Anfänge der Musik. *Bulletin de l'Union Musicologique, 6*, 136–236.

Sachs, C. (1940). *The history of musical instruments*. New York: Norton.

Sachs, C. (1953). *Rhythm and tempo: A study in music history*. New York: Norton.

Sachs, C. (1962). *The wellsprings of music*. The Hague: Martinus Nijhoff.

Savage, P. E., Brown, S., Sakai, E., & Currie, T. E. (2015). Statistical universals reveal the structures and functions of human music. *Proceedings of the National Academy of Sciences, USA*, *112*(29), 8987–8992.

Schachner, A. (2010). Auditory-motor entrainment in vocal mimicking species: Additional ontogenetic and phylogenetic factors. *Communicative and Integrative Biology*, *3*(3), 290–293. doi:10.4161/cib.3.3.11708

Schaller, G. B. (1963). *The mountain gorilla: Ecology and behavior*. Chicago: University of Chicago Press.

Scharff, C., & Petri, J. (2011). Evo-devo, deep homology and FoxP2: Implications for the evolution of speech and language. *Philosophical Transactions of the Royal Society of London B: Biological Sciences*, *366*(1574), 2124–2140. doi:10.1098/rstb.2011.0001

Schellenberg, E. G., & Trehub, S. E. (1999). Culture-general and culture-specific factors in the discrimination of melodies. *Journal of Experimental Child Psychology*, *74*(2), 107–127. doi:10.1006/jecp.1999.2511

Schlaug, G. (2001). The brain of musicians: A model for functional and structural adaptation. *Annals of the New York Academy of Sciences*, *930*, 281–299. doi:10.1111/j.1749-6632.2001.tb05739.x

Schusterman, R. J., Balliet, R., & St. John, S. (1970). Vocal displays under water by the gray seal, the harbor seal and the Steller sea lion. *Psychonomic Science*, *18*(5), 303–305. doi:10.3758/bf03331839

Scott-Phillips, T. D., & Kirby, S. (2010). Language evolution in the laboratory. *Trends in Cognitive Sciences*, *14*(9), 411–417. doi:10.1016/j.tics.2010.06.006

Searcy, W. A., & Nowicki, S. (2005). *The evolution of animal communication: Reliability and deception in signaling systems*. Princeton, NJ: Princeton University Press.

Shubin, N., Tabin, C., & Carroll, S. (1997). Fossils, genes and the evolution of animal limbs. *Nature*, *388*, 639–648.

Shubin, N., Tabin, C., & Carroll, S. (2009). Deep homology and the origins of evolutionary novelty. *Nature*, *457*(7231), 818–823. doi:10.1038/nature07891

Sismondo, E. (1990). Synchronous, alternating and phase-locked stridulation by a tropical katydid. *Science*, *249*(4964), 55–58. doi:10.1126/science.249.4964.55

Smith, K., & Kirby, S. (2008). Cultural evolution: Implications for understanding the human language faculty and its evolution. *Philosophical Transactions of the Royal Society of London B: Biological Sciences*, *363*(1509), 3591–3603. doi:10.1098/rstb.2008.0145

Stark, R. D., Dodenhoff, D. J., & Johnson, E. V. (1998). A quantitative analysis of woodpecker drumming. *Condor*, *100*(2), 350–356. doi:10.2307/1370276

Starnberger, I., Preininger, D., & Hödl, W. (2014). From uni- to multimodality: Towards an integrative view on anuran communication. *Journal of Comparative Physiology A: Neuroethology, Sensory, Neural, and Behavioral Physiology*, *200*(9), 777–787. doi:10.1007/s00359-014-0923-1

Temperley, D. (2004). Communicative pressure and the evolution of musical styles. *Music Perception*, *21*(3), 313–337. doi:10.1525/mp.2004.21.3.313

Thomson, D. H., & Richardson, W. J. (1995). Marine mammal sounds. In W. J. Richardson, C. R. J. Greene, C. I. Malme, & D. H. Thomson (Eds.), *Marine mammals and noise* (pp. 159–204). San Diego, CA: Academic Press; 10.1016/b978-0-08-057303-8.50010-0

Thorpe, W. H. (1963). Antiphonal singing in birds as evidence for avian auditory reaction time. *Nature*, *197*(4869), 774–776. doi:10.1038/197774a0

Thorpe, W. H. (1972). Duetting and antiphonal song in birds: Its extent and significance. *Behaviour* (Suppl. 18), 1–197.

Thorpe, W. H., & North, M. E. W. (1965). Origin and significance of the power of vocal imitation: With special reference to the antiphonal singing of birds. *Nature*, *208*, 219–222. doi:10.1038/208219a0

Tinbergen, N. (1963). On aims and methods of ethology. *Zeitschrift für Tierpsychologie*, *20*(4), 410–433. doi:10.1111/j.1439-0310.1963.tb01161.x

Titze, I. R. (1994). *Principles of voice production*. Englewood Cliffs, NJ: Prentice Hall.

Tolbert, E. (1990). Women cry with words: Symbolization of affect in the Karelian lament. *Yearbook for Traditional Music, 22*, 80–105. doi:10.2307/767933

Trainor, L. J. (2005). Are there critical periods for musical development? *Developmental Psychobiology, 46*(3), 262–278. doi:10.1002/dev.20059

Trainor, L. J., & Trehub, S. E. (1992). A comparison of infants' and adults' sensitivity to Western musical structure. *Journal of Experimental Psychology: Human Perception and Performance, 18*, 394–402. doi:10.1037/0096-1523.18.2.394

Trehub, S. E. (2000). Human processing predispositions and musical universals. In N. L. Wallin, B. Merker, & S. Brown (Eds.), *The origins of music* (pp. 427–448). Cambridge, MA: MIT Press.

Trehub, S. E. (2003). The developmental origins of musicality. *Nature Neuroscience, 6*, 669–673. doi:10.1038/nn1084

Trehub, S. E., & Nakata, T. (2001). Emotion and music in infancy. *Musicae Scientiae, 5*(1), 37–61. doi:10.1177/10298649020050s103

Trehub, S. E., & Trainor, L. J. (1998). Singing to infants: Lullabies and play songs. In C. Rovee-Collier (Ed.), *Advances in infancy research* (pp. 43–77). Greenwich, CT: Ablex.

Vaneechoutte, M. (2014). The origin of articulate language revisited: The potential of a semi-aquatic past of human ancestors to explain the origin of human musicality and articulate language. *Human Evolution, 29*(1–3), 1–33.

von Hornbostel, E. M. (1928). African Negro music. *Africa, 1*(1), 30–62. doi:10.2307/1155862

Weiss, D. J., Garibaldi, B. T., & Hauser, M. D. (2001). The production and perception of long calls by cotton-top tamarins (*Saguinus oedipus*): Acoustic analyses and playback experiments. *Journal of Comparative Psychology, 115*(3), 258–271. doi:10.1037/0735-7036.115.3.258

Wickler, W. (1980). Vocal duetting and the pair bond: 1. Coyness and partner commitment: A hypothesis. *Zeitschrift für Tierpsychologie, 52*(2), 201–209. doi:10.1111/j.1439-0310.1980.tb00711.x

Winkler, H., & Short, L. (1978). A comparative analysis of acoustical signals in pied woodpeckers (Aves, Picoides). *Bulletin of the American Museum of Natural History, 160*, 1–109.

Wold, M., & Cykler, E. (1976). *An introduction to music and art in the Western world.* Dubuque, IA: Wm. C. Brown.

Wood, G. A. (1984). Tool use by the palm cockatoo *Probosciger aterrimus* during display. *Corella, 8*(4), 94–95.

Wood, G. A. (1988). Further field observations of the palm cockatoo *Probosciger aterrimus* in the Cape York Peninsula, Queensland. *Corella, 12*(2), 48–52.

3

Five Fundamental Constraints on Theories of the Origins of Music

Björn Merker, Iain Morley, and Willem Zuidema

Music is a cherished art form and a daily source of inspiration and pleasure, as well as an occasional irritation, for billions. It is also an extraordinarily complex phenomenon that appears to be not only uniquely human but a human universal (Brown, 1991; Brown & Jordania, 2013; von Humboldt, 1836). This uniqueness and universality raises the question of the origins of human musicality and the further issue of its possible biological and evolutionary background. However, as is the case for language and other aspects of human cognition, it is not obvious how to properly constrain our theorizing in this regard so as to avoid producing no more than "just-so stories." Evolutionary biologist Richard Lewontin (1998, p. 130) warned against "the childish notion that everything that is interesting about nature can be understood. History, and evolution is a form of history, simply does not leave sufficient traces, especially when it is the forces that are at issue. Form and even behavior may leave fossil remains, but forces like natural selection do not. It might be interesting to know how cognition (whatever that is) arose and spread and changed, but we cannot know."

Against this blunt pessimism stand those who hold, with Richard Byrne (2000, p. 543), that "comparative analysis of the behaviour of modern primates, in conjunction with an accurate phylogenetic tree of relatedness, has the power to chart the early history of human cognitive evolution." With regard to human musicality, we suspect that neither side of this conceptual divide has rendered good advice to those who would explore its evolutionary origins. Perhaps Lewontin's pessimism might be overcome by casting the comparative and inferential net wide enough. However, to do so we can no longer, as Byrne does, restrict ourselves to the study of primate homologies; rather, we must explore analogies wherever they are found in the animal kingdom. Some traits do, after all, evolve de novo in a lineage. To understand such novelties, analogous developments in unrelated animals provide invaluable information regarding potential selection pressures and ecological conditions favoring their evolution. The fruitfulness of such exercises, whether pursuing homologies or analogies, depends on the extent to which they can be constrained by stubborn facts regarding the phenomenon in search of an evolutionary explanation.

Constraints

Here we focus on a small set of characteristics of human music that should help constrain accounts of its origins. They can be conceived of as basic hurdles that must be cleared along the way to a comprehensive theory of the origins of human music. They were chosen above all for their generality, with the additional desideratum of involving mechanisms that generate consequences for the structural content of music. At present, a number of these constraints are difficult to meet, which means that besides their potential bearing on already proposed theories, they pose challenges for and may perhaps even inspire future ones.

Constraint 1: Cultural Transmission

Music, like language, is a complex product of cultural history. Its present-day patterns rest on traditions extending back over many thousands of years of intergenerational transmission of learned cultural lore (Higham et al., 2012; Le Bomin, Lecointre, & Heyer, 2016; Merker, 2012; Morley, 2013). This simple fact, so obvious that it typically is taken for granted in theories of music origins, nevertheless has profound consequences for any attempt to reconstruct the biological background to human musicality (Honing et al., 2015).

If patterns of cultural goods were only matters of human tastes and preferences—a common misconception regarding the nature of culture—the cultural transmission of musical lore would have no systematic or principled bearing on the reconstruction of music origins. However, when sustained over many generations, intergenerational transmission itself exerts profound and predictable effects on the contents of the transmitted lore, unbeknown to those involved in the transmission, and without any intervention of natural selection or differential reinforcement of outcomes (Bartlett, 1932; Kirby, 1998; Kirby, Cornish, & Smith, 2008; Mesoudi & Whiten, 2004, 2008). Thus, to go in search of evolutionary explanations for aspects of music that result from the transformative powers of cultural transmission would be a serious mistake. As we shall see, major structural features of music are likely to be shaped by the cultural transmission process itself.

The key insight here is that with each generational transfer, the cultural lore (be it language, music, or any similar system transmitted culturally through learning) has to pass the so-called learner bottleneck. Any given learner is exposed to only a portion of the cultural lore extant in the population into which they are born and, moreover, has a limited capacity to absorb even the portion to which he or she is exposed. This means that the many items that make up that lore compete with each other for passage to the next generation through the learner bottleneck. One of the consequences of this competitive filtering is that all aspects of the lore that bear on transmittability, including small differences in learnability and ease of processing, come to transform the cultural corpus in predictable ways, amounting to a cumulative process of informational compression over many generations. This tends to issue in a tight fit between properties of the cultural lore and properties of the learner, introducing commonalities across the lore of different, separated populations, all without horizontal transmission through diffusion or the agency of biological selection.

In the field of language evolution, this mechanism that we refer to as cultural evolution has been extensively studied over the past two decades (e.g., Christiansen & Chater, 2008; Deacon, 1997; Kirby, 2001; Zuidema, 2003; see also Thompson, Kirby, & Smith, 2016 for the interaction between cultural and biological evolution). This work has led to a growing consensus that (1) cultural evolution is a powerful mechanism, (2) many features of languages are potentially best understood as resulting from cultural adaptation to (preexisting) hominin cognitive and physiological features, and (3) theorizing about the evolution of the biological basis of language can sensibly proceed only if we explicitly take into account the possibility that cultural evolution has shaped the linguistic phenotype. There is no reason to believe that any of this is any less applicable to the cultural transmission of music than it is to that of language, because both exhibit the predominantly vertical transmission that is required for the learner bottleneck to do its transformative work (Cavalli-Sforza, 2000; Le Bomin et al., 2016).

Cultural evolution is a gradual, unconscious and obligatory process that extends over many generations and restructures the cultural corpus in ways that increase its salience, expressive economy, communicative generality, and grammatical power, all of which turn on enhanced communicability and learnability in various ways (Christiansen & Chater, 2008; Zuidema, 2003). This permits learners to manage ever larger amounts of cultural content without change in the neural resources devoted to it (through data compression) and lets the cultural products exploit existing peculiarities of neural organization. For instance, Zuidema (2013) discusses the finding of Smith and Lewicki (2006) that the neural code in the auditory nerve of cats appears to be optimized for human speech sounds and argues that this finding makes sense only if the direction of causality is inverted: speech sounds have evolved in a process of cultural evolution to exploit features of a preexisting general mammalian neural code, that is, to achieve maximum discriminability under noise and time constraints.

Another example of the principle may be found in the ubiquitous cross-cultural occurrence of line lengths of around three seconds in poetry (Turner & Pöppel, 1983). This might represent a cultural adaptation to the time span of auditory echoic memory, which is approximately of this length (Darwin, Turvey, & Crowder, 1972), or to a more general psychological time frame of some three seconds (Pöppel, 2004), or even to the length of the human breathing cycle. Whichever physiological regularity is conceived to have driven the process, cultural evolution would have tailored a formal poetic convention to a preexisting physiological constraint.

Some major structural features of music widely distributed across cultures might be a consequence of cultural evolution. Until recently, the failure of most musical tuning systems to conform to the mathematics of small integer ratios was grounds for rejecting Pythagoras's proposal that small integer frequency ratios account for the perception of musical consonance and harmonicity (Burns, 1999). However, recent modeling of the cumulative effects of physiological nonlinearities at each way station of the ascending auditory pathway has disclosed the presence of "resonance neighborhoods" at whole integer ratio spacings on the tonotopic maps of the auditory system (Large, 2011; Large & Almonte, 2012; Lerud, Almonte, Kim, & Large, 2014). This finding not only accommodates a wide range of tuning systems and musical scales found worldwide but appears capable of accounting for human judgments of consonance and tonal stability in terms of elementary functional features of our auditory system (Large, Kim, Flaig, Bharucha, & Krumhansl, 2016; see also Bidelman & Krishnan, 2009; Cousineau, McDermott, & Peretz, 2012; Gill & Purves, 2009; Lots & Stone, 2008; McDermott, Lehr, & Oxenham, 2010).

Note, however, that the pattern of resonance neighborhoods in auditory system tonotopy is a product of elementary and ubiquitous physiological properties of the underlying neural circuitry. It is therefore likely to be shared by all mammals. It did not evolve for purposes of music, in other words, but as an incidental by-product of the interaction of excitation and inhibition in a neural system evolved to process natural sounds efficiently. Not being confined to humans (Kuhl, 1988; Wright, Rivera, Hulse, Shyan, & Neiworth, 2000), these resonances are not likely to represent an adaptation to music. Why, then, are humans the only mammals engaging in musical practices conforming to these subtle but common resonances in auditory physiology?

The fact that our musical practices are indeed cultural patterns subject to intergenerational transmission may supply the needed explanation. In principle, the formal powers of cultural evolution should suffice to allow musical practice to eventually find its way to the ubiquitous and inherent resonant biases of the auditory system, given a long-running cultural tradition of song (Merker, 2006). Through the external loop of intergenerational transmission of learned musical lore, the production of musical patterns would pass through the learner bottleneck to be shaped by preexisting biases on the part of learners, specifically the purely perceptual resonant biases just invoked. This assumes not only a capacity for vocal learning (see constraint 3), but one emancipated from innate song templates, for which there is precedence in the true mimics among the birds (Baylis, 1982; Dowsett-Lemaire, 1979; Mayfield, 1934). From such a beginning, devoid of scales, tonality, and small integer ratio consonances, thousands of generations of cultural transmission would eventually externalize even subtle biases in auditory perception in the musical practices of human cultures (see also Kuhl, 1988).[1] It is possible that some species of avian vocal learners have undergone a similar cultural evolution toward tonal expressiveness (Doolittle & Brumm, 2012; Taylor, 2009).

The principal constraint this process imposes on theories of music origins is that they must provide a nonarbitrary reason for our distant forebears to have engaged in persistent intergenerational transmission of vocal lore lacking the tonal organization of music as we know it for long enough to allow the transmission mechanism to find its way to the resonances already embedded in basic auditory physiology. Since thousands of generations may be required for this to happen (Kirby, 2001; Kirby & Hurford, 2002), the constraint is a real one. Perhaps our forebears, like many bird species, maintained cultural traditions of learned song and became vocal learners as part of the cluster of changes that define the emergence of *Homo* some 2 million years ago (Merker, 2012; Merker & Okanoya, 2007). An increasingly refined vocal communication system for the accurate communication and extraction of emotional information from vocal prosody is likely to have contributed to this process as well (Morley, 2002, 2013, 2014; see also note 3 in Merker, 2000).

More generally, placing the historical nature of the pattern content of human music at the head of the effort to understand its biology greatly facilitates the reconstruction of its evolutionary background. To the extent that its patterns are shaped by the transformative dynamics of persistent intergenerational transmission through the learner bottleneck, there is no need to ask evolutionary selectional mechanisms to equip us with those pattern specifics, even when they happen to be cross-culturally widespread, as in the example of auditory system resonance. That is, even an absolute structural universal of music (see Savage, Brown, Sakai, & Currie, 2015) need not have been present at the first beginnings of music. The formal powers of transgenerational transmission of learned lore, without intervention of horizontal transmission (diffusion), can move initially disparate character states to a final, universally shared one (equifinality).

Constraint 2: Generativity, or Infinite Variety by Finite Means

Music, like language, is generative: it produces infinite pattern variety by finite means (von Humboldt, 1836). The key to that variety in both music and language is not recursion (Hauser, Chomsky, & Fitch, 2002) but combinatorics (Abler, 1989; Merker, 2002; von Humboldt, 1836). By combining a finite set of elements—discrete pitches and durations—music creates composite patterns without limit. For this to be possible, the combining elements must be nonblending in the sense of not producing an average when combined (Fisher, 1930); they must retain their individuality on combining (see figures 3.1A, 3.1B). When that is the case, each such combination "creates something which is not present per se in any of the associated constituents" (von Humboldt, 1836, p. 67), making infinite pattern variety possible (see figure 3.1C). There are four major open-ended generative systems in existence, two of which are natural ones (chemistry and genetics) and two are cultural (music and language; see table 3.1).

In the cases of music and language, the combining elements are conventional, the musical ones arising through a radical reduction in the degrees of freedom available to vocal or instrumental sound production (Merker, 2002). This is accomplished by discretizing two continua, those of pitch and duration, to yield musical notes with determinate pitch (Stumpf, 1911) and—in all rhythmic music—proportional durations based on discretizing

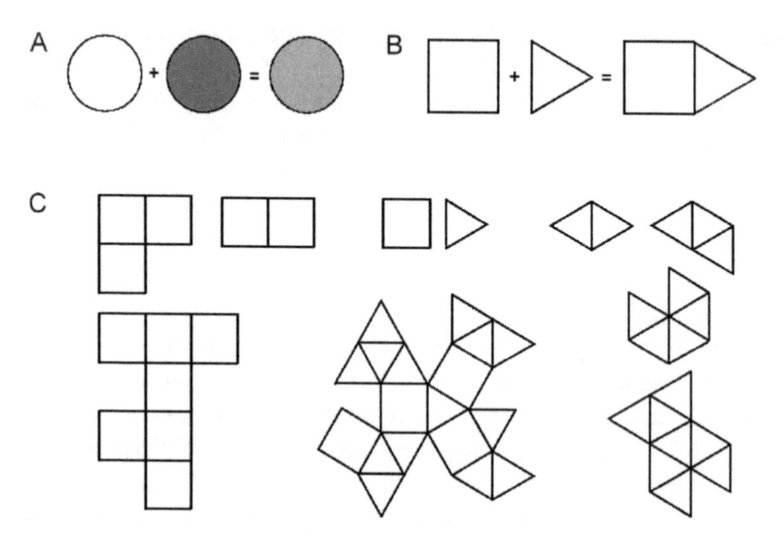

Figure 3.1
Cartoon illustration of the particulate principle of self-diversifying systems, following Abler (1989). (a) A blending system in which combining ingredients average, here exemplified by a drop of ink in water: the combining elements do not generate a qualitatively new entity. Other examples are most mixtures of liquids as well as gases, as in weather systems, patterns of heat conduction, and force fields. (b) A particulate system, in which the combining elements generate a qualitatively new entity by retaining their individuality on combining. (c) A minuscule sample of the infinite generativity of a combinatorics of as few as one or two discrete non-averaging elements.

Table 3.1
The principal open-ended or "self-diversifying" systems

System	Chemistry	Genetics	Music	Language
Product	All molecules	All life forms	All melodies	All sentences
Combining particles	Atoms	Nucleotides	Notes	Phonemes
Particle type	Physical		Conventional	
Domain	Nature		Culture	

time through an isochronous pulse (Arom, 1991). The isochronous pulse of rhythmic music supplies the temporal predictability needed for interindividual entrainment with perfect synchrony. It is covered under constraint 4 and will not be considered further here.

Musical notes are, in other words, not simply pitches. They are individuated pitch locations within a discretized pitch continuum. Even glissandi must travel between these fixed reference points with the same necessity as ordinary notes if they are to be musical. The designation of a specific location on the pitch continuum as a note by a culturally determined pitch standard, applied with a conventionalized margin of tolerance, lifts that location out of its relation of equivalence to its infinitude of pitch neighbors. It breaks its anonymity, as it were, and turns it into an individuated and specific musical note to which a musical figure can return and which can be used repeatedly in the development of a musical pattern.

This discretization of the pitch continuum into determinate pitch sets supplies music with combinable pitch elements featured in musical melodies and chords (Balzano, 1982; Burns, 1999; Krumhansl, 1990). Pitch sets thus supply the particulates, the individuated elements, needed for its combinatorial mill. They are found in all musical traditions cross-culturally (see chapter 14, this volume; Stumpf, 1911). Indeed, one of the distinguishing marks of musicianship anywhere is adherence to the pitch locations designated by a pitch standard during musical performance. Not to do so is to sing or play out of tune, the quintessential demarcation line between musical and other employments of human capacities.

The constraint imposed on theories of music origins by the generativity of music is that no such theory can account for the genesis of music as we know it without giving a credible account of how we came to conquer for ourselves the discretized (particulate) pitches and durations without which there can be no open-ended generativity of music. As we saw in discussing cultural evolution, pitch discretization may be a prime product of a protracted process of cultural transmission exploiting the tonal scaffolding of auditory system resonances. (For the ubiquitous cross-cultural practice of dividing the octave into steps that maximize the individuation of its *intervals,* see Balzano, 1982, and Merker, 2006.)

The ways in which the cultural evolution of musical lore might produce particulate elements needs further research, including the exploration of computational models. Perhaps accounts of how discrete combinatoriality (culturally) evolved in phonology (Oudeyer, 2006; Zuidema & de Boer, 2009) can be adapted to music. In these models, the evolution of a repertoire of continuous trajectories through an acoustic space is studied. Discrete structure emerges in these models as a side effect of the neural encoding (Oudeyer, 2006) or of optimization for discriminability (Zuidema & de Boer, 2009). Both of these proposals would seem applicable to music, providing a potential route to surface combinatorial structure in the musical lore.

If we can make plausible that such a cultural route can lead to productive combinatoriality (generativity) too, there is no need to burden theories of music with Darwinian accounts of the origin of musical notes, scales, and tuning systems. However one conceives of the matter, the point here is only this: that a credible theory of music origins must furnish such an account, short of which the phenomenal pattern richness of human musical culture remains a cipher.

Constraint 3: Vocal Learning

Every song we know how to sing and every word we know how to pronounce is ours through a highly specialized learning capacity that is conspicuous by its absence in other primates, our closest living relatives included. The vocal patterns of song and speech are acquired through motor learning on the basis of heard, culturally transmitted models through a process requiring intact hearing and feedback from one's own voice (Clement, Koopmans-van Beinum, & Pols, 1996; Doupé & Kuhl, 1999; Konishi, 2004; Liu, Wada, & Nottebohm, 2009; Oller & Eilers, 1988; Thorpe, 1961). The process by which they are acquired is technically known as vocal production learning (Janik & Slater, 1997, 2000), a dedicated and highly specialized capacity that has no other common uses in our lives besides song and speech.

Since there can be no human singing without it, the origin of our capacity for vocal production learning bears directly on scenarios for the origins of musicality. The issue is an acute one, since the fact that other primates lack this capacity (Janik & Slater, 1997, 2000) means that we became vocal learners at or after our divergence from the common ancestor we share with chimpanzees. One limb of the comparative method—the tracing of continuities (homologies) with our close evolutionary relatives—is therefore unavailable for reconstructing its origin in this particular case, contrary to Byrne's (2000) model in the quote cited at the start of the chapter.

A variety of context- or learning-based modifications of vocal output do not involve the mechanism of vocal production learning in the technical sense applicable to human song and speech. They include contextual modulation of vocal behavior, socially or environmentally contingent selection among innate calls and their variants, and their learned modification, as Janik and Slater (2000) detailed. There is no dearth of evidence for such vocal phenomena among primates. They are an integral part of the vocal expressiveness that primates share with many other mammals, but they occur without reliance on the specialized mechanism of vocal learning.

Vocal learning proper, by contrast, is the ability to convert heard sound patterns that are not in the species-specific innate vocal repertoire into vocal output, using feedback from one's own voice to achieve the match (Konishi, 2004). It has been studied in detail above all in birds (Thorpe 1961; Zeigler & Marler, 2004), which have a rich assortment of vocal learning phenomena, by no means all the same. They differ along at least six major dimensions of classification, as reviewed by Beecher and Brenowitz (2005). Human vocal learning occupies the more advanced end of several of these dimensions. It is, for example, open ended, allowing new patterns to be added throughout life (though with diminished accuracy after puberty) and is emancipated from dependence on a species-specific vocal template. Given that the human capacity is an advanced one, comprehensive studies of the true mimics among birds (e.g., mynahs, many species of parrots, lyrebirds, butcherbirds, mockingbirds; see Baylis, 1982; Taylor, 2009) are needed to supplement the invaluable knowledge about mechanisms of vocal learning supplied by the bird species typically employed in the study of vocal learning in the laboratory.

As far as is known, vocal production learning proper is found only in humans, cetaceans, pinnipeds, elephants, bats, oscine songbirds, parrots, and hummingbirds (Janik & Slater,

1997; Jarvis, 2006; Kroodsma & Baylis, 1982). Given its absence in nonhuman primates, the process by which our ancestors were equipped with this capacity is a major evolutionary event intervening between the last common ancestor and the first singing or speaking humans. It supplies a major biological constraint, or evolutionary bottleneck (Nottebohm, 1976), on the path to human music. Its origin in our lineage could have been driven by song, speech, or other factors. This virtually forces theorists to come to grips with the order of precedence among song and speech in our ancestry (see Cross et al., 2013, figure 21.1 and accompanying text).

There is currently no good account of how humans evolved the capacity for vocal production learning. As Nottebohm (1976) noted many years ago: "You might find it much harder to explain this first step, vocal learning, than the latter acquisition of language" (p. 645). Nottebohm's warning, we submit, applies to music no less than to language. One possibility in this regard is that we acquired vocal learning, like some of the songbirds, as a means to sustain cultural traditions of learned song (see constraint 1) for purposes best explained by a version of Zahavi's handicap principle known as the developmental stress hypothesis (Hasselquist, Bensch, & von Schantz, 1996; Nowicki, Searcy, & Peters, 2002). Another might be that vocal learning built on comparable abilities for manual imitative learning and variation that were already developed, or developing. (For some difficulties encumbering this gestural alternative, see Merker 2015.) However conceived, our possession of vocal production learning is a fact, and one which any theory of the origins of music leaves unaccounted for at its peril.

Constraint 4: Entrainment

The constraints considered so far apply to both human music (song) and language (speech). They therefore cannot help us home in on evolutionary factors unique to music as such. This is no longer so for the final two constraints we consider, beginning with our capacity to entrain our behavior to one another with perfect synchrony, as defined by Ermentrout (1991). That capacity features persistent phase locking of the cyclical behavior of two or more individuals to one another with zero (or even slightly negative) phase lag at a given tempo, with flexibility to do so at different tempos. This capacity of ours has virtually no uses outside human music, dance, and drill. In trying to understand its biological background, it helps to keep the general lack of biological utility of such behavior in mind. Even without perfect synchrony, the uses for any kind of beat-based entrainment capacity in nature are few and far between (see the survey in Merker, Madison, & Eckerdal, 2009, with further details in Merker, Morley, & Zuidema, 2015). As we shall see, the full capacity under consideration here is a rare one indeed.

In entrainment with perfect synchrony, current behavioral output coincides with the current stimulus event. Because of reaction time limitations, it cannot therefore be based on responding to that stimulus event. Instead it requires a predictive (anticipatory) and cyclical motor timing mechanism that takes an evenly paced stimulus sequence as input.

This stimulus *isochrony* makes the very next beat in the sequence predictable, allowing the timing mechanism to align—or latch—its endogenously produced behavioral isochrony to the stimulus isochrony with zero, or even small negative (anticipatory), phase lag typical of human sensorimotor synchrony (Fraisse, 1982).

Given an evenly paced periodic input, the cardinal characteristic of entrainment with perfect synchrony is that it stays locked to the input sequence despite biologically inevitable noise in the form of local deviations from perfect isochrony, as well as tempo drift. To do so, the predictive timing mechanism must be capable of phase correction, as well as period adjustment (Ermentrout, 1991; Repp, 2001; Semjen, Vorberg, & Schulze, 1998). Thus equipped, it does not, in other words, drift in and out of phase as long as the tempo of the stimulus sequence stays reasonably steady. Rather, it stays locked to the beat as long as it persists (Ermentrout 1991; Merker et al., 2009; Repp & Keller, 2008).

In nature (i.e., outside the laboratory), the mutual entrainment of two or more individuals by such a mechanism obviously cannot occur unless they themselves are capable of producing the evenly paced entraining stimulus. That is, they must be capable of endogenously producing isochronous behavioral output without prior input. A second cardinal characteristic of a species in possession of that mechanism is therefore an expressive capacity, even propensity, to spontaneously produce isochronous behavioral cyclicity of some kind (such as chirping, flashing, hooting, clapping, stomping, or drumming) within the tempo range of its predictive timing mechanism.

This elementary fact supplies an easy-to-apply criterion by which to search the animal kingdom for candidate entraining species. If they do not produce some kind of isochronous behavior spontaneously, they are unlikely to be natural entrainers. To ensure the predictability needed for predictive timing, and hence for perfect synchrony, the endogenously produced isochrony must have low variance in period length. Human isochronous tapping typically exhibits a standard deviation of a few (two to five) percentage points (Madison, 2000; Repp, 2005). Details apart, to be in possession of a natural entrainment capacity is therefore first and foremost a matter of production, not perception.[2]

As far as is currently known, very few species in the entire animal kingdom are capable of entrainment with perfect synchrony. These are human beings, *Homo sapiens*, as attested by a century of sensorimotor synchronization studies (Repp, 2005); three species of tropical firefly, *Pteroptyx malaccae*, *Pteroptyx tener*, and *Luciola pupilla* (Buck, 1988; Ermentrout, 1991; Hanson, 1978, 1982); and a few species of chorusing crickets employing period adjustment in their entrainment behavior (Murphy, Thompson, & Schul, 2016; Nityananda & Balakrishnan, 2007; Walker, 1969). The human capacity is exercised almost exclusively in the domains of music, dance, and drill, while the insects use theirs as a means of mate attraction by either synchronous male chorusing or bioluminescent flashing. Thus, large numbers of lampyrid beetle males gather for synchronous flashing in single rain forest trees during the tropical night. The precise superposition of their signals creates a supernormal mate attraction signal that overcomes the directionally obscuring effects of dense rain forest foliage—the so-called beacon effect (Buck & Buck, 1978).

Firefly entrainment precision falls at the more skilled end of the human range. It is underwritten by the steady isochrony of flash production, which exhibits a standard deviation in period length of less than 3 percent (Ermentrout, 1991). The tempo range over which these fireflies are capable of entraining with perfect synchrony—some 15 percent in either direction around roughly 2 Hz—is narrower than the broad range over which humans do so. It is worth noting, however, that the human range reflects the cognitive strategy of mentally subdividing long periods (so-called subitizing) to maintain precision at low tempos (Madison, 2001; Repp, 2010). Moreover, humans have a preferred tempo centered on 2 Hz (120 beats per minute, henceforth bpm) where precision is highest and around which the tempo of rhythmic music concentrates. In fact, the firefly tempo range of 30 percent covers the central 58 percent of the area of the tempo histogram for 74,000 pieces of Western popular music compiled by Dirk Moelants (2002; see also MacDougall & Moore, 2005). Moreover, it exceeds the range of tempos by which the cockatoo Snowball was tested for synchrony by Schachner and colleagues (Schachner, Brady, Pepperberg, & Hauser, 2009), and falls just short of the plus and minus 20 percent range employed with the same animal by Patel, Iversen, Bregman, and Schulz (2009).

In addition to the insects engaging in entrainment with perfect synchrony already mentioned a number of insect species among the cicadas and crickets have developed synchronous chorusing behavior without the period adjustment mechanism needed to maintain zero phase lag at different tempos (Greenfield 1994, 2005; see also Merker et al., 2009, 2015). Phase resetting alone suffices to keep them steadily locked to the entraining signal over a range of tempos. However, when the tempo of the entrainment signal differs from an individual's spontaneous tempo, this leads to a phase offset whose size is proportional to the difference between an individual's spontaneous tempo and the entraining tempo. When signal tempo is too fast or slow for this to suffice, they nevertheless entrain by various whole integer beat combination ratios familiar from metrical practice in human music (see figures 2 and 3 of Sismondo, 1990, a model of clarity in the presentation of entrainment data).

In light of the foregoing, it should be apparent that none of the nonhuman animals that recently have been assessed for their entrainment or synchronizing capacity give evidence of possessing a mechanism for entrainment with perfect synchrony. Intermittent or episodic synchrony in the course of major phase drifts and repeated loss of entrainment in the presence of a steady beat reported for two species of parrots and a bonobo (Large & Gray, 2015; Patel et al., 2009; Schachner et al., 2009) is not what is expected of an entrainment mechanism, whether human or arthropod. As we have seen, entraining fireflies, cicadas, crickets, and humans, unlike the tested individuals, stay tightly locked to the entraining stimulus as long as the beat stays reasonably steady, and this is true even for species that lack a period adjustment mechanism.

The parrots do, however, show spontaneous responsiveness to and engagement with the rhythmic music to which they are exposed. They are highly encephalized and vocal animals, and presumably they have the auditory capacity to extract the beat from the human music.[3] The videos show them visibly captured by the music and expressing their engagement with it through dancelike bodily movements, albeit without the consistent latching to the beat that is the unique contribution of a predictive entrainment mechanism. The fact that the birds are vocal learners, and even vocal mimics, may be relevant to their performance, though in a different sense from that proposed in the vocal learning hypothesis for rhythmic entrainment by Patel (2006, 2014; Patel et al., 2009).

The high-fidelity duplication by vocal learners of heard patterns with features that in biological terms are arbitrary (*sensu*; Merker 2005) are mastered through lengthy practice entirely without external reinforcement. In previous publications, one of us has proposed that this implies the operation of a motivational mechanism, a "conformal motive," that sustains practice until duplication is achieved. Furthermore, "Around the core of learned control of the voice, the same mimetic ('conformal') motive that ensures copying fidelity in song … [c]ould have been linked to other aspects of … behavioral display. … Such a linkage … extending to bodily (postural and gestural) expressive learning would have turned us into a species of 'imitative generalists'" (Merker, 2009, p. 53; see also Merker 2005, 2012).

Some parrot species appear to possess such a generalized imitative capacity. Many monogamous parrots engage in joint cooperative pair displays, some of which they learn by imitation (Serpell, 1981). White cockatoos are accomplished actors, clowns, and hams, employing a range of human postural attitudes and prosodic gestures in interaction with their keepers.[4] They are also highly susceptible to imprinting on their human keepers, so much so that they must be reared in special isolation containers in order to be capable of breeding with their own kind on sexual maturity. The suggestion here, then, is that the parrots, acculturated to humans, have learned the human ways with rhythmic music and are doing their enthusiastic best to dance, but without the benefit of an endowment for entrainment synchrony that would allow them to keep the beat consistently and with precision. Vocal learning would, on this interpretation, supply the key to the imitative propensity they do possess, not to the synchronizing ability in which they are deficient.

The contrast with a species lacking such ritual imitative capacity is instructive in this regard (see Merker, 2005, 2009; Tennie, Call, & Tomasello, 2006). It is provided by the synchronization study with the bonobo Kuni, in which an experimenter tried to get the animal, equipped with its own drum, to join in human drumming (Large & Gray, 2015). In sharp contrast to the spontaneously performing parrots, the bonobo required various measures of habituation, coaxing, and reinforcement in order to perform. Unfortunately, the hectic human drumming employed in this study, and in particular in experiment 2 designed to assess synchrony, has a pace typical of ape and human noise making and lacks the very characteristics that induce humans (and parrots) to move to a beat (see Janata, Tomic, &

Haberman, 2012; MacDougall & Moore, 2005; Madison, 2006; Merker, 2014; Merker et al., 2009). The design of experiment 1, in which human drumming stopped as soon as Kuni started drumming (presumably to resemble the continuation mode in human tapping studies), may also have discouraged or confused the animal.

To what extent these circumstances explain Kuni's erratic performance, featuring unmotivated tempo changes, frequent pauses, and phase drifts not expected of a species equipped with an entrainment mechanism, is unknown. They do, however, urge caution in treating her performance as indicative of bonobo rhythmic limitations. For one thing, there is a recent report of a common chimpanzee in captivity spontaneously producing a more than five-minute-long solo of untaught drumming (Dufour, Poulin, Curé, & Sterck, 2015). It is not performed to an external beat, but its characteristics are far closer to human musical drumming than anything heard in the samples provided of Kuni's drumming. The drumming is bimanual, rather evenly paced, and as indicated by the most evenly paced passages and stretches of duple meter, has a basic tempo centered on 128 bpm, close to the human preferred tempo of 120 bpm.

Meanwhile, De Waal's (1988) tantalizing report of rhythmic vocal synchrony in captive bonobos still has no follow-up and remains unexplored after close to three decades (see also Merker, 1999, 2000; Merker et al., 2009). In situations of high excitement or tension, two or more bonobos occasionally stage a loud chorus of evenly paced staccato hooting. In doing so, their hoots occur either in synchrony or in alternation at a rhythmic tempo of approximately 2 Hz (120 bpm). This call type has not been reported for common chimpanzees but has been reported under a different name for bonobos in the wild, though without any information regarding group synchrony (Bermejo & Omedes, 1999, figure 47). Accordingly, "few issues would seem to provide more leverage for the comparative study of the biology of human musical rhythmicity than a thorough characterisation of bonobo staccato hooting in the wild. Should it occur, and serve inter-individual entrainment of voices, the genus *Homo* would not be alone among the apes in having evolved a capacity for rhythmic entrainment of voices" (Merker et al., 2009, p. 7).

We now turn to studies that have attempted to explore entrainment in nonhuman animals with a variety of conditioning procedures. Such procedures do not, of course, figure in the ontogeny of either insect or human entrainment. The insect mechanism is innate and automatic, while the human one appears to be largely a matter of maturation. It benefits from both exposure and practice, but to become operative, it is not dependent on formal instruction, let alone instrumental conditioning (Merker et al., 2009).[5] Operant procedures might nevertheless serve to bring out a latent capacity in animals, unresponsive as they are to verbal instructions.

The large positive asynchronies reported for macaques in a continuation task requiring extensive training (Merchant & Honing, 2013; Zarco, Merchant, Prado, & Mendez, 2009) are compatible with reaction time responding (Kauranen & Vanharanta, 1996; macaques, with shorter conduction paths, are likely to be faster than humans). The fact that stimulus presentations with randomized onset intervals produced even greater asynchronies is not evidence of predictive timing in the sense of entrainment synchrony. Reaction times to predictable stimuli are shorter than those to unpredictable ones (cued versus uncued reaction time). A sprinter's dash is initiated by a reaction time limited response to the sound of the start pistol, shortened by the preparatory cues of "ready, steady." The regular stimulus train of the monkeys' task can be construed as a serial cued reaction time signal.

The predictive timing of entrainment synchrony is a very different matter. Its point is to eliminate reliance on the current stimulus event as a trigger for responding, thus making zero asynchrony possible. This is accomplished by aligning an endogenously driven and ongoing behavioral cyclicity (periodicity) with an external isochronous stimulus sequence by small phase and period corrections, a matter unrelated to the shortening of reaction times by a predictive cue. The macaques required an average of sixteen months of training to master their task. They thus had ample time and opportunity to hit on the strategy of targeting the upcoming stimulus itself, perfectly predictable as far as its time of arrival is concerned. That none of them did so is strong evidence against their possession of a mechanism supporting entrainment synchrony by predictive timing.

Similar reaction time considerations may help explain the peculiar results of another study that, like the macaque study, also used operant conditioning procedures to test for entrainment synchrony to short isochronous sequences, this time in budgerigars (Hasegawa, Okanoya, Hasegawa, & Seki, 2011). A wide tempo range spanning interstimulus intervals from 450 to 1800 ms (133 to 33 bpm) was tested. At the faster (apparently preferred) tempos, the birds' interresponse intervals matched those of the stimuli well: they were regular, but asynchronies were large and positive, approaching the birds' reaction time as estimated by responses to randomized intervals. At slow tempos, responses matched interstimulus intervals poorly; they were irregular, yet averaged asynchronies were smaller, though still positive.

Assuming that the reaction time estimate, for the same reasons as in the macaque study, is an overestimate, the birds' fast tempo responses were presumably reaction time based and thus not evidence for entrainment synchrony. For the slow tempos, the paradoxical result that more irregular performance resulted in reduced asynchronies requires an explanation. For humans, synchronizing precision declines substantially above 900 ms interval length (Madison, 2001) unless the strategy of mentally subdividing the interval (subitizing) is adopted (for which see Repp, 2010). The birds cannot be expected to manage long intervals better than humans, and without subitizing, irregular budgerigar performance would be expected, as was indeed the case.

The anomaly of reduced asynchronies with more irregular performance may have been caused as follows. An average phase approximating stimulus onset rather than the point of estimated reaction time (yielding low average asynchrony despite irregular pecking) occurred at 900 ms or longer intervals in six of the forty-six bird × tempo combinations. Premature (anticipatory, jumping the gun) pecking during the long acceptable period before onset, as in the trace of female D in figure 3 of Hasegawa et al. (2011), is more likely to occur at long intervals (slow tempos). Averaged into irregular performance, these early pecks might explain the anomalous result. If so, little remains of a case for entrainment by predictive synchrony in budgerigars as assessed in this study.

Hattori, Tomonaga, and Matsuzawa (2013) used food reinforcement to train three chimpanzees to alternately tap two illuminating and sounding keyboard keys and then assessed the influence of a separate isochronous sound sequence (a distractor sequence) on their tapping performance . Only one of the three chimpanzees (named Ai) showed any systematic relationship between her tapping and the isochronous distractor, and she did so only at one of the three distractor tempos—the tempo that coincided almost perfectly with the self-paced tempo Ai had developed in the course of the conditioning procedure. At that distractor tempo (100 bpm), Ai showed a very broadly tuned assimilation of her tapping to the distractor stimulus, centered on close to zero asynchrony.

Given that Ai showed no influence of the distractor isochrony at any other tempo and that the other two chimpanzees showed no influence at any tempo, there is no way of knowing what the result would have been if one of the three distractor tempos had not happened to coincide with Ai's self-paced tempo.[6] The stimulus matching was, in other words, a quite circumscribed one. Although the tempo coincidence made period adjustment unnecessary, Ai's performance is not explicable as phase correction without period correction, such as that exhibited by a number of insect synchronizers. At other than the tempo at which period correction is not needed, such a mechanism would still generate good entrainment, but with a nonzero asynchrony reflecting the tempo difference between stimulus tempo and preferred tempo (the sea lion study, explored next). No known entrainment mechanism works in the way Ai performed, and certainly not the human one. Perhaps the tempo coincidence allowed the distractor sequence to exert a subliminal influence on Ai's tapping. Subliminal effects are known to occur in human sensorimotor tapping (Madison & Merker, 2004).

We come then to the female sea lion Ronan (Cook, Rouse, Wilson, & Reichmuth, 2013). An operant training regime based on food reinforcement was used to shape her head bobbing to isochronous auditory stimulus trains. Initially this was done on what amounts to a bob-by-bob basis, gradually extended to bobbing sequences. After lengthy training, Ronan bobbed her head reliably to the stimulus train at two tempos. She then readily generalized her performance to five novel tempos but did not generalize to bob in time with a piece of rhythmic popular music. Further training on this piece brought her to criterion performance. She then readily generalized to different tempos of that piece and, without further training, to two different pop tunes. After further training on one of these, she generalized to five novel tempo versions of that tune as well.

A peculiarity of Ronan's performance across this variety of entrainment stimuli, which the authors noted, is an overall tendency to produce positive asynchronies at faster tempos (i.e., to lag the stimulus, and the more so the faster the tempo) and negative asynchronies at lower tempos (analogously, but leading the stimulus). The tendency is not strictly quantitative, and some reversals between adjacent tempos are in evidence (see figures 2B and 2C in Cook et al., 2013), yet as a qualitative trend, the tendency is unmistakable. The same tendency is evident in a subsequent study in which the animal's response to phase and tempo perturbations was studied (Rouse, Cook, Large, & Reichmuth, 2016; see the panels for the 15% and 25% tempo deviations in either direction in figure 1 of that report for striking illustrations of the effect).

The pattern approximates, in other words, the synchronizing mechanism of the less endowed category of insect synchronizers. Their phase delay mechanism generates phase offsets proportional to tempo in both directions from their free-running tempo. It is the most common synchronizing mechanism among a variety of synchronizing insects that lack the mechanism for perfect synchrony (see figure 3 in Buck, Buck, Case, & Hanson, 1981, and the phase response curve, figure 2A in Sismondo, 1990).

Why a lengthy and detailed operant shaping process in an otarid seal should produce an imperfect version of a mode of synchrony possessed by insects on an innate basis is at this point a mystery. It does, however, underscore that what Ronan is doing on the basis of intensive operant training is not what humans do as a matter of course when they synchronize with a beat, quite apart from the fact that they do not require such training to do so.

It is to be regretted that Ronan was not observed while being exposed to highly compelling rhythmic music prior to training and also that no male sea lion has been tested for entrainment. The latter desideratum relates to the point made earlier concerning the centrality of a capacity for isochronous production in any naturally synchronizing species. Sea lion males use a repetitive, evenly paced bark as part of their territorial, herding, and agonistic display behaviors. The tempo of this bark, which in California sea lions has a default around 2 Hz, varies across a considerable range with contextual and motivational factors, with a faster pace signaling higher excitement or stakes (Charrier, Ahonen, & Harcourt, 2011; Schusterman, 1977). At each tempo, the barking exhibits a steady isochrony, save for the first few barks. These animals come equipped, in other words, with a crucial component of natural synchronous chorusing, but are not on record for exhibiting such group behavior. The steady pacing of the male bark at different tempos might nevertheless play a role in sea lion entrainment under appropriate testing conditions.

We conclude this brief review of recent studies of nonhuman entrainment with a note about synchrony of arm movements in macaques (Nagasaka, Chao, Hasegawa, Notoya, & Fujii, 2013). It is claimed that the three animals, paired and facing each other while each engaged in a button press task, spontaneously synchronized their arm movements to one another with asynchronies as low as 1 ms. This occurred despite the fact that the monkeys worked at different speeds whose ranges overlapped in only one of the three pairings and whose phases therefore must have drifted with respect to one another (see figure 2 of that report), a situation bound to generate considerable asynchronies as an inevitable result. As we have pointed out in a previous publication, this inexplicable result is based on a fundamental error in the way the data were analyzed and is thus a methodological artifact (for details, see Merker et al., 2015).

Having thus found no compelling reason to revise our initial supposition that as far as is known, only a few species of insects and humans are equipped with a mechanism for entrainment with perfect synchrony, it is time to ask what might account for this disparate phylogenetic distribution. Humans and fireflies are denizens of different phyla, chordates and arthropods, respectively. This eliminates homology by common descent as a possible reason for their shared capacity. Moreover, the fact that the insects accomplish their perfect synchrony with a brain weighing a fraction of a milligram eliminates both brain size and vertebrate neural circuitry as relevant factors as well. The only remaining causal option is convergent evolution (analogy, homoplasy) under shared selection pressures.

Regarding the nature of those selection pressures, the key, we suggest, is the general uselessness of beat-based synchrony from a biological point of view and the narrow circumstances under which it is capable of serving a function. It is all the more interesting, then, that the behavioral analogy between humans and synchronizing insects is joined by a striking commonality in the behavioral ecology of the otherwise utterly different synchronizers.

The pattern of clumped males needing to attract migrating females in fireflies and other synchronizing insects (see Greenfield, 1994) has a direct parallel in the pattern of male territoriality combined with female exogamy, which, on cladistic grounds, can be assumed to have characterized the social system of the common ancestor we share with chimpanzees (Ghiglieri, 1987; Lovejoy, 2009). In such a setting, in which only females migrate from their natal territory at sexual maturity (Ember, 1978; Pusey, 1979), selection pressures like those that have promoted group synchrony of signaling in synchronizing insects would be present. We might, in other words, be the direct descendants of a subpopulation of the common ancestor that responded to those selection pressures. A detailed scenario for how these pressures may have selected for synchronous chorusing in our ancestry has been presented in prior publications by one of us (Merker, 1999, 2000; Merker et al., 2009). That account will not be recapitulated here, but it will receive some further notice in the next section of this chapter.

Constraint 5: Motivational Basis

Wherever humans live, and however they have organized their societies, they exhibit a behavioral peculiarity of gathering from time to time to sing and dance together in a group (Brown, 1991; Brown & Jordania, 2013; von Humboldt, 1836). By featuring both human learned singing (constraint 3) and entrainment (in the dancing movements and perhaps clapping performed in synchrony with the singing and music—constraint 4) such behavior qualifies as human music. Indeed, the fact that it occurs in every human culture, and subculture too, without exception, unless deliberately suppressed by severe sanctions against it, marks this phenomenon as the most universal human behavior of a musical kind on record.

In its ubiquity, this human propensity for occasional group singing and dancing would seem to constitute a prototypical musical behavior, all the more so since it can be staged entirely without musical instruments as in the traditional trance dance of the hunter-gatherers of the Kalahari Desert (Katz, 1982; see also Trehub, Becker, & Morley, 2015, and chapter 6, this volume). It may in fact represent the motivational core of the human capacity for music from which its many other manifestations may have developed by differentiation, elaboration, and specialization. One is assisted in becoming aware of the peculiarity and specificity of this behavioral propensity by imagining that in exactly those circumstances in which we typically gather to sing and dance together in a group, another human culture would gather in groups to draw pictures together instead.

The ubiquity and specificity of this putatively prototypical musical behavior would seem to require an explanation. In searching for one, we enter for the first time onto the grounds of a possible homology, because certain social displays of our closest living primate relatives may provide a biological background to the human tendency.

As Geissmann (2000) pointed out, there is an association between loud calls (distance calls) and physical display among our closest living primate relatives, the apes. The loud calls that apes use in distance signaling (e.g., long call, pant-hoot, gibbon pair duet) tend to be accompanied by vigorous physical displays such as locomotor excitement, branch shaking, chest beating, and other forms of noise-making that Fitch (2005) called "drumming," though lacking the pulse-based rhythmicity of drumming in the musical sense. These displays do not feature any metrical structure resembling isochrony or any pulse-based rhythmic entrainment between individuals, but they do provide a precedent for the linkage between vocalization and bodily movement that occurs in human group singing and dancing.

Chimpanzees exhibit a social elaboration of this coupling between voice and physical display into an occasional group frenzy called the "carnival display." On irregular occasions, typically when a foraging subgroup discovers a ripe fruit tree or when two subgroups of the same territory meet after a period of separation, the animals launch an excited bout of loud calling, stomping, bursts of running, slapping of tree buttresses, and other means of chaotic noise-making. There are no indications that any kind of interindividual coordination, let alone rhythmic synchrony, forms part of these chimpanzee group displays. They may last for hours, even a whole night, and induce distant subgroups and individuals on the territory, both male and female, to approach and join the fray (Ghiglieri, 1984; Reynolds & Reynolds, 1965; Sugiyama, 1969, 1972; Wrangham, 1975).

Our social-emotional propensity to occasionally gather for excited group displays appears to be shared, in other words, with our closest living relative among the apes, the chimpanzee. We are not alone in sensing a possible connection in this regard. Bayaka pygmy hunter-gatherers inhabit the Congo-Brazzaville rain forest, which they share with chimpanzees. Mokondi massana spirit play, featuring ritual singing and dancing, is a significant aspect of Bayaka culture:

> When BaYaka … hear a chimpanzee "carnival display" from their camp it provokes great hilarity among camp members as one or two of them begin imitating the frenetic actions of the chimpanzees as they pound buttress roots or shriek at the canopy. The camp is launched into laughter as they explicitly ridicule the chimpanzees attempt to stage a ritual (massana), but are incapable of bringing it off properly. Fables such as "Chimpanzee you will die" (sumbu a we) elaborate on this theme describing how chimpanzee tries to get initiated but has to be dissuaded to avoid him being killed during the trials. (Jerome Lewis, personal communication to B.M., 2014, with permission).

If the propensity for an excited social noise-and-movement display is indeed homologous in the two cases, one with musical content and the other without it, this bears directly on theories attributing group or social functions to music. In the case of homology, the social efficacy would derive not from the musical content of the group activity but from its preexisting motivational mechanism shared with chimpanzees by common descent. This group excitement factor has to be controlled for in studies designed to explore the emotional or social significance and consequences of human music.

In our case, the communal display was eventually elaborated by adding metric and melodic structure to the chaotic noise-and-movement display. The refinement takes the form of regularizing the pacing of both voice and bodily display, making the even pace of its tempo (isochrony) the means for entraining the behavior of individuals to one another in an accurately timed group display of rhythmic chanting and dancing. A plausible setting for such a development is the male group territoriality combined with female exogamy, a rare pattern among higher animals, that can be assumed to have characterized the last common ancestor of humans and chimpanzees (Ember, 1978; Ghiglieri, 1987; Lovejoy, 2009; Merker et al., 2009; Pusey, 1979). Merker noted the striking parallelism between this pattern and the male clumping combined with female migration that is the functional and evolutionary key to synchronous chorusing in the insect examples already noted, and proposed it as a selection pressure for the evolution of the human entrainment capacity (Merker, 1999, 2000; Merker et al., 2009). As noted in the latter publication, such a scenario is eminently compatible with the central tenet of the coalition signaling scenario that Hagen and Bryant (2003) proposed.

The constraint we are proposing in this section pertains to the motivational underpinnings of music rather than to its structural content. Something needs to explain the cross-culturally universal human tendency to gather from time to time for group singing and dancing. No theory of music origins can be considered complete without somehow accounting for this tendency. If, as suggested here, the social function and emotional impact of the gatherings, which in our case feature music, far antedate their specifically musical content, then it is not to the musical content but to the decidedly nonmusical social adaptations of our hominoid ancestors that we should look for the secret of the social function and emotional impact of those gatherings.

The Evolutionary Context

In the course of detailing the constraints, we have noted that some distinguishing aspects of music (e.g., scale systems) require no Darwinian explanation for their widespread yet unique occurrence in humans; equifinality can occur as a consequence of characteristics of learning mechanisms and existing constraints providing the necessary frameworks for such development. Other underlying abilities required for musical behaviors, however (e.g., vocal learning, entrainment), are likely to be the product of forms of Darwinian selection. In these cases, the question then arises regarding what modes of selection might be operative and on what specifically they might be operating. There are some general lessons from evolutionary theory that are relevant, but often ignored, in constructing evolutionary scenarios for music (and language).

The first point is that biological evolution is always about genetic change. Although very few genes involved in music have been identified (see chapter 10, this volume), it is important to recognize that evolutionary scenarios implicitly or explicitly assume a sequence of changes in gene frequencies in a population, including the appearance of new genetic variants. Making this assumption explicit helps in avoiding the common fallacies of assuming (implicitly) unrealistic amounts of genetic changes (although that is difficult to quantify), assuming instantaneous adoption of new variants, or ignoring the fact that new variants, arising from mutation, are initially always rare.

A second point is that although evolution involves nonadaptive mechanisms such as random mutation and drift, a series of nonadaptive genetic changes leading to a complex new phenotype is exceedingly improbable. To establish the plausibility of a scenario for a trait shared by all humans, we thus have to show that each new variant conveys a fitness advantage both when it is rare in a population and when it has already become quite common. Eventually, of course, any gene that reliably confers high fitness will go to fixation in the population by its own success (Williams, 1992). With everyone in possession of the gene, it no longer drives differential reproductive success, and the trait it underpins has become part of species-specific nature. Until then, we must show that its advantage applies to the individual who carries it. Simply assuming selection for the benefit of the group is a once widely held fallacy (Williams, 1966). Traits that benefit the group rather than the individual can evolve only under quite specific circumstances described by kin selection and social evolution theory *sensu* Frank (1998).

A third point is that we need to be aware of the fact that the fitness advantage of a trait might not, or not only, come from its contribution to increased success in reproduction through increased survival (natural selection in the narrow sense, though including benefits associated with individuals' ability to establish effective social alliances); it may also come from the trait's effects on increased success in reproduction via attractiveness to potential partners (Darwinian sexual selection; Darwin, 1871). This could be particularly relevant for the evolution of music, as sexual selection is invariably invoked in understanding the evolution of elaborate animal aesthetic displays (where the connection between display and fitness can be very indirect). Music is nothing if not an aesthetic display (although possibly much else besides). Darwin treated it as such and proposed sexual selection as the mechanism behind it.

Finally, a fourth point is that in order to confer a selective advantage, a trait or behavior need not be essential for survival, but need only confer a slightly improved likelihood of survival to procreation or a greater rate of procreation—thus perpetuating and increasing the frequency of that trait or behavior—than would otherwise be the case. There is thus no justification for the common observation that music could hardly be a product of evolution by selection since it is hardly essential for survival. The former observation does not in fact follow from the latter. Furthermore, it does not rule out the possibility that various of the abilities that are used in musical activities may have been initially selectively favored as a consequence of their fulfillment of other purposes (e.g., interpersonal communication and the establishment of interpersonal relationships) and that musical practices may have developed within the context of those uses; musical behaviors have the potential to fulfill some of those same purposes, or other purposes, in even more effective ways. The co-use of these underlying abilities could lead to increasing interdependence among them, uniting them functionally in this new behavioral system, and potentially leading to further selective processes acting on those underlying abilities and the behaviors that use them.

Some of the traits that are essential for musical activities may have been a product of biological (natural or sexual) selection, and this could be by conferring a fitness advantage in either the context of their use in musical activities themselves or a different context of use. Meanwhile, certain properties of music and the traits that support them need not have been the product of biological selection at all.

The constraints we have outlined indicate that some of the abilities prerequisite for music (e.g., entrainment and vocal learning) would appear to have arisen in our lineage, or at the least been adapted from existing mechanisms to take on essentially novel form, in the period between our last common ancestor with chimpanzees (around 6 to 7 million years ago) and the appearance of our own species (around 200,000 years ago). Proposals regarding the emergence of these abilities should be complementary to, and tested against, what we know of the physiology, behavior, and ecology of the hominins in that intervening period. This is no small task, as clearly the knowledge in both paleoanthropology and the study of primate and human cognition is in a state of constant flux. Nevertheless, some aspects of our understanding in both areas are well enough established that we should undertake to ensure that proposals regarding the evolution of these capacities do not contradict core understanding in hominin evolution.

For example, hypotheses regarding the development of collective bodily and vocal display behaviors in early hominins from those exhibited by chimpanzees (and presumably our last common ancestor with them) should be framed in the context of changes in the habitat and group size of successive hominin species. It is now well established that gracile australopithecines exploited more open environments and a more omnivorous diet than higher primates of today but nevertheless continued to exploit wooded environments for shelter and some aspects of subsistence (Elton, 2008; Klein, 2009). Meanwhile, the physiological characteristics and ecological contexts of early *Homo* (*Homo habilis*, early African *Homo erectus*, and their descendants such as *Homo heidelbergensis*) indicate that they were exploiting a far greater range of more open environments, lacustrine and riverine habitats, and that carnivory had increasingly taken the place of arboreal frugivory in their subsistence resource exploitation (Elton, 2008; Klein, 2009). The efficacy of any proposed alterations to ancestral "carnival displays" (Merker et al., 2009), coalitional displays (Hagen & Bryant, 2003), or size-exaggeration vocalizations (Fitch, 2000), for example, needs to be situated within the context of these changes (see also Morley, 2013).

The mating strategies of human ancestors, and the social organization that arises from them, are also relevant to assessing the ecological validity of models regarding foundations of musical behaviors in interpersonal communication and display behaviors. This is because strategies for interpersonal communication, alliance and pair-bond formation, and display behaviors vary according to whether, for example, populations are monogamous, polygynous, or polyandrous. In polygynous species, for example, males compete with other males for access to multiple females, with little or no long-term alliance commitment to any one female. In contrast, monogamous species form pair-wise long-term cooperative bonds between males and females. In each case, the types of cooperation and competition, and with whom cooperation and competition occurs, vary, and the behaviors leading to success in negotiating alliances and long-term bonds, in display directed at the same sex, and directed at the opposite sex, vary accordingly (see, e.g., Lewin & Foley, 2004). Sexual selection is of course not confined to polygynous species; it occurs in monogamous species as well (Miller, 2013; Stewart-Williams & Thomas, 2013; Wachtmeister & Enquist, 2000).

In the case of human ancestors, high levels of sexual dimorphism and rapid developmental life history in australopithecines comparable to that of chimpanzees (Robson & Wood, 2008) has been taken to indicate broadly similar mating strategies and male-female relations as are exhibited by chimpanzees (e.g., Klein, 2009). Meanwhile, trends toward a reduction in sexual dimorphism and increased altriciality in the infants of early *Homo* indicate the development of increased cooperative long-term pair bonding (e.g., Klein, 2009). As noted, these developments have a direct influence on the forms and efficacy of behaviors related to display, sexual selection, pair bonding, and vocalization between adults (see also Merker, 2012). The development of greater altriciality in infants, a longer developmental process and greater dependence on adult care, also have direct impacts on the vocal behaviors between adults and infants (e.g., Falk, 2004) and the learning opportunities of infants and juveniles (e.g., Coqueugniot, Hublin, Veillon, Houët, & Jacob, 2004; O'Connell & DeSilva, 2013; Robson & Wood, 2008).

Similarly, the potential value and form of vocal learning capabilities should be tested and understood against the backdrop of physiological constraints for vocalization capabilities. Humans and higher primates (and our common ancestor with them) exhibit very different levels of innervation, and thus neurological control, of breathing and larynx, and thus pitch, prosody, emphasis, and duration of vocal utterances, with far greater levels of control exhibited by humans in each case. It is clear that these abilities developed over the course of the evolutionary emergence of our genus, *Homo*. For example, recent research (Meyer & Haeusler, 2015) has shown that the cervical and thoracic vertebral canal dimensions of australopithecines fall within the range of those of extant higher primates, while those of *H. erectus* (c.1.8 million years ago, Dmanisi, D2700), *H. antecessor* (ca. 800,000 years ago, Atapuerca Gran Dolina, ATD6-75), and *H. heidelbergensis* (ca. 500,000 years ago, Atapuerca Sima de los Huesos, AT1557) fall within the range of modern *Homo sapiens*. These spinal column dimensions relate to the level of innervation, and thus control of breathing that, alongside laryngeal control via the cranial vagus nerve, is significant for vocal production. In the case of the cervical vertebrae, this relates to control of the diaphragm (via the phrenic nerve at cervical vertebrae C3–C5), influencing pitch intonation, prosody, and emphasis via fine control of inhalation and exhalation (Meyer & Haeusler, 2015). In the case of the thoracic vertebrae, this relates to control of the intercostal and abdominal musculature, contributing to overall tidal volume, length of utterances, and some significant control of amplitude (MacBean et al., 2006).

MacLarnon and Hewitt (1999) had previously argued on the basis of specimen KNM-WT 15000 (Nariokotome Boy) that *H. ergaster/erectus* showed reduced thoracic canal innervation relative to *H. sapiens* and that this likely indicated relatively reduced vocal control. However, Meyer and Haeusler (2015) argue that KNM-WT 15000 appears to be atypical of the species and, indeed, the *Homo* genus, in the case of some of the vertebrae, though for the most part still falling within the lower end of the range of *H. sapiens*. They also argue that the relevance of thoracic canal dimensions for speech control (relative to cervical innervation) has been overstated, as the majority of fine vocal control is achieved with the larynx (via the cranial vagus nerve) in combination with breathing (the diaphragm controlled via the cervical phrenic nerve). However, while perhaps not essential for the intelligibility of modern speech at low amplitude (Meyer & Haeusler, 2015; Tamplin et al., 2013), thoracic control of breathing can have a significant impact on voice projection and phonation length (MacBean et al., 2006; Tamplin et al. 2013) and thus remains important in relation to protomusical and prelinguistic vocal communication. These changes could have served either song or speech, and it is worth noting that on a number of measures, such as tidal volume, range of subglottal pressure, and muscular control, the biomechanics of human song are more demanding than those of conversational speech (Sundberg, 1987). In short, it is with the emergence of our own genus, *Homo*, that we see the first evidence for neurological control of breathing necessary for modern-like vocalizations, and models of the development of complex vocalizations must be situated in the context of this physiological evolution.

The many ways in which evolutionary changes in traits and behaviors relevant to musical behaviors can occur, by biological selection or otherwise, are not mutually exclusive; in contrast, they can interact in important and complex ways, and any or all of them could have operated at various times in the course of human evolution. The distinctions among them have not always been clearly made in the literature discussing evolutionary rationales for musical behaviors, however. It is important that any future proposals do so, and clearly situate such mechanisms within what we know of the social and ecological contexts of human ancestors, and their physiological and neurological capabilities.

Acknowledgments

We thank Henkjan Honing for organizing and the Lorentz Center, Netherlands, for hosting the workshop that resulted in this chapter. Our thanks also to Carel ten Cate, Guy Madison, and an anonymous reviewer for numerous comments and suggestions that have helped us improve its content.

Notes

1. This sketch leaves out the many critical developments a vocal learning tradition must traverse in order to do what we propose it to have done in our case. The vocal learning capacity must first be emancipated from dependence on an innate song template, as in the bird mimics cited in the text. It must also abandon exclusive reliance on the tiny vocal gestures that supply the song elements for most birdsong, learned and unlearned, to include elements more akin to musical notes—those sustained at a given pitch with spectral energy concentrated to the fundamental. There is precedent for this in birds such as the musician wren and the pied butcher-bird of Australia, a mimic and virtuoso singer (Doolittle & Brumm, 2012; Taylor, 2008, 2009; Taylor & Lestel, 2011). Only on the basis of producing such music-like notes is vocal production learning likely to engage auditory system resonances with enough strength and reliability to become a factor in cultural transmission leading to musical tonality. The requirement that all this be in place if the process we have postulated is to get a start may help explain the rarity of actual tonal phenomena in animal song.

2. If isochrony is hidden in stimulus complexity, the requisite capacity for perceptual beat extraction is needed as well, but this is neither a defining aspect of entrainment with perfect synchrony nor part of most studies of sensorimotor synchronization in humans (Repp, 2005). A species that lacks the isochrony production propensity essential for mutual entrainment may nevertheless possess a perceptual beat extraction capacity as a matter of auditory system sophistication alone. That by itself does not make it a synchronizing species. Nor are all cyclical production phenomena based on equal period lengths (isochrony) suitable for interindividual entrainment of behavior. Those employed by entraining species (from human to firefly) typically have a leisurely tempo of 2 Hz or slower. The exceedingly rapid (10 to 60 Hz, i.e., 600 to 3600 bpm) pulse grids extractable from sequence timing in the singing of zebra finches (not known to entrain behaviorally) documented by Norton and Scharff (2016) presumably reflect the temporal granularity of their song system motor control rather than anything related to musical isochrony.

3. See note 2 above.

4. Examples abound on YouTube, of which the following video provides an example ("Cockatoo Finding Out He Is Going to the Vet," https://www.youtube.com/watch?v=5UUjJysUMTw).

5. From the point of view of scenarios for the origin and function of the human capacity for entrainment with perfect synchrony, it is of interest that it appears to be fully in place only by the time of approaching puberty (see further Merker et al., 2009).

6. *Self-paced* is preferable to the term *spontaneous* that occurs in the title and body of the Hattori et al. (2013) paper. The chimpanzees performed within the reinforcement contingencies of the operant regime of the experiment and can hardly be said to have acted spontaneously. The experimental design did not, however, dictate the pace of their responding, which accordingly was self-paced within the multiple constraints of the experimental setup.

References

Abler, W. L. (1989). On the particulate principle of self-diversifying systems. *Journal of Social and Biological Structures, 12*(1), 1–13. doi:10.1016/0140-1750(89)90015-8

Arom, S. (1991). *African polyphony and polyrhythm: Musical structure and methodology.* Cambridge: Cambridge University Press.

Balzano, G. (1982). The pitch set as a level of description for studying musical pitch perception. In M. Clynes (Ed.), *Music, mind, and brain* (pp. 321–351). New York: Plenum Press. doi:10.1007/978-1-4684-8917-0_1

Bartlett, F. C. (1932). *Remembering.* Oxford: Macmillan.

Baylis, J. R. (1982). Avian vocal mimicry: Its function and evolution. In D. E. Kroodsma & E. H. Miller (Eds.), *Acoustic communication in birds* (pp. 51–83). New York: Academic Press.

Beecher, M. D., & Brenowitz, E. A. (2005). Functional aspects of song learning in songbirds. *Trends in Ecology and Evolution, 20*(3), 143–149. doi:10.1016/j.tree.2005.01.004

Bermejo, M., & Omedes, A. (1999). Preliminary vocal repertoire and vocal communication of wild bonobos (*Pan paniscus*) at Lilungu (Democratic Republic of Congo). *Folia Primatologica, 70*(6), 328–357. doi:10.1159/000021717

Bidelman, G. M., & Krishnan, A. (2009). Neural correlates of consonance, dissonance, and the hierarchy of musical pitch in the human brainstem. *Journal of Neuroscience, 29*(42), 13165–13171. doi:10.1523/jneurosci.3900-09.2009

Brown, D. E. (1991). *Human universals.* New York: McGraw-Hill.

Brown, S., & Jordania, J. (2013). Universals in the world's musics. *Psychology of Music, 41*(2), 229–248. doi:10.1177/0305735611425896

Buck, J. (1988). Synchronous rhythmic flashing in fireflies. II. *Quarterly Review of Biology, 63*(3), 265–289. doi:10.1086/415929

Buck, J., & Buck, E. (1978). Towards a functional interpretation of synchronous flashing by fireflies. *American Naturalist, 112*(985), 471–492. doi:10.1086/283291

Buck, J., Buck, E., Case, J. F., & Hanson, F. E. (1981). Control of flashing in fireflies V: Pacemaker synchronization in *Pteroptyx cribellata. Journal of Comparative Physiology, 144*(3), 287–298. doi:10.1007/BF00612560

Burns, E. M. (1999). Intervals, scales, and tuning. In D. Deutsch (Ed.), *The psychology of music* (pp. 215–264). San Diego, CA: Academic Press. doi:10.1016/b978-012213564-4/50008-1

Byrne, R. W. (2000). Evolution of primate cognition. *Cognitive Science, 24*(3), 543–570. doi:10.1207/s15516709cog2403_8

Cavalli-Sforza, L. L. (2000). *Genes, peoples, and languages.* New York: North Point Press.

Charrier, I., Ahonen, H., & Harcourt, R. G. (2011). What makes an Australian sea lion (*Neophoca cinerea*) male's bark threatening? *Journal of Comparative Psychology, 125*(4), 385–392. doi:10.1037/a0024513

Christiansen, M. H., & Chater, N. (2008). Language as shaped by the brain. *Behavioral and Brain Sciences, 31*(5), 489–558. doi:10.1017/s0140525x08004998

Clement, C. J., Koopmans-van Beinum, F. J., & Pols, L. C. W. (1996). Acoustical characteristics of sound production of deaf and normally hearing infants. In *Proceedings of Fourth International Conference on Spoken Language Processing.* Piscataway, NJ: IEEE. doi:10.1109/icslp.1996.607914

Cook, P., Rouse, A., Wilson, M., & Reichmuth, C. (2013). A California sea lion (*Zalophus californianus*) can keep the beat: Motor entrainment to rhythmic auditory stimuli in a nonvocal mimic. *Journal of Comparative Psychology, 127*(4), 412–427. doi:10.1037/a0032345

Coqueugniot, H., Hublin, J.-J., Veillon, F., Houët, F., & Jacob, T. (2004). Early brain growth in *Homo erectus* and implications for cognitive ability. *Nature, 431*, 299–302. doi:10.1038/nature02852

Cousineau, M., McDermott, J. H., & Peretz, I. (2012). The basis of musical consonance as revealed by congenital amusia. *Proceedings of the National Academy of Sciences, 109*(48), 19858–19863. doi:10.1073/pnas.1207989109

Cross, I., Fitch, W. T., Aboitiz, F., Iriki, A., Jarvis, E. D., Lewis, J., … Trehub, S. E. (2013). Culture and evolution. In M. A. Arbib (Ed.), *Language, music, and the brain* (pp. 540–562). Cambridge, MA: MIT Press.

Darwin, C. (1871). *The descent of man and selection in relation to sex*. New York: D. Appleton.

Darwin, C. J., Turvey, M. T., & Crowder, R. G. (1972). An auditory analogue of the Sperling partial report procedure: Evidence for brief auditory storage. *Cognitive Psychology, 3*(2), 255–267. doi:10.1016/0010-0285(72)90007-2

De Waal, F. B. M. (1988). The communicative repertoire of captive bonobos (*Pan paniscus*), compared to that of chimpanzees. *Behaviour, 106*(3), 183–251. doi:10.1163/156853988x00269

Deacon, T. W. (1997). *The symbolic species: The co-evolution of language and the brain*. New York: Norton.

Doolittle, E., & Brumm, H. (2012). O Canto do Uirapuru: Consonant intervals and patterns in the song of the musician wren. *Journal of Interdisciplinary Music Studies, 6*(1), 55–85. doi:10.4407/jims.2013.10.003

Doupé, A. J., & Kuhl, P. K. (1999). Birdsong and human speech: Common themes and mechanisms. *Annual Review of Neuroscience, 22*, 567–631. doi:10.1146/annurev.neuro.22.1.567

Dowsett-Lemaire, F. (1979). The imitation range of the song of the marsh warbler, *Acrocephalus palustris*, with special reference to imitations of African birds. *Ibis, 121*(4), 453–468. doi:10.1111/j.1474-919X.1979.tb06685.x

Dufour, V., Poulin, N., Curé, C., & Sterck, E. H. M. (2015). Chimpanzee drumming: A spontaneous performance with characteristics of human musical drumming. *Scientific Reports, 5*, 1–6. doi:10.1038/srep11320s

Elton, S. (2008). The environmental context of human evolutionary history in Eurasia and Africa. *Journal of Anatomy, 212*(4), 377–393. doi:10.1111/j.1469-7580.2008.00872.x

Ember, C. R. (1978). Myths about hunter-gatherers. *Ethnology, 17*, 439–448. doi:10.2307/3773193

Ermentrout, B. (1991). An adaptive model for synchrony in the firefly *Pteroptyx malaccae*. *Journal of Mathematical Biology, 29*(6), 571–585. doi:10.1007/bf00164052

Falk, D. (2004). Prelinguistic evolution in early hominins: Whence motherese? *Behavioral and Brain Sciences, 27*(4), 491–541. doi:10.1017/s0140525x04000111

Fisher, R. A. (1930). *The genetical theory of natural selection*. Oxford: Clarendon Press.

Fitch, W. T. (2000). The evolution of speech: A comparative review. *Trends in Cognitive Sciences, 4*(7), 258–267. doi:10.1016/s1364-6613(00)01494-7

Fitch, W. T. (2005). The evolution of music in comparative perspective. *Annals of the New York Academy of Sciences, 1060*, 29–49. doi:10.1196/annals.1360.004

Fraisse, P. (1982). Rhythm and tempo. In D. Deutsch (Ed.), *The psychology of music* (pp. 149–180). London: Academic Press.

Frank, S. A. (1998). *Foundations of social evolution*. Princeton, NJ: Princeton University Press.

Geissmann, T. (2000). Gibbon song and human music from an evolutionary perspective. In N. L. Wallin, B. Merker, & S. Brown (Eds.), *The origins of music* (pp. 103–123). Cambridge, MA: MIT Press.

Ghiglieri, M. P. (1984). *The chimpanzees of Kibale forest*. New York: Columbia University Press.

Ghiglieri, M. P. (1987). Sociobiology of the great apes and the hominid ancestor. *Journal of Human Evolution, 16*(4), 319–357. doi:10.1016/0047-2484(87)90065-0

Gill, K. Z., & Purves, D. A. (2009). Biological rationale for musical scales. *PLoS One, 4*(12), 1–9. doi:10.1371/journal.pone.0008144

Greenfield, M. D. (1994). Cooperation and conflict in the evolution of signal interactions. *Annual Review of Ecology and Systematics, 25*, 97–126. doi:10.1146/annurev.ecolsys.25.1.97

Greenfield, M. D. (2005). Mechanisms and evolution of communal sexual displays in arthropods and anurans. *Advances in the Study of Behavior, 35*, 1–62. doi:10.1016/s0065-3454(05)35001-7

Hagen, E. H., & Bryant, G. A. (2003). Music and dance as a coalition signaling system. *Human Nature, 14*(1), 21–51. doi:10.1007/s12110-003-1015-z

Hanson, F. E. (1978). Comparative studies of firefly pacemakers. *Federation Proceedings*, *37*(8), 2158–2164.

Hanson, F. E. (1982). Pacemaker control of rhythmic flashing of fireflies. In D. Carpenter (Ed.), *Cellular pacemakers* (Vol. 2, pp. 81–100). New York: Wiley.

Hasegawa A., Okanoya K., Hasegawa T., & Seki Y. (2011). Rhythmic synchronization tapping to an audio-visual metronome in budgerigars. *Scientific Reports*, *1*, 1–8. doi:10.1038/srep001201

Hasselquist, D., Bensch, S., & von Schantz, T. (1996). Correlation between song repertoire, extra-pair paternity and offspring survival in the great reed warbler. *Nature*, *381*, 229–232. doi:10.1038/381229a0

Hattori, Y., Tomonaga, M., & Matsuzawa, T. (2013). Spontaneous synchronized tapping to an auditory rhythm in a chimpanzee. *Scientific Reports*, *3*, 1–6. doi:10.1038/srep01566

Hauser, M. D., Chomsky, N., & Fitch, W. T. (2002). The faculty of language: What is it, who has it, and how does it evolve? *Science*, *298*(5598), 1569–1579. doi:10.1126/science.298.5598.1569

Higham, T., Basell, L., Jacobi, R., Wood, R., Ramsey, C. B., & Conard, N. J. (2012). Testing models for the beginnings of the Aurignacian and the advent of figurative art and music: The radiocarbon chronology of Geißenklösterle. *Journal of Human Evolution*, *62*(6), 664–676. doi:10.1016/j.jhevol.2012.03.003

Honing, H., ten Cate, C., Peretz, I., & Trehub, S. E. (2015). Without it no music: Cognition, biology and evolution of musicality. *Philosophical Transactions of the Royal Society of London B: Biological Sciences*, *370*(1664), 20140088. doi:10.1098/rstb.2014.0088

Janata, P., Tomic, S. T., & Haberman, J. M. (2012). Sensorimotor coupling in music and the psychology of the groove. *Journal of Experimental Psychology: General*, *141*(1), 54–75. doi:10.1037/a0024208

Janik, V. M., & Slater, P. J. B. (1997). Vocal learning in mammals. *Advances in the Study of Behavior*, *26*, 59–99. doi:10.1016/s0065-3454(08)60377-0

Janik, V. M., & Slater, P. J. B. (2000). The different roles of social learning in vocal communication. *Animal Behaviour*, *60*(1), 1–11. doi:10.1006/anbe.2000.1410

Jarvis, E. D. (2006). Selection for and against vocal learning in birds and mammals. *Ornithological Science*, *5*(1), 5–14. doi:10.2326/osj.5.5

Katz, R. (1982). *Boiling energy: Community healing among the Kalahari Kung*. Cambridge, MA: Harvard University Press.

Kauranen, K., & Vanharanta, H. (1996). Influences of ageing, gender, and handedness on motor performance of upper and lower extremities. *Perceptual and Motor Skills*, *82*(2), 515–525. doi:10.2466/pms.1996.82.2.515

Kirby, S. (1998). *Language evolution without natural selection: From vocabulary to syntax in a population of learners* (Report No. 98-1). Edinburgh, UK: Department of Linguistics, Edinburgh University.

Kirby, S. (2001). Spontaneous evolution of linguistic structure: An iterated learning model of the emergence of regularity and irregularity. *IEEE Transactions on Evolutionary Computation*, *5*(2), 102–110. doi:10.1109/4235.918430

Kirby, S., Cornish, H., & Smith, K. (2008). Cumulative cultural evolution in the laboratory: An experimental approach to the origins of structure in human language. *Proceedings of the National Academy of Sciences*, *105*(31), 10681–10686. doi:10.1073/pnas.0707835105

Kirby, S., & Hurford, J. (2002). The emergence of linguistic structure: An overview of the iterated learning model. In A. Cangelosi & D. Parisi (Eds.), *Simulating the evolution of language* (pp. 121–148). London: Springer-Verlag. 10.1007/978-1-4471-0663-0_6

Klein, R. (2009). *The human career: Human biological and cultural origins*. Chicago: Chicago University Press.

Konishi, M. (2004). The role of auditory feedback in birdsong. *Annals of the New York Academy of Sciences*, *1016*, 463–475. doi:10.1196/annals.1298.010

Kroodsma, D. E., & Baylis, J. R. (1982). A world survey of evidence for vocal learning in birds. In D. E. Kroodsma & E. H. Miller (Eds.), *Acoustic communication in birds* (pp. 311–337). New York: Academic Press.

Krumhansl, C. L. (1990). *Cognitive foundations of musical pitch*. New York: Oxford University Press.

Kuhl, P. K. (1988). Auditory perception and the evolution of speech. *Human Evolution*, *3*, 19–43. doi:10.1007/bf02436589

Large, E. W. (2011). A dynamical systems approach to musical tonality. In R. Huys & V. Jirsa (Eds.), *Nonlinear dynamics in human behavior* (pp. 193–211). New York: Springer-Verlag.

Large, E. W., & Almonte, F. V. (2012). Neurodynamics, tonality, and the auditory brainstem response. *Annals of the New York Academy of Sciences, 1252*(1), E1–E7. doi:10.1111/j.1749-6632.2012.06594.x

Large, E. W., & Gray, P. M. (2015). Spontaneous tempo and rhythmic entrainment in a bonobo (*Pan paniscus*). *Journal of Comparative Psychology, 129*(4), 317–328. doi:10.1037/com0000011

Large, E. W., Kim, J. C., Flaig, N., Bharucha, J., & Krumhansl, C. L. (2016). A neurodynamic account of musical tonality. *Music Perception , 33*(3), 319–331. doi:10.1525/mp.2016.33.3.319

Le Bomin, S., Lecointre, G., & Heyer, E. (2016). The evolution of musical diversity: The key role of vertical transmission. *PLoS One, 11*(3), 1–17. doi:10.1371/journal.pone.0151570

Lerud, K. D., Almonte, F. V., Kim, J. C., & Large, E. W. (2014). Mode-locking neurodynamics predict human auditory brainstem responses to musical intervals. *Hearing Research, 308*, 41–49. doi:10.1016/j.heares.2013.09.010

Lewin, R., & Foley, R. (2004). *Principles of human evolution.* Oxford: Blackwell.

Lewontin, R. C. (1998). The evolution of cognition: Questions we will never answer. In D. Scarborough & S. Sternberg (Eds.), *An invitation to cognitive science: Methods, models, and conceptual issues* (Vol. 4). Cambridge, MA: MIT Press.

Liu, W.-C., Wada, K., & Nottebohm, F. (2009). Variable food begging calls are harbingers of vocal learning. *PLoS One, 4*(6), 1–11. doi:10.1371/journal.pone.0005929

Lots, I. S., & Stone, L. (2008). Perception of musical consonance and dissonance: An outcome of neural synchronization. *Journal of the Royal Society, Interface, 5*(29), 1429–1434. doi:10.1098/rsif.2008.0143

Lovejoy, C. O. (2009). Reexamining human origins in light of *Ardipithecus ramidus. Science , 326*(5949), 74e1–e8. doi:10.1126/science.1175834

MacBean, N., Ward, E., Murdoch, B., Cahill, L., Solley, M., & Geraghty, T. (2006). Characteristics of speech following cervical spinal cord injury. *Journal of Medical Speech-Language Pathology, 14*(3), 167–184.

MacDougall, H. G., & Moore, S. T. (2005). Marching to the beat of the same drummer: The spontaneous tempo of human locomotion. *Journal of Applied Physiology, 99*(3), 1164–1173. doi:10.1152/japplphysiol.00138.2005

MacLarnon, A., & Hewitt, G. (1999). The evolution of human speech: The role of enhanced breathing control. *American Journal of Physical Anthropology, 109*(3), 341–363. doi:10.1002/(SICI)1096-8644(199907)109:3<341::AID-AJPA5>3.0.CO;2-2

Madison, G. (2000). On the nature of variability in isochronous serial interval production. In P. Desain & L. Windsor (Eds.), *Rhythm perception and production* (pp. 95–113). Lisse: Swets and Zeitlinger.

Madison, G. (2001). Variability in isochronous tapping: Higher-order dependencies as a function of inter tap interval. *Journal of Experimental Psychology: Human Perception and Performance, 27*(2), 411–422. doi:10.1037/0096-1523.27.2.411

Madison, G. (2006). Experiencing groove induced by music: Consistency and phenomenology. *Music Perception, 24*(2), 201–208. doi:10.1525/mp.2006.24.2.201

Madison, G., & Merker, B. (2004). Human sensorimotor tracking of continuous subliminal deviations from isochrony. *Neuroscience Letters, 370*(1), 69–73. doi:10.1016/j.neulet.2004.07.094

Mayfield, G. R. (1934). The mockingbird's imitation of other birds. *Migrant, 5*, 17–19.

McDermott, J. H., Lehr, A. J., & Oxenham, A. J. (2010). Individual differences reveal the basis of consonance. *Current Biology, 20*(11), 1035–1041. doi:10.1016/j.cub.2010.04.019

Merchant, H., & Honing, H. (2013). Are non-human primates capable of rhythmic entrainment? Evidence for the gradual audiomotor evolution hypothesis. *Frontiers in Neuroscience, 7*, 1–8. doi:10.3389/fnins.2013.00274

Merker, B. (1999). Synchronous chorusing and the origins of music. *Musicae Scientiae* [Special issue 1999–2000], 59–73. doi:10.1177/10298649000030s105

Merker, B. (2000). Synchronous chorusing and human origins. In N. L. Wallin, B. Merker, & S. Brown (Eds.), *The origins of music* (pp. 315–327). Cambridge, MA: MIT Press.

Merker, B. (2002). Music: The missing Humboldt system. *Musicae Scientiae*, *6*(1), 3–21. doi:10.1177/102986490200600101

Merker, B. (2005). The conformal motive in birdsong, music, and language: An introduction. *Annals of the New York Academy of Sciences*, *1060*(1), 17–28. doi:10.1196/annals.1360.003

Merker, B. (2006). The uneven interface between culture and biology in human music [commentary]. *Music Perception*, *24*(1), 95–98. doi:10.1525/mp.2006.24.1.95

Merker, B. (2009). Ritual foundations of human uniqueness. In S. Malloch & C. Trevarthen (Eds.), *Communicative musicality* (pp. 45–59). Oxford: Oxford University Press.

Merker, B. (2012). The vocal learning constellation: Imitation, ritual culture, encephalization. In N. Bannan (Ed.), *Music, language and human evolution* (pp. 215–260). Oxford: Oxford University Press.

Merker, B. (2014). Groove or swing as distributed rhythmic consonance: Introducing the groove matrix. *Frontiers in Human Neuroscience*, *8*, 1–4. doi:10.3389/fnhum.2014.00454

Merker, B. (2015). Seven theses on the biology of music and language. *Signata*, *6*, 195–213.

Merker, B., Madison, G., & Eckerdal, P. (2009). On the role and origin of isochrony in human rhythmic entrainment. *Cortex*, *45*(1), 4–17. doi:10.1016/j.cortex.2008.06.011

Merker, B., Morley, I., & Zuidema, W. (2015). Five fundamental constraints on theories of the origins of music. *Philosophical Transactions of the Royal Society of London B: Biological Sciences*, *370*(1664), 1–11. doi:10.1098/rstb.2014.0095

Merker, B., & Okanoya, K. (2007). The natural history of human language: Bridging the gaps without magic. In C. Lyon, L. Nehaniv, & A. Cangelosi (Eds.), *Emergence of communication and language* (pp. 403–420). London: Springer-Verlag.

Mesoudi, A., & Whiten, A. (2004). The hierarchical transformation of event knowledge in human cultural transmission. *Journal of Cognition and Culture*, *4*(1), 1–24. doi:10.1163/156853704323074732

Mesoudi, A., & Whiten, A. (2008). The multiple roles of cultural transmission experiments in understanding human cultural evolution. *Philosophical Transactions of the Royal Society of London B: Biological Sciences*, *363*(1509), 3489–3501. doi:10.1098/rstb.2008.0129

Meyer, M. R., & Haeusler, M. (2015). Spinal cord evolution in early *Homo*. *Journal of Human Evolution*, *88*, 43–53.

Miller, G. F. (2013). Mutual mate choice models as the red pill in evolutionary psychology: Long delayed, much needed, ideologically challenging, and hard to swallow. *Psychological Inquiry*, *24*(3), 207–210. doi:10.1080/1047840x.2013.817937

Moelants, D. (2002). Preferred tempo reconsidered. In C. Stevens, D. Burnham, G. McPherson, E. Schubert, & J. Renwick (Eds.), *Proceedings of the Seventh International Conference on Music Perception and Cognition* (pp. 580–583). Adelaide: Causal Productions.

Morley, I. (2002). Evolution of the physiological and neurological capacities for music. *Cambridge Archaeological Journal*, *12*(2), 195–216. doi:10.1017/s0959774302000100

Morley, I. (2013). *The prehistory of music: Human evolution, archaeology, and the origins of musicality*. Oxford: Oxford University Press.

Morley, I. (2014). A multi-disciplinary approach to the origins of music: Perspectives from anthropology, archaeology, cognition and behaviour. *Journal of Anthropological Sciences*, *92*, 147–177. doi:10.4436/JASS.92008

Murphy, M. A., Thompson, N. L., & Schul, J. (2016). Keeping up with the neighbor: A novel mechanism of call synchrony in *Neoconocephalus ensiger* katydids. *Journal of Comparative Physiology A: Neuroethology, Sensory, Neural, and Behavioral Physiology*, *202*(3), 225–234. doi:10.1007/s00359-016-1068-1

Nagasaka, Y., Chao, Z. C., Hasegawa, N., Notoya, T., & Fujii, N. (2013). Spontaneous synchronization of arm motion between Japanese macaques. *Scientific Reports*, *3*, 1–7. doi:10.1038/srep01151

Nityananda, V., & Balakrishnan, R. (2007). Synchrony during acoustic interactions in the bushcricket Mecopoda "Chirper" (Tettigoniidae: Orthoptera) is generated by a combination of chirp-by-chirp resetting and change in intrinsic chirp rate. *Journal of Comparative Physiology A: Neuroethology, Sensory, Neural, and Behavioral Physiology*, *193*, 51–65. doi:10.1007/s00359-006-0170-1

Norton, P., & Scharff, C. (2016). "Bird song metronomics": Isochronous organization of zebra finch song rhythm. *Frontiers in Neuroscience, 10*, 1–15. doi:10.3389/fnins.2016.00309

Nottebohm, F. (1976). Vocal tract and brain: A search for evolutionary bottlenecks. *Annals of the New York Academy of Sciences, 280*, 643–649. doi:10.1111/j.1749-6632.1976.tb25526.x

Nowicki, S., Searcy, W. A., & Peters, S. (2002). Brain development, song learning and mate choice in birds: A review and experimental test of the "nutritional stress hypothesis." *Journal of Comparative Physiology A: Neuroethology, Sensory, Neural, and Behavioral Physiology, 188*(11), 1003–1014. doi:10.1007/s00359-002-0361-3

O'Connell, C., & DeSilva, J. (2013). Mojokerto revisited: Evidence for an intermediate pattern of brain growth in *Homo erectus*. *Journal of Human Evolution, 65*(2), 156–161. doi:10.1016/j.jhevol.2013.04.007

Oller, D. K., & Eilers, R. E. (1988). The role of audition in infant babbling. *Child Development, 59*(2), 441. doi:10.2307/1130323

Oudeyer, P.-Y. (2006). *Self-organization in the evolution of speech: Studies in the evolution of language* (J. R. Hurford, Trans.). Oxford: Oxford University Press.

Patel, A. D. (2006). Musical rhythm, linguistic rhythm, and human evolution. *Music Perception, 24*(1), 99–104. doi:10.1525/mp.2006.24.1.99

Patel, A. D. (2014). The evolutionary biology of musical rhythm: Was Darwin wrong? *PLoS Biology, 12*(5), 1–6. doi:10.1371/journal.pbio.1001821.s001

Patel, A. D., Iversen, J. R., Bregman, M. R., & Schulz, I. (2009). Experimental evidence for synchronization to a musical beat in a nonhuman animal. *Current Biology, 19*(10), 827–830. doi:10.1016/j.cub.2009.05.023

Pöppel, E. (2004). Lost in time: A historical frame, elementary processing units and the 3-second window. *Acta Neurobiologiae Experimentalis, 64*, 295–301.

Pusey, A. (1979). Inter-community transfer of chimpanzees in Gombe National Park. In D. Hamburg & E. McCown (Eds.), *The great apes* (pp. 465–479). Menlo Park, CA: Benjamin/Cummings.

Repp, B. H. (2001). Processes underlying adaptations to tempo changes in sensorimotor synchronization. *Human Movement Science, 20*(3), 277–312. doi:10.1016/s0167-9457(01)00049-5

Repp, B. H. (2005). Sensorimotor synchronization: A review of the tapping literature. *Psychonomic Bulletin and Review, 12*(6), 969–992. doi:10.3758/bf03206433

Repp, B. H. (2010). Self-generated interval subdivision reduces variability of synchronization with a very slow metronome. *Music Perception, 27*(5), 389–397. doi:10.1525/mp.2010.27.5.389

Repp, B. H., & Keller, P. E. (2008). Sensorimotor synchronization with adaptively timed sequences. *Human Movement Science, 27*(3), 423–456. doi:10.1016/j.humov.2008.02.016

Reynolds, V., & Reynolds, R. (1965). Chimpanzees of the Budongo forest. In I. Devore (Ed.), *Primate behavior: Field studies of monkeys and apes* (pp. 368–424). New York: Holt, Rinehart & Winston.

Robson, S., & Wood, B. (2008). Hominin life history: Reconstruction and evolution. *Journal of Anatomy, 212*(4), 394–425. doi:10.1111/j.1469-7580.2008.00867.x

Rouse, A. A., Cook, P. F., Large, E. W., & Reichmuth, C. (2016). Beat keeping in a sea lion as coupled oscillation: Implications for comparative understanding of human rhythm. *Frontiers in Neuroscience, 10*, 1–12. doi:10.3389/fnins.2016.00257

Savage, P. E., Brown, S., Sakai, E., & Currie, T. E. (2015). Statistical universals reveal the structure and function of human music. *Proceedings of the National Academy of Sciences, 112*(29), 8987–8992.

Schachner, A., Brady, T. F., Pepperberg, I. M., & Hauser, M. D. (2009). Spontaneous motor entrainment to music in multiple vocal mimicking species. *Current Biology, 19*(10), 831–836. doi:10.1016/j.cub.2009.03.061

Schusterman, R. J. (1977). Temporal patterning in sea lion barking (*Zalophus californianus*). *Behavioral Biology, 20*(3), 404–408. doi:10.1016/s0091-6773(77)90964-6

Semjen, A., Vorberg, D., & Schulze, H.-H. (1998). Getting synchronized with the metronome: Comparisons between phase and period correction. *Psychological Research, 61*(1), 44–55. doi:10.1007/s004260050012

Serpell, J. (1981). Duets, greetings and triumph ceremonies: Analogous displays in the parrot genus *Trichoglossus*. *Zeitschrift für Tierpsychologie, 55*(3), 268–283. doi:10.1111/j.1439-0310.1981.tb01272.x

Sismondo, E. (1990). Synchronous, alternating, and phase-locked stridulation by a tropical katydid. *Science, 249*(4964), 55–58. doi:10.1126/science.249.4964.55

Smith, E. C., & Lewicki, M. S. (2006). Efficient auditory coding. *Nature, 439*(7079), 978–982. doi:10.1038/nature04485

Stewart-Williams, S., & Thomas, A. G. (2013). The ape that thought it was a peacock: Does evolutionary psychology exaggerate human sex differences? *Psychological Inquiry, 24*(3), 137–168. doi:10.1080/1047840x.2013.804899

Stumpf, C. (1911). *Die Anfänge der Musik* [The beginnings of music]. Leipzig: Barth.

Sugiyama, Y. (1969). Social behavior of chimpanzees in the Budongo Forest, Uganda. *Primates, 9*(3), 225–258. doi:10.1007/bf01730343

Sugiyama, Y. (1972). Social characteristics and socialization of wild chimpanzees. In F. E. Poirer (Ed.), *Primate socialization* (pp. 145–163). New York: Random House.

Sundberg, J. (1987). *The science of the singing voice*. Dekalb: Northern Illinois University Press.

Tamplin, J., Baker, F. A., Grocke, D., Brazzale, D. J., Pretto, J. J., Ruehland, W. R., et al. (2013). Effect of singing on respiratory function, voice, and mood after quadriplegia: A randomized controlled trial. *Archives of Physical Medicine and Rehabilitation, 94*(3), 426–434. doi:10.1016/j.apmr.2012.10.006

Taylor, H. (2008). Decoding the song of the pied butcherbird: An initial survey. *Transcultural Music Review, 12*, 1–30.

Taylor, H. (2009). *Towards a species songbook: Illuminating the vocalisations of the Australian pied butcherbird (Cracticus nigrogularis)* (Unpublished doctoral dissertation). University of Western Sydney, Sydney, Australia.

Taylor, H., & Lestel, D. (2011). The Australian pied butcherbird and the nature-culture continuum. *Journal of Interdisciplinary Music Studies, 5*(1), 57–83.

Tennie, C., Call, J., & Tomasello, M. (2006). Push or pull: Imitation vs. emulation in great apes and human children. *Ethology, 112*(12), 1159–1169. doi:10.1111/j.1439-0310.2006.01269.x

Thompson, B., Kirby, S., & Smith, K. (2016). Culture shapes the evolution of cognition. *Proceedings of the National Academy of Sciences, 113*(16), 4530–4535. doi:10.1072/pnas.1523631113

Thorpe, W. H. (1961). *Bird song*. Cambridge: Cambridge University Press.

Trehub, S. E., Becker, J., & Morley, I. (2015). Cross cultural perspectives on music and musicality. *Philosophical Transactions of the Royal Society of London B: Biological Sciences, 370*(1664), 1–9. doi:10.1098/rstb.2014.0096

Turner, F., & Pöppel, E. (1983). The neural lyre: Poetic meter, the brain, and time. *Poetry, 142*(5), 277–309.

von Humboldt, W. (1836). *Über die Verschiedenheit des Menschlichen Sprachbaues und ihren Einfluss auf die geistige Entwicklung des Menschengeschlechts*. Berlin: Royal Academy of Sciences. English translation, G. C. Buck and F. Raven (Trans.). *Linguistic variability and intellectual development*. Miami: University of Miami Press, 1971.

Wachtmeister, C.-A., & Enquist, M. (2000). The evolution of courtship rituals in monogamous species. *Behavioral Ecology, 11*(4), 405–410. doi:10.1093/beheco/11.4.405

Walker, T. J. (1969). Acoustic synchrony: Two mechanisms in the snowy tree cricket. *Science, 166*(3907), 891–894. doi:10.1126/science.166.3907.891

Williams, G. C. (1966). *Adaptation and natural selection*. Princeton, NJ: Princeton University Press.

Williams, G. C. (1992). *Natural selection: Domains, levels, and challenges*. New York: Oxford University Press.

Wrangham, R. W. (1975). *The behavioural ecology of chimpanzees in Gombe National Park, Tanzania* (Unpublished doctoral dissertation). University of Cambridge, Cambridge, UK.

Wright, A. A., Rivera, J. J., Hulse, S. H., Shyan, M., & Neiworth, J. J. (2000). Music perception and octave generalization in rhesus monkeys. *Journal of Experimental Psychology: General, 129*(3), 291–307. doi:10.1037/0096-3445.129.3.291

Zarco, W., Merchant, H., Prado, L., & Mendez, J. C. (2009). Subsecond timing in primates: Comparison of interval production between human subjects and rhesus monkeys. *Journal of Neurophysiology, 102*(6), 3191–3202. doi:10.1152/jn.00066.2009

Zeigler, H. P., & Marler, P. (Eds.). (2004). Behavioral neurobiology of birdsong [Special issue]. *Annals of the New York Academy of Sciences, 1016.*

Zuidema, W. (2003). How the poverty of the stimulus solves the poverty of the stimulus. In S. Becker, S. Thrun, & K. Obermayer (Eds.), *Advances in neural information processing systems, 15* (pp. 51–58). Cambridge, MA: MIT Press.

Zuidema, W. (2013). Language in nature: On the evolutionary roots of a cultural phenomenon. In P. Binder & K. Smith (Eds.), *The language phenomenon* (pp. 163–189). Berlin: Springer-Verlag.

Zuidema, W., & de Boer, B. (2009). The evolution of combinatorial phonology. *Journal of Phonetics, 37*(2), 125–144. doi:10.1016/j.wocn.2008.10.003

4

The Origins of Music: Auditory Scene Analysis, Evolution, and Culture in Musical Creation

Laurel J. Trainor

The origins of complex behaviors and cognitive abilities are of great interest in the field of evolutionary psychology (Andrews, Gangestad, & Matthews, 2002; Barkow, Cosmides, & Tooby, 1995; Baron-Cohen, 1997). The origin of musical behavior is a particularly interesting example because there is currently no agreement as to whether music was an evolutionary adaptation or a cultural creation. Although the universality and early developmental emergence of musical behavior are consistent with its being an evolutionary adaptation, its adaptive value is not agreed on or, indeed, obvious (Justus & Hutsler, 2005; McDermott & Hauser, 2005; Pinker, 1997; Wallin, Merker, & Brown, 2000). A number of potential evolutionary pressures for music have been proposed and evidence for them discussed (reviewed in Fitch, 2006; Huron, 2003; Justus & Hutsler, 2005; McDermott & Hauser, 2005; Wallin et al., 2000), such as sexual selection (Darwin, 1872; Miller, 2000), social bonding and group cohesion (Brown, 2000; Dunbar, 2004; Roederer, 1984), regulating infant arousal and behavior (Dissanayake, 2000, 2008; Falk, 2004; Trehub & Trainor, 1998), aiding cooperative labor through rhythmic coordination, perceptual and motor practice or skill development (Cross, 2001), conflict resolution, safe time passing, and as a memory aid for preserving important cultural information across generations (Huron, 2003). It has also been proposed that music is not an evolutionary adaptation but, rather, a cultural creation that can stimulate pleasure centers in the brain (e.g., "the auditory cheesecake" hypothesis; Pinker, 1997), a by-product of the evolution of language (e.g., Patel, 2008; Pinker, 1997), or a culturally created "transformative technology" that affects our experience of the world (Patel, 2010; see also chapter 5, this volume).

This chapter is a revision of Trainor (2015). I argue in this chapter that these seemingly opposing views of musical origins—evolutionary adaptation versus a cultural creation—can be reconciled by going beyond simple notions of adaptive processes. Specifically, musical behavior rests on the interaction of adaptations shaped by natural selection and social-cultural forces. A major question is whether adaptations were selected to enhance music specifically or whether the evolutionary pressures were for other traits or capacities related to auditory perception, cognition, and motor skills that, once in place, made music possible. According to the former view, the benefits of musical behavior drove the evolutionary adaptations; according to the latter view, music is a cultural creation that was molded to existing brain structure and capacities that evolved under other pressures.

Evolutionary biologists describe an adaptation as a trait that has been shaped or modified by natural or sexual selection through particular gene-promoting effects (Williams, 1966). An adaptationist hypothesis is therefore a claim about effects that in the ancestral past, were favored by natural or sexual selection and contributed to shape current structure or operation. It is not a claim about current selection pressures that may or may not be maintaining it in populations. Consequently, the study of adaptation is largely a historical science (Mayr, 1983). In evolutionary biology, the term *function* is used for an effect that contributed to the shaping or modification of an adaptation by natural selection.

It is possible for some traits to take on new beneficial effects without being modified by selection for those effects. Such traits are called exaptations for these effects (Gould & Vrba, 1982). The distinction between an exaptation and an adaptation rests on whether the trait has been modified or shaped by selection specifically to facilitate a beneficial effect. For instance, the contour feathers of birds probably evolved first in small dinosaurs for a thermoregulatory function by providing a flat surface over which wind could pass without disturbing the warm air trapped close to the body (Andrews et al., 2002). But the structural organization of contour feathers also proved useful for facilitating flight. Thus, contour feathers were initially an adaptation with respect to thermoregulation, not flight. However, natural selection subsequently lengthened and stiffened the contour feathers located on the forelimbs and tails specifically because of the flight-facilitating effect. Thus, contour feathers were first adapted to thermoregulation, then exapted to flight, and then, finally, some contour feathers underwent secondary adaptation for flight. Note that when a trait does not exhibit any specific modification for a beneficial effect, that effect cannot be said to be a function of the trait. Only adaptations have functions, so facilitating flight is the function of the lengthened and strengthened feathers on the wings and tails of birds but not of contour feathers on the abdomen, unless specific modification for promoting flight could be demonstrated.

Finally, some traits may not be directly favored by natural selection but are inextricably tied (by genetic or developmental constraints) to traits that were the outcome of selection. Such traits are termed *by-products* or *spandrels* (Gould & Lewontin, 1979), after the triangular-shaped spaces between architectural arches. It is impossible to build a row of arches without producing these spaces, although there was no intent to do so. Spandrels can have neutral, beneficial, or even harmful effects. If a spandrel has a beneficial effect, it may also qualify as an exaptation for that effect, provided it has not been modified by selection to promote that effect.

The evaluation of evolutionary hypotheses is difficult, as others have noted (e.g., Andrews et al., 2002; Tooby, Cosmides, & Barrett, 2003; Williams, 1966). Musical behavior does have a number of features consistent with the idea that it was in part an evolutionary adaptation, such as an ancient origin (bone flutes date to at least 36,000 years ago and vocal music likely much earlier; Cross, Zubrow, & Cowan, 2002; d'Errico et al., 2003), universality across human cultures, early ontogenetic emergence without formal instruction, similarities (as well as variations of course) in pitch and rhythmic structures across musical systems, connections between auditory rhythms and entrained movement across cultures, the universal proclivity to respond emotionally to music, and use in ritual and social engagement across societies (e.g., Brown, 2000; Huron, 2003; Justus & Hutsler, 2005; McDermott & Hauser, 2005).

The origins of complex cognitive abilities, such as language, that are highly flexible, generative, and whose manifestations change rapidly over time pose particular challenges for evolutionary theories (Christiansen & Chater, 2008). Just as there are many languages, there are many musical systems, and, like languages, musical systems change over time (Albrecht & Huron, 2014; Huron & Ommen, 2006). Because they carry less conventional sound-meaning mappings, musical systems may change even more rapidly than languages. When different musical systems come into contact, new musical styles can readily emerge. For example, regional folk songs and jazz have influenced classical music, and new styles have emerged from fusions between jazz and rock music. Given that an exclusively evolutionary explanation for the origins of music would have difficulty explaining the variety of musical styles and the rapidity of musical change, there would appear to be a strong cultural component to musical origins.

In the case of music, the evolutionary question has typically been posed as whether musical behavior fits into one of three evolutionary processes. First is *adaptation.* There were selection pressures on the nervous system specifically for musical behavior, such as increased group social cohesion, which led to increased survival, or signaled increased fitness in mate selection. The second is *exaptation.* For example, the evolution of language might be an adaptation, leading to survival benefits for individuals in groups who could use language to communicate specific information. The auditory, memory, and cognitive adaptations needed for language also enabled music, which has survived over the long term because it enriches us culturally, even though music was not directly selected for. The third is *spandrel.* For example, the auditory system evolved under pressure to better sense danger in the environment, and pleasure centers in the brain evolved in order to motivate behaviors needed for survival and procreation. Music just happens to use the auditory system in ways that activate pleasure centers, but the auditory system has not been modified by selection to do so and there is no particular benefit to this.

This chapter takes a somewhat different approach. Rather than starting with the question of what functions music has or had in the past, and therefore what adaptive pressures might have been involved in the emergence of music, this chapter begins by examining the structure of music itself and determining what capabilities are needed for the perception and processing of music. The origins of these capabilities are then examined in light of developmental and cross-species comparisons to determine whether the capabilities in question evolved for functions other than music. Only capabilities necessary for music that did not obviously evolve for any other function are considered as candidates for music-specific adaptations. The three processes of adaptation, exaptation, and spandrels are often intertwined, particularly for the emergence of complex traits and complex cognitive abilities (see Andrews et al., 2002, for a detailed and insightful discussion). In this chapter, I argue that all three processes were likely involved in the emergence of critical structures necessary for music, but for the most part, this occurred through selection pressures for nonmusical functions. Those traits, or inextricably linked traits, may have then enabled musical or protomusical behaviors as cultural creations. However, even if music was largely a cultural creation, it is also possible that to the extent that music itself was beneficial, further music-specific adaptation occurred subsequently. Indeed, for the emergence of something as complex as music, there may have been a number of iterations of adaptive, exaptive, and cultural processes.

Music involves many aspects, such as pitch perception, time perception, pattern perception, rhythm or metrical perception, emotional responses, memory, sound production, and social consequences. It is possible, indeed likely, that different adaptive pressures and histories of adaptive and exaptive processes applied to these different aspects and that in many cases, the adaptive pressures were not for music. In the following sections, I consider pitch-based aspects of music, time- and rhythm-based aspects, and social-emotional aspects. For each, I consider possible evolutionary origins of particular traits or behaviors necessary for music and whether there is evidence for music-specific adaptations.

I also examine evidence from ontogenetic development where available. Ontogenesis is informative, as the early emergence of a trait or ability in development suggests that cultural origins are less likely, or at least that the organism is prepared to learn quickly in that domain. As for cross-species comparisons, in the case of music, it is generally agreed that humans are the only species to produce music (McDermott, 2008). A few other species do engage in music-like behaviors; for example, some vocal learning birds produce generative vocalizations, and some entrain to musical rhythms (Patel, Iversen, Bregman, & Schulz, 2009; Schachner, Brady, Pepperberg, & Hauser, 2009), and sea lions can be trained to move to musical rhythms at different tempos (Cook, Rouse, Wilson, & Reichmuth, 2013; Rouse, Cook, Large, & Reichmuth, 2016). But it is particularly revealing that our genetically closest relatives do not engage in musical activity, nor do musical stimuli appear to interest or engage them (McDermott & Hauser, 2004, 2007; but see Mingle et al., 2014). In any event, neurological structures or processes that play a role in the musical behavior of humans, but are widely conserved across species, are likely to originate from adaptive pressure unrelated to music, and to therefore be exaptations or spandrels with respect to music. Conversely, neurological structures or processes unique to humans represent phenotypic modifications that may have arisen by natural selection for behaviors specific to humans, including musical behavior. These represent candidate adaptations that should be rigorously scrutinized.

This chapter is not intended to provide an exhaustive consideration of the evolutionary and cultural origins of music; rather, it presents hypotheses about how adaptive, exaptive, and cultural processes may have been involved in some aspects of musical emergence in the context of a discussion of how to evaluate hypotheses in this domain. The first sections focus on perceptual prerequisites for musical behavior. In particular, I argue that much of musical spectral (pitch) and temporal (rhythm and meter) structure rests on adaptations of the auditory system for gathering information about what sounding objects are present in the environment and where they are located, a process termed auditory scene analysis (ASA; Bregman, 1990; Pinker, 1997). Specifically, in the next section, "Auditory Scene Analysis," I present a brief overview of ASA and discuss the fact that it is phylogenetically old and emerges early in development. In the following section, "Spectral Analysis and the Origins of Musical Pitch Structure," I consider what aspects of musical pitch structure can and cannot be explained by ASA, and in the fourth section, "Time Processing and the Origins of Musical Rhythm," what aspects of musical temporal structure can and cannot be explained by ASA. I argue that rather than music exhibiting adaptive pressure on the auditory system, it is largely the other way around: pitch and rhythmic structure in music have adapted or conformed to preexisting features of the auditory system. However, there may be some features of music that were evolutionary adaptations, and I consider evidence for these. In the fifth section, "Social and Emotional Functions and the Origins of Music," I examine possible adaptive social and emotional aspects of music and consider whether they might have exerted adaptive pressure for enhanced musical perception and production.

Auditory Scene Analysis

The most basic functions of perception include determining what objects are present in the environment and where they are located (Bregman, 1990; Cherry, 1953), information that is useful for a wide variety of species. Unlike the visual system, where the relative location of objects in space is related to the spatial pattern of activity on the retina and topographic maps in visual pathways, in the auditory system, sound vibration frequency is encoded along the basilar membrane in the inner ear, and this organization is maintained in tonotopic maps throughout subcortical pathways and into primary auditory cortex. Thus, location must be calculated on the basis of complex cues such as interaural time and intensity differences and sound filtering properties of the pinna (Middlebrooks & Green, 1991).

In the visual system, one object may occlude another object, but the corresponding problem in the auditory system is more complex in that (1) most sounds emitted by objects in the environment contain energy across a wide range of frequencies, so different sounds overlap in frequency content, and (2) an auditory environment typically contains many simultaneously sounding objects and the sound waves emitted by these objects (and their echoes) are combined in the air and reach the ear as one complex wave. Thus, ASA involves decomposing the sound input into spectrotemporal components (i.e., the frequency content and how it changes over time) and figuring out how many sound sources there are and which components come from which sound sources. This requires segregation of some components as originating from different sources, as well as the integration of other components as coming from the same sound source. This determination is not an easy problem to solve, and the auditory system relies on a number of cues (Bregman, 1990; Cherry, 1953; Gutschalk & Dykstra, 2014; Snyder & Alain, 2007; Winkler, Denham, & Nelken, 2009).

As Bregman (1990) outlined, ASA in humans has two aspects; bottom-up automatic parsing of the input and top-down controlled processes, which deploy attention and knowledge of familiar sounds. The cues used by the auditory system in automatic ASA have been studied extensively. They can be grouped into two categories: those related to separating *simultaneous* sound sources (e.g., one person's voice from other voices at a cocktail party; see Micheyl & Oxenham, 2010, for a review) and those related to integrating *successive* sounds emitted over time from one object (e.g., integrating successive speech sounds emitted by one talker, or successive notes played by one musical instrument, into a single stream of sound; Moore & Gockel, 2012; van Noorden, 1975). Of course, simultaneous and successive processes occur at the same time. For example, in music written for two voices, the auditory system must determine at any moment in time that there are two voices present and which frequency components (harmonics) belong to each voice, while at the same time following the successive frequency changes within each voice and integrating them into melodic percepts (Huron, 2001).

Bottom-up processes in ASA are sometimes surprisingly opaque to top-down influence (Bregman, 1990), suggesting an evolutionarily ancient origin. Indeed, ASA has been identified across many species (see Fay, 2008, for a review). It also emerges early in human development (Demany, 1982; Fassbender, 1993; Folland, Butler, Payne, & Trainor, 2015; McAdams & Bertoncini, 1997; Smith & Trainor, 2011; Winkler et al., 2003). The cues used to accomplish ASA are complex, but a number have been identified and, in some cases, how they interact when in conflict to produce stable percepts (see Bregman, 1990; Scharine & Letowski, 2009, for reviews). For both simultaneous and successive aspects of ASA, both spectral (frequency) based and temporal (timing) based cues are used. These are discussed in the next sections.

Spectral Analysis and the Origins of Musical Pitch Structure

Pitch perception is fundamental to music, raising the possibility that it might have evolved for musical behavior. However, I will show here that pitch is not given in the stimulus but derived by the brain, and that the perception of pitch is a direct consequence of ASA. Vowel-like vocalizations and musical instrument sounds that are perceived to have a pitch typically have energy at a fundamental frequency, f_0, and at harmonics whose frequencies are at integer multiples of f_0. For example, if $f_0 = 100$ Hz, the harmonic frequencies will be 200, 300, 400, 500, … Hz. The cochlea in the inner ear is stiffer and wider at one end than the other, causing it to vibrate maximally at different points along its length according to the frequency input in a systematic manner. The vibration of the basilar membrane is transduced into electrical signals in the auditory nerve via the inner hair cells along its length, creating a tonotopic representation that is maintained through subcortical nuclei and into primary auditory cortex. Thus, when a complex sound (i.e., one with several frequency components or harmonics) is presented, the basilar membrane performs a sort of Fourier analysis, decomposing it into its frequency components, which are maintained in separate channels. In addition, there is a temporal aspect of frequency coding (e.g., see Cariani, 2004; Cedolin & Delgutte, 2005; Delgutte & Cariani, 1992). Inner hair cells fire at the point of maximal displacement of the basilar membrane, so the timing of populations of neurons also encodes frequency content, and current models of pitch perception combine spectral and temporal cues (Moore & Gockel, 2011; Oxenham, Micheyl, & Keebler, 2009; Santurette & Dau, 2011). Accumulating evidence suggests that it is not until information reaches an area just beyond primary auditory cortex on the lateral side of Heschl's gyrus that the spatial frequency and temporal frequency representations are combined and that the frequency content is integrated into a percept of a single sound (auditory object) with a particular pitch and timbre (Bendor & Wang, 2005; Gockel, Carlyon, Mehta, & Plack, 2011; Hall, Barrett, Akeroyd, & Summerfield, 2005; Patterson, Uppenkamp, Johnsrude, & Griffiths, 2002; Penagos, Melcher, & Oxenham, 2004; Zatorre, 1988).

One might ask why the auditory system decomposes an incoming sound into its frequency components only to reintegrate them once again in cortex. The answer is that the process is necessary for ASA. When two or more sound sources are present in the environment at the same time and their frequency ranges overlap, the only way to determine which frequency components belong to which sound (or, indeed, how many sounds are present) is to decompose the incoming sound wave by frequency and recombine the components according to probable sound sources (figures 4.1A and 4.1B).

One important cue for determining whether a set of simultaneous frequencies should be integrated into a single percept is whether the frequencies are integer multiples of a common fundamental frequency, a common sound structure in human and nonhuman vocalizations. The perception of pitch is one consequence of this process. That pitch is derived in the brain, and not given in the sound input, is clearly demonstrated by the phenomenon known as perception of the pitch of the missing fundamental (figure 4.1D). Specifically, if the energy at f_0 is removed (and masking noise covers any difference tones created by nonlinearities in the ear), the structure of the harmonics leads to the perception of a sound with pitch at f_0, even though there is no energy at that frequency (although timbre will change, of course; Plack, Oxenham, & Fay, 2006). Thus, pitch perception appears to

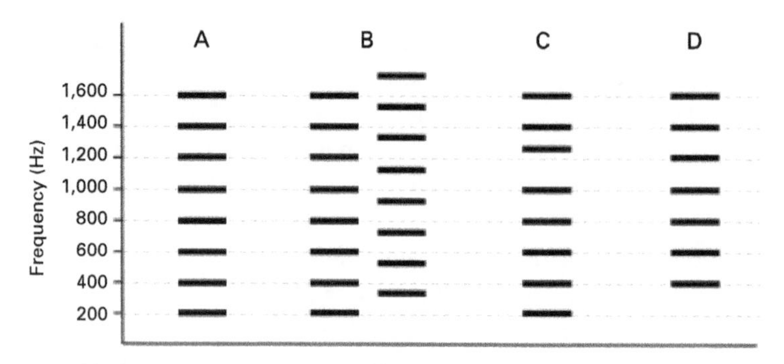

Figure 4.1
Harmonic structure and determining the number of auditory objects with simultaneous sound inputs. (A) A complex tone with fundamental frequency (f_0) at 200 Hz and harmonics at integer multiples of f_0, which is perceived as a single tone (auditory object) with a pitch of 200 Hz. (B) Two complex tones (sound sources) with f_0's at 200 and 260 Hz and their harmonics. Their harmonics overlap in frequency range, so when they simultaneously impinge on the ear, the auditory system must decompose the incoming sound into its frequency components and use its knowledge of harmonic structure to recombine them into representations of the original sound sources. (C) That the brain uses harmonicity to determine the number of auditory objects can be seen by mistuning one harmonic of the 200 Hz complex tone shown in panel A. In this case, two tones are heard. The mistuned harmonic is heard as one auditory object and the remaining components, which are all integer multiples of f_0, fuse into a second auditory object. (D) Pitch of the missing fundamental. That the brain creates the sensation of pitch can be seen clearly in that when f_0 is removed from a complex tone stimulus, the perceived pitch remains at f_0.

have evolved as a consequence of ASA and not specifically for music. Consistent with this idea, many species perceive the pitch of the missing fundamental (e.g., Shofner, 2005). In human infants, perception of the pitch of the missing fundamental emerges at around three months of age as auditory cortex matures and supports information processing (He & Trainor, 2009). Thus, the evidence strongly indicates that pitch perception did not evolve for music but rather was exapted for music. Indeed, it could be considered that in this case, music conformed to the human auditory system rather than the other way around, as has been suggested for language (Christiansen & Chater, 2008; Kirby, Cornish, & Smith, 2008).

Harmonic relations, or their absence, are also used in ASA to separate frequency components into different auditory objects (e.g., Bregman, 1990; Hartmann, 1996). For example, if one harmonic of a complex tone is mistuned, it is no longer integrated with the other frequency components and is perceived as a separate auditory object (Lin & Hartmann, 1998; see figure 4.1C). The ability to hear two objects when a harmonic is mistuned appears to emerge in human infancy at around the same age as the ability to derive pitch from sounds with missing fundamentals (Folland et al., 2015; Folland, Butler, Smith, & Trainor, 2012; He & Trainor, 2009), consistent with the idea that both are part of the same process of ASA. Music often consists of more than one sound at a time. As with the perception of pitch itself, the ability to perceive multiple simultaneous musical lines appears to be based on the evolution of ASA, again consistent with musical structure being a consequence of the human auditory system rather than music driving the evolution of the auditory system.

Other aspects of musical pitch structure also appear to be a consequence of the structure of inner ear. For example, the physical properties of the basilar membrane are such that its frequency tuning increases with increasing frequency (Moore, 2012). Specifically, when two frequencies that differ by less than a critical band are presented simultaneously, their vibration patterns interact on the basilar membrane so that they are not cleanly encoded in different tonotopic channels and it is more difficult to determine which frequencies are present. The size of the critical band increases with increasing frequency up to at least 1,000 Hz (Glasberg & Moore, 1990) and likely well beyond (Shera, Guinan, & Oxenham, 2002, 2010) which means that for lower tones, greater frequency separation is needed in order to clearly perceive the pitches of the tones (Fletcher, 1940; Greenwood, 1990; Zwicker, Flottorp, & Stevens, 1957). Frequency coding on the basilar membrane, in the form of a tonotopic map, is the first step in ASA because only by separating the frequency components in a sound wave can it be determined which components belong to which auditory objects. Critical bands are a direct result of the nature of physical vibrations on the basilar membrane, so they can be considered a by-product of adaptations for ASA. As Huron (2001) points out, music is written with larger pitch differences between, for example, bass and tenor parts than between soprano and alto parts, in a manner that parallels the size of the critical band. It is highly unlikely that music exerted an influence on the evolution of critical band size. Instead, for the pitch content of music to be clear, it must conform to basic constraints of the auditory system that evolved for other functions.

Similarly, when two musical tones are played simultaneously, musicians and nonmusicians and even infants encode the pitch of the higher tone better than that of the lower tone (Marie, Fujioka, Herrington, & Trainor, 2012; Marie & Trainor, 2012; 2014). Interestingly, this effect also originates in interactions between harmonics during frequency coding on the basilar membrane in the cochlea (Trainor, Marie, Bruce, & Bidelman, 2014; see box 4.1). Although there are no animal studies on this effect, its peripheral origin suggests that it will likely also be found in other mammals. Musical composition is consistent with this property of sound encoding, as seen in the widespread placement of the main melody in the highest-pitched voice in polyphonic music (Chon & Huron, 2014; Huron, 2016). It is highly unlikely that the critical band structure in the inner ear was specifically selected for music. Indeed, the effects of critical band structure on frequency encoding and the high-voice superiority effect are likely spandrels (i.e., nonadaptive consequences) of ASA that in turn affect how music is composed and experienced. That said, it is possible that once critical band structure had evolved, music or language, or both, exerted additional pressures to sharpen cochlear tuning. Consistent with this possibility, it has been estimated that human cochlear tuning is better than that of most other mammals by a factor of two to three (Glasberg & Moore, 1990; Gutschalk & Dykstra, 2014; Shera et al., 2010; Snyder & Alain, 2007; Winkler et al., 2009).

Box 4.1

The high-voice superiority effect for pitch and the low-voice superiority effect for timing of simultaneous tones originate in the cochlea of the inner ear. When two simultaneous tones are presented, as in panel A in figure 4.3, the brain responds more strongly to occasional pitch changes of a semitone (1/12 octave) in the higher than the lower tone as measured by the mismatch negativity (MMN) response of the event-related potential (ERP) in electroencephalographic (EEG) recordings, but not when each tone is presented separately. When the high tone or the low tone is passed through a computer model of the auditory periphery (Trainor, Marie, Bruce, & Bidelman, 2014), the harmonics are well represented in the auditory nerve firings (panel B), but when the two tones are presented together, the harmonics of the higher-pitched tone tend to mask the harmonics of the lower-pitched tone (a phenomenon referred to as two-tone masking) largely because the former are more intense than the latter due to the roll-off in intensity with increasing frequency in natural sounds.

When the same tones are presented, but either the higher tone or the lower tone is occasionally presented 50 ms too early, as in panel C (from Hove et al., 2014), the MMN is larger for the timing deviants in the lower-pitched voice. As sounds propagate along the basilar membrane, the high frequencies enervate the basal end up to 10 ms sooner than the low frequencies enervate the apical end, but the low-voice superiority effect for time described here cannot be a consequence of this as time difference is too short and the brain compensates for this difference, perceiving simultaneously presented high and low tones as simultaneous (Wojtczak, Beim, Micheyl, & Oxenham, 2012). The origin of this effect in the inner ear depends instead on the harmonic structure of the tones, as can be seen by the results of passing these stimuli through the model of Ibrahim & Bruce (2010). In panel D, when the two tones come on simultaneously at 50 ms (top), the spike counts in the auditory nerve show a single abrupt onset across all frequency channels. When the lower-pitched tone comes on too early at 0 ms (middle), there is spiking across the frequency range because its fundamental is low and its harmonics therefore cover the frequency range. In this case, there is no clear spike increase when the higher-pitched sound enters at 50 ms and the sound is unambiguously represented as early. However, in the case that the higher tone is too early at 0 ms, there is spiking at this early time for frequencies at its fundamental and above, but a second clear spike increase is seen in the lower-frequency range when the lower tone enters at 50 ms. Thus, the time representation of this stimulus is more ambiguous. These results show that the musical propensity to put the melody in the highest voice and the basic beat in the lowest voice originates in properties of the inner ear.

Another aspect of ASA involves determining when to integrate successive sound events as emanating from one sound source (or stream) versus segregating them as emanating from different sound sources. A number of cues to streaming in ASA have been demonstrated (e.g., see Bregman, 1990; Cherry, 1953; Gutschalk & Dykstra, 2014; Snyder & Alain, 2007; Winkler et al., 2009), and Huron (2001) has outlined how some of them relate to rules of musical composition. Huron's analysis applies to Western music, but it is likely that other musical systems are also greatly influenced by cues evolved for ASA.

For example, one basic ASA cue for integration relates to pitch proximity; the frequency or pitch content of a source is expected to change little over small time periods, reflecting the fact that the sounds emitted by objects do not normally fluctuate rapidly in frequency. That this is a prominent cue in ASA was demonstrated with the gallop rhythm depicted in figure 4.2A (van Noorden, 1975). When the frequencies of the high and low tones are close, all of the tones are integrated into one auditory object, and a gallop rhythm can be heard. The larger the frequency distance is between the high and low tones, the more likely it is that the pattern will be perceived as two auditory objects, one consisting of high tones and the other of low tones, in which case no gallop rhythm is heard (figure 4.2B). Similarly, when the sequence is presented slowly, it is more likely that the tones with different frequencies will be integrated into one auditory object (figure 4.2C), whereas at faster rates, the tones are more likely to separate into individual auditory objects.

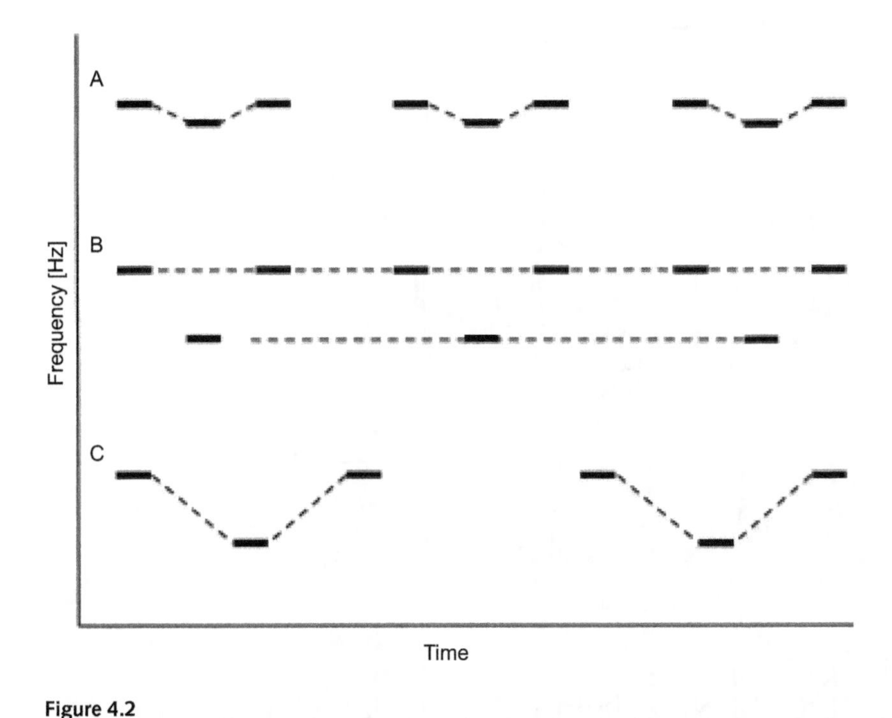

Figure 4.2
The effects of pitch proximity and tempo on determining the number of auditory objects in sequential streams of sounds. (A) When a higher tone repeats at a regular interval and a lower tone repeats at half the tempo of the higher tone and they are arranged as in *a*, all of the tones are perceived to come from a single sound source (as depicted by the dotted lines) and a gallop rhythm is heard. (B) When the higher and lower tones are sufficiently separated in frequency, they can no longer be integrated into a single stream. Two auditory objects are heard, one a repeating high tone and one a repeating low tone, and no gallop rhythm is perceived. This demonstrates that the auditory system expects a single sound source to remain reasonably consistent in pitch. (C) When the tempo of the sequence in *b* is slowed, again the two pitches can be integrated into a single auditory object, and the gallop rhythm is heard again, consistent with the idea that the auditory system expects an auditory object to change pitch slowly (adapted from van Noorden, 1975).

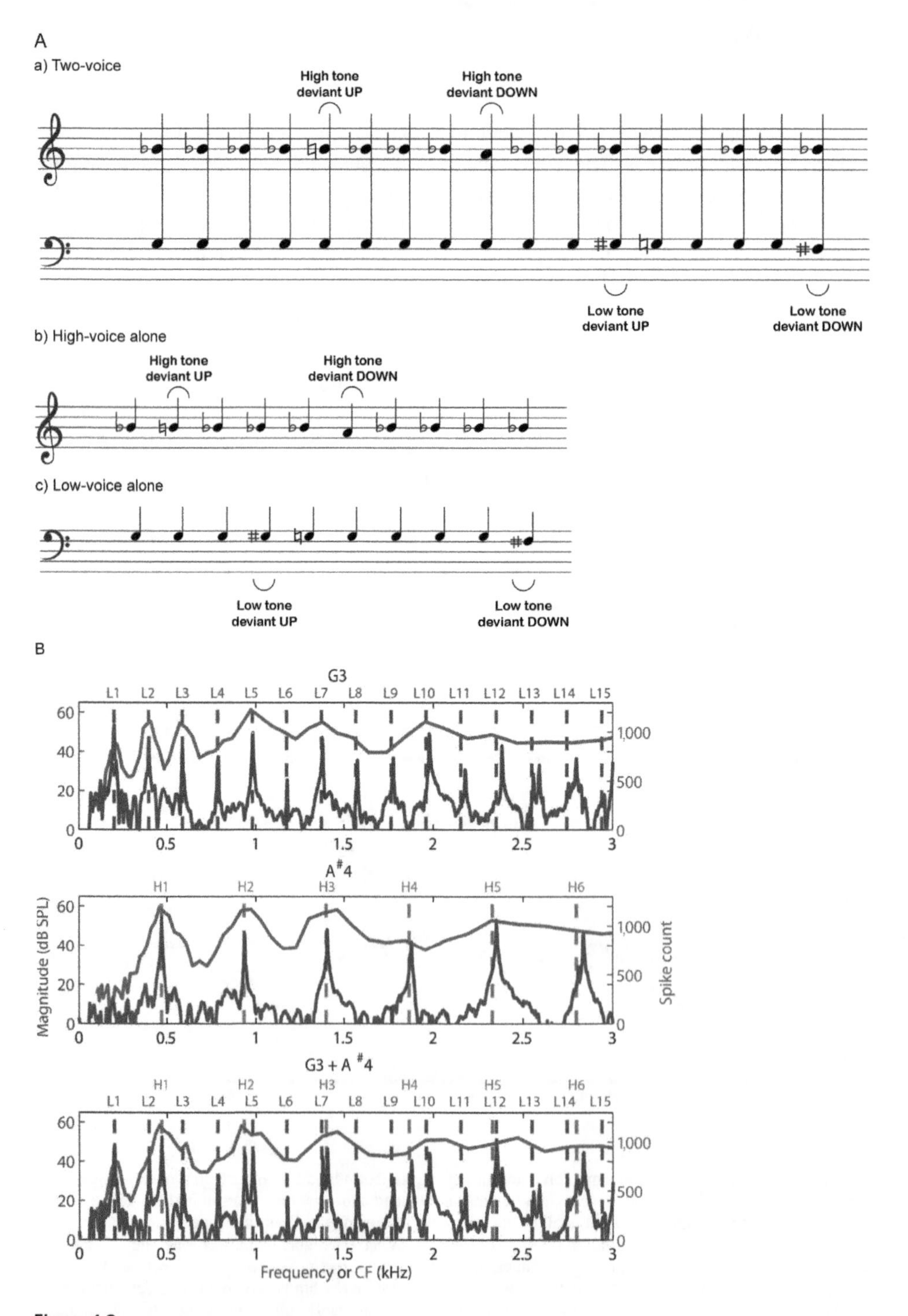

Figure 4.3
The high-voice superiority effect for pitch (panels A and B) and the low-voice superiority effect for timing (panels C and D).

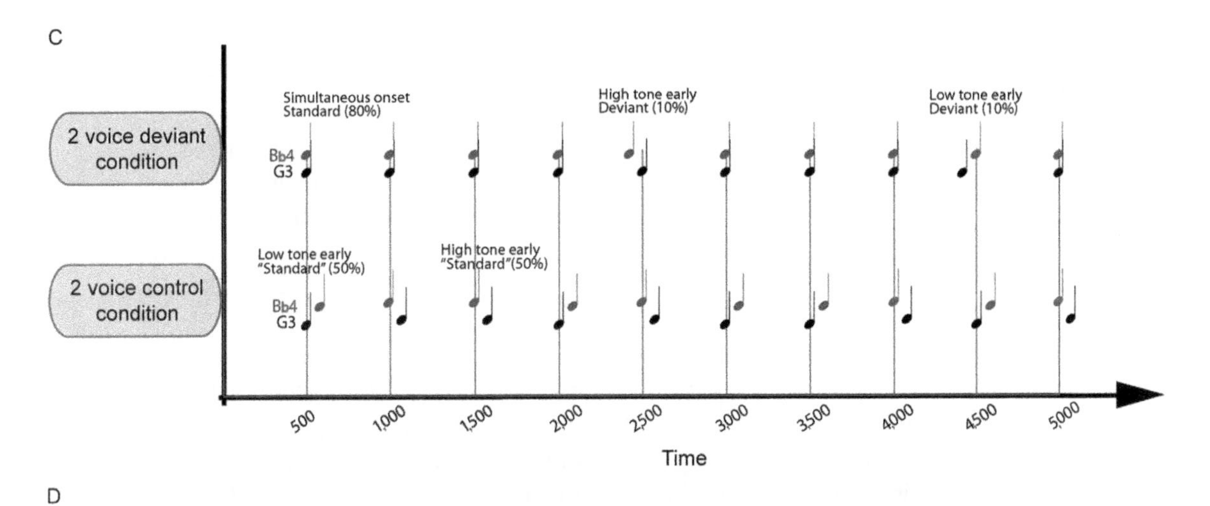

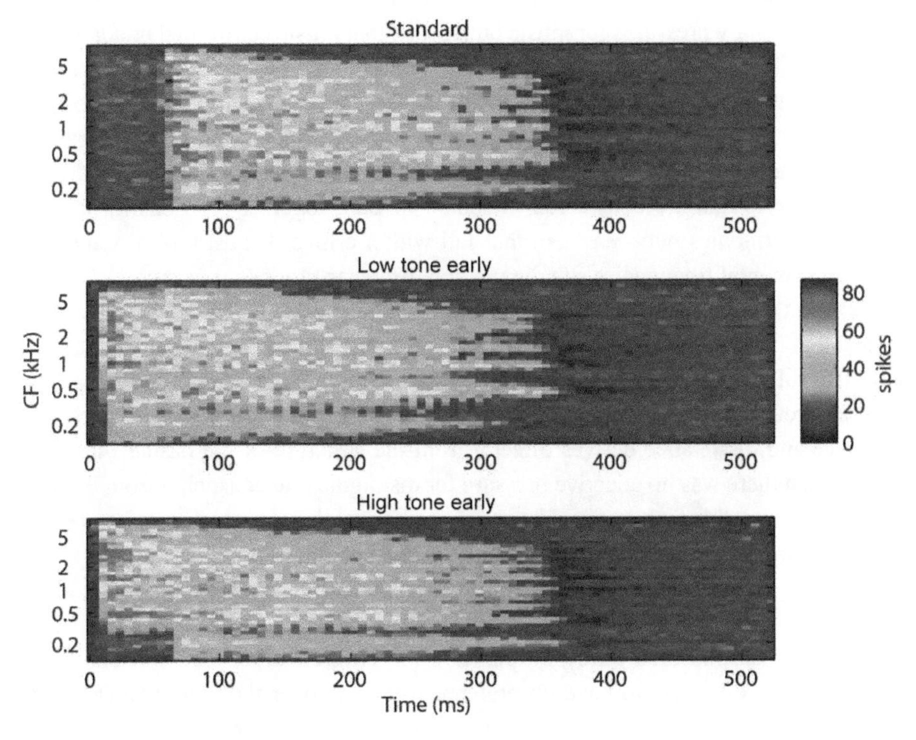

Figure 4.3 (continued)

Huron (2001) showed that most of the Western rules of voice leading (how to compose polyphonic music) are a consequence of cues such as pitch proximity. For example, one set of rules states that when writing successive chords (e.g., in four-part harmony), where it is desirable for the listener to follow each part or stream (e.g., soprano, alto, tenor, bass), keep the same pitch in a particular part from chord to chord if it is possible; if the pitch needs to change, move by the smallest pitch distance possible and, most important, avoid large pitch changes. This enables people to perceive the successive tones from each part as coming from one auditory object and therefore to follow each part over time. Another rule states that it is not a good idea for the different parts to cross pitch so that, for example, the soprano part is higher than the alto part on one chord but lower on the next chord. Again, the principle of pitch proximity dictates that under these conditions, listeners will likely confuse which pitches belong to which part. In sum, the fit between compositional practice and the principles of ASA, and the fact that ASA is phylogenetically more ancient than human music, indicate that much of musical structure was not specifically selected for through evolutionary pressures for music but, rather, that music conformed or adapted to a preexisting auditory system.

Some aspects of musical pitch, however, appear to be specific to music perception, such as the relation between sensory consonance and dissonance and feelings of pleasantness or unpleasantness, and the structure of musical tonality. According to Plomp and Levelt (1965), two tones that are considered to sound pleasant together (consonant) have few nonidentical harmonics between them that fall within critical bands, typically the result of their fundamental frequencies standing in small-integer ratios (e.g., octave 2:1; perfect fifth 3:2). On the other hand, tones that are perceived to sound unpleasant together (dissonant) stand in more complex ratios (e.g., major seventh 15:8; tritone 45:32) and have harmonics that fall within critical bands on the basilar membrane, creating the perception of beating and roughness. According to this theory, the perceptual differentiation of sensory consonance and dissonance derives directly from the structure of the basilar membrane. Assuming that there was no adaptive pressure for distinguishing consonant from dissonant tone combinations, this feature could be considered a spandrel of inner ear structure. Consistent with this notion is evidence that monkeys perceive the difference between sensory consonance and dissonance (Fishman et al., 2001; Izumi, 2000) even though they do not have music.

Interestingly, despite their ability to perceive the difference between consonance and dissonance, monkeys seem to have no preference for one over the other (McDermott & Hauser, 2004). What seems to be special to human music, then, is a preference for consonance over dissonance and the use of dissonance to create musical tension and consonance to resolve that tension. Based on ideas articulated by Stumpf more than one hundred years ago (DeWitt & Crowder, 1987), McDermott, Lehr, and Oxenham (2010) proposed that the perception of consonance, defined as preference, is related to the extent to which all harmonics across the simultaneously presented sounds conformed to a harmonic template consisting of a fundamental frequency and harmonics at integer multiples of that fundamental. Experimentally, they showed that pleasantness has stronger relations to harmonicity than to roughness and beating. It is unknown whether monkeys base their discrimination of consonant and dissonant patterns on beating and roughness or on harmonicity, but it is possible that valenced harmonicity processing is unique to humans. It is clear that musical structure uses preexisting properties of the auditory system that give rise to the distinction between consonance and dissonance, but music appears to add emotional meaning to this distinction. The critical question, then, is whether this assignment of meaning is innate and was specifically selected for, making it a musical adaptation, or whether it is culturally derived. Studies of human infants are potentially informative in this regard, but the results are mixed. Several studies show preferences for consonance early in development (Trainor & Heinmiller, 1998; Trainor, Tsang, & Cheung, 2002; Zentner & Kagan, 1998), but it is unclear whether these early preferences are based on beating and roughness or on harmonicity and whether they are learned or innate (Plantinga & Trehub, 2013). Furthermore, although it is often assumed that the perception of consonance and dissonance in humans is similar around the world, one study suggests that this may not be the case (McDermott, Schultz, Undurraga, & Godoy, 2016). Thus, it can be concluded that human music makes use of the species-general consonance/dissonance distinction but that further research is needed to determine whether the differential assignment of emotional meaning is an adaptation for music or culturally derived.

More broadly than the consonance/dissonance distinction, musical pitch organization has a tonal structure, which dictates which pitch intervals (distances between tones) are used, the functions of different tones within musical scales, and how they are combined sequentially and simultaneously in composition and improvisation (for detailed descriptions of Western tonal pitch space, see Lerdahl, 2001; Lerdahl & Jackendoff, 1983). Just as there are many different languages in the world that share commonalities suggestive of innate biological constraints, there are many different musical systems in the world that share commonalities (e.g., Dowling & Harwood, 1986; Handel, 1989; Jackendoff & Lerdahl, 2006; Justus & Hutsler, 2005; McDermott & Hauser, 2005). Aspects of musical pitch structure that appear to be near universal across musical systems include octave equivalence (musical pitch has several perceptual dimensions, e.g., *chroma*, or notes of a scale, and *octave equivalence*, whereby pitches an octave apart are perceived to be similar and have common note names across octaves); the use of a small number of discrete pitches per octave (e.g., musical *scales*), which is likely a consequence of general memory limitations; and the use of more than one interval size (pitch distance) between notes of musical scales. The latter distinction enables each note of the scale to be related to the other notes in unique ways in terms of pitch relations (Balzano, 1980; 1982). Typically, one note (the tonic) is central, and each other note stands in a unique interval relation to the tonic and to the other notes. Collectively, these relations constitute the pitch space.

Critical questions concern how unique these properties are to human perception and the extent to which they are the direct result of ASA and the basic structure of the auditory system, or whether they have cultural origins. Most of the properties of tonal pitch space already noted do not directly enhance the perception of auditory objects in the environment and are therefore unlikely to reflect direct adaptations for ASA. Furthermore, for the most part, they are not particularly useful for other auditory processing, such as that needed for speech perception. And while the processing of tonal pitch space may rely on faculties such as memory and attention, these cannot fully explain the properties of tonal pitch space (Jackendoff & Lerdahl, 2006). Tonal pitch space and the interval structure of scales appear to be relevant for music alone. Thus, one possibility is that tonal pitch space is a music-specific adaptation. Several genetic studies report that variation in musical ability has a strong genetic component (Drayna, Manichaikul, De Lange, Snieder, & Spector, 2001; Mosing, Madison, Pedersen, Kuja-Halkola, & Ullén, 2014; Oikkonen et al., 2015; Pulli et al., 2008; Ukkola, Onkamo, Raijas, Karma, & Järvelä, 2009; Ukkola-Vuoti et al., 2013; for reviews, see chapter 10, this volume, and Tan, McPherson, Peretz, Berkovic, & Wilson, 2014). However, this tells us little about whether there were selection pressures specifically for music. Although natural selection reduces genetic variability, highly polygenic adaptations, which would characterize music, are expected to show substantial genetic variability as a result of mutation-selection balance (Keller & Miller, 2006). Additionally, the reported genetic differences might actually reflect variation in ASA ability as well and may tell us nothing about music-specific adaptations. In terms of human development, infants and young children learn the specific pitch structure of the music in their environment without formal instruction, just as they learn the language in their environment, suggesting an innate ability to acquire this knowledge, although this ability may or may not be specific to music (e.g., Hannon & Trainor, 2007; Trainor & Corrigall, 2010; Trainor & Hannon, 2013; Trehub, 2003, 2013). A learning mechanism that was selected for one function but is used in a new domain is considered an exapted learning mechanism (Andrews et al., 2002).

Conceiving of tonal pitch space as a music-specific adaptation faces the challenge of different musical systems having somewhat different tonal pitch spaces and the rapidity with which tonal pitch spaces change across time and when different musical systems come into contact, issues that apply equally to adaptationist arguments for language. Recent modeling of language acquisition and language change suggests that it is not necessary, and indeed very difficult, to postulate an innate universal grammar (Christiansen & Chater, 2008). According to this view, rather than language being an evolutionary adaptation, it is a cultural creation molded on preexisting perceptual and cognitive structures adapted for other purposes. It is possible that music behaves similarly and is a cultural creation based on preexisting features of the brain.

Interestingly, while different musical systems use somewhat different scales and have different tonal centers, certain intervals tend to be prominent across musical systems (Gill

& Purves, 2009). Recent work by Large and colleagues (Large, 2010; Large & Almonte, 2012; Lerud, Almonte, Kim, & Large, 2014) shows that neural resonances in the auditory pathway induced by nonlinearities in the system give rise to the intervals prominent across musical systems and that models of such nonlinear oscillation easily learn properties of specific tonal pitch spaces. Thus, the emergence of musical intervals may in fact be a spandrel of basic properties of neural circuits. One difficulty with this argument is that such nonlinear neural circuits are also present in other species, raising the question of why these species have not developed tonal music. Without further research, a definitive answer is impossible. However, it is possible that the potential for tonal pitch space perception is present in other species, but they lack other essential features such as sufficient memory capacity, a link between tonal pitch space and emotional meaning, a cultural means of sustaining such a complex system, or the motor skills to produce music. Indeed, octave equivalence, like the perceptual distinction between consonance and dissonance, has been found in monkeys, at least for simple tonal melodies (Wright, Rivera, Hulse, Shyan, & Neiworth, 2000), although nonhuman species in general have a greater propensity than humans to engage in absolute rather than relative pitch processing.

A further aspect of tonal pitch spaces is important with regard to its origins. Pitch space organization is related to meaning and emotion, as it enables the alternation of tension (moving away from the tonic) and relaxation (moving toward the tonic), and different scales in different musical systems are associated with different meanings. For example, music composed in the Western minor scale tends to convey sadness more than music composed in the major scale. Similarly, many Indian ragas are associated with different meanings and are meant to be played at different times and circumstances (Jairazbhoy, 1995). Just as other species may perceive the distinction between consonance and dissonance but not show preferences in this regard, the mapping of meaning through tonal pitch space is a crucial aspect of human music, and the origin of this mapping must be part of any complete account of the origins of tonal pitch space.

Time Processing and the Origins of Musical Rhythm

Information about the timing of events plays a complementary role to spectral information in ASA (Bregman, 1990). For example, whether frequency component onsets are simultaneous is an important cue for determining whether they originate from the same source, as it is expected that onsets of components emanating from a single auditory object should begin at the same time (Lipp, Kitterick, Summerfield, Bailey, & Paul-Jordanov, 2010). Conversely, components with nonsimultaneous onsets tend to be perceived as belonging to different auditory objects. This principle is central to musical structure. In cases where it is desirable for different simultaneous voices to fuse into a single percept with chordal quality, as in a barbershop quartet, various voices tend to have simultaneous onsets. In polyphonic music, in which it is desirable for each part to be perceived as an independent voice, as in a fugue, each voice tends to change notes at different times (Huron, 2001). As with a number of properties of spectral sound processing, such timing capabilities of the auditory system were likely adaptations for ASA, and musical structure has adapted to these preexisting adaptations rather than driving their existence.

Another basic principle of musical composition is to lay down the basic beat in the lowest-pitched (bass) instruments. Recent research indicates that when two tones are presented simultaneously in a repeating sequence, listeners are better at detecting when the lower tone is occasionally presented 50 ms early (leaving the higher tone on time) compared to when the higher tone is presented 50 ms early (leaving the lower tone on time; Hove, Marie, Bruce, & Trainor, 2014). Furthermore, modeling work suggests that this low-voice superiority effect for time originates in properties of the inner ear (see box 4.1) although the effect is likely sharpened higher in the auditory system (Nelson, Smith, & Young, 2009; Nelson & Young, 2010). Because there is no obvious adaptive reason for this effect, it might simply be a nonadaptive consequence of the structure of the inner ear (spandrel). The important point with respect to music is that music is composed to conform to this preexisting feature of the auditory system.

As with tonal pitch space, aspects of musical rhythm appear to be specific to music (e.g., Fitch, 2006; Patel & Iversen, 2014). Language, for example, has temporal structure, but not the same requirement as music for regularity and temporal precision at the beat level. Musical rhythm has a number of aspects (e.g., Grahn, 2012; Lerdahl & Jackendoff, 1983). The rhythmic surface consists of the sequence of event durations and silences that comprise the music. From this surface, the brain derives the beat, typically a regularly spaced sequence of pulses. That the beat is derived in the brain and not given directly in the stimulus is seen in that beats can be perceived even when there is no physical sound present but the surrounding context implies a beat at that time. Electroencephalogram (EEG) and magnetoencephalogram studies show brain signatures of such "felt" beats (e.g., Fujioka, Trainor, Large, & Ross, 2009; Nozaradan, Peretz, Missal, & Mouraux, 2011). Beats can be mentally subdivided (usually into groups of two or three) or every second or third beat can be perceived as accented, and these levels of beat structure form a metrical hierarchy. In humans, the beat is extracted effortlessly (Janata, Tomic, & Haberman, 2012; Repp & Su, 2013). Furthermore, sensitivity to meter has been shown in young human infants (Hannon & Trehub, 2005; Phillips-Silver & Trainor, 2005; Winkler, Háden, Ladinig, Sziller, & Honing, 2009).

One of the interesting aspects of musical behavior is spontaneous movement to the beat of music (Toiviainen, Luck, & Thompson, 2010). Indeed, most people readily entrain their movements to the beat of music, using various effectors, across tempos from about 1 to 5 Hz. Functional magnetic resonance imaging (fMRI) studies indicate that when listeners perceive musical meter, even in the absence of movement, a wide range of cortical and subcortical (premotor and supplementary cortex and basal ganglia) regions are activated (Chen, Penhune, & Zatorre, 2008; Grahn & Brett, 2007; Kung, Chen, Zatorre, &

Penhune, 2013). EEG studies reveal that slow oscillatory brain responses (in the delta range) reflect both beat and meter frequencies in the stimulus in adults (Arnal, Doelling, & Poeppel, 2014; Lakatos, Karmos, Mehta, Ulbert, & Schroeder, 2008; Lakatos et al., 2013; Nozaradan et al., 2011; Nozaradan, Peretz, & Mouraux, 2012) and infants (Cirelli, Spinelli, Nozaradan, & Trainor, 2016). Furthermore, when isochronous beat patterns are presented, activation in the beta band (15–25 Hz) is modulated at the tempo of the beat (Fujioka, Trainor, Large, & Ross, 2012; Iversen, Repp, & Patel, 2009). Specifically, beta power decreases after each tone onset and rebounds in a predictive manner prior to the onset of the next beat, with the rebound delayed for slower tempos. Interestingly, this same pattern is observed in both auditory and motor regions when people simply listen to the beat, suggesting a strong connection between auditory and motor systems (Fujioka et al., 2012). Furthermore, the influence appears to be bidirectional, in that when people move on either every second or third beat of an ambiguous rhythm pattern (one that can be interpreted as having different metrical structures such as a march or waltz), their movement influences the metrical interpretation of the auditory pattern (Phillips-Silver & Trainor, 2007).

Different timing mechanisms are present in the human brain. Neural circuits for duration-based (absolute) timing can be contrasted with beat-based timing, in which events occur at regular, predictable times (chapter 8, this volume; Teki, Grube, Kumar, & Griffiths, 2011). Musical structure, of course, requires beat-based timing. Developmental and comparative studies are informative about the origins of the ability to perceive beat and meter and the ability to entrain movements to a beat. With respect to nonhuman species, very few seem to entrain to a beat (Schachner et al., 2009). While there are no reports of motoric entrainment to an auditory beat in the wild, some vocal learning birds have demonstrated entrainment in captivity (Patel et al., 2009; Schachner et al., 2009), and one mammal (sea lion) has been trained to move to the beat (Cook et al., 2013; Rouse et al., 2016). Despite these cases, this ability appears to be rare across nonhuman species, and even in cases where it is found, it requires considerable experience or training with humans and their music. Of course, many species produce rhythmic movements, and the advantage of locomotion was likely a major selective pressure for the development of rhythmic movement. But even if some other species are capable of entraining movements to an auditory beat, most do not appear to do so spontaneously, and they lack attentional interest and motivation to do so (Wilson & Cook, 2016). Where humans appear to differ from most other species is in the connections between auditory and motor regions that support metrical perception and motor entrainment to an auditory beat (Patel & Iversen 2014). Studies in nonhuman primates show that duration-based timing is universally present across primate species, but that only rudimentary beat-based timing is present in monkeys and chimpanzees (Merchant & Honing, 2014). Furthermore, the evidence suggests that in monkeys, sensorimotor connections for timing are stronger between vision and movement than between audition and movement (Nagasaka, Chao, Hasegawa, Notoya, & Fujii, 2013; Zarco, Merchant, Prado,

& Mendez, 2009), whereas the reverse is true for humans (Honing & Merchant, 2014). In line with this differentiation across primate species, although human infants are too motorically immature to precisely entrain to the beat (Merker, Madison, & Eckerdal, 2009), they do speed up their movements with increasing beat tempo (Zentner & Eerola, 2010). Moreover, when bounced on either every second or third beat of an ambiguous rhythm pattern, bypassing their motoric immaturity, infants later prefer to listen to the pattern with accents corresponding to how they were bounced (Phillips-Silver & Trainor, 2005). This indicates that motor influence on auditory perception is present in human infants and suggests that the privileged auditory-motor connections for beat and meter that, among primates, are unique to humans are present very early in human development.

Thus, it would appear that the ability for beat-based timing and the privileged connections between auditory and motor systems that enable entrainment to a beat evolved relatively recently within the primate lineage. The question, then, is whether beat-based timing was a music-specific adaptation or whether it emerged for other reasons. A comparison of tonal pitch space with beat-based timing and entrainment in this regard might be useful in addressing this question. Although tonal pitch space appears to be unique to humans, the particular pitch intervals used and their organization may originate in basic properties of nonlinear oscillators that characterize neural circuits. In this case, the neural basis of tonality would be widely conserved across species, and an explanation is necessary for why humans exploited this feature to create music whereas other species did not. Beat-based timing ability and movement entrainment to an auditory beat appear to be substantially different in humans than in other primate species, although a progression of ability in this regard can be seen in the primate lineage (Merchant & Honing, 2014) and may rely on auditory motor circuits that are unique to humans (Patel & Iversen, 2014). Thus, it is possible that these capabilities are not easily explained by nonmusical adaptations. The ability to entrain to an auditory beat has a further important consequence in that when individuals listen to music together and they each entrain their movements to the beat, their ability to synchronize with others is greatly increased. The social consequences of such synchronous movement are discussed in the next section.

Social and Emotional Functions and the Origins of Music

In many cases, musical structure conforms to the properties of an auditory system that evolved for ASA. However, two central features of music cannot be explained completely by ASA: that music induces emotional responses in people and that music is an intensely social activity. The emotional and social aspects of music are likely closely related. With respect to emotion, music not only expresses emotion, but it can induce emotions directly that can be measured physiologically (e.g., by changes in heart rate, galvanic skin responses, EEG and fMRI), behaviorally (e.g., tears), and by verbal reports of emotional experiences (Huron, 2006; Juslin & Sloboda, 2001; Salimpoor, Benovoy, Larcher, Dagher,

& Zatorre, 2011; Trainor & Schmidt, 2003). Common experience of music can therefore instill common emotional reactions in a group of people. This is likely why, even today, people participate in music making or music listening in groups when the goal is to feel a common emotion or to work together to achieve a common goal. For instance, music is almost always present at important social functions such as weddings, funerals, and parties. Fans chant to display their solidarity and offer encouragement at sporting events. Music is used in the military to encourage unity of purpose and present a threatening front to the enemy.

Some properties of nonmusical sounds can induce emotions across a range of species. For example, large menacing animals typically make low, loud sounds, and many species react to such sounds with fear (Morton, 1977). Emotions can also be induced by unexpected events, and music exploits this basic mechanism as well (Huron, 2006; Meyer, 1956; Trainor & Zatorre, 2016). Music exploits these emotional connections to sounds that are conserved across many species, but it appears to go beyond this basic emotional response to sound in using elaborate tonal systems (e.g., Western tonality, Indian ragas) that can express a myriad of emotions, many of them hard to express verbally. Metrical structure also provides a scaffold on which a variety of tempos and rhythmic patterns can induce a range of emotions from peacefulness to agitation and menace. Furthermore, the emotional impact of music in humans is seen early in infancy. For example, mothers sing lullabies to soothe infants and play songs to arouse them and interact playfully (Trehub & Trainor, 1998), and these have differential consequences for infants (Rock, Trainor, & Addison, 1999). Emotional responses to music may be specific to humans and appear to be mediated by specialized physiological mechanisms. In humans, emotional responses to music are mediated by the dopamine system, such that music modulates activation in reward centers in the brain (Salimpoor et al., 2011). More physiological research is needed, but the apparent indifference of other primates to music (McDermott & Hauser, 2004) and very early responses in human infants suggest basic genetically driven differences in the physiology of neural pathways underlying the human emotional response to music and that of other primates. However, this question needs to be informed by more data across species.

With respect to social affiliation, after people move together in synchrony, they rate each other as more likable, trust each other more, and are more likely to cooperate than after moving asynchronously (Anshel & Kippler, 1988; Hove & Risen, 2009; Launay, Dean, & Bailes, 2013; Reddish, Fischer, & Bulbulia, 2013; Tarr, Launay, Cohen, & Dunbar, 2015; Valdesolo, Ouyang, & DeSteno, 2010; Weinstein, Launay, Pearce, Dunbar, & Stewart, 2015; Wiltermuth & Heath, 2009). Because of its predictable beat, music provides an excellent scaffold for synchronized movement with others. Indeed, music and dance are intimately connected, and dance most often involves two or more people. It is notable that dancing is common during courtship, when strong social and emotional bonds are being formed. With respect to development, children who played a game together involving music are more likely to help each other than children who played a game together without music (Kirschner & Tomasello, 2010). Furthermore, recent research indicates that infants as young as fourteen months of age help an experimenter more (e.g., by picking up items she "accidentally" drops) if they were previously bounced to music in synchrony with her movements than if they were bounced at a different tempo (Cirelli, Einarson, & Trainor, 2014; Trainor & Cirelli, 2015). Furthermore, this effect is specific to the person the infant bounced with and does not generalize to neutral strangers (Cirelli, Wan, & Trainor, 2014), although it does generalize to people shown to be friends of the person who moved in sync with the infant (Cirelli, Wan, & Trainor, 2016). Thus, synchronous movement can have powerful effects on social affiliation and cooperation, can help define social groups, and is effective very early in development. Indeed, infants' experiences of being rocked in their mother's arms while being sung to are potentially powerful in strengthening bonds between mother and infant. During adolescence, when the formation of social groups is very important, music is often used to help define individual and group identity (Berns, Capra, Moore, & Noussair, 2010).

Despite the universality and early emergence of entrainment effects (when motor immaturity of young children is bypassed) and associated affiliative consequences, motoric entrainment to an auditory beat has not been found in nonhuman species in the wild (although more research is needed), only a few species spontaneously engage in this behavior when living with humans (Schachner et al., 2009), and it is very difficult, if not impossible, to train this ability in those species that are genetically closest to humans (Honing & Merchant, 2014). Furthermore, there appear to be genetically driven physiological differences between human and nonhuman primates that underlie entrainment (Patel & Iversen, 2014). Thus, unlike many of the features of music that rest on adaptations for other functions such as ASA, emotional responses to music, entrainment, and their affiliative consequences are candidates for music-specific adaptations.

Going back to Darwin (1872), it has been proposed that musical behavior evolved as an indicator of fitness, such that those with good rhythmic entrainment abilities, for example, would be more likely to attract mates (Miller, 2000). This contention is consistent with the observation that, across a wide range of species, elaborate displays such as the peacock tail, which are potentially detrimental to survival by exposing the animal to predators and taking resources away from other activities that might increase survival, are often explained as signals of fitness to conspecifics (Grafen, 1990). According to this hypothesis, musical behavior is an evolutionary adaptation such that the structure and production of music became more and more elaborate through competition as a display of the highest fitness. This view is not without challenges. A full discussion is beyond the scope of this chapter, but the fact that both men and women produce music contrasts with the vast majority of such displays in other species, many of which are specific to males (Searcy & Andersson, 1986). It is possible, however, that music is an outlier on this dimension, and both male and female humans engage in mate selection. Perhaps a more serious challenge is to explain why music is used across a range of situations that seemingly have little to do with mating, such as work songs, parental songs for infants, and children's play songs.

Another proposal is that participating in joint music making increased group cohesion, cooperation, and therefore the survival of individuals who were able to engage in music (e.g., Dissanayake, 2000; Falk, 2004; Fitch, 2006; Huron, 2001). Consistent with this view is evidence that among primates, music engages the dopamine reward system only in humans, and only in humans are there privileged connections between auditory and motor systems underlying beat and metrical processing. On the other hand, music is highly flexible and generative, and it changes rapidly over time, which poses particular challenges for an evolutionary theory of music. Furthermore, it is clear that in large part, musical structure conforms to preexisting features of the auditory system, many of which evolved for ASA and are highly conserved across species, which strongly suggests that music is a cultural creation rather than an evolutionary adaptation. While these two views appear contradictory, they can be reconciled if a complex interaction between evolutionary and cultural processes is considered. For example, music may have originally emerged as a cultural creation made possible by preexisting adaptations related to ASA and other capabilities such as increased memory. However, if benefits arose through increased survival of those who engaged in music making, this could have exerted evolutionary pressure to enhance neural pathways by which music could activate emotional centers in the brain and to enhance pathways linking auditory and motor beat-based timing circuits. In turn, these neurally based adaptations could reinforce the cultural development and sustainability of musical behavior and perhaps explain why humans spend so much time and resources on music and why music is constantly changing.

Conclusion

Both evolutionary adaptation and cultural creation likely played a role in the origins of music. Rather than focusing on an evaluation of different evolutionary versus cultural theories for musical origins, this chapter considers various musical features and whether they were selected to enhance music specifically or were adaptations for nonmusical functions. This analysis shows that many aspects of musical pitch and timing structure conform to features of auditory processing needed for ASA. Given that ASA is much more ancient than music, is highly conserved across many species, and is present early in development, it is concluded that in large part, music has been designed to conform to features of ASA rather than the reverse possibility that effects of music have largely driven the evolution of auditory processing. This lends support to the idea that music may have begun as a cultural creation, exapting preexisting features of the auditory system that had evolved for ASA. However, some aspects of music are not easily explained by ASA or other general capabilities such as increased memory and motor skills. These include emotional and social effects of music. It is possible that engaging in music conferred survival advantages, which led to some music-specific adaptations. For example, the ability to perform beat-based timing and entrain movements to a regular pulse appears to differ between humans and other primates and to be supported by genetically driven brain connections that are present early in human development. Synchronous movement leads to increased group cohesion and potential survival advantages for those who can participate. In this case, music may have conferred survival advantages that led to specific adaptations underlying behaviors such as entrainment, which had advantageous consequences such as social cohesion. Thus, music is likely to have a complex origin involving exaptation of traits evolved for other functions such as ASA, cultural creation, and music-specific adaptations.

Acknowledgments

This chapter was supported by grants from the Natural Sciences and Engineering Research Council of Canada and the Canadian Institutes of Health Research. Thanks to Paul Andrews for insightful comments on an earlier draft and to Susan Marsh-Rollo for help with manuscript preparation.

References

Albrecht, J. D., & Huron, D. (2014). A statistical approach to tracing the historical development of major and minor pitch distributions, 1400–1750. *Music Perception, 31*(3), 223–243. doi:10.1525/mp.2014.31.3.223

Andrews, P. W., Gangestad, S. W., & Matthews, D. (2002). Adaptationism: How to carry out an exaptationist program. *Behavioral and Brain Sciences, 25*(4), 489–504. doi:10.1017/S0140525X02000092

Anshel, A., & Kippler, D. (1988). The influence of group singing on trust and cooperation. *Journal of Music Therapy, 25*(3), 145–155. doi:10.1093/jmt/25.3.145

Arnal, L. H., Doelling, K. B., & Poeppel, D. (2014). Delta-beta coupled oscillations underlie temporal prediction accuracy. *Cerebral Cortex, 25*(9), 3077–3085. doi:10.1093/cercor/bhu103

Balzano, G. J. (1980). The group-theoretic description of 12-fold and microtonal pitch systems. *Computer Music Journal, 4*(4), 66–84. doi:10.2307/3679467

Balzano, G. J. (1982). The pitch set as a level of description for studying musical pitch perception. In M. Clynes (Ed.), *Music, mind and brain* (pp. 321–351). New York: Plenum. doi:10.1007/978-1-4684-8917-0_17

Barkow, J. H., Cosmides, L., & Tooby, J. (1995). *The adapted mind: Evolutionary psychology and the generation of culture*. New York: Oxford University Press.

Baron-Cohen, S. (1997). *Mindblindness: An essay on autism and theory of mind*. Cambridge, MA: MIT Press.

Bendor, D., & Wang, X. (2005). The neuronal representation of pitch in primate auditory cortex. *Nature, 436*, 1161–1165. doi:10.1038/nature03867

Berns, G. S., Capra, C. M., Moore, S., & Noussair, C. (2010). Neural mechanisms of the influence of popularity on adolescent ratings of music. *NeuroImage, 49*(3), 2687–2696. doi:10.1016/j.neuroimage.2009.10.070

Bregman, A. S. (1990). *Auditory scene analysis: The perceptual organization of sound*. Cambridge, MA: MIT Press.

Brown, S. (2000). Evolutionary models of music: From sexual selection to group selection. In F. Tonneau & N. S. Thompson (Eds.), *Perspectives in ethology: Behavior, evolution and culture* (Vol. 13, pp. 231–281). New York: Plenum. doi:10.1007/978-1-4615-1221-9_9

Cariani, P. A. (2004). Temporal codes and computations for sensory representation and scene analysis. *IEEE Transactions on Neural Networks, 15*(5), 1100–1111. doi:10.1109/TNN.2004.833305

Cedolin, L., & Delgutte, B. (2005). Pitch of complex tones: Rate-place and interspike interval representations in the auditory nerve. *Journal of Neurophysiology, 94*(1), 347–362. doi:10.1152/jn.01114.2004

Chen, J. L., Penhune, V. B., & Zatorre, R. J. (2008). Listening to musical rhythms recruits motor regions of the brain. *Cerebral Cortex, 18*(12), 2844–2854. doi:10.1093/cercor/bhn042

Cherry, E. C. (1953). Some experiments on the recognition of speech, with one and with two ears. *Journal of the Acoustical Society of America, 25*(5), 975–979. doi:10.1121/1.1907229

Chon, S. H., & Huron, D. (2014). Does auditory masking explain high voice superiority? In M. K. Song (Ed.), *Proceedings of the International Conference on Music Perception and Cognition.* Seoul, South Korea: Causal Productions.

Christiansen, M. H., & Chater, N. (2008). Language as shaped by the brain. *Behavioral and Brain Sciences, 31*(5), 489–509. doi:10.1017/S0140525X08004998

Cirelli, L. K., Einarson, K. M., & Trainor, L. J. (2014). Interpersonal synchrony increases prosocial behavior in infants. *Developmental Science, 17*(6). doi:10.1111/desc.12193

Cirelli, L. K., Spinelli, C., Nozaradan, S., & Trainor, L. J. (2016). Measuring neural entrainment to beat and meter in infants: Effects of music background. *Frontiers in Neuroscience, 10,* 1–11. doi:10.3389/fnins.2016.00229

Cirelli, L. K., Wan, S. J., & Trainor, L. J. (2014). Fourteen-month-old infants use interpersonal synchrony as a cue to direct helpfulness. *Philosophical Transactions of the Royal Society of London B: Biological Sciences, 369*(1658), 1–8. doi:10.1098/rstb.2013.0400

Cirelli, L. K., Wan, S. J., & Trainor, L. J. (2016). Social effects of movement synchrony: Increased infant helpfulness only transfers to affiliates of synchronously moving partners. *Infancy, 21(6)*, 807–821. doi:10.1111/infa.12140

Cook, P., Rouse, A., Wilson, M., & Reichmuth, C. (2013). A California sea lion (*Zalophus californianus*) can keep the beat: Motor entrainment to rhythmic auditory stimuli in a non vocal mimic. *Journal of Comparative Psychology, 127*(4), 412. doi:10.1037/a0032345

Cross, I. (2001). Music, cognition, culture, and evolution. *Annals of the New York Academy of Sciences, 930,* 28–42. doi:10.1111/j.1749-6632.2001.tb05723.x

Cross, I., Zubrow, E., & Cowan, F. (2002). Musical behaviours and the archaeological record: A preliminary study. In J. Mathieu (Ed.), *British experimental archaeology* (pp. 25–34). Oxford: British Archaeological Reports.

Darwin, C. (1872). *The expression of emotion in man and animals.* London: Murray.

Delgutte, B., & Cariani, P. (1992). Coding of the pitch of harmonic and inharmonic complex tones in the interspike intervals of auditory nerve fibers. In M. E. H. Schouten (Ed.), *The auditory processing of speech* (pp. 37–45). Berlin: Mouton-DeGruyter; 10.1515/9783110879018.37.

Demany, L. (1982). Auditory stream segregation in infancy. *Infant Behavior and Development, 5*(2–4), 261–276. doi:10.1016/S0163-6383(82)80036-2

d'Errico, F., Henshilwood, C., Lawson, G., Vanhaeren, M., Tillier, A. M., Soressi, M., et al. (2003). Archaeological evidence for the emergence of language, symbolism, and music—an alternative multidisciplinary perspective. *Journal of World Prehistory, 17*(1), 1–70. doi:10.1023/A:1023980201043

DeWitt, L. A., & Crowder, R. G. (1987). Tonal fusion of consonant musical intervals: The oomph in Stumpf. *Perception and Psychophysics, 41*(1), 73–84. doi:10.3758/bf03208216

Dissanayake, E. (2000). Antecedents of the temporal arts in early mother-infant interaction. In N. L. Wallin, B. Merker, & S. Brown (Eds.), *The origins of music* (pp. 389–410). Cambridge, MA: MIT Press.

Dissanayake, E. (2008). If music is the food of love, what about survival and reproductive success? *Musicae Scientiae, 12*(1), 169–195. doi:10.1177/1029864908012001081

Dowling, W. J., & Harwood, D. L. (1986). *Music cognition.* Orlando, FL: Academic Press.

Drayna, D., Manichaikul, A., De Lange, M., Snieder, H., & Spector, T. (2001). Genetic correlates of musical pitch recognition in humans. *Science, 291*(5510), 1969–1972. doi:10.1126/science.291.5510.1969

Dunbar, R. (2004). Language, music and laughter in evolutionary perspective. In D. Kimbrough Oller & U. Griebel (Eds.), *Evolution of communication systems: A comparative approach* (pp. 257–274). Cambridge, MA: MIT Press.

Falk, D. (2004). Prelinguistic evolution in early hominins: Whence motherese? *Behavioral and Brain Sciences, 27*(4), 491–503. doi:10.1017/S0140525X04000111

Fassbender, C. (1993). *Auditory grouping and segregation processes in infancy.* Norderstedt: Kaste Verlag.

Fay, R. R. (2008). Sound source perception and stream segregation in non-human vertebrate animals. In W. A. Yost, A. N. Popper, & R. R. Fay (Eds.), *Auditory perception of sound sources* (pp. 307–323). New York: Springer Science+Business Media.

Fishman, Y. I., Volkov, I. O., Noh, M. D., Garell, P. C., Bakken, H., Arezzo, J. C., et al. (2001). Consonance and dissonance of musical chords: Neural correlates in auditory cortex of monkeys and humans. *Journal of Neurophysiology, 86*(6), 2761–2788.

Fitch, W. (2006). The biology and evolution of music: A comparative perspective. *Cognition, 100*(1), 173–215. doi:10.1016/j.cognition.2005.11.009

Fletcher, H. (1940). Auditory patterns. *Reviews of Modern Physics, 12*(1), 47–65. doi:10.1103/RevModPhys.12.47

Folland, N. A., Butler, B. E., Payne, J. E., & Trainor, L. J. (2015). Cortical representations sensitive to the number of perceived auditory objects emerge between 2 and 4 months of age: Electrophysiological evidence. *Journal of Cognitive Neuroscience, 27*(5), 1060–1067. doi:10.1162/jocn_a_00764

Folland, N. A., Butler, B. E., Smith, N. A., & Trainor, L. J. (2012). Processing simultaneous auditory objects: Infants' ability to detect mistunings in harmonic complexes. *Journal of the Acoustical Society of America, 131*, 993–997. doi:10.1121/1.3651254

Fujioka, T., Trainor, L. J., Large, E. W., & Ross, B. (2009). Beta and gamma rhythms in human auditory cortex during musical beat processing. *Annals of the New York Academy of Sciences, 1169*, 89–92. doi:10.1111/j.1749-6632.2009.04779.x

Fujioka, T., Trainor, L. J., Large, E. W., & Ross, B. (2012). Internalized timing of isochronous sounds is represented in neuromagnetic beta oscillations. *Journal of Neuroscience, 32*(5), 1791–1802. doi:10.1523/JNEUROSCI.4107-11.2012

Gill, K. Z., & Purves, D. (2009). A biological rationale for musical scales. *PLoS One, 4*(12), 1–9. doi:10.1371/journal.pone.0008144

Glasberg, B. R., & Moore, B. C. (1990). Derivation of auditory filter shapes from notched-noise data. *Hearing Research, 47*(1–2), 103–138. doi:10.1016/0378-5955(90)90170-t

Gockel, H. E., Carlyon, R. P., Mehta, A., & Plack, C. J. (2011). The frequency following response for dichotic pitch stimuli: No evidence for pitch encoding. *Journal of the Acoustical Society of America, 129*(4), 2592. doi:10.1121/1.3588585

Gould, S. J., & Lewontin, R. C. (1979). The spandrels of San Marco and the Panglossian paradigm: A critique of the adaptationist programme. *Proceedings of the Royal Society of London B: Biological Sciences, 205*(1161), 581–598. doi:10.1098/rspb.1979.0086

Gould, S. J., & Vrba, E. S. (1982). Exaptation: A missing term in the science of form. *Paleobiology, 8*(1), 4–15. doi:10.1017/s0094837300004310

Grafen, A. (1990). Biological signals as handicaps. *Journal of Theoretical Biology, 144*(4), 517–546. doi:10.1016/s0022-5193(05)80088-8

Grahn, J. A. (2012). Neural mechanisms of rhythm perception: Current findings and future perspectives. *Topics in Cognitive Science, 4*(4), 585–606. doi:10.1111/j.1756-8765.2012.01213.x

Grahn, J. A., & Brett, M. (2007). Rhythm perception in motor areas of the brain. *Journal of Cognitive Neuroscience, 19*(5), 893–906. doi:10.1162/jocn.2007.19.5.893

Greenwood, D. D. (1990). A cochlear frequency-position function for several species—29 years later. *Journal of the Acoustical Society of America, 87*(6), 2592–2605. doi:10.1121/1.399052

Gutschalk, A., & Dykstra, A. R. (2014). Functional imaging of auditory scene analysis. *Hearing Research, 307*, 98–110. doi:10.1016/j.heares.2013.08.003

Hall, D. A., Barrett, D. J., Akeroyd, M. A., & Summerfield, A. Q. (2005). Cortical representations of temporal structure in sound. *Journal of Neurophysiology, 94*(5), 3181–3191. doi:10.1152/jn.00271.2005

Hannon, E. E., & Trainor, L. J. (2007). Music acquisition: Effects of enculturation and formal training on development. *Trends in Cognitive Sciences, 11*(11), 466–472. doi:10.1016/j.tics.2007.08.008

Hannon, E. E., & Trehub, S. E. (2005). Tuning in to musical rhythms: Infants learn more readily than adults. *Proceedings of the National Academy of Sciences, 102*(35), 12639–12643. doi:10.1073/pnas.0504254102

Hartmann, W. M. (1996). Pitch, periodicity, and auditory organization. *Journal of the Acoustical Society of America, 100*(6), 3491–3502. doi:10.1121/1.417248

He, C., & Trainor, L. J. (2009). Finding the pitch of the missing fundamental in infants. *Journal of Neuroscience, 29*(24), 7718–7722. doi:10.1523/JNEUROSCI.0157-09.2009

Honing, H., & Merchant, H. (2014). Differences in auditory timing between human and non-human primates. *Behavioral and Brain Sciences, 37*(6), 373–374. doi:10.1017/s0140525x13004056

Hove, M. J., Marie, C., Bruce, I. C., & Trainor, L. J. (2014). Superior time perception for lower musical pitch explains why bass-ranged instruments lay down musical rhythms. *Proceedings of the National Academy of Sciences, 111*(28), 10383–10388. doi:10.1073/pnas.1402039111

Hove, M. J., & Risen, J. L. (2009). It's all in the timing: Interpersonal synchrony increases affiliation. *Social Cognition, 27*(6), 949–960. doi:10.1521/soco.2009.27.6.949

Huron, D. (2001). Tone and voice: A derivation of the rules of voice-leading from perceptual principles. *Music Perception, 19*(1), 1–64. doi:10.1525/mp.2001.19.1.1

Huron, D. (2003). Is music an evolutionary adaptation? In I. Peretz & R. Zatorre (Eds.), *The cognitive neuroscience of music* (pp. 57–78). Oxford: Oxford University Press. doi:10.1093/acprof:oso/9780198525202.003.0005

Huron, D. (2006). *Sweet anticipation: Music and the psychology of expectation.* Cambridge, MA: MIT Press.

Huron, D. (2016). *Voice leading: The science behind a musical art.* Cambridge, MA: MIT Press.

Huron, D., & Ommen, A. (2006). An empirical study of syncopation in American popular music, 1890–1939. *Music Theory Spectrum, 28*(2), 211–231. doi:10.1525/mts.2006.28.2.211

Ibrahim, R. A., & Bruce, I. C. (2010). Effects of peripheral tuning on the auditory nerve's representation of speech envelope and temporal fine structure cues. In E. A. Lopez-Poveda, A. R. Palmer, & R. Meddis (Eds.), *The neurophysiological bases of auditory perception* (pp. 429–438). New York: Springer. doi:10.1007/978-1-4419-5686-6_40

Iversen, J. R., Repp, B. H., & Patel, A. D. (2009). Top-down control of rhythm perception modules early auditory responses. *Annals of the New York Academy of Sciences, 1160*, 58–73. doi:10.1111/j.1749-6632.2009.04579.x

Izumi, A. (2000). Japanese monkeys perceive sensory consonance of chords. *Journal of the Acoustical Society of America, 108*(6), 3073–3078. doi:10.1121/1.1323461

Jackendoff, R., & Lerdahl, F. (2006). The capacity for music: What is it, and what's special about it? *Cognition, 100*(1), 33–72. doi:10.1016/j.cognition.2005.11.005

Jairazbhoy, N. A. (1995). *The rāgs of North Indian music: Their structure and evolution.* Bombay: Popular Press.

Janata, P., Tomic, S. T., & Haberman, J. (2012). Sensorimotor coupling in music and the psychology of the groove. *Journal of Experimental Psychology, 141*(1), 54–75. doi:10.1037/a0024208

Juslin, P. N., & Sloboda, J. A. (2001). *Music and emotion: Theory and research.* Oxford: Oxford University Press.

Justus, T., & Hutsler, J. J. (2005). Fundamental issues in the evolutionary psychology of music: Assessing innateness and domain specificity. *Music Perception, 23*(1), 1–27. doi:10.1525/mp.2005.23.1.1

Keller, M. C., & Miller, G. (2006). Resolving the paradox of common, harmful, heritable mental disorders: Which evolutionary genetic models work best? *Behavioral and Brain Sciences, 29*(4), 385–404. doi:10.1017/s0140525x06009095

Kirby, S., Cornish, H., & Smith, K. (2008). Cumulative cultural evolution in the laboratory: An experimental approach to the origins of structure in human language. *Proceedings of the National Academy of Sciences, 105*(31), 10681–10686. doi:10.1073/pnas.0707835105

Kirschner, S., & Tomasello, M. (2010). Joint music making promotes prosocial behavior in 4-year-old children. *Evolution and Human Behavior, 31*(5), 354–364. doi:10.1016/j.evolhumbehav.2010.04.004

Kung, S. J., Chen, J. L., Zatorre, R. J., & Penhune, V. B. (2013). Interacting cortical and basal ganglia networks underlying finding and tapping to the musical beat. *Journal of Cognitive Neuroscience, 25*(3), 401–420. doi:10.1162/jocn_a_00325

Lakatos, P., Karmos, G., Mehta, A. D., Ulbert, I., & Schroeder, C. E. (2008). Entrainment of neuronal oscillations as a mechanism of attentional selection. *Science, 320*(5872), 110–113. doi:10.1126/science.1154735

Lakatos, P., Musacchia, G., O'Connel, M. N., Falchier, A. Y., Javitt, D. C., & Schroeder, C. E. (2013). The spectrotemporal filter mechanism of auditory selective attention. *Neuron, 77*(4), 750–761. doi:10.1016/j.neuron.2012.11.034

Large, E. W. (2010). A dynamical systems approach to musical tonality. In R. Huys & V. K. Jirsa (Eds.), *Nonlinear dynamics in human behavior* (pp. 193–211). Berlin: Springer-Verlag.

Large, E. W., & Almonte, F. V. (2012). Neurodynamics, tonality, and the auditory brainstem response. *Annals of the New York Academy of Sciences, 1252*, E1–E7. doi:10.1111/j.1749-6632.2012.06594.x

Launay, J., Dean, R. T., & Bailes, F. (2013). Synchronization can influence trust following virtual interaction. *Experimental Psychology, 60*(1), 53–63. doi:10.1027/1618-3169/a000173

Lerdahl, F. (2001). *Tonal pitch space*. New York: Oxford University Press.

Lerdahl, F., & Jackendoff, R. (1983). *A generative grammar of tonal pitch space*. Cambridge, MA: MIT Press.

Lerud, K. D., Almonte, F. V., Kim, J. C., & Large, E. W. (2014). Mode-locking neurodynamics predict human auditory brainstem responses to musical intervals. *Hearing Research, 308*, 41–49. doi:10.1016/j.heares.2013.09.010

Lin, J. Y., & Hartmann, W. M. (1998). The pitch of a mistuned harmonic: Evidence for a template model. *Journal of the Acoustical Society of America, 103*(5), 2608–2617. doi:10.1121/1.422781

Lipp, R., Kitterick, P., Summerfield, Q., Bailey, P. J., & Paul-Jordanov, I. (2010). Concurrent sound segregation based on inharmonicity and onset asynchrony. *Neuropsychologia, 48*(5), 1417–1425. doi:10.1016/j.neuropsychologia.2010.01.009

Marie, C., Fujioka, T., Herrington, L., & Trainor, L. J. (2012). The high-voice superiority effect in polyphonic music is influenced by experience: A comparison of musicians who play soprano-range compared to bass-range instruments. *Psychomusicology: Music, Mind, and Brain, 22*(2), 97–104. doi:10.1037/a0030858

Marie, C., & Trainor, L. J. (2012). Development of simultaneous pitch encoding: Infants show a high voice superiority effect. *Cerebral Cortex, 23*(3), 660–669. doi:10.1093/cercor/bhs050

Marie, C., & Trainor, L. J. (2014). Early development of polyphonic sound encoding and the high voice superiority effect. *Neuropsychology, 57*, 50–58. doi:10.1016/j.neuropsychologia.2014.02.023

Mayr, E. (1983). How to carry out the adaptationist program? *American Naturalist, 121*(3), 324–334. doi:10.1086/284064

McAdams, S., & Bertoncini, J. (1997). Organization and discrimination of repeating sound sequences by newborn infants. *Journal of the Acoustical Society of America, 102*(5), 2945–2953. doi:10.1121/1.420349

McDermott, J. (2008). The evolution of music. *Nature, 453*, 287–288. doi:10.1038/453287a

McDermott, J., & Hauser, M. (2004). Are consonant intervals music to their ears? Spontaneous acoustic preferences in a nonhuman primate. *Cognition, 94*(2), B11–B21. doi:10.1016/j.cognition.2004.04.004

McDermott, J., & Hauser, M. (2005). The origins of music: Innateness, uniqueness, and evolution. *Music Perception, 23*(1), 29–59. doi:10.1525/mp.2005.23.1.29

McDermott, J., & Hauser, M. D. (2007). Nonhuman primates prefer slow tempos but dislike music overall. *Cognition, 104*(3), 654–668. doi:10.1016/j.cognition.2006.07.011

McDermott, J. H., Lehr, A. J., & Oxenham, A. J. (2010). Individual differences reveal the basis of consonance. *Current Biology, 20*(11), 1035–1041. doi:10.1016/j.cub.2010.04.019

McDermott, J. H., Schultz, A. F., Undurraga, E. A., & Godoy, R. A. (2016). Indifference to dissonance in native Amazonians reveals cultural variation in music perception. *Nature, 535*(7613), 547–550. doi:10.1038/nature18635

Merchant, H., & Honing, H. (2014). Are non-human primates capable of rhythmic entrainment? Evidence for the gradual audiomotor evolution hypothesis. *Frontiers in Auditory Cognitive Neuroscience, 7*(274), 1–8. doi:10.3389/fnins.2013.00274

Merker, B., Madison, G., & Eckerdal, P. (2009). On the role and origin of isochrony in human rhythmic entrainment. *Cortex, 45*(1), 4–17. doi:10.1016/j.cortex.2008.06.011

Meyer, L. B. (1956). *Emotion and meaning in music*. Chicago: University of Chicago Press.

Micheyl, C., & Oxenham, A. J. (2010). Pitch, harmonicity and concurrent sound segregation: Psychoacoustical and neurophysiological findings. *Hearing Research, 266*(1–2), 36–51. doi:10.1016/j.heares.2009.09.012

Middlebrooks, J. C., & Green, D. M. (1991). Sound localization by human listeners. *Annual Review of Psychology, 42*, 135–159. doi:10.1146/annurev.psych.42.1.135

Miller, G. (2000). Evolution of human music through sexual selection. In H. L. Wallin, B. Merker, & S. Brown (Eds.), *The origins of music* (pp. 329–360). Cambridge, MA: MIT Press.

Mingle, M. E., Eppley, T. M., Campbell, M. W., Hall, K., Horner, V., & de Waal, F. B. M. (2014). Chimpanzees prefer African and Indian music over silence. *Journal of Experimental Psychology: Animal Learning and Cognition, 40*(4), 502–505. doi:10.1037/xan0000032

Moore, B. C. J. (2012). *An introduction to the psychology of hearing* (6th ed.). Cambridge: Brill.

Moore, B. C. J., & Gockel, H. E. (2011). Resolvability of components in complex tones and implications for theories of pitch perception. *Hearing Research, 276*(1–2), 88–97. doi:10.1016/j.heares.2011.01.003

Moore, B. C. J., & Gockel, H. E. (2012). Properties of auditory stream formation. *Philosophical Transactions of the Royal Society of London B: Biological Sciences, 367*(1591), 919–931. doi:10.1098/rstb.2011.0355

Morton, E. S. (1977). On the occurrence and significance of motivation-structural rules in some bird and mammal sounds. *American Naturalist, 111*(981), 855–869. doi:10.1086/283219

Mosing, M. A., Madison, G., Pedersen, N. L., Kuja-Halkola, R., & Ullen, F. (2014). Practice does not make perfect: No causal effect of music practice on music ability. *Psychological Science, 25*(9), 1795–1803. doi:10.1177/0956797614541990

Nagasaka, Y., Chao, Z. C., Hasegawa, N., Notoya, T., & Fujii, N. (2013). Spontaneous synchronization of arm motion between Japanese macaques. *Scientific Reports, 3*, 1–7. doi:10.1038/srep01151

Nelson, P. C., Smith, Z. M., & Young, E. D. (2009). Wide-dynamic-range forward suppression in marmoset inferior colliculus neurons is generated centrally and accounts for perceptual masking. *Journal of Neuroscience, 29*(8), 2553–2562. doi:10.1523/JNEUROSCI.5359-08.2009

Nelson, P. C., & Young, E. D. (2010). Neural correlates of context-dependent perceptual enhancement in the inferior colliculus. *Journal of Neuroscience, 30*(19), 6577–6587. doi:10.1523/JNEUROSCI.0277-10.2010

Nozaradan, S., Peretz, I., Missal, M., & Mouraux, A. (2011). Tagging the neuronal entrainment to beat and meter. *Journal of Neuroscience, 31*(28), 10234–10240. doi:10.1523/JNEUROSCI.0411-11.2011

Nozaradan, S., Peretz, I., & Mouraux, A. (2012). Selective neuronal entrainment to the beat and meter embedded in a musical rhythm. *Journal of Neuroscience, 32*(49), 17572–17581. doi:10.1523/JNEUROSCI.3203-12.2012

Oikkonen, J., Huang, Y., Onkamo, P., Ukkola-Vuoti, L., Raijas, P., Karma, K., et al. (2015). A genome-wide linkage and association study of musical aptitude identifies loci containing genes related to inner ear development and neurocognitive functions. *Molecular Psychiatry, 20*(2), 275–282. doi:10.1038/mp.2014.8

Oxenham, A. J., Micheyl, C., & Keebler, M. V. (2009). Can temporal fine structure represent the fundamental frequency of unresolved harmonics? *Journal of the Acoustical Society of America, 125*(4), 2189–2199. doi:10.1121/1.3089220

Patel, A. D. (2008). *Music, language, and the brain.* New York: Oxford University Press.

Patel, A. D. (2010). Music, biological evolution, and the brain. In M. Bailar (Ed.), *Emerging disciplines* (pp. 91–144). Houston, TX: Rice University Press.

Patel, A. D., & Iversen, J. (2014). The evolutionary neuroscience of musical beat perception: The action simulation for auditory prediction hypothesis. *Frontiers in Systems Neuroscience, 8.* doi:10.3389/fnsys.2014.00057

Patel, A. D., Iversen, J. R., Bregman, M. R., & Schulz, I. (2009). Experimental evidence for synchronization to a musical beat in a nonhuman animal. *Current Biology, 19*(10), 827–830. doi:10.1016/j.cub.2009.03.038

Patterson, R. D., Uppenkamp, S., Johnsrude, I. S., & Griffiths, T. D. (2002). The processing of temporal pitch and melody information in auditory cortex. *Neuron, 36*(4), 767–776. doi:10.1016/S0896-6273(02)01060-7

Penagos, H., Melcher, J. R., & Oxenham, A. J. (2004). A neural representation of pitch salience in nonprimary human auditory cortex revealed with functional magnetic resonance imaging. *Journal of Neuroscience, 24*(30), 6810–6815. doi:10.1523/JNEUROSCI.0383-04.2004

Phillips-Silver, J., & Trainor, L. J. (2005). Feeling the beat in music: Movement influences rhythm perception in infants. *Science, 308*(5727), 1430. doi:10.1126/science.1110922

Phillips-Silver, J., & Trainor, L. J. (2007). Hearing what the body feels: Auditory encoding of rhythmic movement. *Cognition, 105*(3), 533–546. doi:10.1016/j.cognition.2006.11.006

Pinker, S. (1997). *How the mind works*. New York: Norton.

Plack, C. J., Oxenham, A. J., & Fay, R. R. (2006). *Pitch: Neural coding and perception* (Vol. 24). New York: Springer.

Plantinga, J., & Trehub, S. E. (2013). Revisiting the innate preference for consonance. *Journal of Experimental Psychology: Human Perception and Performance, 40*(1), 40–49. doi:10.1037/a0033471

Plomp, R., & Levelt, W. J. (1965). Tonal consonance and critical bandwidth. *Journal of the Acoustical Society of America, 38*(4), 548–560. doi:10.1121/1.1909741

Pulli, K., Karma, K., Norio, R., Sistonen, P., Göring, H. H. H., & Järvelä, I. (2008). Genome-wide linkage scan for loci of musical aptitude in Finnish families: Evidence for a major locus at 4q22. *Journal of Medical Genetics, 45*(7), 451–456. doi:10.1136/jmg.2007.056366

Reddish, P., Fischer, R., & Bulbulia, J. (2013). Let's dance together: Synchrony, shared intentionality and cooperation. *PLoS One, 8*(8), 1–13. doi:10.1371/journal.pone.0071182

Repp, B. H., & Su, Y. H. (2013). Sensorimotor synchronization: A review of recent research (2006–2012). *Psychonomic Bulletin and Review, 20*(3), 403–452. doi:10.3758/s13423-012-0371-2

Rock, A. M. L., Trainor, L. J., & Addison, T. (1999). Distinctive messages in infant-directed lullabies and play songs. *Developmental Psychology, 35*(2), 527–534. doi:10.1037/0012-1649.35.2.527

Roederer, J. G. (1984). The search for a survival value of music. *Music Perception, 1*(3), 350–356. doi:10.2307/40285265

Rouse, A. A., Cook, P. F., Large, E. W., & Reichmuth, C. (2016). Beat keeping in a sea lion as coupled oscillation: Implications for comparative understanding of human rhythm. *Frontiers in Neuroscience, 10*, 257. doi:10.3389/fnins.2016.00257

Salimpoor, V. N., Benovoy, M., Larcher, K., Dagher, A., & Zatorre, R. J. (2011). Anatomically distinct dopamine release during anticipation and experience of peak emotion to music. *Nature Neuroscience, 14*(2), 257–262. doi:10.1038/nn.2726

Santurette, S., & Dau, T. (2011). The role of temporal fine structure information for the low pitch of high-frequency complex tones. *Journal of the Acoustical Society of America, 129*(1), 282–292. doi:10.1121/1.3518718

Schachner, A., Brady, T. F., Pepperberg, I. M., & Hauser, M. D. (2009). Spontaneous motor entrainment to music in multiple vocal mimicking species. *Current Biology, 19*(10), 831–836. doi:10.1016/j.cub.2009.03.061

Scharine, A. A., & Letowski, T. R. (2009). Auditory conflicts and illusions. In C. E. Rash, M. B. Russo, T. R. Letowski, & E. T. Schmeisser (Eds.), *Helmet-mounted displays: Sensation, perception and cognition issues* (pp. 579–598). Fort Rucker, AL: US Army Aeromedical Research Laboratory.

Searcy, W. A., & Andersson, M. (1986). Sexual selection and the evolution of song. *Annual Review of Ecology and Systematics, 17*, 507–533.

Shera, C. A., Guinan, J. J., & Oxenham, A. J. (2002). Revised estimates of human cochlear tuning from otoacoustic and behavioral measurements. *Proceedings of the National Academy of Sciences, 99*(5), 3318–3323. doi:10.1073/pnas.032675099

Shera, C. A., Guinan, J. J., Jr., & Oxenham, A. J. (2010). Otoacoustic estimation of cochlear tuning: Validation in the chinchilla. *Journal of the Association for Research in Otolaryngology, 11*(3), 343–365. doi:10.1007/s10162-010-0217-4

Shofner, W. P. (2005). Comparative aspects of pitch perception. In C. J. Plack, A. J. Oxenham, R. R. Fay, & A. N. Popper (Eds.), *Pitch* (pp. 56–98). New York: Springer. doi:10.1007/0-387-28958-5_3

Smith, N. A., & Trainor, L. J. (2011). Auditory stream segregation improves infants' selective attention to target tones amid distractors. *Infancy, 16*(6), 655–668. doi:10.1111/j.1532-7078.2011.00067.x

Snyder, J. S., & Alain, C. (2007). Toward a neurophysiological theory of auditory stream segregation. *Psychological Bulletin, 133*(5), 780–799. doi:10.1037/0033-2909.133.5.780

Tan, Y. T., McPherson, G. E., Peretz, I., Berkovic, S. F., & Wilson, S. J. (2014). The genetic basis of music ability. *Frontiers in Psychology, 5*, 1–19. doi:10.3389/fpsyg.2014.00658

Tarr, B., Launay, J., Cohen, E., & Dunbar, R. I. (2015). Synchrony and exertion during dance independently raise pain threshold and encourage social bonding. *Biology Letters, 11*(10), 20150767. doi:10.1098/rsbl.2015.0767

Teki, S., Grube, M., Kumar, S., & Griffiths, T. D. (2011). Distinct neural substrates of duration-based and beat-based auditory timing. *Journal of Neuroscience, 31*(10), 3805–3812. doi:10.1523/JNEUROSCI.5561-10.2011

Toiviainen, P., Luck, G., & Thompson, M. (2010). Embodied meter: Hierarchical eigenmodes in music-induced movement. *Music Perception, 28*(1), 59–70. doi:10.1525/mp.2010.28.1.59

Tooby, J., Cosmides, L., & Barrett, H. C. (2003). The second law of thermodynamics is the first law of psychology: Evolutionary developmental psychology and the theory of tandem, coordinated inheritances: Comment on Lickliter and Honeycutt. *Psychological Bulletin, 129*(6), 858–865. doi:10.1037/0033-2909.129.6.858

Trainor, L. J. (2015). The origins of music in auditory scene analysis and the roles of evolution and culture in musical creation. *Philosophical Transactions of the Royal Society of London B: Biological Sciences, 370*(1664), 20140089. doi:10.1098/rstb.2014.0089

Trainor, L. J., & Cirelli, L. (2015). Rhythm and interpersonal synchrony in early social development. *Annals of the New York Academy of Sciences, 1337*(1), 45–52. doi:10.1111/nyas.12649

Trainor, L. J., & Corrigall, K. A. (2010). Music acquisition and effects of musical experience. In M. Riess-Jones & R. R. Fay (Eds.), *Springer handbook of auditory research: Music perception* (pp. 89–128). Heidelberg: Springer.

Trainor, L. J., & Hannon, E. E. (2013). Musical development. In D. Deutsch (Ed.), *The psychology of music* (3rd ed., pp. 423–498). London: Elsevier.

Trainor, L. J., & Heinmiller, B. M. (1998). The development of evaluative responses to music: Infants prefer to listen to consonance over dissonance. *Infant Behavior and Development, 21*(1), 77–88. doi:10.1016/S0163-6383(98)90055-8

Trainor, L. J., Marie, C., Bruce, I. C., & Bidelman, G. M. (2014). Explaining the high-voice superiority effect in polyphonic music: Evidence from cortical evoked potentials and peripheral auditory models. *Hearing Research, 308*, 60–70. doi:10.1016/j.heares.2013.07.014

Trainor, L. J., & Schmidt, L. A. (2003). Processing emotions induced by music. In I. Peretz & R. Zatorre (Eds.), *The cognitive neuroscience of music* (pp. 310–324). Oxford: Oxford University Press.

Trainor, L. J., Tsang, C. D., & Cheung, V. H. W. (2002). Preference for consonance in 2- and 4-month-old infants. *Music Perception, 20*(2), 187–194. doi:10.1525/mp.2002.20.2.187

Trainor, L. J., & Zatorre, R. J. (2016). The neurobiological basis of musical expectations. In S. Hallam, I. Cross, & M. Thaut (Eds.), *The Oxford handbook of music psychology* (2nd ed.). Oxford: Oxford University Press.

Trehub, S. E. (2003). Musical predispositions in infancy: An update. In R. Zatorre & I. Peretz (Eds.), *The cognitive neuroscience of music* (pp. 3–20). Oxford: Oxford University Press.

Trehub, S. E. (2013). Musical universals: Perspectives from infancy. In J. L. Leroy (Ed.), *Topics in universals in music/Actualité des universaux musicaux* (pp. 5–8). Paris: Editions des Archives Contemporaines.

Trehub, S. E., & Trainor, L. J. (1998). Singing to infants: Lullabies and playsongs. *Advances in Infancy Research, 12*, 43–77.

Ukkola, L. T., Onkamo, P., Raijas, P., Karma, K., & Järvelä, I. (2009). Musical aptitude is associated with AVPR1A-haplotypes. *PLoS One, 4*(5), 1–10. doi:10.1371/journal.pone.0005534

Ukkola-Vuoti, L., Kanduri, C., Oikkonen, J., Buck, G., Blancher, C., Raijas, P., et al. (2013). Genome-wide copy number variation analysis in extended families and unrelated individuals characterized for musical aptitude and creativity in music. *PLoS One, 8*(2), 1–9. doi:10.1371/journal.pone.0056356

Valdesolo, P., Ouyang, J., & DeSteno, D. (2010). The rhythm of joint action: Synchrony promotes cooperative ability. *Journal of Experimental Social Psychology, 46*(4), 693–695. doi:10.1016/j.jesp.2010.03.004

Van Noorden, L. P. A. S. (1975). *Temporal coherence in the perception of tone sequence* (Unpublished doctoral dissertation). Technical University Eindhoven, Eindhoven, Netherlands.

Wallin, N. L., Merker, B., & Brown, S. (2000). *The origins of music*. Cambridge, MA: MIT Press.

Weinstein, D., Launay, J., Pearce, E., Dunbar, R. I., & Stewart, L. (2015). Group music performance causes elevated pain thresholds and social bonding in small and large groups of singers. *Evolution and Human Behavior*, *37*(2), 152. doi:10.1016/j.evolhumbehav.2015.10.002

Williams, G. C. (1966). *Adaptation and natural selection: A critique of some current evolutionary thought.* Princeton, NJ: Princeton University Press.

Wilson, M., & Cook, P. F. (2016). Rhythmic entrainment: Why humans want to, fireflies can't help it, pet birds try, and sea lions have to be bribed. *Psychonomic Bulletin and Review*, *23*(6), 1647–1659. doi:10.3758/s13423-016-1013-x

Wiltermuth, S. S., & Heath, C. (2009). Synchrony and cooperation. *Psychological Science*, *20*(1), 1–5. doi:10.1111/j.1467-9280.2008.02253.x

Winkler, I., Denham, S. L., & Nelken, I. (2009). Modeling the auditory scene: Predictive regularity representations and perceptual objects. *Trends in Cognitive Sciences*, *13*(12), 532–540. doi:10.1016/j.tics.2009.09.003

Winkler, I., Háden, G., Ladinig, O., Sziller, I., & Honing, H. (2009). Newborn infants detect the beat in music. *Proceedings of the National Academy of Sciences*, *106*(7), 2468–2471. doi:10.1073/pnas.0809035106

Winkler, I., Kushnerenko, E., Horvath, J., Ceponiene, R., Fellman, V., & Huotilainen, M., … Sussman, E. (2003). Newborn infants can organize the auditory world. *Proceedings of the National Academy of Sciences*, *100*(20), 11812–11815. doi:10.1073/pnas.2031891100

Wojtczak, M., Beim, J. A., Micheyl, C., & Oxenham, A. J. (2012). Perception of across-frequency asynchrony and the role of cochlear delays. *Journal of the Acoustical Society of America*, *131*(1), 363–377. doi:10.1121/1.3665995

Wright, A. A., Rivera, J. J., Hulse, S. H., Shyan, M., & Neiworth, J. J. (2000). Music perception and octave generalization in rhesus monkeys. *Journal of Experimental Psychology*, *129*(3), 291–307. doi:10.1037/0096-3445.129.3.291

Zarco, W., Merchant, H., Prado, L., & Mendez, J. C. (2009). Subsecond timing in primates: Comparison of interval production between human subjects and rhesus monkeys. *Journal of Neurophysiology*, *102*(6), 3191–3202. doi:10.1152/jn.00066.2009

Zatorre, R. J. (1988). Pitch perception of complex tones and human temporal-lobe function. *Journal of the Acoustical Society of America*, *84*(2), 566–572. doi:10.1121/1.396834

Zentner, M., & Eerola, T. (2010). Rhythmic engagement with music in infancy. *Proceedings of the National Academy of Sciences*, *107*(13), 5768–5773. doi:10.1073/pnas.1000121107

Zentner, M. R., & Kagan, J. (1998). Infants' perception of consonance and dissonance in music. *Infant Behavior and Development*, *21*(3), 483–492. doi:10.1016/s0163-6383(98)90021-2

Zwicker, E. G., Flottorp, G., & Stevens, S. S. (1957). Critical band width in loudness summation. *Journal of the Acoustical Society of America*, *29*(5), 548–557. doi:10.1121/1.1908963

5

Music as a Transformative Technology of the Mind: An Update

Aniruddh D. Patel

Debates over the evolution of human musicality are often framed as a dichotomy between two positions. Either musical behavior originated as a biological adaptation (as first argued by Darwin in 1871 in *The Descent of Man*), or music is a purely cultural invention that builds on brain mechanisms that evolved for other reasons, and thus has little biological significance in its own right (as Pinker held in 1997 in *How the Mind Works*). I have argued that this dichotomy limits our thinking about the biology of music (Patel, 2008). Specifically, I have posited that music is both a human invention and biologically powerful (Patel, 2010). In terms of origins, musical behavior, like the control of fire, is an ancient human invention rather than a product of natural or sexual selection. In this view, music (like fire control) became universal in human societies because what it offered to humans was universally valued. (In the case of music, I suggested these offerings included emotional regulation, a framework for rituals, and a scaffolding for memorizing of long sequences of information, prior to writing.) In terms of biological significance, I have argued that music's rich connections to other cognitive systems and to the brain's reward circuitry give it a powerful ability to shape nonmusical brain functions via mechanisms of neural plasticity, within individual lifetimes. (For an example of this idea applied to relations between musical experience and speech processing, see the OPERA hypothesis, Patel, 2011, 2014a).[1] This beneficial biological impact of music, while not the reason for its origin or maintenance in human societies, makes music a biologically powerful human invention or transformative technology of the mind (TTM; Patel, 2010).[2]

In the years since I proposed this TTM theory, I have begun to reconsider the notion that there is a categorical divide between cultural invention and biological evolution. Inspired by research on gene-culture coevolution within evolutionary biology (e.g., Richerson & Boyd, 2008) and by a growing number of theorists who have proposed interactions between biology and culture in the evolution of music (e.g., Chatterjee, 2013; Cross, 2003; Davies, 2012; Kirschner & Tomasello, 2010; Tomlinson, 2015; Trainor, 2015; van der Schyff, 2013), I believe it is time to update TTM theory to consider the possibility that the invention of musical behavior triggered processes of gene-culture coevolution (GCC). The purpose of this chapter is to show how TTM theory can be integrated with the idea of

GCC and to make specific suggestions for research aimed at exploring whether GCC has indeed been at play in the evolution of human musicality, that is, in shaping biological and cognitive mechanisms essential for perceiving, appreciating, and making music (Honing, ten Cate, Peretz, & Trehub, 2015).

Before embarking on this discussion, it is worth noting that the idea that human inventions have consequences for biological evolution has already been richly theorized in the case of the control of fire. Wrangham (2009) has argued persuasively that the control of fire and the invention of cooking by human ancestors led to coevolutionary changes in physiology, such that modern humans are now biologically adapted to eating cooked food. He argues that cooking makes certain animal proteins, plant carbohydrates, and lipids more digestible and also softens food, all of which reduce the cost of digestion (cf. Groopman, Carmody, & Wrangham, 2015). Consequently, our gut shrank over evolutionary time, allowing valuable metabolic energy to be diverted to our brains, which could then grow larger over evolutionary time since brains are energetically very expensive. Recall that TTM theory draws an analogy between the origin of music and the control of fire. The original purpose of that analogy was to point out how ancient and universal aspects of human culture can originate as inventions rather than as biological adaptations. Yet Wrangham's (2009) argument suggests that ancient human inventions can have biological consequences on evolutionary timescales. This in turn suggests that TTM theory can be updated to include the idea of GCC.

Why Consider Music a Human Invention?

In Patel (2010) I argued that music was an invention because each of the components of musicality (the cognitive foundations of musical behavior) was cognitively linked to some nonmusical mental ability. For some aspects of music cognition, this claim is relatively uncontroversial. For example, the mechanisms involved in auditory scene analysis (the mental parsing of complex auditory patterns into distinct sound sources) appear to play an important role in music cognition (see chapter 4, this volume). For other aspects of music cognition, however, the claim is more controversial. These include tonality processing (which involves the perception of tones in terms of underlying scales, with hierarchies of stability) and the entrainment of rhythmic movement to a periodic beat. These aspects of music cognition are found across cultures (Savage, Brown, Sakai, & Currie, 2015) and have no obvious relationship to nonmusical aspects of human cognition. They are thus logical candidates for modular processes that have been specifically shaped by evolution to support music processing (Patel, 2008; Peretz & Coltheart, 2003). In both cases, however, research points to hidden connections to other cognitive domains (see Patel, 2010, for a detailed discussion). Briefly, there is evidence that the processing of tonality and the hierarchical relations that it entails has links to cognitive mechanisms involved in the processing linguistic structure. (See Kunert & Slevc, 2015, and Patel, 2013, for recent reviews.

For recent empirical evidence, see Kunert, Willems, & Hagoort, 2016; LaCroix, Diaz, & Rogalsky, 2015; and Musso et al., 2015.) There is also evidence that the ability to entrain rhythmic movements to a beat in a predictive and flexible manner has links to the brain mechanisms for vocal learning or vocal control (Patel, Iversen, Bregman, & Schulz, 2009). These more controversial claims are topics of active research and debate (e.g., Cook, Rouse, Wilson, & Reichmuth, 2013; Dalla Bella, Berkowska, & Sowiński, 2015; Hattori, Tomonaga, & Matsuzawa, 2013, 2015; Merchant & Honing, 2014; Patel, 2014b; Peretz, Vuvan, Lagrois, & Armony, 2015; chapter 9, this volume). In light of research on music cognition that has transpired since the original publication of TTM theory, I still believe that music cognition is deeply connected to other cognitive functions (especially language) and that it is possible to explain the origins of human music without invoking natural or sexual selection specifically aimed at musical behavior.

This strategy of using rich cognitive links between music and other mental abilities as evidence for music's origin as a cultural invention is not unique to TTM theory. For example, Pinker (1997) used this strategy when arguing that music is an invented "pleasure technology" that uses the neural machinery of several other adaptive cognitive functions. While I continue to rely on this strategy, I am increasingly cognizant of its limitations. For example, the strategy often overlooks the fact that a behavior that originates as an invention can subsequently lead to evolutionary changes to an organism's biology (see the next section; cf. Fisher & Ridley, 2013; Trainor, 2015). This means that nonmodularity does not rule out that a trait has (at some point after its origin) been subject to natural selection. Another limitation of the strategy is that it tacitly assumes a simple mapping between genes and cognitive processes; it implies that brain specializations for music should involve genes and cognitive processes not shared with other mental domains. Yet evidence from developmental neurobiology suggests that mappings between genes and cognitive traits are usually complex, with most genes acting across multiple cognitive/behavioral traits (Balaban, 2006; Greenspan, 2004). This suggests that even biologically evolved traits need not always show high degrees of genetic or cognitive modularity. A final limitation of the strategy is that does not offer explicit standards for what constitutes evidence for links between music and other cognitive abilities. For example, as Peretz, Vuvan, Lagrois, and Armony point out in chapter 9 (this volume), demonstrating that music and some other cognitive ability (e.g., language) show overlapping brain activations in functional magnetic resonance imaging (fMRI) studies does not prove shared neural circuitry, since distinct neural circuits can coexist in the same brain regions and not be resolvable by fMRI methods due to the limited spatial resolution of this method. Fortunately, there are several other methods for studying cognitive links between music and other mental abilities (e.g., interference experiments; cf. Kunert & Slevc, 2015), so that the search for cognitive links can use multiple approaches.

The broader point is that evidence that music, like the control of fire, originated as a cultural invention (rather than as a biological adaptation) requires more than just demonstrating links between music and other cognitive domains. Seeking evidence for an invention-based origin is an important future topic for TTM theory and will likely require integrating data from genetic and cross-species studies of music cognition. The focus of this chapter, however, is not on music's origins but on considering whether musical behavior, once established, triggered processes of gene-culture coevolution.

Updating TTM Theory in Light of Gene-Culture Coevolution

Gene-culture coevolution (GCC) refers to an interplay between human cultural practices and lasting changes in human biology occurring on evolutionary timescales (Richerson, Boyd, & Henrich, 2010; Ross & Richerson, 2014). Wrangham's (2009) theories regarding the impact of fire control on human evolution provide one example. Wrangham argues that the control of fire dates back 2 million years. This date has been disputed, with some anthropologists arguing that strong evidence of fire control dates back no more than 400,000 years ago. Even this conservative date, however, would allow ample time for biological changes to accrue via GCC. We know this because the best-supported example of GCC comes from a behavior that originated somewhere between 8,000 and 11,000 years ago. This was the invention of dairying, or herding animals in order to drink their milk (Beja Pereira et al., 2003). Before this invention, human consumption of milk was limited to infants and young children, and only they had the enzyme lactase that could break down lactose, or milk sugar. Normally the gene that produces this enzyme is switched off when children mature. Around 8,000 to 11,000 years ago, however, people began herding cattle and making foods (such as cheese) out of their milk. A few thousand years later, a genetic mutation spread through Europe that gave adults the ability to produce lactase. A cultural practice led to a lasting genetic change, and this opened up a new food resource to the early Europeans. This in turn affected the ability of these populations to grow and spread.

Dairying and its impact on lactose metabolism illustrates a feedback loop between human cultural practices and lasting changes in human biology. Given that the oldest known musical instruments are least 40,000 years old (Conard, Malina, & Münzel, 2009) and that nonfossilizing forms of music (such as singing, which is universal in human cultures) may far predate such instruments, there appears to be enough time for musical behavior to have led to GCC.

The original version of TTM theory did not consider GCC and suggested that if music is a human invention, "future research will show that every component of music cognition can either be related to a nonmusical brain function or can be explained via learning in the absence of any evolutionary specialization for music" (Patel, 2010, p. 105). In retrospect, this statement overlooks the possibility that some components of music cognition might originate as secondary uses of other brain functions (i.e., as exaptations; cf. Gould, 1991), but then may become specialized through processes of GCC to support musical behavior. The idea that exaptations can undergo subsequent evolutionary specialization for a new function is well known in biology (Gould & Vrba, 1982). Feathers provide a good example. Feathers are thought to have originated from anatomical structures that originally functioned in heat regulation (or other non-flight-related functions), which only incidentally proved useful in providing aerodynamic lift when hopping or jumping (Prum, 1999). There is no question, however, that as birds evolved, feathers became structurally specialized to support flight. The example of feathers is not a case of GCC, but it does show how a function that starts as a secondary use of another trait can become a target of evolutionary forces. Returning to music, the question is "Did musical behavior have some unintended beneficial consequence for human survival, which could thus trigger processes of GCC?"

Why Might Musical Behavior Have Triggered Gene-Culture Evolution?

If musical behavior originated as a human invention and by chance had a reliable positive impact on survival and reproductive success, this could conceivably trigger processes of GCC. Presumably such processes would act to specialize certain cognitive abilities that were important for musical behavior—for example, by making them develop more robustly and quickly in ontogeny than they would if they were not supported by a biological predisposition. This in turn would promote the acquisition of musical behavior by children. The key question for this section of the chapter is "Which beneficial effects of music might have triggered GCC?" I have argued that musical experience can have positive impacts on other brain functions (such as language processing) through mechanisms of neural plasticity within individual lifetimes (Patel, 2010, 2011, 2014a). This raises the question of whether these beneficial effects would trigger GCC. I see these benefits as enhancing the quality of lives of individuals (e.g., by improving aspects of their speech processing), but it is not clear to me that such benefits could significantly affect survival and reproductive success. Thus, I consider another potential beneficial effect of music related to social behavior in early human groups.

A growing number of theorists are considering the social consequences of temporally coordinated musical behavior such as group singing and dancing (e.g., Dunbar, 2012; Kirschner & Tomasello, 2010; Roederer, 1984). Cross-cultural research shows that singing, playing percussion instruments, and dancing to simple, repetitive music in groups is a universal human tendency (Savage et al., 2015) and is thus likely very ancient. Theorists have argued that such behavior strengthens social bonds and that these bonds are expressed (outside of musical contexts) in prosocial behaviors toward group members. The psychological mechanisms behind these effects are a topic of current interest. Tarr, Launay, and Dunbar (2014), for example, suggest that moving in synchrony with others leads to "self-other merging" and endorphin release, which strengthens social bonds among group members. Whatever the proximate mechanisms for the social effects of music, the key questions from the standpoint of GCC are whether such effects are reliable and whether they could have had consequences for survival and reproduction in early human groups.

In terms of the first question, a growing number of empirical studies have demonstrated that temporally synchronized and simultaneous rhythmic movement between individuals has a positive impact on how they evaluate and treat each other socially after the synchronous interaction (e.g., Hove & Risen, 2009; Kirschner & Tomasello, 2010; Trainor & Cirelli, 2015). In terms of the latter question (whether synchrony could affect survival), theorists have pointed out that since human groups inevitably suffer from internal conflicts and from competition with other groups, greater cohesion within a group could lead to an increased ability of group members to survive these challenges and be reproductively successful (Richerson & Boyd, 2008; Wilson, 2012). Thus, if group music making had a reliable impact on social cohesion in early human groups, it seems plausible that this could trigger processes of GCC to shape the mind in ways that promote engagement in group musical behavior.

Notice that this view is distinct from the idea that musical behavior had its origin in sexual selection, Darwin's hypothesis in *The Descent of Man* (1871). While the sexual selection hypothesis has modern adherents (e.g., Miller, 2011), the social bonding theory of music's origins has been ascending in recent years, perhaps because several empirical studies it has inspired have provided results consistent with the theory. Social bonding theories often imply that musical behavior arose due to its impact on social bonding, yet such theories can be aligned with TTM theory if one assumes that musical behavior first arose as a human invention and then had (unanticipated) beneficial effects on social cohesion. There is an analogy here to the origin of feathers. The beneficial effect of the first feathers on the aerodynamics of hopping or jumping was a fortuitous coincidence. However, by improving survival and reproductive success, this unanticipated benefit ultimately led to the evolutionary specialization of these structures for aerodynamic properties. Of course, the analogy to music is not perfect, because such dermal scales did not arise as an invention among the ancestors of birds, whereas in TTM theory, music originated as a human invention. Nevertheless, the analogy is useful because in both cases, an unanticipated benefit triggers further processes of biological evolution. If one takes this view in the case of music, then a key question is "Where would one look for evidence that the invention of music triggered processes of GCC?"

How to Seek Evidence of Gene-Culture Coevolution for Music Cognition

If, as I have suggested, the invention of musical behavior led to GCC driven by the impact of group music making on social cohesion, then a logical place to look for evidence of GCC is in specialized mental mechanisms that support music making with others. Note that by "specialized mental mechanisms," I do not mean mental capacities that are sealed off (modularized) from other nonmusical aspects of cognition. Indeed, the idea of music as originating as a human invention, which is fundamental to TTM theory, suggests (contra a strong modular view of music cognition) that one will always find links between musical and nonmusical mental processes, since the former have their origins in alternative uses of the latter. The pertinent question is: "Is there evidence that certain mental processes involved in music have undergone further evolutionary specialization to support music cognition (especially for group music making)?" Such mental processes would have a dual life in the brain: they would have a connection to nonmusical brain functions (as already noted), but they would also show a pattern of early or elaborated development that extends beyond what would be expected if they were simply a secondary use of another cognitive ability.

With these ideas in mind, let us briefly consider three aspects of musicality relevant to group music making: pitch control in group singing, synchronizing auditory-motor rhythmic behavior with others, and the use of working memory in learning songs. Is there any evidence that humans have evolutionary neural specializations related to these abilities? The notion of evolutionary specialization is crucial here, since (with extensive training) humans can acquire specialized sensorimotor or cognitive skills that evolution never specialized them to do, such as hitting an accurate backhand in tennis or reading written language. To demonstrate evolutionary (versus purely ontogenetic) specialization, converging evidence is needed from a variety of methods, including developmental, genetic, and cross-species studies (cf. chapters 2, 7, and 10, this volume; Gingras et al., 2015; Hoeschele et al., 2015; Patel & Demorest, 2013; Trehub, 2003). The search for such evidence with regard to pitch control in group singing, synchronizing auditory-motor behavior with others, and the use of working memory in song learning is a topic for future research. This chapter simply highlights these aspects of musicality as relevant areas of research for future work on musicality and GCC.

Pitch Control in Group Singing

Humans have fine voluntary control over their vocal pitch patterns, a trait that is rare (if not absent) in most other primates (Petkov & Jarvis, 2012). This ability is important for speech production, where the regulation of pitch plays an important role in linguistic and affective prosody and (in tone languages) in distinguishing between word forms. In singing as in speech, pitch patterns are systematically regulated, and the ability to regulate pitch accurately in speech and song appears to be related (Mantell & Pfordresher, 2013). Such a relationship is what would be expected if singing is a human invention that builds on speech motor control. Group singing, however, places demands on pitch regulation that go beyond anything required by speech, because during group singing, this regulation must also take into account other (concurrently heard) pitch patterns. That is, group singing requires a particular type of perception-action loop in which one's vocal pitch must be dynamically coordinated with pitch patterns that others simultaneously produce. Throughout the world, this coordination is facilitated by making songs from discrete pitches organized into musical scales (Savage et al., 2015). By using scale pitches as cognitive reference points when making melodies, vocal pitch can be coordinated among individuals in perceptually coherent and precise ways.

The coordination of simultaneous vocal pitch patterns, which may play an important role in creating a sense of group cohesion through music, does not occur in ordinary speech. In ordinary conversation, a speaker and a listener alternate their vocal production, and producing words at the same time leads to a perceptually incoherent jumble. If there has been GCC to support the mechanisms involved in group singing, then one would expect evolutionary neural specializations for the ability to regulate vocal pitch in relation to other simultaneously heard pitch patterns (e.g., the ability to match pitches being sung by others or to sing pitches that perceptually blend with pitches being sung by others, even if not matching their exact pitches). It is interesting to ask whether the tendency to sing in groups in which pitch patterns are (1) produced simultaneously with others and (2) systematically regulated in terms of pitch relations with others might be uniquely human. While the former trait has been documented in some species (e.g., wolves howling together), it is not clear if there are nonhuman animals that show both the former and latter traits. For example, recent work on the Madagascar indri (a type of lemur) has provided evidence for the former trait (group singing involving multiple individuals vocalizing simultaneously), yet no evidence for the latter trait was presented (Gamba et al., 2016). More specifically, quantitative analyses revealed structured temporal interactions between singing indris, but whether they also regulated pitch relations among their voices was not addressed. Indeed, establishing such regulation during animal group singing is nontrivial and will likely require experimental methods, to test if an animal shifts the pitch of its own voice in response to acoustic manipulations that raise or lower the pitches that others produce (cf. Yuan, Rosenberg, & Patel, 2016).

Synchronizing Auditory-Motor Behavior with Others

A core feature of music cognition is the ability to perceive a periodic beat in rhythmic music and synchronize one's movement to it in a predictive and tempo-flexible way. This beat perception and synchronization (BPS) is not part of ordinary speech, yet it may have originated in the brain circuitry for complex vocal learning (Patel, 2006, 2014a; Patel & Iversen, 2014) or vocal control (Fitch, 2015), since vocal learning and control, like BPS, require neuroanatomical specializations for precise and flexible auditory-motor integration (cf. Petkov & Jarvis, 2012). Thus, once again we see a connection between music cognition and a nonmusical brain function. Yet group music making often requires more than synchronization to a common beat. It also requires a dynamic interplay between the rhythmic actions of group members in order to keep sound production synchronized in the face of expressive or unintentional temporal fluctuations produced by individual group members (Keller, Novembre, & Hove, 2014). The ability to adapt quickly to such fluctuations and to stay coordinated in time with others over relatively long periods (i.e., throughout entire songs or dances) is needed in group music making. This type of coordination is distinct from the temporal demands created by other sequential auditory-motor behaviors, such as speech. Of course, speech does involve auditory-motor temporal coordination between individuals, as in the timing of conversational turns (Stivers et al., 2009). However, group music requires temporal coordination with others that are producing sound simultaneously, versus in alternation, with oneself. The accuracy of such simultaneous coordination is likely important in creating a sense of group cohesion through music (cf. Hove & Risen, 2009). Thus, GCC may have acted to create evolutionary neural specializations that support simultaneous auditory-motor rhythmic coordination with others, even in the face of timing variations. In searching for neural specialization for this ability, it would be particularly interesting to investigate if humans show more highly elaborated temporal coordination abilities than other species that engage in simultaneous rhythmic behavior (Ravignani, Bowling, & Fitch, 2014).

The Use of Working Memory in Learning Songs

Making music with others often involves learning songs created by other individuals in the community. Learning songs requires auditory working memory (WM): one must listen, chunk information, and encode it in (and retrieve it from) long-term memory. Of course, music is hardly unique in placing demands on auditory WM. The cognitive mechanisms of auditory sentence comprehension, which require information to be integrated over time, have an important relationship to auditory WM (Just & Carpenter, 1992). Thus, once again, we see that a core aspect of music cognition has important links to other cognitive abilities. And once again, it seems that music places demands on this aspect of cognition that are different from the demands made by cognitive domains. Specifically, in the case of auditory WM, music may demand more capacity than needed for ordinary spoken language. This is because when we are comprehending spoken language, we immediately translate words into semantic concepts and propositions, and the specific words used in sentences do not need to be retained. Thus, a listener who tries to repeat a just-heard sentence will often not repeat back the exact words but will get the gist right. When learning songs for group singing, however, particular word sequences (and pitch sequences) need to be stored and recalled accurately, and this would seem to place higher demands on auditory WM than ordinary language does. In this regard, it is interesting to note that musically trained individuals appear to have larger auditory WM capacity than untrained individuals (Clayton et al., 2016; Kraus, Strait, & Parbery-Clark, 2012; Zuk, Benjamin, Kenyon, & Gaab, 2014). However, for the purposes of this discussion, it is the average human auditory WM capacity that is of interest. Comparative research with other primates suggests that they have very limited auditory WM (e.g., compared to their visual WM, or to the auditory WM of humans; Hashiya & Kojima, 2001; Ng, Plakke, & Poremba, 2014; cf. Carruthers, 2013; Patel, 2016; Scott, Mishkin, & Yin, 2012). This suggests that human auditory WM capacity has expanded significantly over evolutionary time. From the standpoint of research on GCC related to music cognition, the question is whether this expansion far exceeds what would be needed for language alone. If so, this might suggest a neural specialization aimed at supporting not only language processing but also music cognition.

Interestingly, it has been suggested (based on computational modeling) that a relatively limited auditory WM capacity is beneficial in early language acquisition and facilitates the learning of the basic syntactic structures in a language (Elman, 1993), which are then learned in more detail as WM matures to the adult state. This raises the idea that auditory WM capacity in humans has an interesting developmental trajectory. Thus, if music-related GCC has played a role in shaping human auditory WM capacity, one place to look for evidence of this might be in shape of this trajectory from infancy to adulthood, in light of the working memory demands required by linguistic versus musical processing.

Conclusion

In this chapter I have argued that music is a biologically powerful human invention (a transformative technology of the mind) that may have triggered processes of gene-culture coevolution. This view resonates with evolutionary theories of music that argue for an interplay between biological and cultural evolution. If the current proposal is correct, then certain core processes involved in music cognition will have important links to other domains of cognition, yet also show evidence of neural specialization specifically aimed at supporting musical behavior, particularly group music making.

Acknowledgments

I thank Simon Fisher for insightful comments and Richard Wrangham for discussions regarding his theory of how the control of fire impacted human evolution.

Notes

1. The OPERA hypothesis posits that music training results in enhanced speech processing when five conditions are met: (1) Overlap: there is anatomical overlap in the brain networks that process an acoustic feature used in both music and speech (e.g., waveform periodicity, amplitude envelope); (2) precision: music places higher demands on these shared networks than does speech, in terms of the precision of processing; (3) emotion: the musical activities that engage this network elicit strong positive emotion; (4) repetition: the musical activities that engage this network are frequently repeated; and (5) attention: the musical activities that engage this network are associated with focused attention. According to the OPERA hypothesis, when these conditions are met, neural plasticity drives the networks in question to function with higher precision than needed for ordinary speech communication. Yet since speech shares these networks with music, speech processing benefits. The OPERA hypothesis was introduced in Patel (2011) and focused on sensory processing; for an updated version of the hypothesis that considers both sensory and cognitive processing, see Patel (2014a).

2. The original paper on TTM (Patel, 2010) is available online at https://mitpress.mit.edu/books/origins-musicality.

References

Balaban, E. (2006). Cognitive developmental biology: History, process, and fortune's wheel. *Cognition, 101*(2), 298–332. doi:10.1016/j.cognition.2006.04.006

Beja-Pereira, A., Luikart, G., England, P. R., Bradley, D. G., Jann, O. C., Bertorelle, G., et al. (2003). Gene-culture coevolution between cattle milk protein genes and human lactase genes. *Nature Genetics, 35*(4), 311–313. doi:10.1038/ng1263

Carruthers, P. (2013). Evolution of working memory. *Proceedings of the National Academy of Sciences, 110*(Suppl. 2), 10371–10378. doi:10.1073/pnas.1301195110

Clayton K. K., Swaminathan, J., Yazdanbakhsh, A., Zuk, J., Patel, A. D., & Kidd, G., Jr. (2016). Executive function, visual attention and the cocktail party problem in musicians and non-musicians. *PLoS One, 11*(7): 1–17. doi:10.1371/journal. pone.0157638

Conard, N. J., Malina, M., & Münzel, S. C. (2009). New flutes document the earliest musical tradition in southwestern Germany. *Nature, 460*, 737–740. doi:10.1038/nature08169

Cook, P., Rouse, A., Wilson, M., & Reichmuth, C. (2013). A California sea lion (*Zalophus californianus*) can keep the beat: Motor entrainment to rhythmic auditory stimuli in a non vocal mimic. *Journal of Comparative Psychology, 127*(4), 412–427. doi:10.1037/a0032345

Cross, I. (2003). Music and biocultural evolution. In M. Clayton, T. Herbert, & R. Middleton (Eds.), *The cultural study of music: A critical introduction* (pp. 19–30). New York: Routledge.

Dalla Bella, S., Berkowska, M., & Sowiński, J. (2015). Moving to the beat and singing are linked in humans. *Frontiers in Human Neuroscience, 9*, 1–13. doi:10.3389/fnhum.2015.00663

Darwin, C. (1871). *The descent of man and selection in relation to sex*. London: John Murray.

Davies, S. (2012). *The artful species: Aesthetics, art, and evolution*. Oxford: Oxford University Press.

Dunbar, R. I. M. (2012). On the evolutionary function of song and dance. In N. Bannan (Ed.), *Music, language, and human evolution* (pp. 201–214). New York: Oxford University Press. doi:10.1093/acprof:osobl/9780199227341.003.0008

Elman, J. L. (1993). Learning and development in neural networks: The importance of starting small. *Cognition, 48*(1), 71–99.

Fisher, S. E., & Ridley, M. (2013). Culture, genes, and the human revolution. *Science, 340*(6135), 929–930.

Fitch, W. T. (2015). The biology and evolution of musical rhythm: An update. In I. Toivonen, P. Csúri, & E. van der Zee (Eds.), *Structures in the mind: Essays on language, music, and cognition in honor of Ray Jackendoff* (pp. 293–324). Cambridge, MA: MIT Press.

Gamba, M., Torti, V., Estienne, V., Randrianarison, R. M., Valente, D., Rovara, P., et al. (2016). The Indris have got rhythm! Timing and pitch variation of a primate song examined between sexes and age classes. *Frontiers in Neuroscience, 10*, 1–12. doi:10.3389/fnins.2016.00249

Gingras, B., Honing, H., Peretz, I., Trainor, L. J., & Fisher, S. E. (2015). Defining the biological bases of individual differences in musicality. *Philosophical Transactions of the Royal Society of London B: Biological Sciences, 370*(1664), 1–15. doi:10.1098/rstb.2014.0092

Gould, S. J. (1991). Exaptation: A crucial tool for an evolutionary psychology. *Journal of Social Issues, 47*(3), 43–65. doi:10.1111/j.1540-4560.1991.tb01822.x

Gould, S. J., & Vrba, E. S. (1982). Exaptation—a missing term in the science of form. *Paleobiology, 8*(1), 4–15. doi:10.1017/S0094837300004310

Greenspan, R. (2004). *E pluribus unum, ex uno plura*: Quantitative- and single-gene perspectives on the study of behavior. *Annual Review of Neuroscience, 27*, 79–105. doi:10.1146/annurev.neuro.27.070203.144323

Groopman, E. D., Carmody, R. N., & Wrangham, R. W. (2015). Cooking increases net energy gain from a lipid-rich food. *American Journal of Physical Anthropology, 156*(1), 11–18. doi:10.1002/ajpa.22622

Hashiya, K., & Kojima, S. (2001). Acquisition of auditory-visual intermodal matching-to-sample by a chimpanzee (*Pan troglodytes*): Comparison with visual-visual intramodal matching. *Animal Cognition, 4*(3), 231–239. doi:10.1007/s10071-001-0118-3

Hattori, Y., Tomonaga, M., & Matsuzawa, T. (2013). Spontaneous synchronized tapping to an auditory rhythm in a chimpanzee. *Scientific Reports, 3*, 1–6. doi:10.1038/srep01566

Hattori, Y., Tomonaga, M., & Matsuzawa, T. (2015). Distractor effect of auditory rhythms on self-paced tapping in chimpanzees and humans. *PLoS One, 10*(7), 1–17. doi:10.1371/journal.pone.0130682

Hoeschele, M., Merchant, H., Kikuchi, Y., Hattori, Y., & ten Cate, C. (2015). Searching for the origins of musicality across species. *Philosophical Transactions of the Royal Society of London B: Biological Sciences, 370*(1664), 1–9. doi:10.1098/rstb.2014.0094

Honing H., ten Cate, C., Peretz, I., & Trehub, S.E. (2015). Without it no music: Cognition, biology and evolution of musicality. *Philosophical Transactions of the Royal Society of London B: Biological Sciences, 370*(1664), 1–8. doi:10.1098/rstb.2014.0088

Hove, M. J., & Risen, J. L. (2009). It's all in the timing: Interpersonal synchrony increases affiliation. *Social Cognition, 27*(6), 949–960. doi:10.1521/soco.2009.27.6.949

Just, M. A., & Carpenter, P. A. (1992). A capacity theory of comprehension: Individual differences in working memory. *Psychological Review, 99*(1), 122–149. doi:10.1037/0033-295x.99.1.122

Keller, P. E., Novembre, G., & Hove, M. J. (2014). Rhythm in joint action: Psychological and neurophysiological mechanisms for real-time interpersonal coordination. *Philosophical Transactions of the Royal Society of London B: Biological Sciences, 369*(1658), 1–12. doi:10.1098/rstb.2013.0394

Kirschner, S., & Tomasello, M. (2010). Joint music making promotes prosocial behavior in 4-year-old children. *Evolution and Human Behavior, 31*(5), 354–364. doi:10.1016/j.evolhumbehav.2010.04.004

Kraus, N., Strait, D. L., & Parbery-Clark, A. (2012). Cognitive factors shape brain networks for auditory skills: Spotlight on auditory working memory. *Annals of the New York Academy of Sciences, 1252*(1), 100–107. doi:10.1111/j.1749-6632.2012.06463.x

Kunert, R., & Slevc, L. R. (2015). Evidence beyond neuroimaging: A commentary on "Neural overlap in processing music and speech (Peretz et al., 2015)." *Frontiers in Human Neuroscience, 9*, 1–3. doi:10.3389/fnhum.2015.00330

Kunert, R., Willems, R. M., & Hagoort, P. (2016). Language influences music harmony perception: Effects of shared syntactic integration resources beyond attention. *Royal Society Open Science, 3*(2), 1–29. doi:10.1098/rsos.150685

LaCroix, A. N., Diaz, A. F., & Rogalsky, C. (2015). The relationship between the neural computations for speech and music perception is context-dependent: An activation likelihood estimate study. *Frontiers in Psychology, 6*, 1–19. doi:10.3389/fpsyg.2015.01138

Mantell, J. T., & Pfordresher, P. Q. (2013). Vocal imitation of song and speech. *Cognition, 127*(2), 177–202. doi:10.1016/j.cognition.2012.12.008

Merchant, H., & Honing, H. (2014). Are non-human primates capable of rhythmic entrainment? Evidence for the gradual audiomotor evolution hypothesis. *Frontiers in Neuroscience, 7*, 1–8. doi:10.3389/fnins.2013.00274

Miller, G. (2011). *The mating mind: How sexual choice shaped the evolution of human nature*. New York: Knopf.

Musso, M., Weiller, C., Horn, A., Glauche, V., Umarova, R., Hennig, J., et al. (2015). A single dual-stream framework for syntactic computations in music and language. *NeuroImage, 117*, 267–283. doi:10.1016/j.neuroimage.2015.05.020

Ng, C.-W., Plakke, B., & Poremba, A. (2014). Neural correlates of auditory recognition memory in the primate dorsal temporal pole. *Journal of Neurophysiology, 111*(3), 455–469. doi:10.1152/jn.00401.2012

Patel, A. D. (2006). Musical rhythm, linguistic rhythm, and human evolution. *Music Perception, 24*(1), 99–104. doi:10.1525/mp.2006.24.1.99

Patel, A. D. (2008). *Music, language, and the brain*. New York: Oxford University Press.

Patel, A. D. (2010). Music, biological evolution, and the brain. In M. Bailar (Ed.), *Emerging disciplines* (pp. 91–144). Houston, TX: Rice University Press.

Patel, A. D. (2011). Why would musical training benefit the neural encoding of speech? The OPERA hypothesis. *Frontiers in Psychology, 2*, 1–14. doi:10.3389/ fpsyg.2011.00142

Patel, A. D. (2013). Sharing and nonsharing of brain resources for language and music. In M. Arbib (Ed.), *Language, music, and the brain: A mysterious relationship* (pp. 329–355). Cambridge, MA: MIT Press.

Patel, A. D. (2014a). Can nonlinguistic musical training change the way the brain processes speech? The expanded OPERA hypothesis. *Hearing Research, 308*, 98–108. doi:10.1016/j.heares.2013.08.011.

Patel, A. D. (2014b). The evolutionary biology of musical rhythm: Was Darwin wrong? *PLoS Biology, 12*(3), 1–6. doi:10.1371/journal.pbio.1001821

Patel, A. D. (2016). Using music to study the evolution of cognitive mechanisms relevant to language. *Psychonomic Bulletin and Review*. doi:10.3758/s13423-016-1088-4

Patel, A. D., & Demorest, S. (2013). Comparative music cognition: Cross-species and cross-cultural studies. In D. Deutsch (Ed.), *The psychology of music* (3rd ed., pp. 647–681). London: Academic Press/Elsevier.

Patel, A. D., & Iversen, J. R. (2014). The evolutionary neuroscience of musical beat perception: The action simulation for auditory prediction (ASAP) hypothesis. *Frontiers in Systems Neuroscience, 8*, 1–14. doi:10.3389/fnsys.2014.00057

Patel, A. D., Iversen, J. R., Bregman, M. R., & Schulz, I. (2009). Experimental evidence for synchronization to a musical beat in a nonhuman animal. *Current Biology, 19*(10), 827–830. doi:10.1016/j.cub.2009.03.038

Peretz, I., & Coltheart, M. (2003). Modularity of music processing. *Nature Neuroscience, 6*(7), 688–691. doi:10.1038/nn1083

Peretz, I., Vuvan, D., Lagrois, M. É., & Armony, J. L. (2015). Neural overlap in processing music and speech. *Philosophical Transactions of the Royal Society of London B: Biological Sciences, 370*(1664), 1–8. doi:10.1098/rstb.2014.0090

Petkov, C. I., & Jarvis, E. D. (2012). Birds, primates, and spoken language origins: Behavioral phenotypes and neurobiological substrates. *Frontiers in Evolutionary Neuroscience, 4*, 1–24. doi:10.3389/fnevo. 2012.00012

Pinker, S. (1997). *How the mind works*. London: Allen Lane.

Prum, R. O. (1999). Development and evolutionary origin of feathers. *Journal of Experimental Zoology, 285*(4), 291–306. doi:10.1002/(SICI)1097-010X(19991215)285:4<291::AID-JEZ1>3.0.CO;2–9

Richerson, P. J., & Boyd, R. (2008). *Not by genes alone: How culture transformed human evolution*. Chicago: University of Chicago Press.

Richerson, P. J., Boyd, R., & Henrich, J. (2010). Gene-culture coevolution in the age of genomics. *Proceedings of the National Academy of Sciences, 107*(Suppl. 2), 8985–8992. doi:10.1073/pnas.0914631107

Ravignani, A., Bowling, D. L., & Fitch, W. T. (2014). Chorusing, synchrony, and the evolutionary functions of rhythm. *Frontiers in Psychology, 5*, 1–15. doi:10.3389/fpsyg.2014.01118

Roederer, J. G. (1984). The search for a survival value of music. *Music Perception, 1*(3), 350–356. doi:10.2307/40285265

Ross, C. T., & Richerson, P. J. (2014). New frontiers in the study of human cultural and genetic evolution. *Current Opinion in Genetics and Development, 29*, 103–109. doi:10.1016/j.gde.2014.08.014

Savage, P. E., Brown, S., Sakai, E., & Currie, T. E. (2015). Statistical universals reveal the structures and functions of human music. *Proceedings of the National Academy of Sciences, 112*(29), 8987–8992. doi:10.1073/pnas.1414495112

Scott, B. H., Mishkin, M., & Yin, P. (2012). Monkeys have a limited form of short-term memory in audition. *Proceedings of the National Academy of Sciences, 109*(30), 12237–12241. doi:10.1073/pnas.1209685109

Stivers, T., Enfield, N. J., Brown, P., Englert, C., Hayashi, M., & Heinemann, T., … Levinson, S. C. (2009). Universals and cultural variation in turn-taking in conversation. *Proceedings of the National Academy of Sciences, 106*(26), 10587–10592. doi:10.1073/pnas.0903616106

Tarr, B., Launay, J., & Dunbar, R. I. (2014). Music and social bonding: “Self-other” merging and neurohormonal mechanisms. *Frontiers in Psychology, 5*, 1–10. doi:10.3389/fpsyg.2014.01096

Tomlinson, G. (2015). *A million years of music: The emergence of human modernity*. Cambridge, MA: MIT Press.

Trainor, L. J. (2015). The origins of music in auditory scene analysis and the roles of evolution and culture in musical creation. *Philosophical Transactions of the Royal Society of London B: Biological Sciences, 370*(1664), 1–14. doi:10.1098/rstb.2014.0089

Trainor, L. J., & Cirelli, L. (2015). Rhythm and interpersonal synchrony in early social development. *Annals of the New York Academy of Sciences, 1337*, 45–52. doi:10.1111/nyas.12649

Trehub, S. E. (2003). The developmental origins of musicality. *Nature Neuroscience, 6*(7), 669–673.

van der Schyff, D. (2013). Music, culture and the evolution of the human mind: Looking beyond dichotomies. *Hellenic Journal of Music, Education and Culture, 4*, http://hejmec.eu/journal/index.php/HeJMEC/article/viewFile/40/37.

Wilson, E. O. (2012). *The social conquest of earth.* New York: Norton.

Wrangham, R. (2009). *Catching fire: How cooking made us human.* New York: Basic Books.

Yuan, J., Rosenberg, J. R., & Patel, A. D. (2016). An empirical study of dogs howling to music. In T. Zanto (Ed.), *Proceedings of the 14th International Conference on Music Perception and Cognition* (pp. 557–560). Adelaide: Causal Productions.

Zuk, J., Benjamin, C., Kenyon, A., & Gaab, N. (2014). Behavioral and neural correlates of executive functioning in musicians and non-musicians. *PLoS One, 9*(6), 1–14. doi:10.1371/journal.pone.0099868

CROSS-CULTURAL, CROSS-SPECIES, AND CROSS-DOMAIN STUDIES

6

Cross-Cultural Perspectives on Music and Musicality

Sandra E. Trehub, Judith Becker, and Iain Morley

Universals and Contrasts

Music is universal, transmitted through generations, usually performed in the presence of others, and of extreme antiquity. Although there is no interculturally valid definition of music, and the term *music* is found only in selected cultures, a number of presumptive universals indicate that musicality is a prominent and distinctive characteristic of humankind. People in all cultures engage in activities that we would call music, often in relation to play and everywhere in relation to ritual. They sing, an activity recognized on the basis of context or by cultural consensus as different from speech. They also have some form of instrumental music, however rudimentary.

Music making is necessarily a cultural performance because conventions about the structure of music, instrumentation, context of performance, and meaning are learned. It is a system of communication transmitted through ongoing transgenerational interaction. Music making is a social performance, even when performed or listened to alone. The solitary performer often has an audience in mind. For example, when music students at North American universities practice alone in hermetic practice rooms, they are often imagining an audience. Solitary music making for personal pleasure is likely to evoke memories of learning the piece (when, where, and from whom) and previous experiences of playing and hearing it. Musical pieces, like performers, are saturated with contextual, social memory.

In the Western industrialized world, music listening is commonly solitary, and iPods, earbuds, and headphones are ubiquitous. The listener seems alone, but the music has an implied social context such as a symphony orchestra, a rapper before a live audience, a DJ at a club, or imagined friends. Even a solitary basement musician producing a multitrack composite performance is likely to invoke or imitate a social context.

An exception to the truism that music making is a social performance is the deliberate use of music to shut out the world and create an alternate reality. At times, the iPod-solitary-listening reality can produce something close to a deeply spiritual or mystical experience.

I've walked that way for—I don't know how many years … and it's very boring, so having the music makes me see things that I would see everyday in a kind of new way—like a leaf falling or something. It might be like, "Wow, a leaf has fallen!" (Herbert, 2011, p. 58)

Unlike solitary, self-directed performances, music performances are political in the sense that they are situated and embedded within structures of power and influence. Where they are performed, by whom, and for whom reveal a great deal about the cultural and social status of the performer and the performance:

Who has creative and economic control?

Which people and assumptions dictate the terms, not only of the creative work produced by artists but also of the critical and scholarly writing about it?

What are the ways that the race, gender, class, sexual orientation, and generation of an artist shape his or her professional experiences and aesthetic choices? (Mahon, 2014, p. 328)

Because musical performances are socially and culturally situated, they come to be ethically saturated as well. Our deepest values may be implied by participation in a particular genre of music such as a hootenanny (populist), heavy metal concert (male machismo), rave (youth culture plus drugs), or European classical concert (upper-middle-class values and lifestyle). There are exceptions, of course, but the stereotypes are more likely than not to apply to any given listener at a musical event. As Wittgenstein (2001) claimed, "Ethics and aesthetics are one and the same."

Music is believed to affect our emotions, to involve some kind of arousal, ranging all the way from mild pleasure or displeasure to profoundly transformed states of consciousness such as trancing (Becker, 2004; Juslin & Sloboda, 2010). Nevertheless, the scientific study of music and emotion has focused primarily on solitary listening, which is but one of many modes of listening across cultures (Becker, 2010). Cross-cultural research in this domain is limited and largely restricted to listeners' categorization of intended and felt emotions by means of a small set of emotion words (e.g., *happy*, *sad*, *fearful*, *angry*) or facial expressions (Balkwill & Thompson, 1999; Balkwill, Thompson, & Matsunaga, 2003; Fritz et al., 2009). Even within a culture, listeners' perception of emotion differs markedly from their felt emotion, and the latter varies considerably with differences in listening context, experimenter- versus self-selected music, and number of response alternatives (Liljeström, Juslin, & Västfäll, 2013; Salimpoor, Benovoy, Longo, Cooperstock, & Zatorre, 2009; Zentner, Grandjean, & Scherer, 2008).

What Is Called *Music*?

Many languages, including most North American Indian languages (Leon Portilla, 2007) and several African languages such as that of the Basongye of Zaire (Merriam, 1964) and the Tiv of Nigeria (Keil, 1979) have names for individual genres of music but no term that includes all of their musical genres. An important commonality among musical systems of the world is the conjoining of music and ritual, often including dance and speech. The term for music frequently includes other activities. For example, Sanskrit *sangīta*, Thai *wai khruu* (Bohlman, 2002), and *nkwa* of the Igbo of Nigeria (Balkwill & Thompson, 1999) encompass music and dance as facets of the same activity, making no clear terminological distinction among them:

> Honest observers are hard pressed to find a single indigenous group in Africa that has a term congruent to the usual western notion of "music." There are terms for more specific acts like singing, playing instruments, and more broadly performing (dance, games, music); but the isolation of musical sound from other arts proves a western abstraction, of which we should be aware when we approach the study of performance in Africa. (Stone, 1982, p. 7)

The synonymity of music and dance may be recognized by the participants, as in the ritual dance of the Maring of Papua New Guinea at the Kaiku festival, in which the dance and ritual pledge are interdependent in transforming the natural order (Rappaport, 1999), and the Candomblé Afro-Brazilian religion, in which music encompasses what we recognize as music and religion (Bohlman, 2002).

In contrast, the chanting of religious texts in Judaism (*ta'ameh ha-mikrah*) and Islam (*qirā'ah*) would be classified as musical by external observers but as recitation by the practitioners (Bohlman, 2002). As with the preceding examples, however, the musical and ritual elements are mutually dependent facets of the performative activity.

Statistical Universals

No structural aspects of music are implemented identically in all known musical systems. Nonetheless, some characteristics have a wide distribution globally and are considered statistical universals (Brown & Jordania, 2013; Savage, Brown, Sakai, & Currie, 2015).

Perhaps foremost among statistical universals is the idea of musicality itself—that everyone has the capacity or potential for engaging in a range of musical activities. Moreover, one does not just sing; one sings *something*, and that *something* can be identified in one way or another (Nettl, 2000). What is performed has an identity, a name or moniker that sets it apart from other musical acts and from any other kind of utterance. For example, what is sung may be considered a unit in a ritual, the creation of someone, performed by someone, or performed at some particular place. Another statistical universal is the presence of musical units or phrases, identifiable by repetition or some form of redundancy. Repetition of units or phrases, whether identical or with variation, is widespread in musical systems across the globe (Huron, 2006; Margulis, 2014). Musical units or phrases can be transposed and still retain their identity (Brown & Jordania, 2013). The musical phrase, or combination of phrases, frequently consists of unequal intervals, often major seconds and thirds, and commonly combining to produce pentatonic scales (Brown & Jordania, 2013; Nettl, 2005; Trehub, 2000). While the idea of a scale is by no means universal, the analytical construction of scale indicates that most musical systems in the world do not exceed seven notes within an octave.

The sense of octave equivalence is found wherever men, women, and children sing together in unison. In Bali, Indonesia, the sense of octave equivalence is underlined by being deliberately undermined. Equivalent pairs of keys of Balinese bronze xylophones are deliberately tuned so that the octave across each pair is "out of tune" or "stretched" to produce the desired beats when both keys are struck simultaneously (Tenzner, 1991). The use of beats for aesthetic purposes in Bali and other musical cultures (Jordania, 2006) is inconsistent with the claim that beats, which are typically avoided in Western music, are innately unpleasant (Trainor & Heinmiller, 1998; Zentner & Kagan, 1998).

Most musical systems are predominantly isorhythmic, which means that the same rhythmic configuration, once established, tends to continue throughout. A given beat within an isorhythmic configuration is commonly subdivided into two or three units (Brown & Jordania, 2013).

Certain ideas about music are also frequent across the globe. One of these is the belief that one's own system of music is natural. To consider one's music system as natural is to endow it with a kind of necessity or power that it might not claim otherwise. But naturalness can be located in many different realms. Naturalness is often identified in Western music, as well as in India, the Middle East, and parts of sub-Saharan Africa, on the basis of acoustics—the building of scales on the overtone structure of any single tone (Gill & Purves, 2009; Mukhergee, 2004).

A different kind of link with nature, and therefore with spiritual power, is found in ideas about music of the Kaluli, an ethnic group of the highlands of Papua New Guinea. The Kaluli believe that human composers are reworking birdsongs. Their most important ceremonial song genre consists of four descending tones in imitation of the call of the *muni* bird, which can be heard as the voices of their deceased ancestors (Feld, 2012).

Another system of ideas concerning the naturalness of musical systems is in the lore surrounding Central Javanese court gamelan ensembles. Indic conceptions of power were imported into Java in the first millennium CE and were drawn on by the sultans of central Java to enhance their status. According to Indic cosmology, female energy, *sakti*, which permeates the universe, is found everywhere for all time, and is morally neutral. While we would consider this system of ideas a construct, it was a description of the natural world and the universe for its practitioners. To be an effective ruler in Java, the male sultan, through meditation, added to his effectiveness by attracting and assimilating cosmic *sakti*. In modern Indonesian, with its Indic past receding, *sakti* is translated as power. Any gamelan ensemble, but especially old palace gamelans, are believed to possess or embody *sakti*, in part because they are made of bronze, an alloy of copper and tin that comes from the earth and must be transformed by fire. In Java, blacksmiths and gong makers are believed to embody *sakti*, a power needed to transform substances of the earth, through fire, into wondrous objects such as a *kris* (dagger) or a gamelan instrument. A large palace ensemble always includes two large gongs, one female and one male. The female gong, which is considered older and more primal than the male gong, is always used to end a piece. The division of a pair of male and female instruments extends to other paired instruments, the bronze xylophones of the ensemble as well as the gongs. Thus, the naturalness and power of a gamelan ensemble is attributed in part to its metaphoric iconicity with cosmic power (Becker, 1988).

Scholarly and Everyday Ideas about the Origins of Music

Music is often ascribed to spiritual or supernatural aspects of the natural world. In the ancient and medieval worlds, East and West, music was considered a primal source of the phenomenal world. Pythagoras and Plato formulated the occidental theory of cosmic music, the relations between music intervals and numbers, the character of scales, the "harmony of the spheres," and their influence on nature and society (James, 1993). Scholars in ancient China created a complex system of relations among social phenomena, elements of the calendar, and relations between tones. "The origin of music is in the very remote past. It was … rooted in the Grand Unity. The Grand Unity gave birth to … heaven and earth" (ca. 239 BCE; DeWoskin, 1982).

In ancient India, sound was considered sacred in origin (Rowell, 1992), and Hindu aspiring for liberation or for association with a chosen deity performed a "'sonic act" informed by a "sonic theology" (Sullivan, 1997). Ancient theories of the relation of music to the cosmos have continuing influence, but commonplace notions about music and the supernatural also abound, such as Native American and African beliefs that dreams or visions are the sources of specific songs (Boas, 1888/1964; McAllester, 1954; Nettl, 1989; Owomoyele, 2002). A related and widespread belief is that music has profound effects on our minds and bodies (see figure 6.1) and can ameliorate physical and mental problems (DeNora, 2013; Owomoyele, 2002; Sacks, 2008; Sullivan, 1997).

The Antiquity of Musical Activities

A deep history to musical activities is suggested by their existence in all known human societies. They are key components of ritual activities, cosmologies, and the management of social relationships—all core elements of maintaining human groups. This ubiquity suggests shared mechanisms by which musical activities inevitably emerge in human societies despite divergent histories. Alternatively, musical activities could be part of a shared history of human groups that predates their divergence from one another.

The earliest direct evidence of musical activities is necessarily limited to musical instrument use. In many musical traditions, however, musical behaviors need not involve instrumentation, and much instrumentation would not preserve archaeologically. For example, among the Blackfoot and Sioux Native Americans (McAllester, 1954; Nettl, 1992), the

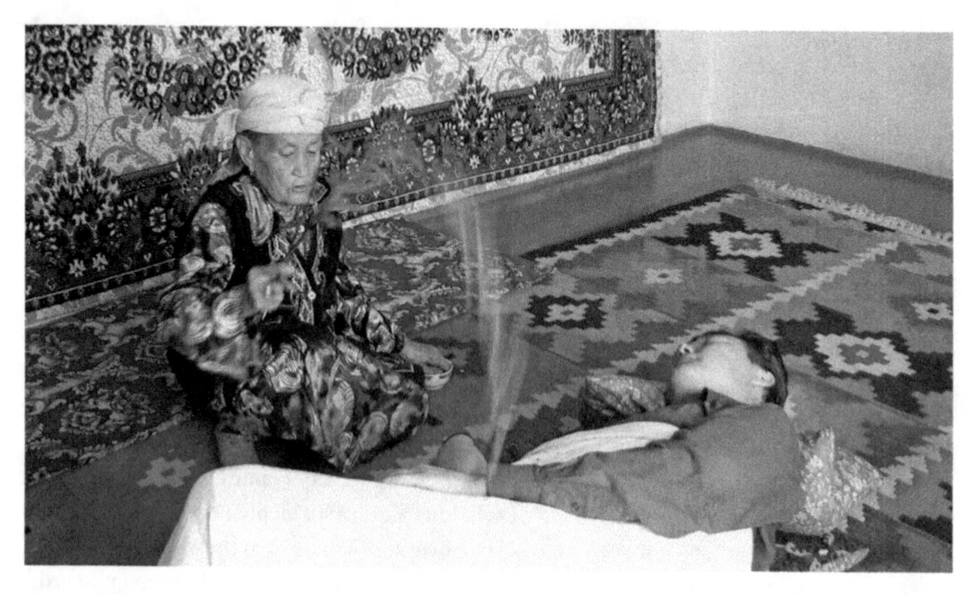

Figure 6.1
Karakalpak woman performing a trance-inducing ritual to speed recovery from illness (Autonomous Republic of Karakalpakstan, Uzbekistan). Photographer: Frédéric Léotar.

Aka and Mbuti African pygmies (Ichikawa, 1999; Locke, 1996), the Yupik of southwest Alaska (Johnston, 1989), and the Pintupi-speaking Australian Aborigines (Breen, 1994), the melodic content of music is largely provided by the voice, with instrumentation primarily being percussive. When instruments are used, they are often made from natural, biodegradable materials that would not preserve archaeologically under most conditions and are supplemental to the use of the body for producing melodic and rhythmic content.

This means that archaeological evidence for musical activities, in the form of preserved instruments, is unlikely to represent the earliest musical activities or their extent. Nevertheless, it is clear that the earliest populations of modern humans (*Homo sapiens*) to enter Europe, more than 40,000 years ago, were engaged in musical activities. The oldest known pipes or flutes come from the sites of Geissenklösterle, Hohle Fels, and Vogelherd in the Swabian Jura of Germany (Conard, Malina, & Münzel, 2009; Conard, Malina, Münzel, & Seeberger, 2004; d'Errico et al., 2003; Hahn & Münzel, 1995). These instruments are made from swan bone, vulture bone, and mammoth ivory. All come from layers associated with Aurignacian technologies, among the earliest tool types produced by modern humans in Europe, and, in the case of the Geissenklösterle examples, the layers have recently been redated to 43,150 to 39,370 calibrated (or calendar) years before present (Higham et al., 2012). The bird bone examples are made from modified radius and ulna bones (lower arm/wing bones). Bird bones are light and naturally hollow, so are relatively easily worked.

The lower wing bones of large birds, such as vultures, eagles, swans, and some geese, are of sufficient size to function as multipitch pipes. These have had the ends (epiphyses) removed and finger holes prepared by thinning the bone surface and piercing or boring a hole with a sharp tool (Buisson, 1990; Lawson & d'Errico, 2002). For the mammoth ivory examples, a more laborious process was used to produce a somewhat larger equivalent to the bird bones. The ivory was soaked to separate its laminar layers, cut in half along its length, the central core removed, and the two halves resealed with airtight resin glue, having had finger holes bored along the length (Conard et al., 2004, 2009).

Further examples associated with Aurignacian technologies are known from Spy, Belgium (what may be panpipes; Otte, 1979; Scothern, 1992), Abri Blanchard (Harrold, 1988; Jelinek, 1990), and Isturitz, France (Buisson, 1990; Lawson & d'Errico, 2002; Passemard, 1923, 1944), which suggest widespread musical activities at that time. Given the sophistication of the production techniques and the association with some of the earliest *H. sapiens* populations in Europe, it seems likely that these represent part of a behavioral repertoire that predates their arrival in Europe. These instruments constitute the first of an extensive record of pipes and other instruments from throughout the Upper Paleolithic period (from around 45,000 years ago until around 12,000 years ago, at the end of the last ice age). In several cases (e.g., Isturitz, Mas d'Azil), these instruments were found at sites that were focal points for large gatherings at particular times of the year (Bahn, 1983; Scothern, 1992).

As noted, among many traditional hunter-gatherer societies, melodic instruments were a small part of the sound-producing traditions, with the melody being primarily vocal and instruments being primarily percussive. Were this the case among Paleolithic human populations as well, then the record of bone pipes represents a small proportion of the instruments produced and used in a small proportion of musical activities.

Although pipes are perhaps the most readily recognized and best-preserved instruments from Upper Paleolithic contexts, other possible sound producers are known from this period. These include rasps (scraped idiophones; Dauvois, 1989, 1999; Huyge, 1990, 1991; Kuhn & Stiner, 1998), bullroarers ("free aerophones"; Bahn, 1983; Dauvois, 1989), struck bones (osseophones; Bibikov, 1978; Hadingham, 1980), rocks, and stalactites and stalagmites (lithophones; Dams, 1984, 1985; Glory, 1964, 1965; Glory, Vaultier, & Farinha Dos Santos, 1965). All are well-known forms of sound production in ethnographic contexts. For example, bullroarers have been used in Australia, Africa, North America, New Guinea, and among the Maori of New Zealand and the Sami of Scandinavia. Their use often has sacred and religious associations, and their powerful religious roles in ancient Greece are well documented (Haddon, 1898; Harding, 1973; Lang, 1884; Maioli, 1991; Montagu, 2007). Similarly, the use of "rock gongs" or "singing stones" has been documented in historical and contemporary contexts in the Canary Islands (Álvarez & Siemens, 1988), Sweden (Henschen-Nyman, 1988), India (Boivin, 2004; Boivin, Brumm, Lewis, Robinson, & Korisettar, 2007), Bolivia, Southeast Asia, Australia, and Africa (Fagg, 1997;

Montagu, 2007). Rock gongs produce a single tone or multiple tones depending on where they are struck. As expected from their extensive geographical and temporal range, their uses are varied, including signaling (the sounds can be audible for several miles) and accompanying ritual, singing, and dancing (Boivin, 2004; Boivin et al., 2007; Montagu, 2007).

Music and Social Organization

Conjoining music, dance, and ritual language within an event that addresses the existential concerns of the community is the most universally valued of musical activities. The entrainment of human bodies in a group transforms a musical event such as a ceremony or a ritual, even a small communal or family gathering, into an enhancer of community spirit. Making music together builds a community together, which may be the most adaptive and evolutionarily significant aspect of musical experience worldwide (Benzon, 2001; Dunbar, 2012; Freeman, 2000; Loersch & Arbuckle, 2013).

Among the Pintupi-speaking Aborigines of Australia (Myers, 1999), musical activities are an important part of interactions with neighboring groups during the dry seasons (Bahuchet, 1999; Kehoe, 1999). The Yupik of southwest Alaska achieve "socially lubricating" functions through music whose lyrics and dance actions diffuse tension through teasing or comic means (Johnston, 1989). Both of these peoples, hunter-gatherers in very different environments and separated by many thousands of miles, believe that their musical activities directly influence the world around them, that they have come from the land, and that they are akin to the other fauna in their environment (Ichikawa, 1999; Johnston, 1989; Myers, 1999; Nettl, 1992).

Although there is variation in the permissible roles of individuals and genders in musical activities, the activities are often inclusive, with little distinction between performers and audience. All who are present participate in the activity in some capacity. For the Aka and Mbuti equatorial African pygmies, music is a communal, cooperative activity, with no individual considered a specialist musician (Ichikawa, 1999). Songs related to hunting are typically performed by men (Bahuchet, 1999), and performances related to rites of passage are typically performed by women (Ichikawa, 1999). For all other music and dance performances, persons of any age or gender may participate (Turino, 1992). Similarly, for the Blackfoot and Sioux Native Americans, gender roles and specific responsibilities were delineated in music used for ritual purposes, but a wide cross-section of the community participated in the second-most-common use, social dancing (Nettl, 1992).

Songs also function as a repository of knowledge and cultural values that can be transmitted across persons and generations, as with the Alaskan Yupik, the Australian Pintupi-speaking Aborigines, and many other groups (Johnston, 1989; Myers, 1999). In rural Vietnam, for example, several traditional songs provide detailed guidance for planting and harvesting crops (Cong-Huyen-Ton-Nu, 1976). The "Alphabet Song" and counting songs in many cultures continue to play an important didactic role in childhood.

Music and Group Cohesion

Although it is likely that the social aspects of musical engagement account for the perpetuation of music across cultures and millennia (Dunbar, 2012; Loersch & Arbuckle, 2013), these aspects have received considerably less attention in psychology and neuroscience than the perceptual, cognitive, and emotional processes of solitary listeners and the neural correlates of these processes (Deutsch, 2013; Juslin & Sloboda, 2013; Patel, 2008; Zatorre & Salimpoor, 2013). Progress in documenting the everyday music listening habits of Western adults (DeNora, 2013; Herbert, 2011; Juslin & Laukka, 2004; North, Hargreaves, & Hargreaves, 2004) has not been matched by progress in delineating the consequences of various listening contexts for arousal, emotion regulation, and well-being. Even when emotional responsiveness is of principal interest, music that evokes episodic memories or extramusical associations is commonly excluded (Salimpoor et al., 2009) despite the impact of such associations on listeners' emotional experience.

Musical Caregiving

Music is ubiquitous in caregiving. Caregivers across cultures sing to infants and have done so from time immemorial (Trehub & Trainor, 1998). Their lullabies to soothe infants and induce sleep are usually melodically, rhythmically, and lyrically simple (Brown, 1980; Hawes, 1974; Sakata, 1987; Unyk, Trehub, Trainor, & Schellenberg, 1992), and they are readily identifiable across cultures (Trehub, Unyk, & Trainor, 1993a). Lullabies are commonly performed with rhythmic movement (e.g., rocking) and touch (Ayres, 1973) so the child's introduction to music is multimodal even in cultures where solitary listening is the norm.

Lullabies are the songs of choice in caregiving contexts with almost constant physical contact between caregiver and infant, which often involves limited face-to-face contact (Trehub & Gudmundsdottir, 2015). High levels of contact maximize infants' comfort and safety in the challenging conditions of the developing world (Hrdy, 2009; LeVine, 1988). An exception to this practice involves the Bayaka pygmies of Central Africa, where a mother typically responds to a howling baby by yodeling even louder while rhythmically patting the baby's back (Lewis, 2013). Lullaby lyrics in most cultures express praise and affection for infants, but at times, stressed caregivers use the privacy of the dyadic context and noncomprehending listener to give lyrical expression to their personal difficulties (Masuyama, 1989; Trehub & Trainor, 1998), with potentially soothing consequences for singers and listeners.

Middle-class European and American mothers, having the luxury of focusing on intellectual and social stimulation, place infants in secure devices (e.g., seats, swings) and engage in face-to-face play (Richman, Miller, & Solomon, 1988). Their singing and melodious speech enable them to "keep in touch" with infants (Falk, 2004). The songs of choice in such contexts are lively play songs, with lullabies commonly reserved for bedtime routines (Trehub & Trainor, 1998). Regardless of caregivers' choice of songs, their manner of performance, featuring higher-than-usual pitch level, slower-than-usual tempo, warm vocal tone, and smiling, is discernible to naive listeners within and across cultures (Trainor, 1996; Trainor, Clark, Huntley, & Adams, 1997; Trehub, Plantinga, & Russo, 2016; Trehub et al., 1997; Trehub, Unyk, & Trainor, 1993b). The maternal performing style is also individually distinctive. Mothers sing a few songs repeatedly and in a highly stereotyped manner—typically with identical pitch level and tempo (Bergeson & Trehub, 2002). The result is a distinctive dyadic ritual that eases the burdens of caregiving, enhances infants' sense of security, and cements caregiver-infant bonds.

Aspects of the maternal singing style are similar to the intimate, group-involving style of singing that Lomax (1968) identified in highly integrated communities around the world. That style has been linked to the blurring of boundaries between the self and others (Pantaleoni, 1985; Zuckerkandl, 1973). Mothers' face-to-face singing modulates the arousal levels of contented infants (Shenfield, Trehub, & Nakata, 2003) and is more effective than maternal speech at reducing the high arousal of distressed infants (Ghazban, 2013). Even with auditory-only materials from unfamiliar mothers, songs recorded in an infant's presence engage infants more readily than comparable recordings that lack an infant audience (Masataka, 1999; Trainor, 1996). In fact, infants maintain their composure for roughly twice as long when listening to audio recordings of infant-directed singing (unfamiliar song, unfamiliar language) than to comparable recordings of speech (Corbeil, Trehub, & Peretz, 2016).

Music in Communal Contexts

Music listening evokes pleasure even in the solitary, unnatural context of the laboratory or scanner (Blood & Zatorre, 2001; Janata, Tomic, & Haberman, 2012; Zatorre & Salimpoor, 2013). The pleasure and social consequences are amplified, however, when music is experienced with family or peers (Boer & Abubakar, 2014) and even more so when music features active involvement as in singing, drumming, or dancing (Dunbar, Kaskatis, MacDonald, & Barra, 2012). Communal musical experiences and collective rituals in general enhance social cohesion and prosocial behavior (Dunbar, 2012; Sosis & Ruffle, 2004; Whitehouse, 2004), presumably through jointly experienced elation and synchronized action (Durkheim, 1995; Freeman, 2000; Janata & Parsons, 2013).

Synchronous Arousal

Heightened arousal strengthens social bonds among those who engage in collaborative rituals regardless of the presence or absence of music, positive affect, or overt physical coordination. For example, extreme rituals that include beating, mutilation, or excessive exertion intensify social bonds among participants and empathic observers, as reflected in the extent of their identification, cooperation, and coordinated physiological activity (Atkinson & Whitehouse, 2011; Konvalinka et al., 2011; Xygalatas et al., 2013). In the annual fire-walking ritual of a small rural village in Spain, fire walkers walk barefoot across glowing red coals as friends, relatives, and thousands of visiting spectators look on. The arousal of the fire walkers' friends and relatives becomes synchronized with that of the fire walkers, as reflected in the microstructure of their cardiac activity (Konvalinka et al., 2011). The arousal of unrelated spectators, although elevated, does not exhibit the signature cardiac features of the fire walkers and related observers. Incidentally, music may well prime participants (fire walkers and audience) for the main event because it is preceded by the fire walkers' dancing around the glowing coals, and each fire walker is summoned to the challenge by a trumpet call.

Communal musical experiences are typically of moderate emotional intensity, but examples of intense performances with positive valence are common among the Bayaka pygmies of the Central African Republic (Lewis, 2013). They engage in extended bouts of (Aka) vigorous singing, dancing, and storytelling, in contrast to the reported low frequency of high-intensity rituals and high frequency of low-intensity rituals (Atkinson & Whitehouse, 2011; Whitehouse, 2004). For enjoyable musical rituals rather than punitive nonmusical rituals, intensity may be compatible with high frequency. For the Bayaka pygmies, musical activities play a critical role in the maintenance and intergenerational transmission of their egalitarian and cooperative values and practices (Lewis, 2013).

High-intensity rituals, including loud, long-sustained singing and trancing are also found in charismatic Christian congregations, especially Pentecostals. Originally from the United States, Pentecostal congregations, with hours-long musical services, are now found worldwide (Anderson, 2013).

Synchronous Action

Although synchronous action is not a prerequisite for the formation or maintenance of social bonds, synchronous activity in groups of Buddhist chanters and Hindu devotional singers results in greater identification, trust, and generosity among participants than does the nonsynchronized activity of cross-country running groups (Fischer, Callander, Reddish, & Bulbulia, 2013). The lesser prosocial consequences of cross-country running are inconsistent with the contention that social bonding arises from the energetic aspects of music making (Dunbar et al., 2012).

Prosocial consequences are evident even when synchronous activity occurs without music. For example, three initially unfamiliar individuals who walk in step evaluate one another more favorably than similar threesomes who walk together but not in synchrony (Wiltermuth & Heath, 2009). According to historian William McNeill (1995), the military practice of marching in step no longer has the significance it once had for battle formation, but it is maintained because of the sense of unity it fosters among marchers from diverse political, religious, and socioeconomic backgrounds.

The prosocial consequences of synchronous activity are evident in early childhood. After four-year-old children engage in cooperative play that includes singing and playing percussion instruments, they exhibit more prosocial behavior than four-year-olds who engage in similar cooperative play without music (Kirschner & Tomasello, 2010). Infants move rhythmically but not synchronously to rhythmic music (Zentner & Eerola, 2010). By seven months of age, however, they are sensitive to the metrical patterns of movement experienced while being bounced to music but not to comparable bouncing that they simply observe (Phillips-Silver & Trainor, 2005). By fourteen months of age, they are also sensitive to movement patterns that they observe. After watching an adult bouncing synchronously with their own bouncing, they offer more assistance to that adult than to one who engaged in nonsynchronous bouncing (Cirelli, Einarson, & Trainor, 2014).

Imitation

Synchronous action is one route to social bonding and prosocial behavior in childhood and adulthood. Imitation is another. An important pillar of our musicality and our ritual propensity in general is our ability to imitate or provide a high-fidelity copy of the vocal and bodily actions of others (Donald, 1991; Merker, 2009). Ordinary interpersonal interaction often involves unconscious mimicry or low-fidelity copying that is prompted by prosocial attitudes of the mimic and has prosocial consequences for the person mimicked (van Baaren, Decety, Dijksterhuis, van der Leij, & van Leeuwen, 2011). Such mimicry is presumed to arise from our motivation to conform or be like others (Meltzoff, 2007; Merker, 2009; Tomasello, 2003). Similar affiliative motivations seem to underlie young children's production of exact imitations of the actions of others (Over & Carpenter, 2013) and their enhanced responsiveness to individuals who imitate their actions (Agnetta & Rochat, 2004). For example, eighteen-month-old toddlers initiate more play with adults who imitate their actions than with those who interact without action imitation (Fawcett & Liszkowski, 2012). The motivation to act like others reaches a high plane in musical and ritual activity where the form is primary and the immediate functions are secondary or inconsequential (Merker, 2009).

Music, Meaning, and Communication

For members of many small communities, past and present, musical rituals have meanings that are transparent within the native community but opaque to others. There may be different levels of meaning as well, for example, holistic meanings about particular rituals, like a call for rain or for animals to present themselves for the hunt, and specific meanings linked to individual elements, like the gender of a gong. Music theorists and scientists consider these meanings extramusical, according them lesser importance than intramusical meanings involving the internal structure of music. However, these so-called extramusical meanings can be highly conventionalized, unlike the idiosyncratic extramusical meanings (e.g., "our song") that are typically subsumed under this category. Although music lacks communicative specificity in comparison with language, its power sometimes exceeds that of language in social, emotional, and spiritual domains. In live contexts, moreover, music can communicate to greater numbers of individuals and over greater distances than language can.

Laboratory studies of isolated listeners and music makers have yielded important insights into sensorimotor and cognitive skills and their neural underpinnings, but they have illuminated little about the significance of music for individuals, peer groups, and communities, and for cultural evolution. Some experimental anthropologists, evolutionary biologists, and evolutionary psychologists are beginning to make inroads into these issues by using precise measurement techniques with naturally occurring groups of music makers and listeners (Dunbar et al., 2012; Fischer et al., 2013).

Finally, the commonalities and differences in musical forms and functions across cultures suggest new directions for music cognition, ethnomusicology, and neuroscience and a pivot away from the predominant scientific focus on instrumental music in the Western European tradition. Empirical and field studies of ecologically valid music in diverse settings can complement traditional laboratory research, generating richer conceptions of human musicality than those currently available (see chapter 1, this volume). Social cognition (Seyfarth & Cheney, 2014) and iterated (person-to-person) learning (Kirby, Griffiths, & Smith, 2014) are considered to make vital contributions to the evolution of language. Their role may be even greater in the evolution of music (Trehub, 2015).

Acknowledgments

The preparation of this chapter was assisted by funding from the Natural Sciences and Engineering Council of Canada, the Social Sciences and Humanities Research Council of Canada, and Advancing Interdisciplinary Research in Singing (AIRS) to S.E.T.

References

Agnetta, B., & Rochat, P. (2004). Imitative games by 9-, 14-, and 18-month old infants. *Infancy*, *6*(1), 1–36. doi:10.1207/s15327078in0601_1

Álvarez, R., & Siemens, L. (1988). The lithophonic use of large natural rocks in the prehistoric Canary Islands. In E. Hickman & D. Hughes (Eds.), *The archaeology of early music cultures: Third International Meeting of the ICTM Study Group on Music Archaeology* (pp. 1–10). Bonn, Germany: Verlag für systematische Musikwissenschaft.

Anderson, A. H. (2013). *To the ends of the earth: Pentecostalism and the transformation of world Christianity*. Oxford: Oxford University Press.

Atkinson, Q. D., & Whitehouse, H. (2011). The cultural morphospace of ritual form: Examining modes of religiosity cross-culturally. *Evolution and Human Behavior*, *32*(1), 50–62. doi:10.1016/j.evolhumbehav.2010.09.002

Ayres, B. (1973). Effects of infant carrying practices on rhythm in music. *Ethos*, *1*(4), 381–404. doi:10.1525/eth.1973.1.4.02a00020

Bahn, P. (1983). Late Pleistocene economies in the French Pyrenees. In G. Bailey (Ed.), *Hunter-gatherer economy in prehistory: A European perspective* (pp. 167–185). Cambridge: Cambridge University Press.

Bahuchet, S. (1999). Aka Pygmies. In R. B. Lee & R. Daly (Eds.), *The Cambridge encyclopedia of hunters and gatherers* (pp. 190–194). Cambridge: Cambridge University Press.

Balkwill, L.-L., & Thompson, W. B. (1999). A cross-cultural investigation of emotion in music: Psychophysical and cultural cues. *Music Perception*, *17*(1), 43–64. doi:10.2307/40285811

Balkwill, L.-L., Thompson, W. B., & Matsunaga, R. (2003). Recognition of emotion in Japanese, Western, and Hindustani music by Japanese listeners. *Japanese Psychological Research*, *46*(4), 337–349. doi:10.1111/j.1468-5584.2004.00265.x

Becker, J. (1988). Earth, fire, *sakti*, and the Javanese gamelan. *Ethnomusicology*, *32*(3), 388–391. doi:10.2307/851938

Becker, J. (2004). *Deep listeners*. Bloomington: University of Indiana Press.

Becker, J. (2010). Exploring the habitus of listening: Anthropological perspectives. In P. N. Juslin & J. A. Sloboda (Eds.), *Handbook of music and emotion: Theory, research, applications* (pp. 127–157). Oxford: Oxford University Press.

Benzon, W. L. (2001). *Beethoven's anvil: Music in mind and culture*. New York: Basic Books.

Bergeson, T. R., & Trehub, S. E. (2002). Absolute pitch and tempo in mothers' songs to infants. *Psychological Science*, *13*(1), 72–75. doi:10.1111/1467-9280.00413

Bibikov, S. (1978). A Stone Age orchestra. In D. Hunter & P. Whitten (Eds.), *Readings in physical anthropology and archaeology* (pp. 134–148). London: Harper & Row.

Blood, A. J., & Zatorre, R. J. (2001). Intensely pleasurable responses to music correlate with activity in brain regions implicated in reward and emotion. *Proceedings of the National Academy of Sciences*, *98*(20), 11818–11823. doi:10.1073/pnas.191355898

Boas, F. (1964). *The central Eskimo*. Washington, DC: Smithsonian Institution. (Original work published 1888.)

Boer, B., & Abubakar, A. (2014). Music listening in families and peer groups: Benefits for young people's social cohesion and emotional well-being across four cultures. *Frontiers in Psychology*, *5*, 1–15. doi:10.3389/fpsyg.2014.00392

Bohlman, P. (2002). *World music*. Oxford: Oxford University Press.

Boivin, N. (2004). Rock art and rock music: Petroglyphs of the South Indian Neolithic. *Antiquity*, *78*(299), 38–53. doi:10.1017/S0003598X00092917

Boivin, N., Brumm, A., Lewis, H., Robinson, D., & Korisettar, R. (2007). Sensual, material, and technological understanding: Exploring prehistoric soundscapes in south India. *Journal of the Royal Anthropological Institute*, *13*(2), 267–294. doi:10.1111/j.1467-9655.2007.00428.x

Breen, M. (1994). I have a dreamtime: Aboriginal music and black rights in Australia. In S. Broughton, M. Ellingham, D. Muddyman, & R. Trillo (Eds.), *World music: The rough guide* (pp. 655–662). London: Rough Guides.

Brown, M. J. E. (1980). Lullaby. In S. Sadie (Ed.), *The new Grove dictionary of music and musicians* (pp. 313–314). London: Macmillan.

Brown, S., & Jordania, J. (2013). Universals in the world's musics. *Psychology of Music*, *41*(2), 229–248. doi:10.1177/0305735611425896

Buisson, D. (1990). Les flûtes paléolithiques d'Isturitz [Paleolithic flutes of Isturitz]. *Bulletin de la Société Préhistorique Française*, *87*(10), 420–433. doi:10.3406/bspf.1990.9925

Cirelli, L. K., Einarson, K. M., & Trainor, L. J. (2014). Interpersonal synchrony increases prosocial behavior in infants. *Developmental Science*, *17*(6), 1003–1011. doi:10.1111/desc.12193

Conard, N., Malina, M., & Münzel, S. (2009). New flutes document the earliest musical tradition in southwestern Germany. *Nature, 460*, 737–740. doi:10.1038/nature08169

Conard, N., Malina, M., Münzel, S., & Seeberger, F. (2004). Eine mammutelfenbeinflöte aus dem Aurignacien des Geissenklösterle: Neue belege für eine musikalische tradition im Frühen Jungpaläolithikum auf der Schwäbischen Alb [A mammoth ivory flute from the Aurignacian goats of Geissenklösterle: New evidence of musical traditions in Early Upper Palaeolithic in the Swabian Alb]. *Archäologisches Korrespondenzblatt, 34*, 447–462.

Cong-Huyen-Ton-Nu, N. T. (1976). The function of folk songs in Vietnam. In J. Blacking (Ed.), *The performing arts: Music and dance* (pp. 141–151). The Hague: Mouton.

Corbeil, M., Trehub, S. E., & Peretz, I. (2016). Singing delays the onset of infant distress. *Infancy, 21*(3), 373–391. doi:10.1111/infa.12114

Dams, L. (1984). Preliminary findings et the "Organ Sanctuary" in the cave of Nerja, Malaga, Spain. *Oxford Journal of Archaeology, 3*(1), 1–14. doi:10.1111/j.1468-0092.1984.tb00112.x

Dams, L. (1985). Palaeolithic lithophones: Descriptions and comparisons. *Oxford Journal of Archaeology, 4*(1), 31–46. doi:10.1111/j.1468-0092.1985.tb00229.x

Dauvois, M. (1989). Son et musique Paléolithiques [Paleolithic sound and music]. *Dossiers d'Archéologie, 142*, 2–11.

Dauvois, M. (1999). Mesures acoustiques et témoins sonores osseux Paléolithiques [Acoustic measurements of Paleolithic bone sounds]. In H. Camps-Febrer (Ed.), *Préhistoire d'os, recuil d'études sur l'industrie osseuse préhistorique offert à Mme Henriette Camps-Febrer*. Aix-en-Provence: Publications de l'Université de Provence.

DeNora, T. (2013). *Music asylums: Wellbeing through music in everyday life*. Surrey: Ashgate.

d'Errico, F., Henshilwood, C., Lawson, G., Vanhaeren, M., Tillier, A.-M., Soressi, M., ... Julien, M. (2003). Archaeological evidence for the emergence of language, symbolism and music: An alternative multidisciplinary perspective. *Journal of World Prehistory, 17*, 1–70. doi:10.1023/A:1023980201043

Deutsch, D. (Ed.). (2013). *The psychology of music* (3rd ed.). San Diego, CA: Elsevier.

DeWoskin, K. (1982). *A song for one or two: Music and the concept of art in early China* (Michigan Papers in Chinese Studies, No. 42). Ann Arbor, MI.

Donald, M. (1991). *Origins of the modern mind.* Cambridge, MA: Harvard University Press.

Dunbar, R. I. M. (2012). On the evolutionary function of song and dance. In N. Bannan (Ed.), *Music, language, and human evolution* (pp. 201–214). Oxford: Oxford University Press.

Dunbar, R. I. M., Kaskatis, K., MacDonald, I., & Barra, V. (2012). Performance of music elevates pain threshold and positive affect: Implications for the evolutionary function of music. *Evolutionary Psychology, 10*(4), 688–702. doi:10.1177/147470491201000403

Durkheim, E. (1995). *The elementary forms of religious life*. New York: Free Press.

Fagg, M. (1997). *Rock music* (Pitt-Rivers Museum Occasional Papers 14). Oxford: Pitt-Rivers Museum.

Falk, D. (2004). Prelinguistic evolution in early hominins: Whence motherese? *Behavioral and Brain Sciences, 27*(4), 491–503. doi:10.1017/S0140525X04000111

Fawcett, C., & Liszkowski, U. (2012). Mimicry and play initiation in 18-month-old infants. *Infant Behavior and Development, 35*(4), 689–696. doi:10.1016/j.infbeh.2012.07.014

Feld, S. (2012). *Sound and sentiment: Birds, weeping, poetics, and song in Kaluli expression*. Durham, NC: Duke University Press.

Fischer, R., Callander, R., Reddish, P., & Bulbulia, J. (2013). How do rituals affect cooperation? *Human Nature, 24*, 115–125. doi:10.1007/s12110-013-9167-y

Freeman, W. (2000). A neurobiological role of music in social bonding. In N. L. Wallin, B. Merker, & S. Brown (Eds.), *The origins of music* (pp. 411–424). Cambridge, MA: MIT Press.

Fritz, T., Jentschke, S., Gosselin, N., Sammler, D., Peretz, I., Turner, I., et al. (2009). Universal recognition of three basic emotions in music. *Current Biology, 19*(7), 573–576. doi:10.1016/j.cub.2009.02.058

Ghazban, N. (2013). *Emotion regulation in infants using maternal singing and speech* (Unpublished doctoral dissertation). Ryerson University, Toronto, Ontario, Canada.

Gill, K. Z., & Purves, D. (2009). A biological rationale for musical scales. *PLoS One, 4*(12), 1–9. doi:10.1371/journal.pone.0008144

Glory, A. (1964). La Grotte de Roucador [Roucador Cave]. *Bulletin de la Société Préhistorique Française, 61*, clxvi–clxix.

Glory, A. (1965). Nouvelles découvertes de dessins rupestres sur le causse de Gramat [New discoveries of rock paintings on the Causse de Gramat]. *Bulletin de la Société Préhistorique Française, 62*, 528–536.

Glory, A., Vaultier, M., & Farinha Dos Santos, M. (1965). La grotte ornée d'Escoural (Portugal) [The decorated cave of Escoural]. *Bulletin de la Société Préhistorique Française, 62*, 110–117.

Haddon, A. (1898). *The study of man*. New York: John Murray.

Hadingham, E. (1980). *Secrets of the ice age: The world of the cave artists*. London: Heinemann.

Hahn, J., & Münzel, S. (1995). Knochenflöten aus den Aurignacien des Geissenklösterle bei Blaubeuren, Alb-Donau-Kreis [Bone flutes of Aurignacian goats from Geissenklösterle at Blaubeuren, Alb-Donau-Kreis]. *Fundberichte aus Baden-Würtemberg, 20*, 1–12.

Harding, J. (1973). The bull-roarer in history and in antiquity. *African Music, 5*(3), 40–42.

Harrold, F. (1988). The Chatelperronian and the early Aurignacian in France. In J. Hoffecker & C. Wolf (Eds.), *The early Upper Palaeolithic* (pp. 157–191). Oxford: Archaeopress.

Hawes, B. L. (1974). Folksongs and function: Some thoughts on the American lullaby. *Journal of American Folklore, 87*(344), 140–148. doi:10.2307/539474

Henschen-Nyman, O. (1988). Cup-marked standing stones in Sweden. In E. Hickman & D. Hughes (Eds.), *The archaeology of early music cultures: Third International Meeting of the ICTM Study Group on Music Archaeology* (pp. 11–16). Bonn: Verlag für systematische Musikwissenschaft GmbH.

Herbert, R. (2011). *Everyday musical listening: Absorption, dissociation and trancing*. Surrey: Ashgate.

Higham, T., Basell, L., Jacobi, R., Wood., R., Bronk Ramsey, C., & Conard, N. (2012). Testing models for the beginnings of the Aurignacian and the advent of the figurative art and music: The radiocarbon chronology of Geißenklösterle. *Journal of Human Evolution, 62*(6), 664–676. doi:10.1016/j.jhevol2012.03.003

Hrdy, S. B. (2009). *Mothers and others: The evolutionary origins of mutual understanding*. Cambridge, MA: Harvard University Press.

Huron, D. (2006). *Sweet anticipation: Music and the psychology of expectation*. Cambridge, MA: MIT Press.

Huyge, D. (1990). Mousterian skiffle? Note on a Middle Palaeolithic engraved bone from Schulen, Belgium. *Rock Art Research, 7*, 125–132.

Huyge, D. (1991). The "Venus" of Laussel in the light of ethnomusicology. *Archeologie in Vlaanderen, 1*, 11–18.

Ichikawa, M. (1999). The Mbuti of northern Congo. In R. B. Lee & R. Daly (Eds.), *The Cambridge encyclopedia of hunters and gatherers* (pp. 210–214). Cambridge: Cambridge University Press.

James, J. (1993). *The music of the spheres: Music, science and the natural order of the universe*. New York: Grove.

Janata, P., & Parsons, L. M. (2013). Neural mechanisms of music, singing, and dancing. In M. Arbib (Ed.), *Language, music, and the brain: A mysterious relationship* (pp. 307–328). Cambridge, MA: MIT Press.

Janata, P., Tomic, S. T., & Haberman, J. M. (2012). Sensorimotor coupling in music and the psychology of the groove. *Journal of Experimental Psychology. General, 141*(1), 54–75. doi:10.1037/a0024208

Jelinek, J. (1990). *Art in the mirror of ages: The beginnings of artistic activities*. Brno: Moravian Museum, Anthropos Institute.

Johnston, T. (1989). Song categories and musical style of the Yupik Eskimo. *Anthropos, 84*, 423–431.

Jordania, J. (2006). Who asked the first question? The origins of human choral singing, intelligence, language and speech. Tblisi: *Logos*.

Juslin, P. N., & Laukka, P. (2004). Expression, perception, and induction of musical emotions: A review and a questionnaire study of everyday listening. *Journal of New Music Research, 33*(3), 217–238. doi:10.1080/0929821042000317813

Juslin, P. N., & Sloboda, J. A. (2010). *Handbook of music and emotion: Theory, research, applications*. Oxford: Oxford University Press.

Juslin, P. N., & Sloboda, J. A. (2013). Music and emotion. In D. Deutsch (Ed.), *The psychology of music* (3rd ed., pp. 583–645). San Diego, CA: Elsevier. 10.1016/B978-0-12-381460-9.00015-8

Kehoe, A. (1999). Blackfoot and other hunters of the North American Plains. In R. B. Lee & R. Daly (Eds.), *The Cambridge encyclopedia of hunters and gatherers* (pp. 36–40). Cambridge: Cambridge University Press.

Keil, C. (1979). *Tiv song*. Chicago: University of Chicago Press.

Kirby, S., Griffiths, T., & Smith, K. (2014). Iterated learning and the evolution of language. *Current Opinion in Neurobiology, 28*, 108–114. doi:10.1016/j.conb.2014.07.004

Kirschner, S., & Tomasello, M. (2010). Joint music making promotes prosocial behavior in 4-year-old children. *Evolution and Human Behavior, 31*(5), 354–364. doi:10.1016/j.evolhumbehav.2010.04.004

Konvalinka, I., Xygalatas, D., Bulbulia, J., Schjødt, U., Jegindø, E., & Wallot, S., … Roepstorff, A. (2011). Synchronized arousal between performers and related spectators in a fire-walking ritual. *Proceedings of the National Academy of Sciences of the United States of America, 108*(20), 8514–8519. doi:10.1073/pnas.1016955108

Kuhn, S., & Stiner, M. (1998). The earliest Aurignacian of Riparo Mochi (Liguria, Italy). *Current Anthropology, 39*(Suppl. 3), 175–189. doi:10.1086/204694

Lang, A. (1884). *Custom and myth*. London: Harper.

Lawson, G., & d'Errico, F. (2002). Microscopic, experimental and theoretical re-assessment of Upper Palaeolithic bird-bone pipes from Isturitz, France: Ergonomics of design, systems of notation and the origins of musical traditions. In E. Hickman, A. Kilmer, & R. Eichman (Eds.), *Studien zur Musikarchäologie III* (pp. 119–142). Rahden: Verlag Marie Leidorf.

Leon Portilla, M. (2007). La musica de los aztecas [Music of the Aztecs]. *Pauta, 103*, 7–19.

LeVine, R. A. (1988). Human parental care: Universal goals, cultural strategies, individual behavior. *New Directions for Child and Adolescent Development*, no. 40, 3–12. doi:10.1002/cd.23219884003

Lewis, J. (2013). A cross-cultural perspective on the significance of music and dance to culture and society: Insights from BaYaka Pygmies. In M. Arbib (Ed.), *Language, music, and the brain: A mysterious relationship* (pp. 45–65). Cambridge, MA: MIT Press.

Liljeström, S., Juslin, P. N., & Västfäll, D. (2013). Experimental evidence of the roles of music choice, social context, and listener personality in emotional reactions to music. *Psychology of Music, 41*(5), 579–599. doi:10.1177/0305735612440615

Locke, D. (1996). Africa: Ewe, Mande, Dagbamba, Shona and BaAka. In J. Titon (Ed.), *Worlds of music: An introduction to the music of the world's people* (3rd ed., pp. 83–144). New York: Schirmer.

Loersch, C., & Arbuckle, N. L. (2013). Unraveling the mystery of music: Music as an evolved group process. *Journal of Personality and Social Psychology, 105*(5), 777–798. doi:10.1037/a0033691

Lomax, A. (1968). *Folk song style and culture*. Washington, DC: American Association for the Advancement of Science.

Mahon, M. (2014). Music, power, and practice. *Ethnomusicology, 58*(2), 327–333. doi:10.5406/ethnomusicology.58.2.0327

Maioli, W. (1991). *Le origini il suono e la musica* [The origins of sound and music]. Milan: Jaca Book.

Margulis, E. H. (2014). *On repeat: How music plays the mind*. New York: Oxford University Press.

Masataka, N. (1999). Preference for infant-directed singing in 2-day-old hearing infants of deaf parents. *Developmental Psychology, 35*(4), 1001–1005. doi:10.1037/0012-1649.35.4.1001

Masuyama, E. E. (1989). Desire and discontent in Japanese lullabies. *Western Folklore, 48*(2), 144–148. doi:10.2307/1499687

McAllester, D. (1954). *Enemy way music: A study of social and esthetic values as seen in Navaho music*. Cambridge, MA: Peabody Museum of American Archeology and Ethnology.

McNeill, W. (1995). *Keeping together in time: Dance and drill in human history*. Cambridge, MA: Harvard University Press.

Meltzoff, A. N. (2007). "Like me": A foundation for social cognition. *Developmental Science, 10*(1), 126–134. doi:10.1111/j.1467-7687.2007.00574.x

Merker, B. (2009). Ritual foundations of human uniqueness. In S. Malloch & C. Trevarthen (Eds.), *Communicative musicality: Exploring the basis of human uniqueness* (pp. 45–59). New York: Oxford University Press.

Merriam, A. (1964). *The anthropology of music.* Evanston, IL: Northwestern University Press.

Montagu, J. (2007). *Origins and development of musical instruments.* Lanham, MD: Scarecrow Press.

Mukhergee, P. (2004). *The scales of Indian music.* New Delhi: Aryan Books International.

Myers, F. (1999). Pintupi-speaking Aboriginals of the Western Desert. In R. B. Lee & R. Daly (Eds.), *The Cambridge encyclopedia of hunters and gatherers* (pp. 348–357). Cambridge: Cambridge University Press.

Nettl, B. (1989). *Blackfoot musical thought: Comparative perspectives.* Kent, OH: Kent State University Press.

Nettl, B. (1992). North American Indian music. In B. Nettl, C. Capwell, P. Bohlman, I. Wong, & T. Turino (Eds.), *Excursions in world music* (pp. 260–277). Englewood Cliffs, NJ: Prentice Hall.

Nettl, B. (2000). An ethnomusicologist contemplates musical universals. In N. L. Wallin, B. Merker, & S. Brown (Eds.), *The origins of music* (pp. 464–472). Cambridge, MA: MIT Press.

Nettl, B. (2005). *The study of ethnomusicology.* Champaign: University of Illinois Press.

North, A. C., Hargreaves, D. J., & Hargreaves, J. J. (2004). Uses of music in everyday life. *Music Perception, 22*(1), 41–77. doi:10.1525/mp.2004.22.1.41

Otte, M. (1979). Le Paléolithique Supérieur Ancien en Belgique [The Upper Paleolithic in Belgium]. *Monographies d'Archéologie Nationale, 5.* Brussels, Belgium: Musées Royaux d'Art et d'Histoire.

Over, H., & Carpenter, M. (2013). The social side of imitation. *Child Development Perspectives, 7*(1), 6–11. doi:10.1111/cdep.12006

Owomoyele, O. (2002). *Cultures and customs of Zimbabwe.* Westport, CT: Greenwood Press.

Pantaleoni, H. (1985). *On the nature of music.* Oneonta, NY: Welkin.

Passemard, E. (1923). Une flûte Aurignacienne d'Isturitz [An Aurignacian flute from Isturitz]. *Proceedings of the Congrès de l'Association Français pour l'Avancement des Sciences, Compte rendu de la 46e session, Montpelier*, 474–476.

Passemard, E. (1944). La cavern d'Isturitz en Pays Basque [Isturitz cave in the Basque Country]. *Préhistoire, 9*, 7–95.

Patel, A. D. (2008). *Music, language, and the brain.* New York: Oxford University Press.

Phillips-Silver, J., & Trainor, L. J. (2005). Feeling the beat: Movement influences infant rhythm perception. *Science, 308*, 1430. doi:10.1126/science.1110922

Rappaport, R. (1999). *Ritual and religion in the making of humanity.* Cambridge: Cambridge University Press.

Richman, A. L., Miller, P. M., & Solomon, M. J. (1988). The socialization of infants in suburban Boston. *New Directions for Child and Adolescent Development*, no. *40*, 65–74. doi:10.1002/cd.23219884008

Rowell, L. (1992). *Music and musical thought in early India.* Chicago: University of Chicago Press.

Sacks, O. (2008). *Musicophilia: Tales of music and the brain.* New York: Knopf.

Sakata, H. L. (1987). Hazara women in Afghanistan: Innovators and preservers of musical tradition. In E. Koskoff (Ed.), *Women and music in cross-cultural perspective* (pp. 85–95). Westport, CT: Greenwood Press.

Salimpoor, V. N., Benovoy, N., Longo, G., Cooperstock, J. R., & Zatorre, R. J. (2009). The rewarding acts of music listening are related to degree of emotional arousal. *PLoS One, 4*, 1–14. doi:10.1371/journal.pone.0007487

Savage, P. E., Brown, S., Sakai, E., & Currie, T. E. (2015). Statistical universals reveal the structures and functions of human music. *Proceedings of the National Academy of Sciences, 112*(29), 8987–8992. doi:10.1073/pnas.1414495112

Scothern, P. (1992). *The music-archeology of the Palaeolithic within its cultural setting* (Unpublished doctoral dissertation). University of Cambridge, Cambridge, UK.

Seyfarth, R. M., & Cheney, D. L. (2014). The evolution of language from social cognition. *Current Opinion in Neurobiology, 28*, 5–9. doi:10.1016/j.conb.2014.04.003

Shenfield, T., Trehub, S. E., & Nakata, T. (2003). Maternal signing modulates infant arousal. *Psychology of Music, 31*(4), 365–375. doi:10.1177/03057356030314002

Sosis, R., & Ruffle, B. (2004). Religious ritual and cooperation: Testing for a relationship on Israeli religious and secular kibbutzim. *Current Anthropology, 44*(5), 713–722. doi:10.1086/379260

Stone, R. (1982). *Let the inside be sweet: The interpretation of music event among the Kpelle of Liberia.* Bloomington: Indiana University Press.

Sullivan, L. (1997). *Enchanting powers: Music in the world's religions.* Cambridge, MA: Harvard University Press.

Tenzner, M. (1991). *Balinese music.* Seattle: University of Washington Press.

Tomasello, M. (2003). *Constructing a language: A usage based theory of language acquision.* Cambridge, MA: Harvard University Press.

Trainor, L. J. (1996). Infant preferences for infant-directed versus noninfant-directed playsongs and lullabies. *Infant Behavior and Development, 19*(1), 83–92. doi:10.1016/S0163-6383(96)90046-6

Trainor, L. J., Clark, E. D., Huntley, A., & Adams, B. A. (1997). The acoustic basis of preferences for infant directed singing. *Infant Behavior and Development, 20*(3), 383–396. doi:10.1016/S0163-6383(97)90009-6

Trainor, L. J., & Heinmiller, B. M. (1998). The development of evaluative responses to music: Infants prefer to listen to consonance over dissonance. *Infant Behavior and Development, 21*(3), 77–88. doi:10.1016/S0163-6383(98)90055-8

Trehub, S. E. (2000). Human processing predispositions and musical universals. In N. L. Wallin, B. Merker, & S. Brown (Eds.), *The origins of music* (pp. 427–448). Cambridge, MA: MIT Press.

Trehub, S. E. (2015). Cross-cultural convergence of musical features. *Proceedings of the National Academy of Sciences, 112*(29), 8809–8810. doi:10.1073/pnas.1510724112

Trehub, S. E., & Gudmundsdottir, H. R. (2015). Mothers as singing mentors for infants. In G. Welch & J. Nix (Eds.), *Oxford handbook of singing.* Oxford: Oxford University Press. doi:10.1093/oxfordhb/9780199660773.013.25

Trehub, S. E., Plantinga, J., & Russo, F. A. (2016). Maternal vocal interactions with infants: Reciprocal visual influences. *Social Development, 25*(3), 665–683. doi:10.1111/sode.12164

Trehub, S. E., & Trainor, L. J. (1998). Singing to infants: Lullabies and play songs. *Advances in Infancy Research, 12*, 43–77.

Trehub, S. E., Unyk, A. M., Kamenetsky, S. B., Hill, D. S., Trainor, L. J., Henderson, J. L., et al. (1997). Mothers' and fathers' singing to infants. *Developmental Psychology, 33*(3), 500–507. doi:10.1037/0012-1649.33.3.500

Trehub, S. E., Unyk, A. M., & Trainor, L. J. (1993a). Adults identify infant-directed music across cultures. *Infant Behavior and Development, 16*(2), 193–211. doi:10.1016/0163-6383(93)80017-3

Trehub, S. E., Unyk, A. M., & Trainor, L. J. (1993b). Maternal singing in cross-cultural perspective. *Infant Behavior and Development, 16*(3), 285–295. doi:10.1016/0163-6383(93)80036-8

Turino, T. (1992). The music of sub-Saharan Africa. In B. Nettl, C. Capwell, P. Bohlman, I. Wong, & T. Turino (Eds.), *Excursions in world music* (pp. 165–195). Englewood Cliffs, NJ: Prentice Hall.

Unyk, A. M., Trehub, S. E., Trainor, L. J., & Schellenberg, E. G. (1992). Lullabies and simplicity: A cross-cultural perspective. *Psychology of Music, 20*(1), 15–28. doi:10.1177/0305735692201002

van Baaren, R. B., Decety, J., Dijksterhuis, A., van der Leij, A., & van Leeuwen, M. L. (2011). Being imitated: Consequences of nonconsciously showing empathy. In J. Decety & W. J. Ickes (Eds.), *The social neuroscience of empathy* (pp. 31–42). Cambridge, MA: MIT Press.

Whitehouse, H. (2004). *Modes of religiosity: A cognitive theory of religious transmission.* Lanham, MD: AltaMira Press.

Wiltermuth, S. S., & Heath, C. (2009). Synchrony and cooperation. *Psychological Science, 20*(1), 1–5. doi:10.1111/j.1467-9280.2008.02253.x

Wittgenstein, L. (2001). *Tractatus logico-philosophicus.* London: Routledge.

Xygalatas, D., Mitkidis, P., Fischer, R., Reddish, P., Skewes, J., & Geertz, A., … Bulbulia, J. (2013). Extreme rituals promote prosociality. *Psychological Science, 24*(8), 1602–1605. doi:10.1177/0956797612472910

Zatorre, R. J., & Salimpoor, V. N. (2013). From perception to pleasure: Music and its neural substrates. *Proceedings of the National Academy of Sciences of the United States of America, 110*(Suppl. 2), 10430–10438. doi:10.1073/pnas.1301228110

Zentner, M., & Eerola, T. (2010). Rhythmic engagement with music in infancy. *Proceedings of the National Academy of Sciences, 107*(13), 5768–5773. doi:10.1073/pnas.1000121107

Zentner, M., Grandjean, D., & Scherer, K. R. (2008). Emotions evoked by the sounds of music: Characterization, classification, and measurement. *Emotion, 8*(4), 492–521. doi:10.1037/1528-3542.8.4.494

Zentner, M., & Kagan, J. (1998). Infants' perception of consonance and dissonance in music. *Infant Behavior and Development, 21*(3), 483–492. doi:10.1016/s0163-6383(98)90021-2

Zuckerkandl, V. (1973). *Sound and symbol: Man the musician* (Vol. 2). Princeton, NJ: Princeton University Press.

7

Searching for the Origins of Musicality across Species

Marisa Hoeschele, Hugo Merchant, Yukiko Kikuchi, Yuko Hattori, and Carel ten Cate

Honing, ten Cate, Peretz, and Trehub (2015) suggest that the origins of musicality—the capacity that makes it possible for us to perceive, appreciate, and produce music—can be pursued productively by searching for components of musicality in other species. Perhaps the most obvious starting point in this endeavor is the examination of animal responses to music. In 1984, Porter and Neuringer were the first to conduct an experiment from this perspective by training pigeons (*Columba livia*) to discriminate the music of different composers. The authors used an operant paradigm, where pigeons received a food reward after pecking one of two discs during presentation of excerpts from several Bach pieces for organ and Stravinsky's *Rite of Spring*. Pigeons were trained to respond to the left disc during Bach and a right disc during Stravinsky excerpts. With time, they learned this discrimination. Once the pigeons were making few errors, they were presented with novel excerpts from the same composers and similar excerpts from other composers. The pigeons generalized to all of these novel stimuli through their responses to the two choice discs in a way that mirrored that of human participants.

A more recent study was performed using a similar operant paradigm with carp (*Cyprinus carpio*) using blues and classical stimuli and found comparable results (Chase, 2001); after initial training with a small set of blues and classical music stimuli, carp were able to correctly classify stimuli from these genres that they had never heard before. How can we interpret the fact that distantly related animal species have human-like boundaries for the categorization of such complex auditory stimuli? Moreover, what can we learn from such studies?

Key Problems in Studying Biomusicology

The genre classification performance of pigeons and carp has an analogue in studies on visual categorization. In their seminal study, Herrnstein and Loveland (1964) successfully trained pigeons to discriminate photos that contained humans from photos that did not, with all photos exhibiting considerable variability. Subsequent debate has centered on whether the pigeons were detecting humans or simply using local features (e.g., large flesh-colored area) to solve the task (Weisman & Spetch, 2010). Similarly, pigeons and carp in the music categorization tasks may use specific local features (e.g., presence or absence of a specific frequency) rather than global or abstract features to solve the task. One way to determine what features are controlling the behavior in each species would be to present altered stimuli that are missing some of the features from the rich training stimuli or present some features of the rich stimuli in isolation.

A second, and equally important, issue is motivational. Aside from the music categorization abilities demonstrated in pigeons and carp, do they have any music preferences? They might. Groups of gilthead seabreams (*Sparus aurata*) that were exposed to different pieces of classical music showed physiological differences after prolonged exposure (Papoutsoglou et al., 2015). We also know that chimpanzees prefer at least some types of music to silence, as they spent more time close to a speaker when music was being played than when it was not (Mingle et al., 2014). But another primate species not as closely related to humans, the cotton-top tamarins (*Saguinus oedipus*), showed the opposite result (McDermott & Hauser, 2007). Several other studies have looked at animal musical preferences using similar place preference paradigms. Chiandetti and Vallortigara (2011) put newborn chicks (*Gallus gallus*) in an environment with consonant music playing on one side and dissonant music on the other. They found that the chicks spent more time on the side with consonant music. When McDermott and Hauser (2004) presented cotton-top tamarins with consonant stimuli in one arm of a y-maze and dissonant stimuli in the other arm, the animals showed no preference. Western adults confronted with similar contingencies spent more time listening to consonant than to dissonant sounds. Although such a preference paradigm suggests that a feature such as consonance and dissonance is relevant to a given species, it tells us little about the mechanisms underlying these preferences. The human preference for consonance over dissonance is at least partially based on the physical properties of sound and is evident across many cultures (Cook & Fujisawa, 2006). It is equally important to ascertain the factors contributing to music-related preferences in nonhuman species.

Another problem is the selection of appropriate species for study. Using a traditional laboratory approach, it is possible to take virtually any species with sound-sensing capacity and rudimentary learning capacities and measure its physiology and train it to discriminate different sounds. But which species are relevant for biomusicology? One approach is to study species based on shared ancestry. Species that are closely related to humans are likely to share some of our abilities and are therefore good models for experiments that would be difficult to conduct in humans. For traits that are not shared among closely related species, it is easier to pinpoint the differences in underlying mechanisms. Another approach is to study species based on shared traits. Species that are more distantly related sometimes share traits without sharing a common ancestor with those traits. By examining the evolutionary convergence between these nonrelated species, we can identify biological constraints or mechanisms required for that trait and the selection pressures giving rise to it. For example, some of the traits that are considered highly relevant for biomusicology to date are vocal learning and entrainment. Vocal learning involves the ability to produce vocalizations based on auditory input, and entrainment involves the ability to synchronize movements with an external stimulus (usually sound). Both traits are uncommon in nonhuman species but shared across some unrelated species, so their study could provide clues to their nature and possible interaction (Schachner, Brady, Pepperberg, & Hauser, 2009; Tyack, 2008). Thus, both approaches, examining closely related and distantly related species, can be quite useful for probing the biological basis of musicality.

Thus, we need to consider the perceptual abilities of animals, their natural preferences, and their similarity to humans in terms of phylogeny or shared traits. This task necessitates the combined fruits of traditional laboratory studies with artificial stimuli (focused on the analysis of perceptual mechanisms) and studies with naturalistic stimuli (that are biologically relevant to the animals) or spontaneous behaviors in order to break down the elements of musicality. There is much debate about the relative utility of these two approaches. Proponents of naturalistic studies argue that training animals to perform "unnatural" behaviors, or using stimuli that differ markedly from those in their natural environment, does not constitute an appropriate comparison for human behaviors that emerge without training. They also point out that artificial stimuli can result in underestimation of animal abilities. Proponents of experimental perceptual laboratory research note that experimental control and systematic comparisons across species may reveal underlying abilities and potential that are not apparent in natural behavior. As a result, laboratory studies can shed light on the presence of cognitive abilities that support such behavior, as well as their biological basis. Both sides have valid insights about the limitations of the other approach, but they fail to appreciate its strengths and the utility of combined approaches. We hope this chapter will demonstrate the complementarity of both approaches. For definitions of musical terms, see table 7.1.

Experimental Laboratory Studies of Auditory Perception

Rhythm

The processing of auditory rhythms—both the underlying pulse or beat and the organization of the beats into repetitive groups—relies on basic features of the auditory system. The use of operant conditioning procedures has revealed greater temporal sensitivity in birds than in humans (Dooling, Leek, Gleich, & Dent, 2002) when evaluating perceptual differences among brief stimuli. However, this does not mean they are necessarily able to detect rhythmic patterns. In one study (Hagmann & Cook, 2010), pigeons successfully learned to differentiate two metrical patterns (8/4 versus 3/4) but showed limited transfer of the discrimination to different tempos. Their learning did not generalize to metrical patterns in a different timbre, and they had difficulty differentiating rhythmic sequences from random sequences. European starlings, however, learned to differentiate rhythmic from

Table 7.1
Glossary of relevant musical terms

Term	Definition
Beat	The underlying pulse, or unit of time, in music
Entrainment	The ability to perceive a beat in music and align bodily movement with it
Melody	A sequences of tones defined by its pitch patterning and rhythm
Meter	The recurring pattern of stressed and unstressed beats in music
Musicality	The capacity that underlies the human ability to perceive, appreciate, and produce music
Pitch	A perceptual attribute related to the fundamental frequency that enables comparisons of sounds as higher or lower
Prosody	Rhythm, loudness, pitch, and tempo of speech
Rhythm	A nonrandom repetitive temporal auditory pattern
Timbre	The quality of musical sound that distinguishes different sound sources such as voices and specific musical instruments
Vocal learning	Long-term modification of vocal production by imitation

nonrhythmic sequences and showed a broader range of generalization (Hulse, Humpal, & Cynx, 1984). Zebra finches (*Taeniopygia guttata*) learned to discriminate isochronous from irregular stimuli but were unable to generalize to novel tempos (van der Aa, Honing, & ten Cate, 2015). Another experiment showed that both zebra finches and budgerigars (*Melopsittacus undulates*) could discriminate a regular from an irregular beat pattern, but showed only limited generalization of this discrimination to modified versions of the stimuli (ten Cate, Spierings, Hubert, & Honing, 2016) Reviewing all evidence for rhythm perception in birds, ten Cate et al. (2016) concluded that birds of different species attend primarily to absolute time intervals (i.e., exact durations of elements and pauses), but that some species can also attend to more global rhythmic patterns.

For primates, the use of neurophysiological (e.g., single-cell electrophysiology and functional magnetic resonance imaging) measures has revealed that the neural substrates of sequencing and timing behaviors overlap with those related to human music perception and performance (for a review, see Janata & Grafton, 2003) and that the motor cortico-basal ganglia-thalamo-cortical circuit (mCBGT) plays an important role (Grahn, 2009; Grahn & Brett, 2007; Zatorre, Chen, & Penhune, 2007). Trained rhesus macaques (*Macaca mulatta*) are not only capable of interval timing and motor sequencing tasks performed by humans (Hikosaka, Rand et al., 2002; Mendez, Prado, Mendoza, & Merchant, 2010; Merchant, Battaglia-Mayer, & Georgopoulos, 2003; Merchant, Zarco, Pérez, Prado, & Bartolo, 2011; Zarco, Merchant, Prado, & Mendez, 2009), but they also show similar neural activation in mCBGT in sequential tasks (Hikosaka, Nakamura, Sakai, & Nakahara, 2002; Tanji, 2001), single-interval timing tasks (Macar, Coull, & Vidal, 2006; Merchant, Pérez, Zarco, & Gámez, 2013; Mita, Mushiake, Shima, Matsuzaka, & Tanji, 2009), and tasks involving synchronized tapping to an isochronous metronome (Crowe, Zarco, Bartolo, & Merchant, 2014; Merchant et al., 2013). These results suggest that the role of mCBGT in auditory rhythm processing is shared across these two species of primates (Merchant, Grahn, Trainor, Rohrmeier & Fitch, 2015; chapter 8, this volume). Uncovering findings such as these was possible only with the use of artificial laboratory probing of the limits of the perceptual systems of these two species.

Timbre

Timbre perception has not received much attention in the animal literature, but laboratory studies have begun to provide some insights. As with temporal intervals, avian species seem to detect more finely grained timbral differences than humans do (Amagai, Dooling, Shamma, Kidd, & Lohr, 1999; Cynx, Williams, & Nottebohm, 1990; Lohr & Dooling, 1998). Timbre is considered a surface feature (Warker & Halpern, 2005) because humans recognize the same musical patterns regardless of the timbre of presentation (e.g., vocal, piano, flute). This ability to generalize across timbres is reportedly present from the newborn period (Háden et al., 2009). It is also present in at least one songbird. A jackdaw trained to discriminate two different sound patterns was able to maintain this discrimination over a wide diversity of sounds (Reinert, 1965). In one study, humans' responses to chords readily generalized across timbres, but that was not the case with black-capped chickadees (*Poecile atricapillus*; Hoeschele, Cook, Guillette, Hahn, & Sturdy, 2014). Zebra finches exhibit generalization across timbres (Ohms, Gill, Van Heijningen, Beckers, & ten Cate, 2010; Spierings & ten Cate, 2014). When trained to discriminate between two words produced by male (or female) speakers, they showed generalization across speaker gender (i.e., fundamental frequency and spectral differences). Clearly more research is needed to clarify the nature and extent of timbre generalization across species.

Pitch

The pitch of a sound is typically based on the fundamental frequency of that sound (Dowling & Harwood, 1986), although human listeners can perceive the pitch of a sound in which the fundamental frequency is missing (Oxenham, 2012). The ability to perceive the so-called missing fundamental is present in infants as young as three months of age (Lau & Werner, 2012) and is also demonstrable in cats (*Felis catus*; Heffner & Whitfield, 1976), rhesus monkeys (Tomlinson, 1988), starlings (Cynx & Shapiro, 1986), and zebra finches (Uno, Maekawa, & Kaneko, 1997). The suggestion is that this ability is shared across species, but its generality has not been established empirically.

Listeners sometimes evaluate fundamental frequency or pitch in an absolute manner, as when musicians with absolute pitch correctly name the pitch class of musical notes (e.g., 440 Hz as A; Ward, 1999) or nonmusicians distinguish the original pitch level of highly familiar recorded music from versions that have been shifted by one or two semitones (Schellenberg & Trehub, 2003). In general, birds are superior to mammals at detecting absolute pitch (Friedrich, Zentall, & Weisman, 2007; Weisman et al. 1998; Weisman, Njegovan, Williams, Cohen, & Sturdy, 2004; but see Weisman et al., 2010). In most cases, however, human listeners focus on relations among pitches rather than absolute pitch levels while listening to music. In musical contexts, moreover, human listeners exhibit octave generalization, perceiving the similarity of notes that are one or more octaves apart (Burns, 1999; Patel, 2003).

The evidence for octave generalization in nonhuman species is both limited and controversial. Blackwell and Schlosberg (1943) claimed that rats (*Rattus norvegicus*) generalized from training stimuli in one octave to test stimuli in another octave. However, there are alternative explanations of the findings because the stimuli may have contained harmonics that provided common cues across octaves (Burns, 1999). Suggestive evidence for octave generalization comes from a bottlenose dolphin (*Tursiops truncatus*) that mimicked sounds outside her vocal range by reproducing them an octave apart from the original (Richards, Wolz, & Herman, 1984). Interestingly, rhesus monkeys trained to differentiate melodies in a same-different task responded to octave-transposed melodies as "same" for Western tonal but not atonal melodies (Wright, Rivera, Hulse, Shyan, & Neiworth, 2000).

To date, there is no evidence of octave generalization in avian species. Cynx (1993) trained starlings to discriminate between two tones and then tested whether they generalized this discrimination to the octave. They did not. The failure of human listeners to exhibit octave generalization on the same task (Hoeschele, Weisman, & Sturdy, 2012) called the starling findings into question. In a similar operant training task, humans exhibited octave generalization (Hoeschele, Weisman, & Sturdy, 2012), but an adaptation of the task for black-capped chickadees revealed no octave generalization (Hoeschele, Weisman, Guillette, Hahn, & Sturdy, 2013). The evidence is consistent with the absence of octave generalization in birds, but more research is needed with a wider range of species before the question can be resolved definitively.

With regard to relative pitch, several studies have found that nonhuman animals could be trained to discriminate among chords (i.e., simultaneous combinations of tones): European starlings (Hulse, Bernard, & Braaten, 1995), Java sparrows (*Lonchura oryzivora*; Watanabe, Uozumi, & Tanaka, 2005), Japanese monkeys (*Macaca fuscata*; Izumi, 2000), pigeons (Brooks & Cook, 2010) and black-capped chickadees (Hoeschele et al., 2014; Hoeschele, Cook, Guillette, Brooks, & Sturdy, 2012), and rats (Crespo-Bojorque & Toro, 2015). All of these studies ensured the animals were not simply memorizing the absolute properties of the sounds by presenting novel stimuli with identical or similar relative pitch properties but different absolute pitches. Interestingly, all species except rats were able to transfer what they learned to these novel stimuli.

Evidence of relative pitch processing with sounds presented sequentially rather than simultaneously is less promising. Starlings, brown-headed cowbirds (*Molothrus ater*), and northern mockingbirds (*Mimus polyglottos*) were trained to discriminate ascending from descending note patterns (Hulse & Cynx, 1985, 1986; Hulse, Cynx, & Humpal,

1984; MacDougall-Shackleton & Hulse, 1996; Page, Hulse, & Cynx, 1989). However, they failed to generalize these patterns to novel pitch levels when the altered patterns were outside the training range, although they could quickly learn to do so. In general, it appeared that the birds encoded both absolute and relative pitch information in discriminating the patterns, but they depended more on absolute information. Another set of studies trained zebra finches and black-capped chickadees to discriminate sets of pitches based on either their pitch ratios (i.e., relative pitch) or their absolute frequencies. Both species learned the discrimination more quickly when there was a simple relative pitch rule that they could use, although the discrimination was quite difficult for the birds in comparison to learning a simple absolute pitch rule. In short, these birds can engage in relative pitch processing, although they rely primarily on the absolute pitch of sounds (Njegovan & Weisman, 1997; Weisman, Njegovan, & Ito, 1994). In one study, a bottlenose dolphin learned to respond to descending pitch contours and after extensive training generalized that response to descending pitch contours regardless of the component pitches (Ralston & Herman, 1995).

In general, nonhuman species recognize the relative pitch patterns of single chords more readily than those of note sequences. Three factors may be implicated. First, the component frequencies of chords give rise to qualities such as sensory consonance and dissonance (Helmholtz, 1877) that contribute to their distinctiveness. Second, chords, as single events, pose fewer memory demands than sequences of notes. Third, there are suggestions that harmonic (simultaneous) pitch ratios are processed at early stages of the auditory cortical pathway in rhesus macaques (Kikuchi, Horwitz, Mishkin, & Rauschecker, 2014). As a result, the pitch ratios of chords or simultaneously presented notes may be processed more automatically and therefore compared more readily than the pitch ratios of melodic sequences. However, another possibility is that the animals are using spectral shape, in the case of chords, rather than pitch. A recent study with starlings showed that these birds had trouble generalizing melodies across transpositions of pitch and timbre, but not transpositions of spectral shape (Bregman, Patel, & Gentner, 2016). As such, some animals may attend to pitch only indirectly as it relates to spectral shape, and breaking down a sound into "pitch" and "timbre" may not reflect the processing that occurs with these species.

High-Order Acoustic Patterns

Building on the foundations of the auditory system and interval timing is the perception of grouping. Gestalt psychologists noted long ago that a group of visual or auditory elements has qualities that are more than the sum of its parts. A repeated series of tones of equal frequency and amplitude, with one tone having longer duration than the others, is perceived as an iambic pattern in which the long sound marks the end of a sound unit. A repeated series of tones in which one tone has higher frequency or amplitude than the others is perceived as a trochaic pattern in which the higher-pitched or louder sound marks the beginning of the sound unit (Woodrow, 1909). This type of patterning is common in music as well as speech, and young infants seem to spontaneously recognize trochaic patterns (Bion, Benavides-Varela, & Nespor, 2011). Naive rats seem to group tones according to trochaic but not iambic rules (de la Mora, Nespor, & Toro, 2013), but with experience, they can develop a grouping bias for both short-long and long-short tone pairs (Toro & Nespor, 2015), which indicates that such grouping abilities are not exclusive to human listeners. Further research is needed to explore the nature of auditory grouping abilities across species.

The ability to perceive higher-order temporal patterns in a stream of sounds is relevant to speech as well as music perception. The perception of speech prosody, or the melody of speech, is relevant to music perception. In many languages, for example, statements end with a falling terminal pitch contour, and yes/no questions end with a rising terminal pitch contour (Cruttenden, 1981). Different languages also have different prosodic patterns (Cutler, Dahan, & van Donselaar, 1997). Tamarins (Tincoff et al., 2005), rats (Toro, Trobalon, & Sebastián-Gallés, 2003, 2005) and Java sparrows (Naoi, Watanabe, Maekawa, & Hibiya, 2012) have shown the ability to discriminate between spoken sentences in different languages and to generalize this discrimination to novel sentences. Zebra finches use pitch, duration, and amplitude to discriminate prosodic patterns, and they can generalize specific prosodic patterns of speech syllables to novel syllables (Spierings & ten Cate, 2014). The same was shown for budgerigars, including timbre as a further prosodic cue (Hoeschele & Fitch, 2016). Further exploration of nonhuman species' sensitivity to melodic aspects of speech may be a fruitful approach to the study of some aspects of musicality.

Criticisms

When nonhuman animals are trained to discriminate auditory patterns, they typically take a lot longer to learn the task than their human counterparts. A critic may ask, for example, whether a bird trained for hundreds of trials to discriminate chords can really be compared to a human who discriminates the chords with minimal or no training. That situation does not negate the value of comparisons of music perception in human and nonhuman listeners. Although human listeners may require little training for specific tasks, they have had years of exposure to music and have a wealth of implicit musical knowledge. Moreover, the ability of nonhuman listeners to perform certain tasks, even after extensive training, can provide insights into the mechanisms underlying that ability.

Consider the studies of interval timing in rhesus macaques and humans. As the interaction between the auditory stimulus and required motor output becomes more complex, monkeys' performance lags increasingly behind that of humans. In one study, monkeys and humans were required to tap on a push button to produce six isochronous intervals in a sequence. An auditory stimulus was present to guide tapping during the first three taps but not the last three, which required internal timing based on the preceding auditory stimulus or taps (Donnet et al., 2014; Zarco et al., 2009). Although monkeys produced rhythmic movements with appropriate tempo matching, their movements lagged by approximately 250 ms after each auditory stimulus, even after long periods of training (close to a year). In contrast, humans easily perform the same task with no training, showing stimulus movement asynchronies approaching zero or with negative values (Repp, 2005; Zarco et al., 2009). Such differences in two closely related species make it possible to predict that the mCBGT may have subtle but critical differences that gradually evolved in order to process complex auditory information and use it in a predictive fashion to control the temporal and sequential organization of movement, as recently stated in the gradual audiomotor evolution hypothesis (Merchant & Honing, 2014).

Even if humans and monkeys had comparable experience with the stimuli in such experimental tasks, which they do not, both have very different interpretations of the experimental context and the experimenter's intentions, even where efforts have been made to minimize differences in training requirements and outcomes (Hoeschele, Cook, Guillette, Hahn, & Sturdy, 2014; Hoeschele, Cook et al., 2012; Hoeschele et al., 2013;Weisman et al., 2010).

Summary

Overall, the evidence indicates the enormous potential of systematic experimental laboratory studies on the perception of some components of musicality with nonhuman species. Operant conditioning studies have the potential to reveal skills that are not part of an animal's natural repertoire. Animals' performance in these tasks is deeply rooted in the limitations and adaptive plasticity of their nervous system (Mendoza & Merchant, 2014). By using these animals as models, we can also gain information about the neural activity (e.g., through electrophysiological recordings) as well as manipulations (e.g., pharmacology, electric stimulation, optogenetics) that can alter brain mechanisms and corresponding behavior, facilitating our understanding of the neural underpinnings of musicality in humans.

Studies Linked to Natural Behavior

Laboratory experiments with artificial stimuli have been helpful in revealing perceptual skills and perceptual-motor coordination in nonhuman species. It is possible, however, that their use may lead to underestimates of ability. One alternative or supplementary approach is to use biologically relevant stimuli in laboratory studies. Another is to study music-like features in the natural behavior (e.g., vocalizations) of animals.

Incorporating Naturalistic Stimuli into Experimental Work

As noted, laboratory research with artificial stimuli has revealed that birds focus on absolute aspects of pitch rather than relative pitch (Hulse & Cynx, 1985, 1986; Hulse, Cynx, & Humpal, 1984; MacDougall-Shackleton & Hulse, 1996; Njegovan & Weisman,

1997; Page et al., 1989; Weisman et al., 1994), but evidence from field studies suggests otherwise. For example, fieldwork with black-capped chickadees has shown that they produce a simple two-note tonal song that can be sung at different absolute pitches but maintains its relative pitch ratio (Weisman, Ratcliffe, Johnsrude, & Hurly, 1990). Moreover, this relative pitch ratio is produced more accurately by dominant males (Christie, Mennill, & Ratcliffe, 2004), and accurately produced song pitch ratios are preferred by females (Weisman & Ratcliffe, 2004). These findings prompted laboratory research on this issue with chickadees (Hoeschele, Guillette, & Sturdy, 2012).

Chickadees were trained to discriminate pitch ratios presented at different absolute frequencies and made use of this relevant song pitch ratio. An experimental group was trained to respond to the pitch ratio from chickadee song and not to respond to two nonchickadee-song pitch ratios. A control group was trained to respond to a nonchickadee-song pitch ratio and not respond to two different nonchickadee-song pitch ratios. The chickadees that were required to identify the pitch ratio of their song learned the task more quickly than the control group, suggesting that it was easier for the chickadees to learn to discriminate the natural song pitch ratio than other pitch ratios (Hoeschele, Guillette, & Sturdy, 2012). A related study showed that starlings that were trained to discriminate conspecific vocalizations were able to maintain that discrimination even when the songs were transposed (i.e., pitches changed but pitch relations preserved; Bregman, Patel, & Gentner, 2012), raising the possibility that absolute pitch processing has priority over relative pitch processing only with stimuli lacking in ecological validity.

Rhythm Perception

There is also evidence indicating a greater sensitivity to natural stimuli in the realm of rhythm perception. Although pigeons have difficulty with rhythm perception tasks involving artificial stimuli (Hagmann & Cook, 2010), the natural coo vocalizations of pigeons and doves, neither of them vocal learners, are rhythmic. The collared dove produces a coo that consists of five elements of different duration: three notes separated by two silences (Ballintijn & ten Cate, 1999). Playback experiments in the field show that replacing the second or third note of the coo by silence caused little change in the behavioral response to the coo. When the removed note was not replaced by silence, shortening the duration of the coo, or when the pauses before and after the second note were reversed, the response was significantly reduced. This suggested a sensitivity to the overall rhythmic structure of the coos (Slabbekoorn & ten Cate, 1999).

Although rhythm perception in doves may be closely tied to properties of their natural coos, this suggests that it is important to explore sensitivity to rhythms in patterns that share at least some properties with natural vocalizations. In another study, zebra finches were trained to discriminate conspecific songs and subsequently tested with novel versions that altered amplitude, fundamental frequency, or duration (Nagel, McLendon, & Doupe,

2010). Although performance decreased substantially with changes in amplitude or fundamental frequency, it was maintained with duration changes of over 25 percent, well beyond zebra finches' reported sensitivity to temporal changes (Dooling et al., 2002). The implication may be that zebra finches are better at recognizing the rhythmic patterning of songs across tempo generalizations than they are in recognizing that of artificial stimuli (but see ten Cate et al., 2016, for alternative interpretations).

Vocal Learning

The study of vocal learning has traditionally been tied more to its relevance to language acquisition (Doupe & Kuhl, 1999) than to musicality. However, vocal learning can also be relevant to understanding more about musical abilities. For example, consider the extensive research of Nicolai (Güttinger, Turner, Dobmeyer, & Nicolai, 2002; Nicolai, Gundacker, Teeselink, & Güttinger, 2014) on vocal learning in the bullfinch (*Pyrrhula pyrrhula*), a songbird. Although bullfinches normally learn their species-specific songs from conspecifics, they were trained to sing folk melodies whistled to them. One bullfinch learned a forty-five-note tune from a human tutor and sang it in transposition (i.e., at a different pitch level), indicating exceptional vocal learning and relative pitch processing skills, also incorporating appropriate rhythm. Other bullfinches alternated parts with the human tutor, as in antiphonal singing, indicating that they anticipated as well as followed the notes of a learned melody. Experiments such as these build on the natural abilities of animals, productively extending them to controlled contexts.

Entrainment and Beat Induction

The two best-known features of musicality found in distantly related species are vocal learning and entrainment (see table 7.1). There are suggestions that the two abilities are related (Patel, Iverson, Bregman, & Schulz, 2009b). To date, the species that have been shown to exhibit both vocal learning and entrainment are distantly related to humans. Figure 7.1 shows the relatedness of various vertebrate species, indicating which have vocal learning and entrainment abilities. Snowball, the sulfur-crested cockatoo (*Cacatua galerita eleonora*) whose dance video became a YouTube sensation, helped renew scientific interest in entrainment in nonhuman species. Systematic study revealed that Snowball could synchronize his movements to the beat of music and adjust his rate of movement to changes in tempo (Patel, Iverson, Bregman, & Schulz, 2009a), challenging the notion that entrainment is uniquely human. The authors suggested, moreover, that such entrainment might be evident in other species that, like humans, are vocal learners. In fact, a study of YouTube videos featuring animal "dancing" provided confirmation of entrainment to music in vocal learning species but not in other species (e.g., dogs; Schachner et al., 2009). Another consistent finding was successful training of a budgerigar to tap along with a beat (Hasegawa, Okanoya, Hasegawa, & Seki, 2011), although there is so far no evidence that

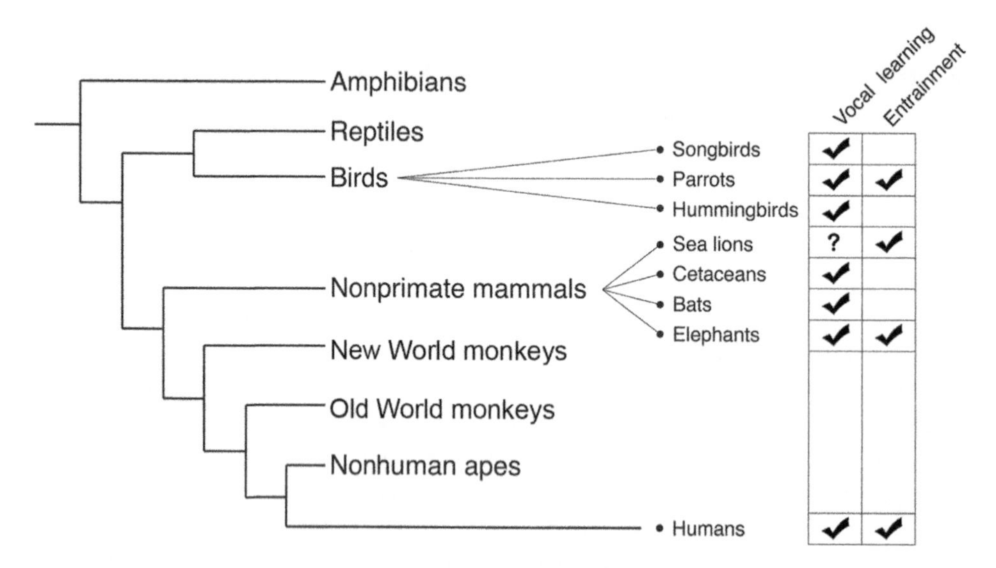

Figure 7.1
Species with vocal learning and entrainment abilities and their relationship in a phylogenetic tree.

budgerigars can spontaneously generalize across tempo transformations of a beat pattern (ten Cate et al., 2016).

That only vocal learners have the capacity for entrainment seems reasonable, given that the three avian groups in which vocal learning has evolved independently have similar functional neural pathways that are not shared with vocal nonlearners and are comparable to humans (Jarvis, 2004). That is, they have a direct connection between auditory perception areas and motor areas (Feenders et al., 2008). Entrainment may necessitate this kind of neural architecture (Patel et al., 2009b). At the same time, Schachner et al. (2009) found evidence for entrainment in only one of the avian vocal learning subgroups, the parrot species, and not in songbirds, and the relationship between vocal learning and beat perception in birds is still far from clear (ten Cate et al., 2016). The only nonparrot species in which entrainment has been detected to date is elephants. Although elephants show evidence of vocal learning (Stoeger et al. 2012), their vocal learning mechanism is unknown, but it is likely to differ from that of parrots. A compelling recent study showed that a California sea lion (*Zalophus californianus*) could also be trained to synchronize with a beat and then spontaneously generalized to music (Cook, Rouse, Wilson, & Reichmuth, 2013). This species is not thought to be a vocal learner, although some other pinnipeds are vocal learners (Tyack, 2008). It is possible that sea lions have vocal learning abilities that are as yet unknown. However, it could also be that the ability to synchronize with a beat requires only part of what is required for vocal learning or even that entrainment abilities can occur without any of the components for vocal learning. Another study showed that chimpanzees, one of the closest nonvocal learning relatives of humans, spontaneously entrained to a beat while completing a motor tapping task (Hattori, Tomonaga, & Matsuzawa, 2013, 2015), while another one showed regular sequences of spontaneous drumming (Dufour, Poulin, Curé, & Sterck, 2015). Also a bonobo showed spontaneous tempo matching in a social drumming interaction with a human drummer (Large & Gray 2015). Clearly, the proposed connection between vocal learning and entrainment (Patel et al., 2009b) requires further research with species that are not vocal learners.

If entrainment is defined more broadly, it could include many nonvocal learning species such as several species of fireflies synchronizing flashing with one another (Buck, 1988) and several species of frogs (Gerhardt & Huber, 2002) and katydids (Greenfield & Schul, 2008) synchronizing their chorusing. Identifying a pulse and locking in phase with it is a simpler task than detecting and entraining to a beat within a stream of music, where the beat is not always marked with an acoustic event and other acoustic events are present between beats (see Patel et al., 2009b, for review). Understanding the range of natural abilities related to entrainment could clarify what is relevant for musicality.

Natural Behavior

There are other potentially productive means of studying the precursors of musicality in nonhuman species. One approach is to search for music-like features in animal vocalizations. For example, Araya-Salas (2012) examined whether the pitch ratios created by adjacent notes of the song of nightingale wrens (*Microcerculus philomela*) conform to harmonic pitch ratios. From 243 comparisons, only six were significantly close to harmonic pitch ratios, suggesting no consistent use of harmonic pitch ratios. However, musician wrens favored consonant rather than dissonant intervals in their songs, with a clear preference for perfect rather than imperfect consonances (Doolittle & Brumm, 2012). Also, Doolittle, Gingras, Endres, and Fitch (2014) showed that the hermit thrush (*Catharus guttatus*), a bird that has long been thought to be especially "musical," does appear to use simple pitch ratios that follow the harmonic series the way humans do. Another approach builds on the studies by Hartshorne (1973) and others in seeking "aesthetic" features in natural birdsong that might have arousing or emotive consequences, as music does for human listeners. This notion was met with considerable skepticism, but Rothenberg, Roeske, Voss, Naguib, and Tchernichovski (2014) posed a similar question with thrush nightingales (*Luscinia luscinia*). According to their analysis, the songs of thrush nightingales have similar patterns of tension and resolution to those of music, which create expectation and anticipation in human listeners. If a certain level of familiarity and novelty is valued across species that produce complex songs, this could lead to insights on the origins of our motivation for music. Finally, another relevant recent discovery is the synchronization observed in the multimodal courtship behavior of birds. Java sparrows coordinate bill-click sounds with the syntax of their songs (Soma & Mori 2015), while male blue-capped cordon-bleus (*Uraeginthus cyanocephalus*) not only sing to attract a mate but also perform a visual display that could be described as a tap dance (Ota, Gahr, & Soma, 2015). Such displays appear at least superficially analogous to humans singing and clapping or dancing together and would lend themselves well to further study in the context of musicality.

Another route to discovering music-like behaviors in nonhuman species is to make predictions from the natural behavior of humans. For example, humans generalize across timbres, recognizing a melody regardless of the instrument on which it is played. In most cases, it makes sense not to generalize across timbres. Different spectral information can change the meaning of vocalizations not only in human speech with different vowels but in animal vocalizations as well (e.g., see Templeton, Greene, & Davis, 2005). A study species that may be more fruitful for timbre generalization research would be one that mimics the vocalizations of other species. For satin bowerbirds (*Ptilonorhynchus violaceus*), female preference for mates and male mate success may depend on the accuracy with which males imitate heterospecific vocalizations (Coleman, Patricelli, Coyle, Siani, & Borgia, 2007). If the mimetic accuracy is what is important, and not simply how well learned a song is (as has been shown to be important in other species, e.g., Nowicki, Searcy, & Peters, 2002), female bowerbirds would need to assess the original heterospecific vocalization, and the conspecific imitation, in a way that is similar to a human evaluating a singer's performance in comparison to a pianist. She would need to be able to distinguish the two but also generalize between them in the sense that she is aware that they are meant to be the same thing. In short, reflecting on natural human and nonhuman behaviors that are musically relevant can provide ideas about species and abilities that offer promising directions for comparative study.

Conclusion

Naturalistic studies have revealed important abilities and questions related to the biological basis of music such as vocal learning and entrainment. They have also suggested new directions for experimental research on perceptual abilities.

Experimental studies on perception often reveal abilities that are not used by nonhuman species under natural conditions. Knowledge of the underlying capacity for those abilities can contribute to an understanding of their evolutionary, developmental, and physiological foundations. The capacity for a particular ability, even if it is unrealized in nature, may arise from the evolutionary history of the species. Identifying their evolutionary pressures may be facilitated by studying the limits of these abilities.

Research on aspects of musicality in various nonhuman species is increasing, but it is rare to find naturalistic studies of musically relevant abilities and studies of the limits of those abilities in the same species. An example of such a combined approach involves the chickadee, which has been the subject of extensive field and laboratory research. This blend of research methods made it possible to understand the relative pitch processing skills of this species (Christie, Mennill, & Ratcliffe, 2004; Hoeschele, Guillette, & Sturdy, 2012; Njegovan & Weisman, 1997; Weisman & Ratcliffe, 2004; Weisman et al., 1990). Hence, experimental laboratory studies on artificial stimuli and more naturalistic studies can provide equally important and complementary insights into musically relevant skills.

Currently, vocal learning and entrainment are the principal focus of research on musically related behaviors and their underpinnings in nonhuman species. There are other potentially productive questions that could be pursued. For example, what kinds of behavior require relative pitch preferences like those present in chickadees (Hoeschele, Guillette, & Sturdy, 2012)? What kinds of behavior require timbre generalization like that observed in zebra finches (Ohms et al. 2010; Spierings & ten Cate, 2014)? Why is auditory grouping relevant in some species? One way forward is to search for relevant behavioral and perceptual abilities of species less studied in the laboratory and to examine the natural behaviors of species commonly studied in the laboratory.

Acknowledgments

This chapter is a slightly modified and updated version of a paper originally published as "Searching for the Origins of Musicality across Species" in *Philosophical Transactions of the Royal Society B* (see Hoeschele, Merchant, Kikuchi, Hattori, & ten Cate, 2015). M.H. was funded by a European Research Council advanced grant (No. 230604 "SOMACCA") awarded to W. Tecumseh Fitch at the University of Vienna and a Banting Postdoctoral Fellowship awarded by the Natural Sciences and Engineering Research Council of Canada hosted by the University of Vienna during the writing of the original article that this chapter is based on (see Hoeschele et al., 2015). M.H. is currently funded by a Lise Meitner Postdoctoral Fellowship (M 1732-B19) from the Austrian Science Fund. H.M. is supported by PAPIIT IN201214-25, CONACYT 236836, CONACYT Fronteras 196. Y.K. is currently supported by a contract from the National Institutes of Health, BBSRC (BB/J009849/1), and Wellcome Trust (WT102961/Z/13/Z) grants to C. Petkov at Newcastle University. Y.H. is supported by a Grant-in-Aid for Young Scientists (B)(26730074) from the Japan Society for the Promotion of Science (JSPS). Thank you to everyone who attended the "What Makes Us Musical Animals? Cognition, Biology and the Origins of Musicality" workshop for the stimulating meeting that led to this book and our special issue in *Philosophical Transactions B*, especially Henkjan Honing for organizing the meeting and the other attendees who participated in the discussion relating directly to this chapter on natural VS artificial studies of animal music perception, including Peter Tyack and Aniruddh Patel. Thank you to Sandra Trehub for her very detailed revisions as editor of the *Philosophical Transactions B* version of this chapter and to the anonymous reviewers for their insights on its organization. Thank you to Daniel L. Bowling for help drawing the figure.

References

Amagai, S., Dooling, R. J., Shamma, S., Kidd, T. L., & Lohr, B. (1999). Detection of modulation in spectral envelopes and linear-rippled noises by budgerigars (*Melopsittacus undulatus*). *Journal of the Acoustical Society of America, 105*(3), 2029–2035. doi:10.1121/1.426736

Araya-Salas, M. (2012). Is birdsong music? Evaluating harmonic intervals in songs of a neotropical songbird. *Animal Behaviour, 84*(2), 309–313. doi:10.1016/j.anbehav.2012.04.038

Ballintijn, M. R., & ten Cate, C. (1999). Variation in number of elements in the perch-coo vocalization of the collared dove (*Streptopelia decaocto*) and what it may tell about the sender. *Behaviour, 136*(7), 847–864. doi:10.1163/156853999501603

Bion, R., Benavides-Varela, S., & Nespor, M. (2011). Acoustic markers of prominence influence infants' and adults' segmentation of speech sequences. *Language and Speech, 54*(1), 123–140. doi:10.1177/0023830910388018

Blackwell, H. R., & Schlosberg, H. (1943). Octave generalization, pitch discrimination, and loudness thresholds in the white rat. *Journal of Experimental Psychology, 33*(5), 407–419. doi:10.1037/h0057863

Bregman, M. R., Patel, A. D., & Gentner, T. Q. (2012). Stimulus-dependent flexibility in non-human auditory pitch processing. *Cognition, 122*(1), 51–60. doi:10.1016/j.cognition.2011.08.008

Bregman, M. R., Patel, A. D., & Gentner, T. Q. (2016). Songbirds use spectral shape, not pitch, for sound pattern recognition. *Proceedings of the National Academy of Sciences, 113*(6), 1666–1671. doi:10.1073/pnas.1515380113

Brooks, D. I., & Cook, R. G. (2010). Chord discrimination by pigeons. *Music Perception, 27*(3), 183–196. doi:10.1525/mp.2010.27.3.183

Buck, J. (1988). Synchronous rhythmic flashing in fireflies. II. *Quarterly Review of Biology, 63*(3), 265–289. doi:10.1086/415929

Burns, E. M. (1999). Intervals, scales, and tuning. In D. Deutsch (Ed.), *The psychology of music* (2nd ed., pp. 215–264). San Diego, CA: Academic Press.

Chase, A. R. (2001). Music discriminations by carp (*Cyprinus carpio*). *Animal Learning and Behavior, 29*(4), 336–353. doi:10.3758/BF03192900

Chiandetti, C., & Vallortigara, G. (2011). Chicks like consonant music. *Psychological Science, 22*(10), 1270–1273. doi:10.1177/0956797611418244

Christie, P. J., Mennill, D. J., & Ratcliffe, L. M. (2004). Pitch shifts and song structure indicate male quality in the dawn chorus of black-capped chickadees. *Behavioral Ecology and Sociobiology, 55*(4), 341–348. doi:10.1007/s00265-003-0711-3

Coleman, S. W., Patricelli, G. L., Coyle, B., Siani, J., & Borgia, G. (2007). Female preferences drive the evolution of mimetic accuracy in male sexual displays. *Biology Letters, 3*(5), 463–466. doi:10.1098/rsbl.2007.0234

Cook, N. D., & Fujisawa, T. X. (2006). The psychophysics of harmony perception: Harmony is a three-tone phenomenon. *Empirical Musicology Review, 1*(2), 106–126.

Cook, P., Rouse, A., Wilson, M., & Reichmuth, C. (2013). A California sea lion (*Zalophus californianus*) can keep the beat: Motor entrainment to rhythmic auditory stimuli in a non vocal mimic. *Journal of Comparative Psychology, 127*(4), 412–427. doi:10.1037/a0032345

Crespo-Bojorque, P., & Toro, J. M. (2015). The use of interval ratios in consonance perception by rats (*Rattus norvegicus*) and humans (*Homo sapiens*). *Journal of Comparative Psychology, 129*(1), 42–51. doi:10.1037/a0037991

Crowe, D. A., Zarco, W., Bartolo, R., & Merchant, H. (2014). Dynamic representation of the temporal and sequential structure of rhythmic movements in the primate medial premotor cortex. *Journal of Neuroscience, 34*(36), 11972–11983. doi:10.1523/ jneurosci.2177–14.2014

Cruttenden, A. (1981). Falls and rises: Meanings and universals. *Journal of Linguistics, 17*(1), 77. doi:10.1017/s0022226700006782

Cutler, A., Dahan, D., & van Donselaar, W. (1997). Prosody in the comprehension of spoken language: A literature review. *Language and Speech, 40*(2), 140–201. doi:10.1177/002383099704000203

Cynx, J. (1993). Auditory frequency generalization and a failure to find octave generalization in a songbird, the European starling (*Sturnus vulgaris*). *Journal of Comparative Psychology, 107*(2), 140–146. doi:10.1037/0735-7036.107.2.140

Cynx, J., & Shapiro, M. (1986). Perception of missing fundamental by a species of songbird (*Sturnus vulgaris*). *Journal of Comparative Psychology, 100*(4), 356–360. doi:10.1037/0735-7036.100.4.356

Cynx, J., Williams, H., & Nottebohm, F. (1990). Timbre discrimination in zebra finch (*Taeniopygia guttata*) song syllables. *Journal of Comparative Psychology, 104*(4), 303–308. doi:10.1037/0735-7036.104.4.303

De la Mora, D. M., Nespor, M., & Toro, J. M. (2013). Do humans and nonhuman animals share the grouping principles of the iambic-trochaic law? *Attention, Perception and Psychophysics, 75*(1), 92–100. doi:10.3758/s13414-012-0371-3

Donnet, S., Bartolo, R., Fernandes, J. M., Cunha, J. P. S., Prado, L., & Merchant, H. (2014). Monkeys time their pauses of movement and not their movement-kinematics during a synchronization-continuation rhythmic task. *Journal of Neurophysiology, 111*(10), 2138–2149. doi:10.1152/jn.00802.2013

Dooling, R. J., Leek, M. R., Gleich, O., & Dent, M. L. (2002). Auditory temporal resolution in birds: Discrimination of harmonic complexes. *Journal of the Acoustical Society of America, 112*(2), 748–759. doi:10.1121/1.1494447

Doolittle, E., & Brumm, H. (2012). O Canto do Uirapuru: Consonant intervals and patterns in the song of the musician wren. *Journal of Interdisciplinary Music Studies, 6*(1), 55–85.

Doolittle, E. L., Gingras, B., Endres, D. M., & Fitch, W. T. (2014). Overtone-based pitch selection in hermit thrush song: Unexpected convergence with scale construction in human music. *Proceedings of the National Academy of Sciences, 111*(46), 16616–16621. doi:10.1073/pnas.1406023111

Doupe, A. J., & Kuhl, P. K. (1999). Birdsong and human speech: Common themes and mechanisms. *Annual Review of Neuroscience, 22*, 567–631. doi:10.1146/annurev.neuro.22.1.567

Dowling, W. J., & Harwood, D. L. (1986). *Music cognition*. Orlando, FL: Academic Press.

Dufour, V., Poulin, N., Curé, C., & Sterck, E. H. M. (2015). Chimpanzee drumming: A spontaneous performance with characteristics of human musical drumming. *Scientific Reports, 5*, 11320. doi:10.1038/srep11320

Feenders, G., Liedvogel, M., Rivas, M., Zapka, M., Horita, H., Hara, E., et al. (2008). Molecular mapping of movement-associated areas in the avian brain: A motor theory for vocal learning origin. *PLoS One, 3*(3), 1–27. doi:10.1371/journal.pone.0001768

Friedrich, A., Zentall, T., & Weisman, R. (2007). Absolute pitch: Frequency-range discriminations in pigeons (*Columba livia*)—comparisons with zebra finches (*Taeniopygia guttata*) and humans (*Homo sapiens*). *Journal of Comparative Psychology, 121*, 95–105. doi:10.1037/0735-7036.121.1.95

Gerhardt, H. C., & Huber, F. (2002). *Acoustic communication in insects and anurans*. Chicago: University of Chicago Press.

Grahn, J. A. (2009). The role of the basal ganglia in beat perception. *Annals of the New York Academy of Sciences, 1169*, 35–45. doi:10.1111/j.1749-6632.2009.04553.x

Grahn, J. A., & Brett, M. (2007). Rhythm and beat perception in motor areas of the brain. *Journal of Cognitive Neuroscience, 19*(5), 893–906. doi:10.1162/jocn.2007.19.5.893

Greenfield, M. D., & Schul, J. (2008). Mechanisms and evolution of synchronous chorusing: Emergent properties and adaptive functions in Neoconocephalus katydids (*Orthoptera: Tettigoniidae*). *Journal of Comparative Psychology, 122*(3), 289–297. doi:10.1037/ 0735–7036.122.3.289

Güttinger, H. R., Turner, T., Dobmeyer, S., & Nicolai, J. (2002). Melodiewahrnehmung und Wiedergabe beim Gimpel: Untersuchungen an liederpfeifenden und Kanariengesang imitierenden Gimpeln (*Pyrrhula pyrrhula*). *Journal für Ornithologie, 143*(3), 303–318. doi:10.1046/j.1439-0361.2002.02017.x

Háden, G. P., Stefanics, G., Vestergaard, M. D., Denham, S. L., Sziller, I., & Winkler, I. (2009). Timbre-independent extraction of pitch in newborn infants. *Psychophysiology, 46*(1), 69–74. doi:10.1111/j.1469-8986.2008.00749.x

Hagmann, C. E., & Cook, R. G. (2010). Testing meter, rhythm, and tempo discriminations in pigeons. *Behavioural Processes, 85*(2), 99–110. doi:10.1016/j.beproc.2010.06.015

Hartshorne, C. (1973). *Born to sing: An interpretation and world survey of bird song*. Bloomington: Indiana University Press.

Hasegawa, A., Okanoya, K., Hasegawa, T., & Seki, Y. (2011). Rhythmic synchronization tapping to an audio-visual metronome in budgerigars. *Scientific Reports, 1*, 120. doi:10.1038/srep00120

Hattori, Y., Tomonaga, M., & Matsuzawa, T. (2013). Spontaneous synchronized tapping to an auditory rhythm in a chimpanzee. *Scientific Reports, 3*, 1566. doi:10.1038/ srep01566

Hattori, Y., Tomonaga, M., & Matsuzawa, T. (2015). Distractor effect of auditory rhythms on self-paced tapping in chimpanzees and humans. *PLoS One, 10*(7), 1–17. doi:10.1371/ journal.pone.0130682

Heffner, H., & Whitfield, I. C. (1976). Perception of the missing fundamental by cats. *Journal of the Acoustical Society of America, 59*(4), 915–919. doi:10.1121/1.380951

Helmholtz, H. L. F. (1877). *On the sensations of tone as a physiological basis for the theory of music.* New York: Dover.

Herrnstein, R. J., & Loveland, D. H. (1964). Complex visual concept in a pigeon. *Science, 146*(3643), 549–551. doi:10.1126/science.146.3643.549

Hikosaka, O., Nakamura, K., Sakai, K., & Nakahara, H. (2002). Central mechanisms of motor skill learning. *Current Opinion in Neurobiology, 12*(2), 217–222. doi:10.1016/s0959-4388(02)00307-0

Hikosaka, O., Rand, M., Nakamura, K., Miyachi, S., Kitaguchi, K., Sakai, K., et al. (2002). Long-term retention of motor skill in macaque monkeys and humans. *Experimental Brain Research, 147*(4), 494–504. doi:10.1007/ s00221-002-1258-7

Hoeschele, M., Cook, R. G., Guillette, L. M., Brooks, D. I., & Sturdy, C. B. (2012). Black-capped chickadee (*Poecile atricapillus*) and human (*Homo sapiens*) chord discrimination. *Journal of Comparative Psychology, 126*(1), 57–67. doi:10.1037/ a0024627

Hoeschele, M., Cook, R. G., Guillette, L. M., Hahn, A. H., & Sturdy, C. B. (2014). Timbre influences chord discrimination in black-capped chickadees (*Poecile atricapillus*) but not humans (*Homo sapiens*). *Journal of Comparative Psychology, 128*(4), 387–401. doi:10.1037/a0037159

Hoeschele, M., & Fitch, W. T. (2016). Phonological perception by birds: Budgerigars can perceive lexical stress. *Animal Cognition, 19*(3), 643–654. doi:10.1007/s10071-016-0968-3

Hoeschele, M., Guillette, L. M., & Sturdy, C. B. (2012). Biological relevance of acoustic signal affects discrimination performance in a songbird. *Animal Cognition, 15*(4), 677–688. doi:10.1007/s10071-012-0496-8

Hoeschele, M., Merchant, H., Kikuchi, Y., Hattori, Y., ten Cate, C., & Hoeschele, M. (2015). Searching for the origins of musicality across species. *Philosophical Transactions of the Royal Society B: Biological Sciences, 370*(1664), 1–9. doi:10.1098/rstb.2014.0094

Hoeschele, M., Weisman, R. G., Guillette, L. M., Hahn, A. H., & Sturdy, C. B. (2013). Chickadees fail standardized operant tests for octave equivalence. *Animal Cognition, 16*(4), 599–609. doi:10.1007/s10071-013-0597-z

Hoeschele, M., Weisman, R. G., & Sturdy, C. B. (2012). Pitch chroma discrimination, generalization, and transfer tests of octave equivalence in humans. *Attention, Perception, and Psychophysics, 74*(8), 1742–1760. doi:10.3758/s13414-012-0364-2

Honing, H., ten Cate, C., Peretz, I., & Trehub, S. E. (2015). Without it no music: Cognition, biology, and evolution of musicality. *Philosophical Transactions of the Royal Society B: Biological Sciences, 370*(1664), 1–8. doi:10.1098/rstb.2014.0088

Hulse, S. H., Bernard, D. J., & Braaten, R. F. (1995). Auditory discrimination of chord-based spectral structures by European starlings (*Sturnus vulgaris*). *Journal of Experimental Psychology: General, 124*(4), 409–423. doi:10.1037/0096-3445.124.4.409

Hulse, S. H., & Cynx, J. (1985). Relative pitch perception is constrained by absolute pitch in songbirds (*Mimus, Molothrus*, and *Sturnus*). *Journal of Comparative Psychology, 99*(2), 176–196. doi:10.1037//0735-7036.99 .2.176

Hulse, S. H., & Cynx, J. (1986). Interval and contour in serial pitch perception by a passerine bird, the European starling (*Sturnus vulgaris*). *Journal of Comparative Psychology, 100*(3), 215–228. doi:10.1037//0735-7036.100 .3.215

Hulse, S. H., Cynx, J., & Humpal, J. (1984). Absolute and relative pitch discrimination in serial pitch perception by birds. *Journal of Experimental Psychology: General, 113*(1), 38–54. doi:10.1037/0096-3445.113.1.38

Hulse, S. H., Humpal, J., & Cynx, J. (1984). Discrimination and generalization of rhythmic and arrhythmic sound patterns by European starlings (*Sturnus vulgaris*). *Music Perception, 1*(4), 442–464. doi:10.2307/ 40285272

Izumi, A. (2000). Japanese monkeys perceive sensory consonance of chords. *Journal of the Acoustical Society of America, 108*(6), 3073–3078. doi:10.1121/1.1323461

Janata, P., & Grafton, S. T. (2003). Swinging in the brain: Shared neural substrates for behaviors related to sequencing and music. *Nature Neuroscience, 6*(7), 682–687. doi:10.1038/nn1081

Jarvis, E. D. (2004). Learned birdsong and the neurobiology of human language. *Annals of the New York Academy of Sciences, 1016*, 749–777. doi:10.1196/annals.1298.038

Kikuchi, Y., Horwitz, B., Mishkin, M., & Rauschecker, J. P. (2014). Processing of harmonics in the lateral belt of macaque auditory cortex. *Frontiers in Neuroscience, 8*, 204. doi:10.3389/fnins.2014.00204

Large, E. W., & Gray, P. M. (2015). Spontaneous tempo and rhythmic entrainment in a bonobo (*Pan paniscus*). *Journal of Comparative Psychology, 129*(4), 317–328. doi:10.1037/com0000011

Lau, B. K., & Werner, L. A. (2012). Perception of missing fundamental pitch by 3- and 4-month-old human infants. *Journal of the Acoustical Society of America, 132*(6), 3874–3882. doi:10.1121/1.4763991

Lohr, B., & Dooling, R. J. (1998). Detection of changes in timbre and harmonicity in complex sounds by zebra finches (*Taeniopygia guttata*) and budgerigars (*Melopsittacus undulatus*). *Journal of Comparative Psychology, 112*(1), 36–47. doi:10.1037/0735-7036.112.1.36

Macar, F., Coull, J., & Vidal, F. (2006). The supplementary motor area in motor and perceptual time processing: fMRI studies. *Cognitive Processing, 7*(2), 89–94. doi:10.1007/s10339-005-0025-7

MacDougall-Shackleton, S. A., & Hulse, S. H. (1996). Concurrent absolute and relative pitch processing by European starlings (*Sturnus vulgaris*). *Journal of Comparative Psychology, 110*(2), 139–146. doi:10.1037//0735-7036.110.2.139

McDermott, J., & Hauser, M. (2004). Are consonant intervals music to their ears? Spontaneous acoustic preferences in a nonhuman primate. *Cognition, 94*(2), B11–B21. doi:10.1016/j.cognition.2004.04.004

McDermott, J., & Hauser, M. D. (2007). Nonhuman primates prefer slow tempos but dislike music overall. *Cognition, 104*(3), 654–668. doi:10.1016/j.cognition.2006.07.011

Mendez, J. C., Prado, L., Mendoza, G., & Merchant, H. (2010). Temporal and spatial categorization in human and non-human primates. *Frontiers in Integrative Neuroscience, 5*, 1–10. doi:10.3389/fnint.2011.00050

Mendoza, G., & Merchant, H. (2014). Motor system evolution and the emergence of high cognitive functions. *Progress in Neurobiology, 122*, 73–93. doi:10.1016/j.pneurobio. 2014.09.001

Merchant, H., Battaglia-Mayer, A., & Georgopoulos, A. P. (2003). Interception of real and apparent motion targets: Psychophysics in humans and monkeys. *Experimental Brain Research, 152*(1), 106–112. doi:10.1007/s00221-003-1514-5

Merchant, H., Grahn, J., Trainer, L., Rohrmeier, M., & Fitch, T. W. (2015). Finding the beat: A neural perspective across humans and non-human primates. *Philosophical Transactions of the Royal Society B: Biological Sciences, 370*(1664), 1–16. doi:10.1098/rstb.2014.0093

Merchant, H., & Honing, H. (2014). Are non-human primates capable of rhythmic entrainment? Evidence for the gradual audiomotor evolution hypothesis. *Frontiers in Neuroscience, 7*, 1–8. doi:10.3389/fnins.2013.00274

Merchant, H., Pérez, O., Zarco, W., & Gámez, J. (2013). Interval tuning in the primate medial premotor cortex as a general timing mechanism. *Journal of Neuroscience, 33*(21), 9082–9096. doi:10.1523/jneurosci.5513-12.2013

Mingle, M. E., Eppley, T. M., Campbell, M. W., Hall, K., Horner, V., & de Waal, F. B. M. (2014). Chimpanzees prefer African and Indian music over silence. *Journal of Experimental Psychology: Animal Behavior Processes, 40*(4), 502–505. doi:10.1037/xan0000032

Mita, A., Mushiake, H., Shima, K., Matsuzaka, Y., & Tanji, J. (2009). Interval time coding by neurons in the presupplementary and supplementary motor areas. *Nature Neuroscience, 12*(4), 502–507. doi:10.1038/nn.2272

Nagel, K. I., McLendon, H. M., & Doupe, A. J. (2010). Differential influence of frequency, timing, and intensity cues in a complex acoustic categorization task. *Journal of Neurophysiology, 104*(3), 1426–1437. doi:10.1152/jn.00028.2010

Naoi, N., Watanabe, S., Maekawa, K., & Hibiya, J. (2012). Prosody discrimination by songbirds (*Padda oryzivora*). *PLoS One, 7*(10), 1–7. doi:10.1371/journal.pone.0047446

Nicolai, J., Gundacker, C., Teeselink, K., & Güttinger, H. R. (2014). Human melody singing by bullfinches (*Pyrrhula pyrrula*) gives hints about a cognitive note sequence processing. *Animal Cognition*, *17*(1), 143–155. doi:10.1007/s10071-013-0647-6

Njegovan, M., & Weisman, R. (1997). Pitch discrimination in field- and isolation-reared black-capped chickadees (*Parus atricapillus*). *Journal of Comparative Psychology*, *111*(3), 294–301. doi:10.1037/0735-7036.111.3.294

Nowicki, S., Searcy, W. A., & Peters, S. (2002). Quality of song learning affects female response to male bird song. *Proceedings of the Royal Society B: Biological Sciences*, *269*(1503), 1949–1954. doi:10.1098/rspb.2002.2124

Ohms, V. R., Gill, A., Van Heijningen, C. A. A., Beckers, G. J. L., & ten Cate, C. (2010). Zebra finches exhibit speaker-independent phonetic perception of human speech. *Proceedings of the Royal Society B: Biological Sciences*, *277*(1684), 1003–1009. doi:10.1098/rspb.2009.1788

Ota, N., Gahr, M., & Soma, M. (2015). Tap dancing birds: The multimodal mutual courtship display of males and females in a socially monogamous songbird. *Scientific Reports*, *5*, 1–6. doi:10.1038/srep16614

Oxenham, A. J. (2012). Pitch perception. *Journal of Neuroscience*, *32*(39), 13335–13338. doi:10.1523/jneurosci.3815-12.2012

Page, S. C., Hulse, S. H., & Cynx, J. (1989). Relative pitch perception in the European starling (*Sturnus vulgaris*): Further evidence for an elusive phenomenon. *Journal of Experimental Psychology: Animal Behavior Processes*, *15*(2), 137–146. doi:10.1037/0097-7403.15.2.137

Papoutsoglou, S. E., Karakatsouli, N., Psarrou, A., Apostolidou, S., Papoutsoglou, E. S., Batzina, A., et al. (2015). Gilthead seabream (*Sparus aurata*) response to three music stimuli (Mozart—"Eine kleine Nachtmusik," Anonymous—"Romanza," Bach—"Violin Concerto No. 1") and white noise under recirculating water conditions. *Fish Physiology and Biochemistry*, *42*(1), 219–232. doi:10.1007/s10695-014-0018-5

Patel, A. D. (2003). Language, music, syntax and the brain. *Nature Neuroscience*, *6*(7), 674–681. doi:10.1038/nn1082

Patel, A. D., Iversen, J. R., Bregman, M. R., & Schulz, I. (2009a). Experimental evidence for synchronization to a musical beat in a nonhuman animal. *Current Biology*, *19*(10), 827–830. doi:10.1016/j.cub.2009.03.038

Patel, A. D., Iversen, J. R., Bregman, M. R., & Schulz, I. (2009b). Studying synchronization to a musical beat in nonhuman animals. *Annals of the New York Academy of Sciences*, *1169*, 459–469. doi:10.1111/j.1749-6632.2009.04581.x

>Porter, D., & Neuringer, A. (1984). Music discriminations by pigeons. *Journal of Experimental Psychology: Animal Behavior Processes*, *10*(2), 138–148. doi:10.1037//0097-7403.10.2.138

Ralston, J. V., & Herman, L. M. (1995). Perception and generalization of frequency contours by a bottlenose dolphin (*Tursiops truncatus*). *Journal of Comparative Psychology*, *109*(3), 268–277. doi:10.1037/0735-7036.109.3.268

Reinert, J. (1965). Takt und Rhythmusunterscheidung bei Dohlen. *Zeitschrift für Tierpsychologie*, *22*(6), 623–671. doi:10.1111/j.1439-0310.1965.tb01683.x

Repp, B. H. (2005). Sensorimotor synchronization: A review of the tapping literature. *Psychonomic Bulletin and Review*, *12*(6), 969–992. doi:10.3758/bf03206433

Richards, D. G., Wolz, J. P., & Herman, L. M. (1984). Vocal mimicry of computer-generated sounds and vocal labeling of objects by a bottlenosed dolphin, *Tursiops truncatus*. *Journal of Comparative Psychology*, *98*(1), 10–28. doi:10.1037/0735-7036.98.1.10

Rothenberg, D., Roeske, T. C., Voss, H. U., Naguib, M., & Tchernichovski, O. (2014). Investigation of musicality in birdsong. *Hearing Research*, *308*, 71–83. doi:10.1016/j.heares.2013.08.016

Schachner, A., Brady, T. F., Pepperberg, I. M., & Hauser, M. D. (2009). Spontaneous motor entrainment to music in multiple vocal mimicking species. *Current Biology*, *19*(10), 831–866. doi:10.1016/j.cub.2009.03.061

Schellenberg, E. G., & Trehub, S. E. (2003). Good pitch memory is widespread. *Psychological Science*, *14*(3), 262–266. doi:10.1111/1467-9280.03432

Slabbekoorn, H., & ten Cate, C. (1999). Collared dove responses to playback: Slaves to the rhythm. *Ethology*, *105*(5), 377–391. doi:10.1046/j.1439-0310.1999.00420.x

Soma, M., & Mori, C. (2015). The songbird as a percussionist: Syntactic rules for non-vocal sound and song production in Java sparrows. *PLoS One*, *10*(5), 1–10. doi:10.1371/journal.pone.0124876

Spierings, M. J., & ten Cate, C. (2014). Zebra finches are sensitive to prosodic features of human speech. *Proceedings of the Royal Society B: Biological Sciences*, *281*(1787), 1–7. doi:10.1098/rspb.2014.0480

Stoeger, A. S., Mietchen, D., Oh, S., de Silva, S., Herbst, C. T., Kwon, S., et al. (2012). An Asian elephant imitates human speech. *Current Biology*, *22*, 2144–2148. doi:10.1016/j.cub.2012.09.022

Tanji, J. (2001). Sequential organization of multiple movements: Involvement of cortical motor areas. *Annual Review of Neuroscience*, *24*, 631–651. doi:10.1146/annurev.neuro.24.1.631

Templeton, C. N., Greene, E., & Davis, K. (2005). Allometry of alarm calls: Black-capped chickadees encode information about predator size. *Science*, *308*(5730), 1934–1937. doi:10.1126/science.1108841

ten Cate, C., Spierings, M., Hubert, J., & Honing, H. (2016). Can birds perceive rhythmic patterns? A review and experiments on a songbird and a parrot species. *Frontiers in Psychology*, *7*, 1–14. doi:10.3389/fpsyg.2016.00730

Tincoff, R., Hauser, M., Tsao, F., Spaepen, G., Ramus, F., & Mehler, J. (2005). The role of speech rhythm in language discrimination: Further tests with a non-human primate. *Developmental Science*, *8*(1), 26–35. doi:10.1111/j.1467-7687.2005.00390.x

Tomlinson, R. W. W. (1988). Perception of the missing fundamental in nonhuman primates. *Journal of the Acoustical Society of America*, *84*(2), 560–565. doi:10.1121/1.396833

Toro, J. M., & Nespor, M. (2015). Experience-dependent emergence of a grouping bias. *Biology Letters*, *11*(9), 1–4. doi:10.1098/rsbl.2015.0374.

Toro, J. M., Trobalon, J. B., & Sebastián-Gallés, N. (2003). The use of prosodic cues in language discrimination tasks by rats. *Animal Cognition*, *6*(2), 131–136. doi:10.1007/s10071-003-0172-0

Toro, J. M., Trobalon, J. B., & Sebastian-Galles, N. (2005). Effects of backward speech and speaker variability in language discrimination by rats. *Journal of Experimental Psychology: Animal Behavior Processes*, *31*(1), 95–100. doi:10.1037/0097-7403.31.1.95

Tyack, P. L. (2008). Convergence of calls as animals form social bonds, active compensation for noisy communication channels, and the evolution of vocal learning in mammals. *Journal of Comparative Psychology*, *122*(3), 319–331. doi:10.1037/a0013087

Uno, H., Maekawa, M., & Kaneko, H. (1997). Strategies for harmonic structure discrimination by (*Taeniopygia guttata*). *Behavioural Brain Research*, *89*(1–2), 225–228. doi:10.1016/S0166-4328(97)00064-8

van der Aa, J., Honing, H., & ten Cate, C. (2015). The perception of regularity in an isochronous stimulus in zebra finches (*Taeniopygia guttata*) and humans. *Behavioural Processes*, *115*, 37–45. doi:10.1016/j.beproc.2015.02.018

Ward, W. D. (1999). Absolute pitch. In D. Deutsch (Ed.), *The psychology of music* (pp. 265–298). San Diego, CA: Academic Press.

Warker, J. A., & Halpern, A. R. (2005). Musical stem completion: Humming that note. *American Journal of Psychology*, *118*(4), 567–585.

Watanabe, S., Uozumi, M., & Tanaka, N. (2005). Discrimination of consonance and dissonance in Java sparrows. *Behavioural Processes*, *70*(2), 203–208. doi:10.1016/ j.beproc.2005.06.001

Weisman, R. G., Balkwill, L.-L., Hoeschele, M., Moscicki, M. K., Bloomfield, L. L., & Sturdy, C. B. (2010). Absolute pitch in boreal chickadees and humans: Exceptions that test a phylogenetic rule. *Learning and Motivation*, *41*(3), 156–173. doi:10.1016/ j.lmot.2010.04.002

Weisman, R. G., Njegovan, M., & Ito, S. (1994). Frequency ratio discrimination by zebra finches (*Taeniopygia guttata*) and humans (*Homo sapiens*). *Journal of Comparative Psychology*, *108*, 363–372.

Weisman, R., Njegovan, M., Sturdy, C. B., Phillmore, L., Coyle, J., & Mewhort, D. (1998). Frequency-range discriminations: Special and general abilities in zebra finches (*Taeniopygia guttata*) and humans (*Homo sapiens*). *Journal of Comparative Psychology*, *112*(3), 244–258. doi:10.1037/0735-7036.112.3.244

Weisman, R. G., Njegovan, M. G., Williams, M. T., Cohen, J. S., & Sturdy, C. B. (2004). A behavior analysis of absolute pitch: Sex, experience, and species. *Behavioural Processes*, *66*(3), 289–307. doi:10.1016/j.beproc.2004.03.010

Weisman, R., & Ratcliffe, L. (2004). Relative pitch and the song of black-capped chickadees. *American Scientist, 92*(6), 532–539. doi:10.1511/2004.6.532

Weisman, R., Ratcliffe, L., Johnsrude, I., & Hurly, T. A. (1990). Absolute and relative pitch production in the song of the black-capped chickadee. *Condor, 92*(1), 118–124. doi:10.2307/1368390

Weisman, R. G., & Spetch, M. L. (2010). Determining when birds perceive correspondence between pictures and objects: A critique. *Comparative Cognition and Behavior Reviews, 5*, 117–131. doi:10.3819/ccbr.2010.50006

Woodrow, H. (1909). A quantitative study of rhythm: The effect of variations in intensity, rate and duration. *Archives de Psychologie, 14*, 1–66.

Wright, A. A., Rivera, J. J., Hulse, S. H., Shyan, M., & Neiworth, J. J. (2000). Music perception and octave generalization in rhesus monkeys. *Journal of Experimental Psychology: General, 129*(3), 291–307. doi:10.1037/0096-3445.129.3.291

Zarco, W., Merchant, H., Prado, L., & Mendez, J. C. (2009). Subsecond timing in primates: Comparison of interval production between human subjects and rhesus monkeys. *Journal of Neurophysiology, 102*(6), 3191–3202. doi:10.1152/jn.00066.2009

Zatorre, R. J., Chen, J. L., & Penhune, V. B. (2007). When the brain plays music: Auditory-motor interactions in music perception and production. *Nature Reviews: Neuroscience, 8*(7), 547–558. doi:10.1038/nrn2152

8

Finding the Beat: A Neural Perspective across Humans and Nonhuman Primates

Hugo Merchant, Jessica Grahn, Laurel J. Trainor, Martin Rohrmeier, and W. Tecumseh Fitch

Beat perception is a cognitive ability that allows the detection of a regular pulse (or beat) in music and permits synchronous responding to this pulse during dancing and musical ensemble playing (Honing, 2013; Large & Palmer, 2002). Most people can recognize and reproduce a large number of rhythms, and they can move in synchrony to the beat by rhythmically timed movements of different body parts (such as finger or foot taps or body sway). Beat perception and synchronization can be considered fundamental musical traits that arguably played a decisive role in the origins of music (Honing, 2013). A large proportion of human music is organized by a quasi-isochronous pulse and frequently also in a metrical hierarchy in which the beats of one level are typically spaced at two or three times those of a faster level (in the simplest Western cases, the tempo of one level is 1/2 [march meter] or 1/3 [waltz meter] that of the other), and human listeners can typically synchronize at more than one level of the metrical hierarchy (Drake, Jones, & Baruch, 2000; Large & Jones, 1999). Furthermore, movement on every second versus every third beat of an ambiguous rhythm pattern (one, for example, that can be interpreted as either a march or a waltz) biases listeners to interpret it as either a march or a waltz, respectively (Phillips-Silver & Trainor, 2007). Therefore, the concept of beat perception and synchronization implies both that the beat does not always need to be physically present in order to be "perceived" and the pulse evokes a particular perceptual pattern in the subject via active cognitive processes. Interestingly, humans do not need special training to perceive and motorically entrain to the beat in musical rhythms; rather, it appears to be a robust, ubiquitous and intuitive behavior. Indeed, even young infants perceive metrical structure (Winkler, Háden, Ladinig, Sziller, & Honing, 2009), and an infant who is bounced on every second beat or on every third beat of an ambiguous rhythm pattern is biased to interpret the meter of the auditory rhythm in a manner consistent with how he or she was moved to it. Thus, although rhythmic entrainment is a complex phenomenon that depends on a dynamic interaction between the auditory and motor systems in the brain (Merchant & Honing, 2014; Zatorre, Chen, & Penhune, 2007), it emerges very early in development without special training (Phillips-Silver & Trainor, 2005).

Recent studies support the notion that the timing mechanisms used in the brain depend on whether the time intervals in a sequence can be timed relative to a steady beat (relative, or beat-based, timing) or not (absolute, or duration-based) timing (Povel & Essens, 1985; Teki, Grube, Kumar, & Griffiths, 2011; Yee, Holleran, & Jones, 1994; (Schubotz, Friederici, Yves von Cramon, 2000). In relative timing, time intervals are measured relative to a regular perceived beat (Teki et al., 2011), to which individuals are able to entrain. In absolute timing, the absolute duration of individual time intervals is encoded discretely, like a stopwatch, and no entrainment is possible. In this regard, the recent gradual audiomotor hypothesis suggests that the complex entrainment abilities of humans seem to have evolved gradually across primates, with a duration-based timing mechanism present across the entire primate order (Mendez, Prado, Mendoza, & Merchant., 2011; Merchant, Battaglia-Mayer, & Georgopoulos, 2003; Zarco, Merchant, Prado, & Mendez, 2009), and a beat-based mechanism that is most developed in humans, shows some of the properties in monkeys, and is present at an intermediate level in chimpanzees (Merchant & Honing, 2014). For example, myriad studies have demonstrated that humans rhythmically entrain to isochronous stimuli with almost perfect tempo and phase matching (Repp & Su, 2013). *Tempo* or *period matching* means that the period of movement precisely equals the musical beat period. *Phase matching* means that rhythmic movements occur near or at the onset times of musical beats.

Macaques are able to produce rhythmic movements with proper tempo matching during a synchronization-continuation task (SCT), where they tapped on a push-button to produce six isochronous intervals in a sequence, three guided by stimuli, followed by three internally timed (without the sound) intervals (Zarco et al., 2009). Macaques reproduce the intervals with only slight underestimations (about 50 ms), and their inter-tap interval variability increases as a function of the target interval, as does human subjects' variability in the same task (Merchant, Zarco, Pérez, Prado, & Bartolo, 2011; Zarco et al., 2009). Crucially, these monkeys produce isochronous rhythmic movements by temporalizing the pause between movements (called dwell) and not the movements' duration (Donnet et al., 2014), reminiscent of human results (Doumas & Wing, 2007). These observations suggest that monkeys use an explicit timing strategy to perform the SCT, where the timing mechanism controls the duration of the movement pauses, which also triggers the execution of stereotyped pushing movements across each produced interval in the rhythmic sequence. However, the taps of macaques typically occur about 250 ms after stimulus onset, whereas humans show asynchronies close to zero (perfect phase matching) or even anticipate stimulus onset, moving slightly ahead of the beat (Zarco et al., 2009). The positive asynchronies in monkeys are shorter than their reaction times in a control task with random interstimulus intervals, suggesting that monkeys do have some temporal prediction capabilities during SCT but that these abilities are not as finely developed as in humans (Zarco et al., 2009). Subsequent studies have shown that monkeys' asynchronies can be reduced to about 100 ms with a different training strategy (Merchant & Honing, 2014) and that monkeys show tempo matching to periods between 350 and 1000 ms, similar to what has been seen in humans (Konoike, Mikami, & Miyachi, 2012; Merchant & Honing, 2014). Importantly, a recent study demonstrated that Japanese macaques are able to generate periodic saccadic movements towards visual targets in two positions with negative asynchronies (Takeya et al., 2017). Hence, it appears that macaques possess some of the components of the brain machinery used in humans for beat perception and synchronization (Honing & Merchant 2014; Patel, 2014; García-Garibay, Cadena-Valencia, Merchant & de Lafuente, 2016).

In order to understand how motoric entrainment to a musical beat is accomplished in the brain, it is important to compare neurophysiological and behavioral data across humans and nonhuman primates. Because humans appear to have uniquely good abilities for beat perception and synchronization, it is important to examine the basis of these abilities using functional magnetic resonance imaging (fMRI), electroencephalogram (EEG), and magnetoencephalography (MEG) data. When these techniques have been used, a complementary set of studies (which we discuss in more detail) has described the neural circuits and the potential neural mechanisms engaged during both rhythmic entrainment and beat perception in humans. In addition, the recording of multiple-site extracellular signals in behaving monkeys has provided important clues about the neural underpinnings of temporal and sequential aspects of rhythmic entrainment. Most of the neural data in monkeys have been collected during the SCT task. Hence, spiking responses of cells and local field potential (LFP) recordings of monkeys during the synchronization phase of the SCT must be compared and contrasted to neural studies of beat perception and synchronization in humans, which use more macroscopic techniques such as EEG and brain imaging. In this chapter, we review both human and monkey data and attempt to synthesize these different neuronal levels of explanation. We end by providing a set of desiderata for neurally grounded computational models of beat perception and synchronization.

Functional Imaging of Beat Perception and Entrainment in Humans

Although studies with humans have used the SCT task, generally with isochronous sequences, humans can spontaneously entrain to nonisochronous sequences as well if they have a temporal structure that induces beat perception (Fitch & Rosenfeld, 2007; Patel, Iversen, Chen, & Repp, 2005; Repp, Iversen, & Patel, 2008). Sequences that induce a beat are often termed metric, and activity to these sequences can be compared with activity to sequences that do not (nonmetric). Different researchers use somewhat different heuristics and terminology for creating metric and nonmetric sequences, but the underlying idea is similar: simple metric rhythms induce clear beat perception, complex metric rhythms less so, and nonmetric rhythms not at all.

During beat perception and synchronization, activity is consistently observed in several brain areas. Subcortical structures include the cerebellum, the basal ganglia (most often the putamen, but also caudate nucleus and globus pallidus), and thalamus, and cortical areas include the supplementary motor area (SMA) and pre-SMA, premotor cortex (PMC), as well as auditory cortex (Bengtsson et al., 2009; Chen, Penhune, & Zatorre, 2008a, 2008b; Chen, Zatorre, & Penhune, 2006; Grahn, 2009; Grahn & Brett, 2007; Grahn & Rowe, 2009, 2013; Lewis, Wing, Pope, Praamstra, & Miall, 2004; Teki et al., 2011). Less frequently, ventrolateral prefrontal cortex (VLPFC, sometimes labeled as anterior insula) and inferior parietal cortex activations are observed (Grahn & Brett, 2007; Kung, Chen, Zatorre, & Penhune, 2013; Vuust, Ostergaard, Pallesen, Bailey, & Roepstorff, 2009; Vuust, Roepstorff, Wallentin, Mouridsen, & Ostergaard, 2006). These areas are depicted in figure 8.1. Although the specific role of each area is still emerging, evidence is accumulating for distinctions between particular areas and networks. The basal ganglia and SMA appear to be involved in beat perception (Cameron, Pickett, Earhart, & Grahn, 2016; Grahn & Brett, 2009), whereas the cerebellum does not. Visual beat perception may be mediated by similar mechanisms to auditory beat perception. Beat perception also leads to greater functional connectivity, or interaction, between auditory and motor areas, particularly for musicians. We now consider this evidence in more detail.

Several fMRI studies indicate that the cerebellum is more important for absolute than relative timing. During various perceptual judgment tasks, the cerebellum is more active

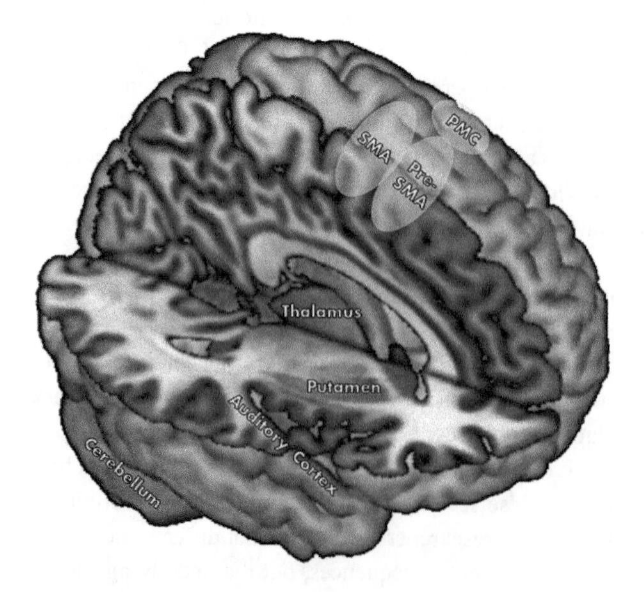

Figure 8.1
Brain areas commonly activated in functional magnetic resonance studies of rhythm perception (cut out to show mid-sagittal and horizontal planes, tinted gray). Besides auditory cortex and the thalamus, many of the brain areas of the rhythm network are traditionally thought to be part of the motor system. SMA = supplementary motor area, PMC = premotor cortex. (Online version in color.)

for nonmetric than metric rhythms (Grahn & Rowe, 2009; Teki et al., 2011). Similarly, when participants tap along to sequences, cerebellar activity increases as metric complexity increases and the beat becomes more difficult to perceive (Kung et al., 2013). The cerebellum also responds more during learning of nonmetric than metric rhythms (Ramnani & Passingham, 2001). The fMRI findings are supported by findings from other methods: deficits in encoding of single durations, but not of metric sequences, occur when cerebellar function is disrupted by disease (Grube, Cooper, Chinnery, & Griffiths, 2010) or through transcranial magnetic stimulation (Grube, Lee, Griffiths, Barker, & Woodruff, 2010). Thus, although the cerebellum is commonly activated during rhythm tasks, the evidence indicates it is involved in absolute, not relative, timing, and therefore does not play a significant role in beat perception or entrainment.

In contrast, relative timing appears to rely on the basal ganglia, specifically the putamen, and the SMA, as simple metric rhythms, compared to complex or nonmetric rhythms, elicit greater putamen and SMA activity across various perception and production tasks (Geiser, Notter, & Gabrieli, 2012; Grahn & Brett, 2007; Grahn, Henry, & McAuley, 2011; Grahn & Rowe, 2013; Teki et al., 2011). For complex rhythms, SMA and putamen activity can be observed when a beat is eventually induced by several repetitions of the rhythm (Chapin et al., 2010). Importantly, increases in putamen and SMA activity during simple metric compared to nonmetric rhythms cannot be attributed to nonmetric rhythms simply being more difficult; even when task difficulty is manipulated to equate performance, greater putamen and SMA activity is still evident for simple metric rhythms (Grahn & Brett, 2007). Furthermore, the greater activity occurs even when participants are not instructed to attend to the rhythms (Bengtsson et al., 2009) or when they attend to nontemporal aspects of the stimuli such as loudness (Geiser et al., 2012) or pitch (Grahn & Rowe, 2009). Thus, greater putamen activity cannot be attributed to metric rhythms' simply facilitating performance on temporal tasks. Finally, when the basal ganglia are compromised, as in Parkinson's disease, discrimination performance with simple metric rhythms is selectively impaired. Overall, these findings indicate that the basal ganglia not only respond during beat perception but are crucial for normal beat perception to occur.

Interestingly, basal ganglia activity does not appear to correlate with the speed of the beat that is perceived (Chen et al., 2008a), instead showing maximal activity around 500 to 700 ms (Riecker, Kassubek, Gröschel, Grodd, & Ackermann, 2006; Riecker, Wildgruber, Mathiak, Grodd, & Ackermann, 2003). Findings from behavioral work suggest that beat perception is maximal at a preferred beat period near 500 ms (Fraisse, 1984; van Noorden & Moelants, 1999). Therefore, the basal ganglia are not simply responding equally to any perceived temporal regularity in the stimuli, but are most responsive to regularity at the frequency that best induces a sense of the beat.

Most fMRI studies of beat perception use auditory stimuli, since beat perception and synchronization occur more readily with auditory than visual stimuli (Chen, Repp, & Patel, 2002; Grahn & Schuit, 2012; Hove, Fairhurst, Kotz, & Keller, 2013; McAuley & Henry,

2010; Merchant, Zarco, & Prado, 2008; Patel et al., 2005; Repp & Penel, 2002). Thus, it is unclear whether visual beat perception would also be mediated by basal ganglia structures. One study found that presenting visual sequences with isochronous timing elicited greater basal ganglia activity than randomly timed stimuli (Marchant & Driver, 2012), although it is unclear if participants truly perceived a beat in the visual condition. Another way to induce visual beat perception is for visual rhythms to be primed by earlier presentations of the same rhythm in the auditory modality. When metric visual rhythms are perceived after auditory rhythms, putamen activity is greater than when visual rhythms are presented without auditory priming. Moreover, the amount of that increase predicts whether a beat is perceived in the visual rhythm (Grahn et al., 2011). This suggests that when an internal representation of the beat is induced during the auditory presentation, the beat can be continued in subsequently presented visual rhythms, and this visual beat perception is mediated by the basal ganglia.

In addition to measuring regional activity, fMRI can be used to assess functional connectivity, or interactions, between brain areas. During beat perception, greater connectivity is observed between the basal ganglia and cortical motor areas, such as the SMA and PMC (Grahn & Rowe, 2009). Furthermore, connectivity between PMC and auditory cortex was found in one study to increase as the salience of the beat in an isochronous sequence increased (Chen et al., 2006). However, a later study using metric, not isochronous, sequences found that auditory-premotor activity increased as metric complexity increased (i.e., connectivity increased as perception of the beat decreased; Kung et al., 2013). Thus, further clarification is needed about the role of auditory-premotor connectivity in isochronous versus metric sequences. Musical training is also associated with greater connectivity between motor and auditory areas. During a synchronization task, musicians showed more bilateral patterns of auditory-premotor connectivity than nonmusicians did (Chen et al., 2008b). Musicians also showed greater auditory-premotor connectivity than nonmusicians did during passive listening, when no movement was required (Grahn & Rowe, 2009). Thus, beat perception increases functional connectivity both within the motor system and between motor and auditory systems. One hypothesis is that increased auditory-premotor connectivity might be important for integrating auditory perception with a motor response (Chen et al., 2008a; Zatorre et al., 2007), and perhaps this occurs even if the response is not executed. Interestingly, this hypothesis has been confirmed using EEG and MEG techniques, as we review below, suggesting a fundamental role of the audiomotor system in beat perception and synchronization.

Beat perception unfolds over time. When a rhythm is first heard, the beat must be discovered. After beat finding occurs, an internal representation of the beat rate can be formed, allowing prediction of future beats as the rhythm continues. Two fMRI studies have attempted to determine whether the role of the basal ganglia is finding the beat, predicting future beats, or both (Grahn & Rowe, 2013; Kung et al., 2013). In Grahn and Rowe's study, participants heard multiple rhythms in a row that either did or did not have a beat. Putamen activity was low during the initial presentation of a beat-based rhythm, during which participants were engaged in finding the beat. Activity was high when beat-based rhythms followed one after the other, during which participants had a strong sense of the beat, suggesting that the putamen is more involved in predicting than finding the beat. The suggestion that the putamen and SMA are involved in maintaining an internal representation of beat intervals is supported by findings of greater putamen and SMA activation during the continuation phase, and not the synchronization phase, during synchronization-continuation tasks (Lewis et al., 2004; Rao et al., 1997) similar to those described for macaques in the next section (Merchant et al., 2011). Patients with SMA lesions also show a selective deficit in the continuation phase but not the synchronization phase of the synchronization-continuation task (Halsband, Ito, & Freund, 1993). However, a second fMRI study compared activation during the initial hearing of a rhythm, during which participants were engaged in beat finding, to subsequent tapping of the beat as they heard the rhythm again. In contrast to the previous study, putamen activity was similar during finding and synchronized tapping (Kung et al., 2013). These different fMRI findings may result from different experimental paradigms, stimuli, or analyses, but more research will be needed to determine the role of the basal ganglia in beat finding and beat prediction.

In summary, a network of motor and auditory areas responds not only during tapping in synchrony with a beat, but also during the perception of sequences that have a beat, to which one could synchronize, or entrain. Beat perception elicits greater activity than perception of rhythms without a beat in subsets of these motor areas, including the putamen and supplementary motor area. Motor and auditory areas also exhibit greater coupling during synchronization to and perception of the beat. A viable mechanism for this coupling may be oscillatory responses, not directly measurable with fMRI, but readily observed using EEG or MEG.

Oscillatory Mechanisms Underlying Rhythmic Behavior in Humans: Evidence from EEG and MEG

The perception of a regular beat in music can be studied in human adults and newborns (Honing, Bouwer, & Háden, 2014; Winkler et al., 2009), as well as in nonhuman primates (Honing, Merchant, Háden, Prado, & Bartolo, 2012; Honing, Bouwer, Prado, & Merchant, 2017) using a particular event-related brain potential (ERP) called mismatch negativity (MMN). MMN is a preattentive brain response reflecting cortical processing of rare unexpected ("deviant") events in a series of ongoing "standard" events (Näätänen, Paavilainen, Rinne, & Alho, 2007). When a complex repeating beat pattern was used, MMN was observed in human adults in response to changes in the pattern at strong beat locations, suggesting extraction of metrical beat structure. Similar results were found in newborn infants, but unfortunately the stimuli used with infants confounded MMN responses to beat changes with a general response to a change in the number of instruments sounding (Winkler et al., 2009), so future studies are needed with newborns in order to verify these results. Honing et al. (2012) also recorded ERPs from the scalp of macaque monkeys. This study demonstrated that an MMN-like ERP component could be measured in rhesus monkeys for both pitch deviants and unexpected omissions from an isochronous tone sequence. However, the study also showed that macaques are not able to detect the beat induced by a varying complex rhythm while being sensitive to the rhythmic grouping structure. Consequently, these results support the notion of different neural networks being active for interval-based and beat-based timing, with a shared interval-based timing mechanism across primates, as discussed in the introduction (Merchant & Honing, 2014). These findings also suggest that monkeys have some capabilities for beat perception, particularly when the stimuli are isochronous, corroborating the hypothesis that the complex entrainment abilities of humans have evolved gradually across primates and with a primordial beat-based mechanism already present in macaques.

MMN has also been used to examine another interesting feature of musical rhythm perception—that across cultures, the basic beat tends to be laid down by bass-range instruments. Hove, Marie, Bruce, and Trainor (2014) showed that when two tones are played simultaneously in a repeated isochronous rhythm, deviations in the timing of the lower-pitched tone elicit a larger MMN response compared to deviations in the timing of the higher-pitched tone. Furthermore, the results are consistent with behavioral responses. When one tone in an otherwise isochronous sequence occurs too early and people are tapping along to the beat, they tend to tap too early on tone following the early one, presumably as a form of error correction. Hove et al. (2014) found that people adjusted the timing of their tapping more when the lower tone was too early compared to when the higher tone was too early. Finally, using a computer model of the peripheral auditory system, these authors showed that this low-voice superiority effect for timing has a peripheral origin in the cochlear of the inner ear. Interestingly, this effect is opposite that for detecting pitch changes in two simultaneous tones or melodies (Fujioka, Trainor, Ross, Kakigi, & Pantev, 2005; Marie, Fujioka, Herrington, & Trainor, 2012), where larger MMN responses are found for pitch deviants in the higher than in the lower tone or melody, an effect found in infants as well as adults (Marie & Trainor, 2012, 2014). The high-voice superiority effect also appears to depend on nonlinear dynamics in the cochlea (Trainor, Marie, Bruce, & Bidelman, 2014). It remains unknown as to whether macaques also show a high-voice superiority effect for pitch and a low-voice superiority effect for timing.

In humans, not only do we extract the beat structure from musical rhythms, but doing so propels us to move in time to the extracted beat. In order to move in time to a musical rhythm, the brain must extract the regularity in the incoming temporal information and predict when the next beat will occur. There is increasing evidence across sensory systems, using a range of tasks, that predictive timing involves interactions between sensory and motor systems and that these interactions are accomplished by oscillatory behavior of neuronal circuits across a number of frequency bands, particularly the delta (1–3 Hz) and beta (15–30 Hz) bands (Arnal, 2012; Fujioka, Trainor, Large, & Ross, 2009, 2012; Lakatos, Karmos, Mehta, Ulbert, & Schroeder, 2008; Morillon & Schroeder, 2015). In general, oscillatory behavior is thought to reflect communication between different brain regions and, in particular, influences of endogenous "top-down" processes of attention and expectation on perception and response selection (Arnal, Wyart, & Giraud, 2011; Besle et al., 2011; Engel, Fries, & Singer, 2001; Giraud & Poeppel, 2012; Schroeder & Lakatos, 2009; Schroeder, Wilson, Radman, Scharfman, & Lakatos, 2010). Importantly, such oscillatory behavior can be measured in humans using EEG and MEG.

The optimal range for the perception of musical tempos is 1 to 3 Hz (Large, 2008), which coincides with the frequency range of neural delta oscillations. Indeed, neural oscillations in the delta range have been shown to frequency (Nozaradan, Peretz, Missal, & Mouraux, 2011; Nozaradan, Peretz, & Mouraux, 2012) and phase (Arnal & Giraud, 2012) align with the tempo of incoming stimuli for musical rhythms, as well as for stimuli with less regular rhythms, such as speech (Giraud et al., 2007). Several researchers have recently suggested that temporal prediction is actually accomplished in the motor system, perhaps through some sort of movement simulation (Arnal, 2012; Arnal, Doelling, & Poeppel, 2014; Arnal & Giraud, 2012; Kotz & Schwartze, 2010; Merchant & Yarrow, 2016; Patel & Iversen, 2014; Schubotz, 2007; Tian & Poeppel, 2010) and that this information feeds back to sensory areas (corollary discharges or efference copies) to enhance processing of incoming information at particular points in time. Temporal expectations appear to align the phase of delta oscillations in sensory cortical areas, such that response times to stimuli that happen to be presented at large delta oscillation peaks are processed more quickly (Lakatos et al., 2008) and more accurately (Arnal et al., 2014) than stimuli presented at other times. This phase alignment of delta rhythms appears to provide a neural instantiation of dynamic attending theory proposed by Large and Jones (1999), whereby attention is drawn to particular points in time and stimulus processing at those points is enhanced.

Metrical structure is derived in the brain based not only on periodicities in the input rhythm but also on expectations for regularity. Thus, in isochronous sequences of identical tones, some tones (e.g., every second or every third tone) can be perceived as accented. And beat locations can be perceived as metrically strong even in the presence of syncopation, in which the most prominent physical stimuli occur off the beat. Interestingly, phase alignment of delta oscillations in auditory cortex reflects the metrical interpretation of input rhythms rather than simply the periodicities in the stimulus (Nozaradan et al., 2011, 2012) as predicted by resonance theory (Large & Snyder, 2009). For example, when listeners are presented with a 2.4 Hz sequence of tones, a component at 2.4 Hz can be seen in the measured EEG response (Nozaradan et al., 2011). But when listeners imagine strong beats every second tone, oscillatory activity in the EEG at 1.2 Hz is also evident, whereas when they imagine strong beats every third tone, EEG oscillations at 0.8 Hz and 1.6 Hz (harmonic of the ternary meter) are evident.

Neural oscillations in the beta range have also been shown to play an important role in predictive timing (Fujioka, Ross, & Trainor, 2015; Fujioka et al., 2009, 2012; Iversen, Repp, & Patel, 2009). For example, Iversen et al. (2009) showed that beta (15–30 Hz) but not gamma (30–50 Hz) oscillations evoked by tones in a repeating pattern were affected by whether listeners imagined them as being on strong or weak beats. Fujioka et al. (2015) found that imagining every second versus every third beat as accented when listening to an isochronous sequence of identical tones resulted in different patterns of power fluctuations in beta, reflecting endogenous influences. Interestingly, beta oscillations have long been associated with motor processes (Engel & Fries, 2010). For example, beta amplitude decreases during motor planning and movement, recovering to baseline once the movement is completed (Gerloff et al., 1998; Pollok, Südmeyer, Gross, & Schnitzler, 2005; Pfurtscheller, 1981; Salmelin, Hámáaláinen, Kajola, & Hari, 1995; Toma et al., 2002).

Of importance in the present context are findings that beta oscillations are crucial for predictive timing in auditory beat processing and that they involve interactions between auditory and motor regions. Fujioka et al. (2012) recorded MEG while participants listened without attending (while watching a silent movie) to isochronous beat sequences at three different tempos in the delta range (2.5, 1.7, and 1.3 Hz). Examining responses from auditory cortex, they found that the power of induced (non-phase-locked) activity in the beta band decreased after the onset of each beat and reached a minimum at approximately 200 ms after beat onset, regardless of tempo. However, the timing of the beta power rebound depended on the tempo, such that maximum beta band power was reached just prior to the onset of the next beat (figure 8.2). Thus, the beta rebound tracked the tempo of the beat and predicted the onset of the following beat.

While little developmental work has yet been done, one recent study indicates that similar beta band fluctuations can be seen in children, at least for slower tempos (Cirelli

Figure 8.2
Induced neuromagnetic responses to isochronous beat sequences at three different tempos. (A) Time-frequency plots of induced oscillatory activity in right auditory cortex between 10 and 40 Hz in response to a fast (390 ms onset-to-onset; upper plot), moderate (585 ms; middle plot) and slow (780 ms; lower plot) tempo ($n = 12$). (B) The time courses of oscillatory activity in the beta band (20–22 Hz) for the three tempos, showing beta desynchronization immediately after stimulus onset (shown by arrows), followed by a rebound with timing predictive of the onset of the next beat. The dashed horizontal lines indicate the 99 percent confidence limits for the group mean. (C) The time at which the beta desynchronization reaches half power (squares) and minimum power (triangles), and the time at which the rebound (resynchronization) reaches half power (circles). The timing of the desynchronization is similar across the three stimulus tempos, but the time of the rebound resynchronization depends on the stimulus tempo in a predictive manner. (D) Areas across the brain in which beta activity is modulated by the auditory stimulus, showing involvement of both auditory cortices and a number of motor regions. Data from Fujioka et al. (2012). (Online version in color.)

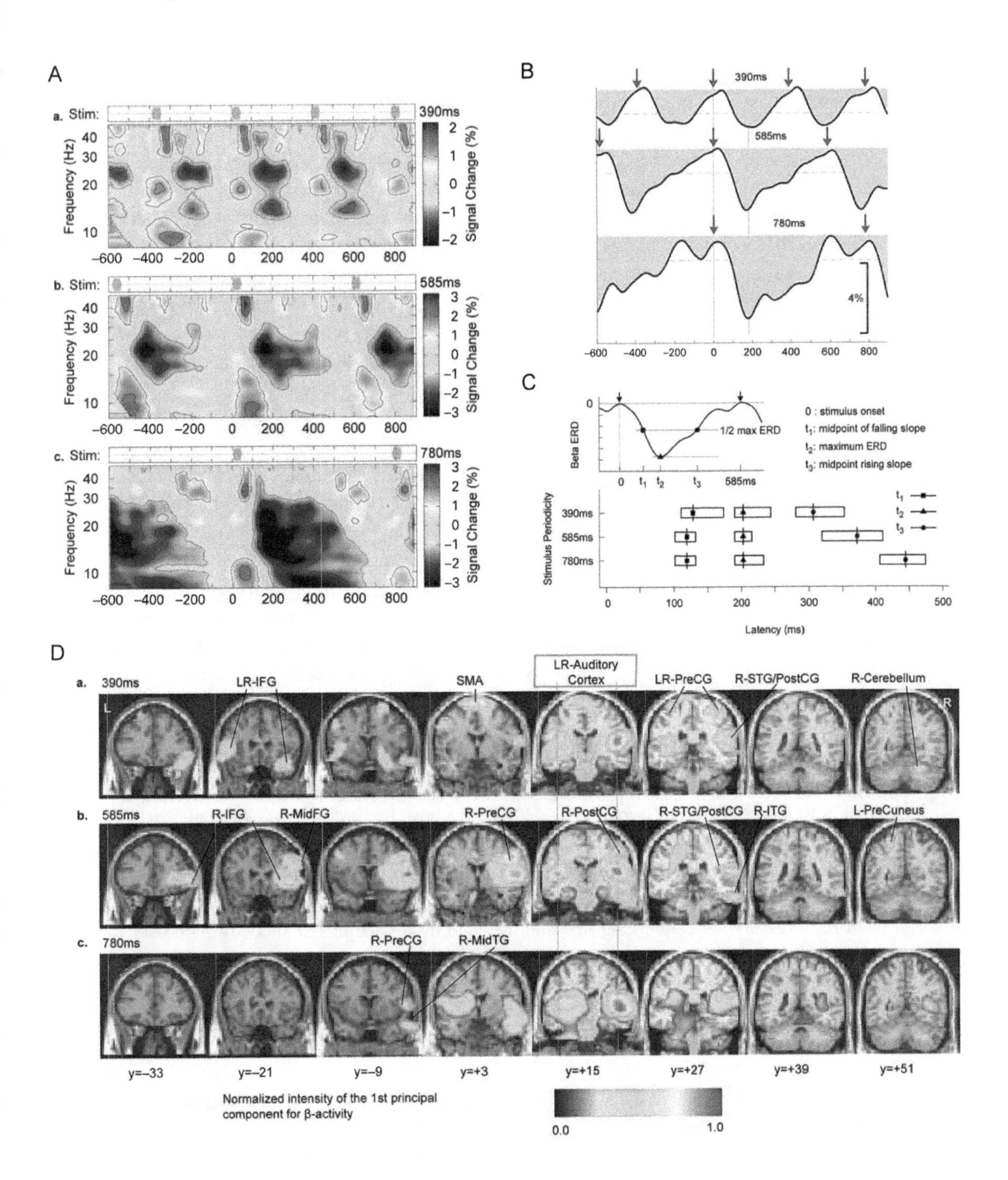
A
a. Stim:
390ms
Frequency (Hz)
Signal Change (%)
b. Stim:
585ms
c. Stim:
780ms
B
390ms
585ms
780ms
4%
C
Beta ERD
0 : stimulus onset
t_1: midpoint of falling slope
t_2: maximum ERD
t_3: midpoint rising slope
1/2 max ERD
585ms
Stimulus Periodicity
390ms
585ms
780ms
Latency (ms)
D
a. 390ms
LR-IFG
SMA
LR-Auditory Cortex
LR-PreCG
R-STG/PostCG
R-Cerebellum
b. 585ms
R-IFG
R-MidFG
R-PreCG
R-PostCG
R-STG/PostCG
R-ITG
L-PreCuneus
c. 780ms
R-PreCG
R-MidTG
y=−33
y=−21
y=−9
y=+3
y=+15
y=+27
y=+39
y=+51
Normalized intensity of the 1st principal component for β-activity
0.0
1.0

et al., 2014). Another study (Fujioka et al., 2009) found that if a beat in a sequence is omitted, the decrease in beta power does not occur, suggesting that the decrease in beta power is tied to the sensory input, whereas the rebound in beta power is predictive of the expected onset of the next beat and is internally generated. These results suggest that oscillations in beta and delta frequencies are connected, with beta power fluctuating according to the phase of delta oscillations, consistent with previous work (Cravo, Rohenkohl, Wyart, & Nobre, 2011). Finally, fluctuations in beta band power reflect not only temporal expectations (predictive timing) but also expectations for content (predictive coding). Specifically, occasional presentation of an unexpected deviant pitch in an isochronous sequence of standard-pitched tones perturbed beta oscillations around 200 to 300 ms after deviant tone onsets (Chang, Bosnyak, & Trainor, 2016). This suggests that beta oscillations reflect attentional processes related to predictions for both timing and content.

To examine the role of the motor cortex in beta power fluctuations, Fujioka et al. (2012) looked across the brain for regions that showed similar fluctuations in beta power as in the auditory cortices and found that this pattern of beta power fluctuation occurred across a wide range of motor areas. Interestingly, the beta power fluctuations appear to be in opposite phase in auditory and motor areas, suggestive that the activity might reflect some kind of sensorimotor loop connecting auditory and motor regions. This is consistent with a recent study in the mouse showing that axons from M2 synapse with neurons in deep and superficial layers of auditory cortex and have a largely inhibitory effect on auditory cortex (Nelson et al., 2013). This is interesting in that interactions between excitatory and inhibitory neurons often give rise to neural oscillations (Hoppensteadt & Izhikevich, 1997). Furthermore, axons from these same M2 neurons also extend to several subcortical areas important for auditory processing. Finally, Nelson et al. (2013) found that stimulation of M2 neurons affected the activity of neurons in auditory cortex. Thus, the connections they described likely reflect the circuits giving rise to the delta-frequency oscillations in beta band power described by Fujioka et al. (2012). Because the task of Fujioka et al. (2012) involved listening without attention or motor movement, the results indicate that motor involvement in beat processing is obligatory and may provide a rationale for why music makes people want to move to the beat.

In sum, research is converging that auditory and motor regions connect through oscillatory activity, particularly at delta and beta frequencies, with motor regions providing the predictive timing needed for the perception of, and entrainment to, musical rhythms.

Neurophysiology of Rhythmic Behavior in Monkeys

The study of the neural basis of beat perception in monkeys and its comparison with humans started just recently, such that the first attempts are the previously described MMN experiments in macaques. As far as we know, no neurophysiological study has been performed on a beat perception task in behaving monkeys. In contrast, a complete set of single-cell and LFP experiments has been carried out in the monkeys during the execution of the SCT. A critical aspect of these studies is that they have been performed in cortical and subcortical areas of the circuit for beat perception and synchronization described in fMRI and MEG studies in humans, specifically in the SMA/pre-SMA, as well as in the putamen of behaving monkeys. These areas are deeply interconnected and are fundamental processing nodes of the motor cortico-basal-ganglia-thalamo-cortical (mCBGT) circuit across all primates (Mendoza & Merchant, 2014). Hence, direct inferences of the functional associations between the different levels of organization measured in this circuit with diverse techniques can be performed across species. This is particularly true for data collected during the synchronization phase of the SCT in monkeys and for human studies of beat synchronization. In addition, cautious generalizations can also be made to the beat perception experiments in humans, since its mechanisms seem to have a large overlap with beat synchronization, as described in the previous two sections.

The signals recorded from extracellular recordings in behaving animals can be filtered to obtain either single cell action potentials or LFPs. Action potentials last approximately1 ms and are emitted by cells in spike trains, whereas LFPs are complex signals determined by the input activity of an area in terms of population excitatory and inhibitory postsynaptic potentials (Buzsáki, Anastassiou, & Koch, 2012). Hence, LFPs and spike trains can be considered the input and output stages of information processing, respectively. Furthermore, LFPs are a local version of the EEG that is not distorted by the meninges and the scalp and that is a signal that provides important information about the input–output processing inside local networks (Buzsáki et al., 2012). The analysis of putaminal LFPs in monkeys performing the SCT revealed an orderly change in the power of transient modulations in the gamma (30–70 Hz) and beta (15–30 Hz) band as a function of the duration of the intervals produced in the SCT (figure 8.3; Bartolo & Merchant, 2015; Bartolo, Prado, & Merchant, 2014). The burst of LFP oscillations showed different preferred intervals so that a range of recording locations represented all the tested durations. These results suggest that the putamen contains a representation of interval duration, where different local cell populations oscillate in the beta or gamma bands for specific intervals during the SCT. Therefore, the transient modulations in the oscillatory activity of different cell ensembles in the putamen as a function of tempo can be part of the neural underpinnings for beat synchronization (Merchant & Bartolo, 2017). Indeed, LFPs tuned to the interval duration in the synchronization phase of the SCT could be considered an empirical substrate of the neural resonance hypothesis that suggests that perception of pulse and meter results from rhythmic bursts of high-frequency neural activity in response to music (Large, 2008). Accordingly, high-frequency bursts in the gamma and beta bands may enable communication between neural areas in humans, such as auditory and motor cortices, during rhythm perception and production (Merchant & Yarrow, 2016; Snyder & Large, 2005; Zanto, Large, Fuchs, & Kelso, 2005).

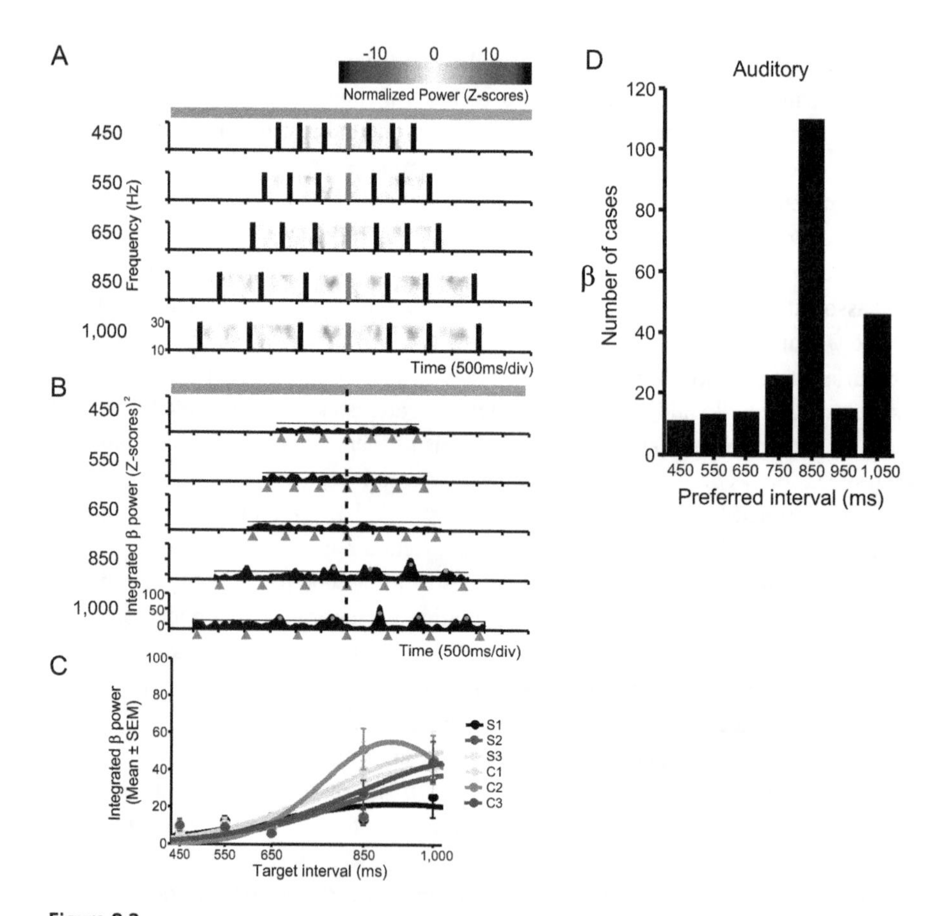

Figure 8.3
Time-varying modulations in the beta power of an LFP signal that show selectivity to interval duration. (A) Normalized (*z*-scores, color-code) spectrograms in the beta-band. Each plot corresponds to the target interval indicated on the left. The horizontal axis is the time during the performance of the task. The times at which the monkey tapped on the push-button are indicated by black vertical bars, and all the spectrograms are aligned to the last tap of the synchronization phase (gray vertical bar). Light and dark gray horizontal bars at the top of the panel represent the synchronization and continuation phases, respectively. (B) Plots of the integrated power time series for each target duration. Tap times are indicated by gray triangles below the time axis. Dots correspond to power modulations above the 1SD threshold (black solid line) for a minimum of 50 ms across trials. The vertical dotted line indicates the last tap of the synchronization phase. (C) Interval tuning in the integrated beta power. Dots are the mean ±SEM and lines correspond to the fitted Gaussian functions. Tuning functions were calculated for each of the six elements of the task sequence (S –S3 for synchronization and C1–C3 for continuation) and are color-coded (see inset). (D) Distribution of preferred interval durations for all the recorded LFPs in the beta band with significant tuning for the auditory condition. Data from Bartolo et al. (2014). (Online version in color.)

In a series of elegant studies, Schroeder and colleagues (2010) have shown that when attention is allocated to auditory or visual events in a rhythmic sequence, delta oscillations of primary visual and auditory cortices of monkeys are entrained (i.e., phase-locked) to the attended modality (Lakatos et al., 2008, 2009). As a result of this entrainment, the neural excitability and spiking responses of the circuit have a tendency to coincide with the attended sensory events (Lakatos et al., 2008). Furthermore, the magnitude of the spiking responses and the reaction time of monkeys responding to the attended modality in the rhythmic patterns of visual and auditory events are correlated with the delta phase entrainment in a trial-by-trial basis (Lakatos et al., 2008, 2009). Finally, attentional modulations for one of the two modalities is accompanied by large-scale neuronal excitability shifts in the delta band across a large circuit of cortical areas in the human brain, including primary sensory and multisensory areas, as with all premotor and prefrontal areas (Besle et al., 2011). Therefore, these authors hypothesized that the attention modulation of delta phase locking across a large cortical circuit is a fundamental mechanism for sensory selection (Schroeder & Lakatos, 2009). Needless to say, the delta entrainment across this large cortical circuit should be present during the SCT; however, the putaminal LFPs in monkeys did not show modulations in the delta band during this task. It is known that delta oscillations have their origin in thalamocortical interactions (Steriade, Dossi, & Nunez, 1991) and that they play a critical role in the long-range interactions between cortical areas in behaving monkeys (Nácher, Ledberg, Deco, & Romo, 2013). Hence, a possible explanation for this apparent discrepancy is that delta oscillations are part of the mechanism of information flow within cortical areas but not across the mCBGT circuit, which in turn could use interval tuning in the beta and gamma bands to represent the beat. Subcortical recordings in the relay nuclei of the mCBGT circuit in humans performing beat perception and synchronization could test this hypothesis.

Surprisingly, the transient changes in oscillatory activity also showed an orderly change as a function of the phase of the task, with a strong bias toward the synchronization phase for the gamma band (when aligned to the stimuli), whereas there is a strong bias toward the continuation phase for the beta band (when aligned to the tapping movements; Bartolo et al., 2014). These results are consistent with the notion that gamma band oscillations predominate during sensory processing (Fries, 2009) or when changes in the sensory input or cognitive set are expected (Engel & Fries, 2010). Thus, the local processing of visual or auditory cues in the gamma band during the SCT may serve for binding neural ensembles that process sensory and motor information within the putamen during beat synchronization (Bartolo et al., 2014). However, these findings seem to be at odds with the role of the beta oscillations in connecting the predictive timing signals from human motor regions to auditory areas needed for the perception of and entrainment to musical rhythms. In this case, it is also critical to consider that gamma oscillations in the putamen during beat synchronization seem to be local and more related to sensory cue selection for the execution of movement sequences (Kimura, 1992; Romo, Merchant, Ruiz, Crespo &

Zainos, 1995; Merchant, Zainos, Hernadez, Salinas, & Romo, 1997) rather than associated with the establishment of top-down predictive signal between motor and auditory cortical areas.

The beta band activity in the putamen of nonhuman primates showed a preference for the continuation phase of the SCT, which is characterized by the internal generation of isochronous movements. Consequently, the global beta oscillations could be associated with the maintenance of a rhythmic set and the dominance of an endogenous reverberation in the mCBGT circuit, which in turn could generate internally timed movements and override the effects of external inputs (figure 8.4; Bartolo et al., 2014). It is clear from comparing the experiments in humans and macaques that in both species, the beta band is deeply involved in prediction and linked with the processing regular events (isochronous or with a particular metric) across large portions of the brain.

The functional properties of cell spiking responses in the putamen and SMA were also characterized in monkeys performing the SCT (Mendoza, Peyrache, et al., 2016) Neurons in these areas show a graded modulation in discharge rate as a function of interval duration in the SCT (Bartolo et al., 2014; Merchant et al., 2014; Merchant, Harrington, & Meck, 2013; Merchant, Pérez, Zarco, & Gamez, 2013). Psychophysical studies on learning and generalization of time intervals predicted the existence of cells tuned to specific interval durations (Bartolo & Merchant, 2009; Nagarajan, Blake, Wright, Byl, & Merzenich, 1998). Indeed, large populations of cells in both areas of the mCBGT are tuned to different interval durations during the SCT, with a distribution of preferred intervals that covers all durations in the hundreds of milliseconds, although there was a bias toward long preferred intervals (Merchant, Pérez, et al., 2013; figure 8.3D). The bias toward the 800 ms interval in the neural population could be associated with the preferred tempo in rhesus monkeys, similar to what has been reported in humans for the putamen activation at the preferred human tempo of 500 ms (Fraisse, 1984; van Noorden & Moelants, 1999). Thus, the interval tuning observed in the gamma and beta oscillatory activity of putaminal LFPs is also present in the discharge rate of neurons in SMA and the putamen, and probably across all the mCBGT circuit. The tuning association between spiking activity and LFPs is not always obvious across the SNC, since the latter is a complex signal that depends on the following factors: the input activity of an area in terms of population excitatory and inhibitory postsynaptic potentials; the regional processing of the microcircuit surrounding the recording electrode; the cytoarchitecture of the studied area; and the temporally synchronous fluctuations of the membrane potential in large neuronal aggregates (Buzsáki et al., 2012). Consequently, the fact that interval tuning is ubiquitous across areas and neural signals during the SCT underlines the role of this population signal in encoding tempo during beat synchronization, and probably during beat perception too.

Beat perception and synchronization not only have a predictive timing component; they are also immersed in a sequence of sensory and motor events. In fact, the regular and repeated cycles of sound and movement during beat synchronization can be fully

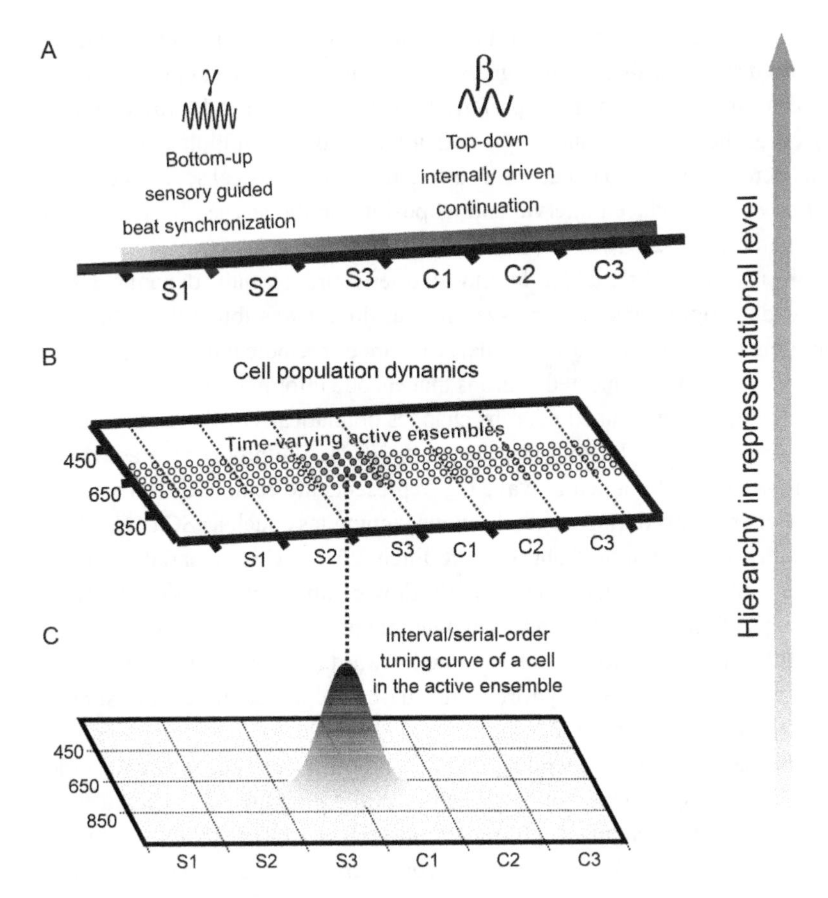

Figure 8.4
Multiple layers of neuronal representation for duration, serial order, and context in which the synchronization-continuation task is performed. (A) LFP activity that switches from gamma activity during sensory driven (bottom-up) beat synchronization to beta activity during the internally generated (top-down) continuation phase. (B) Dynamic representation of the sequential and temporal structure of the SCT. Small ensembles of interconnected cells are activated in a consecutive chain of neural events. (C) A duration/serial-order tuning curve for a cell that is part of the dynamically activated ensemble on top. Dotted horizontal line links the tuning curve of the cell with the neural chain of events in 1.4b. Data from Merchant et al. (2013). (Online version in color.)

described by their sequential and temporal information. The neurophysiological recordings in SMA underscored the importance of sequential encoding during the SCT. A large population of cells in this area showed response selectivity to the sequential organization of the SCT: they were tuned to one of the six serial orders of the SCT (three in the synchronization and three in the continuation phase; figure 8.3C; Merchant, Pérez, et al., 2013). Again, all the possible preferred serial orders are represented across cell populations. Cell tuning is an encoding mechanism that the cerebral cortex uses to represent different sensory, motor, and cognitive features (Merchant, de Lafuente, Peña-Ortega, &

Larriva-Sahd, 2012), which include the duration of the intervals and the serial order of movements produced rhythmically. These signals must be integrated as a population code, where the cells can vote in favor of their preferred interval or serial order to generate a neural "tag." Hence, the temporal and sequential information is multiplexed in a cell population signal across the mCBGT that works as the notes of a musical score in order to define the duration of the produced interval and its position in the learned SCT sequence (Merchant, Harrington et al., 2013).

Interestingly, the multiplexed signal for duration and serial order is quite dynamic. Using encoding and decoding algorithms in a time-varying fashion, it was found that SMA cell populations represent the temporal and sequential structure of periodic movements by activating small ensembles of interconnected neurons that encode information in rapid succession, so that the pattern of active tuned neurons changes dramatically within each interval (figure 8.3B top; Crowe, Zarco, Bartolo, & Merchant, 2014). The progressive activation of different ensembles generates a neural wave that represents the complete sequence and duration of produced intervals during an isochronous tapping task such as SCT (Crowe et al., 2014; Merchant, Grahn, Trainor, Rohrmeier, & Fitch, 2015). This potential anatomo-functional arrangement should include the dynamic flow of information inside the SMA and through the loops of the mCBGT. Thus, each neuronal pool represents a specific value of both features and is tightly connected, through feedforward synaptic connections, to the next pool of neurons, as in the case of synfire chains described in neural network simulations (Gewaltig, Diesmann, & Aertsen, 2001).

A key feature of the SMA neural encoding and decoding for duration and serial order is that its width increases as a function of the target interval during the SCT (Crowe et al., 2014). The time window in which tuned cells represent serial order, for example, is narrow for short interval durations and becomes wider as the interval increases. Thus, duration and serial order are represented in relative terms in SMA, which, as far as we know, is the first single cell neural correlate of beat-based timing and is congruent with the functional imaging studies in the mCBGT circuit already described (Teki et al., 2011).

The investigation of the neural correlates of beat perception and synchronization in humans has stressed the importance of the top-down predictive connection between motor areas and the auditory cortex (Merchant & Yarrow, 2016). A natural following step in the monkey neurophysiological studies will be the simultaneous recording of premotor and auditory areas during tapping synchronization and beat perception tasks. Initial efforts have been started in this direction. However, it is important to emphasize that monkeys show a large bias toward visual rather than auditory cues to drive their tapping behavior (Merchant & Honing, 2014; Nagasaka, Chao, Hasegawa, Notoya, & Fujii, 2013; Zarco et al., 2009), which contrast with the strong bias toward the auditory modality in music and dance (Honing, 2013; Patel et al., 2005). In fact, during the SCT, humans show smaller temporal variability and better accuracy with auditory than visual interval markers (Repp & Penel, 2002; Merchant et al., 2008). It has been suggested that the human perceptual system abstracts the rhythmic-temporal structure of visual stimuli into an auditory representation that is automatic and obligatory (Brodsky, Henik, Rubinstein, & Zorman, 2003; Guttman, Gilroy, & Blake, 2005). Thus, it is quite possible that the human auditory system has a privileged access to the temporal and sequential mechanisms working inside mCBGT circuit in order to determine the exquisite rhythmic abilities of the Homo sapiens (Honing & Merchant, 2014; Merchant & Honing, 2014). The monkey brain seems to not have this strong audio-motor network, as revealed in comparative DTI experiments (Mendoza & Merchant, 2014; Rilling et al., 2008). Hence, it is possible that the audio-motor circuit could be less important than the visuo-motor network in monkeys during beat perception and synchronization. The tuning properties of monkey SMA cells multiplexing the temporal and sequential structure of the SCT were very similar across visual and auditory metronomes (Merchant, Harrington, & Meck, 2013). However, recently, Merchant, Perez et al. (2015) showed that a larger number of SMA cells respond specifically to visual rather than to auditory cues during the SCT, given the first support to the monkeys' stronger association in the visuo-motor system during rhythmic behaviors. Importantly, this study also showed that a substantial group of statistically sensory-aligned cells (Pérez, Kass, & Merchant, 2013) predict the appearance of the next stimulus in the sequence instead of simply responding to the previous stimulus (Merchant, Pérez, et al., 2015; Merchant & Averbeck, 2017). Therefore, the activity of this neural subpopulation can serve as a top-down predictive signal to create the indispensable sensory-motor coupling that permits an animal to execute the SCT. A final point regarding the modality issue is that human subjects improve their beat synchronization when static visual cues are replaced by moving visual stimuli (Hove, Fairhurst et al., 2013; Hove, Iversen, Zhang, & Repp, 2013), stressing the need to use visual moving stimuli to cue beat perception and synchronization in nonhuman primates (Mendoza & Merchant, 2014; Patel, 2014; but see García-Garibay, Cadena, Merchant, & de Lafuente, 2016).

Taken together, the evidence we have discussed supports the notion that beat synchronization is intrinsically linked to the dynamics of cell populations that are tuned for duration and serial order through the mCBGT and encodes this information in relative, rather than absolute, timing terms. Furthermore, the oscillatory activity of the putamen measured with LFPs showed that gamma activity reflects local computations associated with sensory-motor (bottom-up) processing during beat synchronization, whereas beta activity involves the entrainment of large putaminal circuits, probably in conjunction with other elements of mCBGT, during internally driven (top-down) rhythmic tapping. A critical question regarding these empirical observations is how the different levels of neural representation interact and are coordinated during beat perception and synchronization. One approach to answering this question would involve integrating the data we have reviewed with the results obtained from neurally grounded computational models. Indeed, a recent paper indicates that the predictive timing behavior during the SCT, as well as the neural decoding of elapsed time using the activity of simultaneously recorded neurons in

MPC, can be explained by a set of concatenated drift-diffusion models (Merchant & Averbeck, 2017).

Implications for Computational Models of Beat Induction

A considerable variety of models have been developed that are in some way relevant to the core neural and comparative issues in rhythm perception. Broadly, they differ with respect to their purpose and the level of cognitive modeling (in David Marr's sense of algorithmic, computational, and implementational levels; Marr, 1982). These models include rule-based cognitive models, musical information retrieval systems designed to process and categorize music files, and dynamical systems models. Although we make no attempt to review all such models here in detail, we will make a few general observations about the usefulness of each model type for our central problem in this review: understanding the neural basis and comparative distribution of rhythmic abilities.

Rule-based models (e.g., Jackendoff & Lerdahl, 1983; Longuet-Higgins & Lee, 1982) are intended to provide a very high-level computational description of what rhythmic cognition entails in terms of both pulse and meter induction (reviewed in Drake et al., 2000). Such models make little attempt to specify how, computationally or neurally, this is accomplished, but they do provide a clear description of what any neurally grounded model of human BPS and meter perception should be able to explain. Thus, Jackendoff and Lerdahl (1983) emphasize the need to attribute both pulse and meter to a musical surface, and they propose various abstract principles in terms of well-formedness and preference rules to accomplish these goals. Longuet-Higgins and Lee (1982) emphasize that any model of musical cognition must be able to cope with syncopated rhythms, in which some musical events do not coincide with the pulse or a series of events appear to conflict with the established pulse.

A central point made by many rule-based modelers concerns the importance of both bottom-up and top-down processes in determining the rhythmic inferences made by a listener (Desain & Honing, 1999). As a simple example, the vast majority of drum patterns in popular music have bass drum hits on the downbeat (e.g., beats 1 and 3 of a 4/4 pattern) and snare hits on the upbeats (beats 2 and 4). While not inviolable, this statistical generalization leads listeners familiar with these genres to have strong expectations about how to assign musical events to a particular metrical position, thus using learned top-down cognitive processes to reduce the inherent ambiguity of the musical surface. Similarly, many genres have particular rhythmic tropes (e.g., the clave pattern typical of salsa music) that, once recognized, provide an immediate orientation to the pulse and meter in this style. This presumably allows experienced listeners to infer pulse and meter more rapidly (e.g., after a single measure) than would be possible for a listener relying solely on bottom-up processing. It would be desirable for models to allow this type of top-down processing to occur and to provide mechanisms whereby learning about a style can influence more stimulus-drive bottom-up processing.

A second class of potential models for rhythmic processing comes from the practical world of musical information retrieval (MIR) systems. Despite their commercial orientation, such models must efficiently solve problems similar to those addressed in cognitively orientated models. Particularly relevant are beat-finding and tempo-estimation algorithms that aim to process recordings and accurately pinpoint the rate and phase of pulses (e.g., Dixon, 2007; Gouyon & Dixon, 2005). Because such models are designed to deal with real recorded music files, successful MIR systems have higher ecological validity than rule-based systems designed to deal only with notated scores or MIDI files. Furthermore, because their goal is practical and performance driven, and backed by considerable research money, engineering expertise, and competitive evaluation (such as MIREX), we can expect this range of models to provide a menu of systems able to do in practice what any normal human listener easily accomplishes. While we cannot necessarily expect such systems to show any principled connection to the computations humans in fact use to find and track a musical beat, such models should eventually provide insight into which computational approaches and algorithms do and do not succeed, and which challenges are successfully addressed by which techniques. Some of these findings should be relevant to cognitive and neural researchers. Unfortunately, and perhaps surprisingly, no MIR system currently available can fully reproduce the rhythmic processing capabilities of a normal human listener (Gouyon & Dixon, 2005). At present this class of models supports the contention that despite its intuitive simplicity, human rhythmic processing is by no means computationally trivial.

The third class of models—dynamical-systems-based approaches—is most relevant to our purposes in this chapter. Such models attempt to be mathematically explicit and engage with real musical excerpts (not just scores or other symbolic abstractions). Compared to the other two classes, there is considerable variety within this class, and we will not attempt a comprehensive review. Rather we will focus on a class of models that Large and his colleagues developed over several decades and represent the current state of the art for this model category.

Large and his colleagues (Large, 2008; Large & Jones, 1999; Large & Kolen, 1994; Velasco & Large, 2011) have introduced and developed a class of mathematical models of beat induction that can reproduce many of the core behavioral characteristics of human beat and meter perception. Because the Large (2008) model has served as the basis for subsequent improvements and is comprehensively and accessibly described, we describe it briefly here. Large's model is based on a set of nonlinear Hopf oscillators. Such oscillators have two main states, damped oscillation and self-sustained oscillation, where an energy parameter alpha determines which of these states the system is in. Sustained oscillation is the state of interest, because only in this state can an oscillator maintain an implicit beat in the face of temporary silence or conflicting information (e.g., syncopation). In Large's

(2008) model, a set of such nonlinear Hopf oscillators is assembled into a network in which each oscillator has inhibitory connections with the others. This leads to a network where each oscillator competes with the others for activation and the winner(s) are determined by the rhythmic input signal fed into the entire network.

The Large model had ninety-six such oscillators with fixed preferred periods spaced logarithmically from 100 ms (600 BPM) to 1500 ms (40 BPM), thus covering the effective span of human tempo perception. Because each oscillator has its own preferred oscillation rate and a coupling strength to the input, as well as a specific inhibitory connection to the other N–1 oscillators, this model overall has more than N^2 (in this case, more than nine thousand) free parameters. However, these parameters are mostly set based on a priori considerations of human rhythmic perception (e.g., for preferred pulse rate around 600 ms or the tempo range described above) to avoid tweaking parameters to fit a given data set. In Large (2008), this model was tested using ragtime piano excerpts, and the results were compared to experimental data produced by humans tapping to these same excerpts (Snyder & Krumhansl, 2001). A good fit was found between human performance and the model's behavior.

Nonlinear oscillator networks of this sort exhibit a number of desirable properties important for any algorithm-level model of human beat and meter perception. First and foremost, the notion of self-sustained oscillation allows the network, once stimulated with rhythmic input, to persist in its oscillation in the face of noise, syncopation, or silence. This is a crucial feature for any cognitive model of beat perception that is both robust and predictive. Second, the networks that Large and colleagues explored can exhibit hysteresis (meaning that the current behavior of the system depends not only on the current input but on the system's past behavior). Thus, once a given oscillator or group of oscillators is active, they tend to stay that way unless outcompeted by other oscillators, again providing resistance to premature resetting of tempo or phase inference due on syncopation or short-term deviations.

Beyond these beat-based factors, these models also provide for meter perception, in that multiple related oscillators can be, and typically are, active simultaneously. Thus, the system as a whole represents not just the tactus (typically tapping rate) but also harmonics or subharmonics of that rate. Finally, these models have been tested and vetted against multiple sources of data, including both behavioral and, more recently, neural studies, and generally they perform quite well at matching the specific patterns in data derived from humans (Velasco & Large, 2011).

Despite these many virtues, these models also have certain drawbacks for those seeking to understand the neural basis and comparative distribution of rhythmic capabilities. Most fundamentally, these models are implemented at a rather generic mathematical level. Because they have been designed to mirror cognitive and behavioral data, neither the oscillator nor the network properties have any direct connection to properties of specific brain regions, neural circuits, or measurable neuronal properties. For example, both the human preference for a tactus period around 600 ms, and our species' preference for small integer ratios in meter, are hand-coded into the model rather than deriving from any more fundamental properties of the brain systems involved in rhythm perception (e.g., refractory periods of neurons or oscillatory time constants of cortical or basal ganglia circuits). Large (2008) explicitly considers this generic mathematical framework, implemented "without worrying too much about the details of the neural system" (p. 533), to be a desirable property. However, it makes the model and its parameters difficult to integrate with the converging body of neuroscientific evidence we have reviewed. Thus, for example, we would like to know why humans (and perhaps some other animals) prefer to have a tactus around 600 ms. Does this follow from some external property of the body (e.g., resonance frequency of the motor effectors) or some internal neural factor (e.g., preferred frequency for cortico-cortical sensorimotor oscillations, or preferred firing rates of basal ganglia neurons)? In case of the latter, are the observed behavioral preferences a result of properties of the oscillators themselves (as implied in Velasco & Large, 2011), the network connectivity (Large, 2008), or both? Similarly, a neurally grounded model should eventually be able to derive the human preference for small integer ratios in meter perception from some more fundamental neural or cognitive constraints (e.g., properties of coupling between cortical and basal ganglia circuits) rather than hard-coding them into the model.

Ultimately, human cognitive neuroscientists need models that make more specific testable predictions about the neural circuitry, predictions that can be evaluated using modern brain imaging data. Finally, the apparently patchy comparative distribution of rhythmic abilities among different animal species remains mysterious from the viewpoint of a generic model that in principle applies to the nervous system of any species possessing a set of self-sustaining nonlinear oscillators (e.g., virtually any vertebrate species). Why is it that macaques, despite their many similarities with humans, apparently find it so difficult to predictively time their taps to a beat and show no evidence for grouping events into metrical structures (Merchant & Honing, 2014)? Why is it that the one chimpanzee (out of three tested) that shows any evidence at all for BPS does so only for a single preferred tempo and does not generalize this behavior to other tempos (Hattori, Tomonaga, & Matsuzawa, 2013)? What is it about the brain of parrots or sea lions that, in contrast to these primates, enables them to quite flexibly entrain to rhythmic musical stimuli (Cook, Rouse, Wilson, & Reichmuth, 2013; Hasegawa, Okanoya, Hasegawa, & Seki, 2011; Schachner, 2010)? Why is it that some very small and apparently simple brains (e.g., of fireflies or crickets) can accomplish flexible entrainment, while those of dogs apparently cannot (Ravignani, Bowling, & Fitch, 2014)? Addressing any of these questions requires models more closely tied to measurable properties of animal brains than those currently available.

Another desideratum for the next generation of models of rhythmic cognition would include more explicit treatment of the motor output component observed in entrainment behavior to help understand when and why such movements occur. Humans can tap their fingers, nod their heads, tap their feet, sway their bodies, or combine such movements during dancing, and all of these output behaviors appear to be to some extent cognitively interchangeable. Musicians can do the same with their voices or their hands and feet. Does this remarkable flexibility mean that rhythmic cognition is primarily a sensory and central phenomenon, with no essential connection to the details of the motor output? With different species, does it matter that in some species (e.g., nonhuman primates), we study finger taps as the motor output, whereas in others (e.g., birds or pinnipeds), we use head bobs or beak taps? Could other species do better with a vocal response? To what extent do different models predict that output mode should play an important role in determining success or failure in BPS tasks or in preferred entrainment tempos? While incorporating a motor component in perceptual models does not pose a major computational challenge, it would allow us to evaluate such issues, and perhaps help us understand the different roles of cortical, and, basal ganglia circuits in human and animal rhythmic cognition.

Several features outlined in this section could provide fertile inspiration for implementational models at the neural level. First, the several independent loops involved in rhythmic cognition have different fundamental properties. Cortico-cortical loops (e.g., connecting premotor and auditory regions) clearly play an important role in rhythm perception, and a key characteristic of long-range connections in cortex is that they are excitatory—pyramidal cells exciting other pyramidal cells in a feedback loop that in principle can lead to uncontrolled positive feedback and seizure. This is avoided in cortex via local inhibitory neurons. Such inhibitory interneurons in turn provide a target for excitatory feedback projections to "sculpt" ongoing local activity by stimulating local inhibition. This can be conceptualized as high-level motor systems making predictions that bias information processing in "lower" auditory regions via local inhibitory interneurons that tune local circuit oscillations. In sharp contrast, basal ganglia loops are characterized by pervasive inhibitory connections, where inhibition of inhibition is the rule. The computational character of these two loops is thus fundamentally different in ways that are likely, if properly modeled, to have predictable consequences in terms of both temporal integration windows and the way in which cortico-cortical and basal ganglia loops interact with and influence one another. The basal ganglia are also a main locus for dopaminergic reward circuitry projections from the midbrain, which inject learning signals into the forebrain loops. Thus, this circuit may be a preferred locus for the rewarding effects of rhythmic entrainment or learning of rhythmic patterns.

Clearly each of these classes of models addresses interesting questions, and one cannot expect any single model to span all of these levels. Models at the rule-based (computational) and algorithmic levels are currently the most mature and have already provided a good scaffolding for empirical research in cognition and behavior. However, our progress in understanding the neurocomputational basis of rhythm perception in humans and other animals will require a more concerted focus on the computational properties of actual neural circuits, at an implementational level. We hope that the brief review and critique helps to encourage more attention to modeling at Marr's (1982) implementational level.

Conclusion

This chapter described and compared current knowledge concerning the anatomical and functional basis of beat perception and synchronization in human and nonhuman primates, ending by delineating how this knowledge could be used to build neurally grounded models. Hence, our review provides an integrated panorama across fields that has only been treated separately before (Ackermann, Hage, & Ziegler, 2014; Merchant & Honing, 2014; Patel & Iversen, 2014; Patel et al., 2005). It is clear that the human mCBGT circuit is engaged not only during motoric entrainment to a musical beat, but also during the perception of simple metric rhythms. This indicates that the motor system is involved in the representation of the metrical structure of auditory stimuli. Furthermore, this motoric representation is predictive and can induce in auditory cortex an expectation process for metrical stimuli. The predictive signals are conveyed to the sensory areas via oscillatory activity, particularly at delta and beta frequencies. Noninvasive data from humans are complemented by direct recordings of single cell and microcircuit activity in behaving macaques, showing that SMA, the putamen, and probably all of the relay nuclei of the mCBGT circuit use different encoding strategies to represent the temporal and sequential structure of beat synchronization. Indeed, interval tuning could be a mechanism used by the mCBGT to represent the beat tempo during synchronization. As for humans, oscillatory activity in the beta band is deeply involved in generating the internal set used to process regular events.

There is a strong consensus that the motor system makes use of multiple levels of neural representation during beat perception and synchronization. However, implementation-level models, more tightly tied to properties of cells and neural circuits, are urgently needed to help describe and make sense of this (still incomplete) empirical information. Dynamical system approaches to model the neural representations of beat are successful at an algorithmic level, but incorporating single cell intrinsic properties, cell tuning and ramping activity, microcircuit organization and connectivity, and the dynamic communication between cortical and subcortical areas in realistic models would be welcome. The predictions generated by neurally grounded models would help drive further empirical research to bridge the gap across different levels of brain organization during beat perception and synchronization.

Acknowledgments

H.M. is supported by PAPIIT IN201214-25, CONACYT 236836, and CONACYT Fronteras en la Ciencia 196. J.A.G. is supported by the Natural Sciences and Engineering Research Council of Canada. L.J.T. is supported by grants from the Nature Sciences and Engineering Research Council of Canada and the Canadian Institutes of Health Research. W.T.F. thanks the ERC (Advanced Grant SOMACCA 230604) for support.

References

Ackermann, H., Hage, S. R., & Ziegler, W. (2014). Brain mechanisms of acoustic communication in humans and nonhuman primates: An evolutionary perspective. *Behavioral and Brain Sciences*, *37*(6), 529–546. doi:10.1017/s0140525x13003099

Arnal, L. H. (2012). Predicting "when" using the motor system's beta-band oscillations. *Frontiers in Human Neuroscience*, *6*, 1–3. doi:10.3389/fnhum.2012.00225

Arnal, L. H., Doelling, K. B., & Poeppel, D. (2014). Delta-beta coupled oscillations underlie temporal prediction accuracy. *Cerebral Cortex*, *25*(9), 3077–3085. doi:10.1093/cercor/bhu103

Arnal, L. H., & Giraud, A.-L. (2012). Cortical oscillations and sensory predictions. *Trends in Cognitive Sciences*, *16*(7), 390–398. doi:10.1016/j.tics.2012.05.003

Arnal, L. H., Wyart, V., & Giraud, A.-L. (2011). Transitions in neural oscillations reflect prediction errors generated in audiovisual speech. *Nature Neuroscience*, *14*(6), 797–801. doi:10.1038/nn.2810

Bartolo, R., & Merchant, H. (2009). Learning and generalization of time production in humans: Rules of transfer across modalities and interval durations. *Experimental Brain Research*, *197*(1), 91–100. doi:10.1007/s00221-009-1895-1

Bartolo, R., & Merchant, H. (2015). Oscillations are linked to the initiation of sensory-cued movement sequences and the internal guidance of regular tapping in the monkey. *Journal of Neuroscience*, *35*(11), 4635–4640. doi:10.1523/jneurosci.4570-14.2015

Bartolo, R., Prado, L., & Merchant, H. (2014). Information processing in the primate basal ganglia during sensory-guided and internally driven rhythmic tapping. *Journal of Neuroscience*, *34*(11), 3910–3923. doi:10.1523/jneurosci.2679-13.2014

Bengtsson, S. L., Ullén, F., Henrik Ehrsson, H., Hashimoto, T., Kito, T., Naito, E., et al. (2009). Listening to rhythms activates motor and premotor cortices. *Cortex*, *45*(1), 62–71. doi:10.1016/j.cortex.2008.07.002

Besle, J., Schevon, C. A., Mehta, A. D., Lakatos, P., Goodman, R. R., McKhann, G. M., et al. (2011). Tuning of the human neocortex to the temporal dynamics of attended events. *Journal of Neuroscience*, *31*(9), 3176–3185. doi:10.1523/ jneurosci.4518–10.2011

Brodsky, W., Henik, A., Rubinstein, B.-S., & Zorman, M. (2003). Auditory imagery from musical notation in expert musicians. *Perception and Psychophysics*, *65*(4), 602–612. doi:10.3758/bf03194586

Buzsáki, G., Anastassiou, C. A., & Koch, C. (2012). The origin of extracellular fields and currents—EEG, ECoG, LFP and spikes. *Nature Reviews: Neuroscience*, *13*, 407–420. doi:10.1038/nrn3241

Cameron, D. J., Pickett, K. A., Earhart, G. M., & Grahn, J. A. (2016). The effect of dopaminergic medication on beat-based auditory timing in Parkinson's disease. *Frontiers in Neurology*, *7*, 1–8. doi:10.3389/fneur.2016.00019

Chang, A., Bosnyak, D. J., & Trainor, L. J. (2016). Unpredicted pitch modulates beta oscillatory power during rhythmic entrainment to a tone sequence. *Frontiers in Psychology*, *7*, 1–13. doi:10.3389/fpsyg.2016.00327

Chapin, H. L., Zanto, T., Jantzen, K. J., Kelso, S. J. A., Steinberg, F., & Large, E. W. (2010). Neural responses to complex auditory rhythms: The role of attending. *Frontiers in Psychology*, *1*, 1–18. doi:10.3389/fpsyg.2010.00224

Chen, J. L., Penhune, V. B., & Zatorre, R. J. (2008a). Listening to musical rhythms recruits motor regions of the brain. *Cerebral Cortex*, *18*(12), 2844–2854. doi:10.1093/ cercor/bhn042

Chen, J. L., Penhune, V. B., & Zatorre, R. J. (2008b). Moving on time: Brain network for auditory-motor synchronization is modulated by rhythm complexity and musical training. *Journal of Cognitive Neuroscience*, *20*(2), 226–239. doi:10.1162/jocn.2008.20018

Chen, Y., Repp, B. H., & Patel, A. D. (2002). Spectral decomposition of variability in synchronization and continuation tapping: Comparisons between auditory and visual pacing and feedback conditions. *Human Movement Science*, *21*(4), 515–532. doi:10.1016/s0167-9457(02)00138-0

Chen, J. L., Zatorre, R. J., & Penhune, V. B. (2006). Interactions between auditory and dorsal premotor cortex during synchronization to musical rhythms. *NeuroImage*, *32*(4), 1771–1781. doi:10.1016/j.neuroimage.2006.04.207

Cirelli, L. K., Bosnyak, D., Manning, F. C., Spinelli, C., Marie, C., Fujioka, T., et al. (2014). Beat-induced fluctuations in auditory cortical beta-band activity: Using EEG to measure age-related changes. *Frontiers in Psychology, 5*, 1–9. doi:10.3389/ fpsyg.2014.00742

Cook, P., Rouse, A., Wilson, M., & Reichmuth, C. (2013). A California sea lion (*Zalophus californianus*) can keep the beat: Motor entrainment to rhythmic auditory stimuli in a non vocal mimic. *Journal of Comparative Psychology, 127*(4), 412–427. doi:10.1037/a0032345

Cravo, A. M., Rohenkohl, G., Wyart, V., & Nobre, A. C. (2011). Endogenous modulation of low frequency oscillations by temporal expectations. *Journal of Neurophysiology, 106*(6), 2964–2972. doi:10.1152/jn.00157.2011

Crowe, D. A., Zarco, W., Bartolo, R., & Merchant, H. (2014). Dynamic representation of the temporal and sequential structure of rhythmic movements in the primate medial premotor cortex. *Journal of Neuroscience, 34*(36), 11972–11983. doi:10.1523/jneurosci.2177-14.2014

Desain, P., & Honing, H. (1999). Computational models of beat induction: The rule-based approach. *Journal of New Music Research, 28*(1), 29–42. doi:10.1076/jnmr.28.1.29.3123

Dixon, S. (2007). Evaluation of the audio beat tracking system BeatRoot. *Journal of New Music Research, 36*(1), 39–50. doi:10.1080/09298210701653310

Donnet, S., Bartolo, R., Fernandes, J. M., Cunha, J. P. S., Prado, L., & Merchant, H. (2014). Monkeys time their pauses of movement and not their movement-kinematics during a synchronization-continuation rhythmic task. *Journal of Neurophysiology, 111*(10), 2138–2149. doi:10.1152/jn.00802.2013

Doumas, M., & Wing, A. M. (2007). Timing and trajectory in rhythm production. *Journal of Experimental Psychology: Human Perception and Performance, 33*(2), 442–455. doi:10.1037/0096-1523.33.2.442

Drake, C., Jones, M. R., & Baruch, C. (2000). The development of rhythmic attending in auditory sequences: Attunement, referent period, focal attending. *Cognition, 77*(3), 251–288. doi:10.1016/s0010-0277(00)00106-2

Engel, A. K., & Fries, P. (2010). Beta-band oscillations—signalling the status quo? *Current Opinion in Neurobiology, 20*(2), 156–165. doi:10.1016/j.conb.2010.02.015

Engel, A. K., Fries, P., & Singer, W. (2001). Dynamic predictions: Oscillations and synchrony in top-down processing. *Nature Reviews: Neuroscience, 2*, 704–716. doi:10.1038/35094565

Fitch, W. T., & Rosenfeld, A. J. (2007). Perception and production of syncopated rhythms. *Music Perception, 25*(1), 43–58. doi:10.1525/mp.2007.25.1.43

Fraisse, P. (1984). Perception and estimation of time. *Annual Review of Psychology, 35*, 1–36

Fries, P. (2009). Neuronal gamma-band synchronization as a fundamental process in cortical computation. *Annual Review of Neuroscience, 32*, 209–224. doi:10.1146/annurev.neuro.051508.135603

Fujioka, T., Ross, B., & Trainor, L. J. (2015). Beta-band oscillations represent auditory beat and its metrical hierarchy in perception and imagery. *Journal of Neuroscience, 35*(45), 15187–15198. doi:10.1523/jneurosci.2397-15.2015

Fujioka, T., Trainor, L. J., Large, E. W., & Ross, B. (2009). Beta and gamma rhythms in human auditory cortex during musical beat processing. *Annals of the New York Academy of Sciences, 1169*, 89–92. doi:10.1111/j.1749-6632.2009.04779.x

Fujioka, T., Trainor, L. J., Large, E. W., & Ross, B. (2012). Internalized timing of isochronous sounds is represented in neuromagnetic beta oscillations. *Journal of Neuroscience, 32*(5), 1791–1802. doi:10.1523/jneurosci.4107-11.2012

Fujioka, T., Trainor, L. J., Ross, B., Kakigi, R., & Pantev, C. (2005). Automatic encoding of polyphonic melodies in musicians and nonmusicians. *Journal of Cognitive Neuroscience, 17*(10), 1578–1592. doi:10.1162/089892905774597263

García-Garibay, O., Cadena, J., Merchant, H., & de Lafuente, V. (2016). Monkeys share the human ability to internally maintain a temporal rhythm. *Frontiers in Psychology, 7*, 1971. doi:10.3389/fpsyg.2016.01971

Geiser, E., Notter, M., & Gabrieli, J. D. E. (2012). A corticostriatal neural system enhances auditory perception through temporal context processing. *Journal of Neuroscience, 32*(18), 6177–6182. doi:10.1523/jneurosci.5153-11.2012

Gerloff, C., Richard, J., Hadley, J., Schulman, A. E., Honda, M., & Hallett, M. (1998). Functional coupling and regional activation of human cortical motor areas during simple, internally paced and externally paced finger movements. *Brain, 121*, 1513–1531. doi:10.1093/brain/121.8.1513

Gewaltig, M.-O., Diesmann, M., & Aertsen, A. (2001). Propagation of cortical synfire activity: Survival probability in single trials and stability in the mean. *Neural Networks, 14*(6–7), 657–673. doi:10.1016/s0893-6080(01)00070-3

Giraud, A. L., Kleinschmidt, A., Poeppel, D., Lund, T. E., Frackowiak, R. S., & Laufs, H. (2007). Endogenous cortical rhythms determine cerebral specialization for speech perception and production. *Neuron, 56*(6), 1127–1134. doi:10.1016/j.neuron.2007.09.038

Giraud, A. L., & Poeppel, D. (2012). Cortical oscillations and speech processing: Emerging computational principles and operations. *Nature Neuroscience, 15*, 511–517. doi:10.1038/nn.3063

Gouyon, F., & Dixon, S. (2005). A review of automatic rhythm description systems. *Computer Music Journal, 29*(1), 34–54. doi:10.1162/comj.2005.29.1.34

Grahn, J. A. (2009). The role of the basal ganglia in beat perception: Neuroimaging and neuropsychological investigations. *Annals of the New York Academy of Sciences, 1169*, 35–45. doi:10.1111/j.1749-6632.2009.04553.x

Grahn, J. A., & Brett, M. (2007). Rhythm perception in motor areas of the brain. *Journal of Cognitive Neuroscience, 19*(5), 893–906. doi:10.1162/jocn.2007.19.5.893

Grahn, J. A., & Brett, M. (2009). Impairment of beat-based rhythm discrimination in Parkinson's disease. *Cortex, 45*(1), 54–61. doi:10.1016/j.cortex.2008.01.005

Grahn, J. A., Henry, M. J., & McAuley, J. D. (2011). FMRI investigation of cross-modal interactions in beat perception: Audition primes vision, but not vice versa. *NeuroImage, 54*(2), 1231–1243. doi:10.1016/j.neuroimage.2010.09.033

Grahn, J. A., & Rowe, J. B. (2009). Feeling the beat: Premotor and striatal interactions in musicians and nonmusicians during beat perception. *Journal of Neuroscience, 29*(23), 7540–7548. doi:10.1523/JNEUROSCI.2018-08.2009

Grahn, J. A., & Rowe, J. B. (2013). Finding and feeling the musical beat: Striatal dissociations between detection and prediction of regularity. *Cerebral Cortex, 23*(4), 913–921. doi:10.1093/cercor/bhs083

Grahn, J. A., & Schuit, D. (2012). Individual differences in rhythmic abilities: Behavioural and neuroimaging investigations. *Psychomusicology: Music, Mind, and Brain, 22*(2), 105–121. doi:10.1037/a0031188

Grube, M., Cooper, F. E., Chinnery, P. F., & Griffiths, T. D. (2010). Dissociation of duration-based and beat-based auditory timing in cerebellar degeneration. *Proceedings of the National Academy of Sciences, 107*(25), 11597–11601. doi:10.1073/pnas.0910473107

Grube, M., Lee, K. H., Griffiths, T. D., Barker, A. T., & Woodruff, P. W. (2010). Transcranial magnetic theta-burst stimulation of the human cerebellum distinguishes absolute, duration-based from relative, beat-based perception of subsecond time intervals. *Frontiers in Psychology, 1*, 1–8. doi:10.3389/ fpsyg.2010.00171

Guttman, S. E., Gilroy, L. A., & Blake, R. (2005). Hearing what the eyes see: Auditory encoding of visual temporal sequences. *Psychological Science, 16*(3), 228–235. doi:10.1111/ j.0956–7976.2005.00808

Halsband, U., Ito, N., & Freund, H. J. (1993). The role of premotor cortex and the supplementary motor area in the temporal control of movement in man. *Brain, 116*, 243–266. doi:10.1093/brain/116.1.243

Hasegawa, A., Okanoya, K., Hasegawa, T., & Seki, Y. (2011). Rhythmic synchronization tapping to an audio-visual metronome in budgerigars. *Scientific Reports, 1*, 1–8. doi:10.1038/srep00120

Hattori, Y., Tomonaga, M., & Matsuzawa, T. (2013). Spontaneous synchronized tapping to an auditory rhythm in a chimpanzee. *Scientific Reports, 3*, 1–6. doi:10.1038/ srep01566

Honing, H. (2013). Structure and interpretation of rhythm in music. In D. Deutsch (Ed.), *Psychology of Music* (3rd ed., pp. 369–404). London: Academic Press; 10.1016/b978-0-12-381460-9.00009-2

Honing, H., Bouwer, F. L., Prado, L., and Merchant, H. (submitted). Rhesus monkeys (*Macaca mulatta*) sense isochrony in rhythm, but not the beat. *Cortex.*

Honing, H., Bouwer, F., & Háden, G. P. (2014). Perceiving temporal regularity in music: The role of auditory event-related potentials (ERPs) in probing beat perception. In H. Merchant & V. de Lafuente (Eds.), *Neurobiology of interval timing.* Berlin: Springer. doi:10.1007/978-1-4939-1782-2_16

Honing, H., & Merchant, H. (2014). Differences in auditory timing between human and nonhuman primates. *Behavioral and Brain Sciences, 37*(6), 557–558. doi:10.1017 /s0140525x13004056

Honing, H., Merchant, H., Háden, G. P., Prado, L., & Bartolo, R. (2012). Rhesus monkeys (*Macaca mulatta*) detect rhythmic groups in music, but not the beat. *PLoS One*, *7*(12), e51369. doi:10.1371/journal.pone.0051369

Hoppensteadt, F. C., & Izhikevich, E. M. (1997). *Weakly connected neural networks*. New York: Springer-Verlag.

Hove, M. J., Fairhurst, M. T., Kotz, S. A., & Keller, P. E. (2013). Synchronizing with auditory and visual rhythms: An fMRI assessment of modality differences and modality appropriateness. *NeuroImage*, *67*, 313–321. doi:10.1016/j.neuroimage.2012.11.032

Hove, M. J., Marie, C., Bruce, I. C., & Trainor, L. J. (2014). Superior time perception for lower musical pitch explains why bass-ranged instruments lay down musical rhythms. *Proceedings of the National Academy of Sciences*, *111*(28), 10383–10388. doi:10.1073/pnas.1402039111

Hove, M. J., Iversen, J. R., Zhang, A., & Repp, B. H. (2013). Synchronization with competing visual and auditory rhythms: Bouncing ball meets metronome. *Psychological Research*, *77*(4), 388–398. doi:10.1007/ s00426-012-0441-0

Iversen, J. R., Repp, B. H., & Patel, A. D. (2009). Top-down control of rhythm perception modulates early auditory responses. *Annals of the New York Academy of Sciences*, *1169*, 58–73. doi:10.1111/j.1749-6632.2009.0457 9.x

Jackendoff, F., & Lerdahl, R. (1983). *A generative theory of tonal music*. Cambridge, MA: MIT Press.

Kimura, M. (1992). Behavioral modulation of sensory responses of primate putamen neurons. *Brain Research*, *578*(1–2), 204–214. doi:10.1016/0006-8993(92)90249-9

Konoike, N., Mikami, A., & Miyachi, S. (2012). The influence of tempo upon the rhythmic motor control in macaque monkeys. *Neuroscience Research*, *74*(1), 64–67. doi:10.1016/j.neures.2012.06.002

Kotz, S. A., & Schwartze, M. (2010). Cortical speech processing unplugged: A timely subcortico-cortical framework. *Trends in Cognitive Sciences*, *14*(9), 392–399. doi:10.1016/j.tics.2010.06.005

Kung, S.-J., Chen, J. L., Zatorre, R. J., & Penhune, V. B. (2013). Interacting cortical and basal ganglia networks underlying finding and tapping to the musical beat. *Journal of Cognitive Neuroscience*, *25*(3), 401–420. doi:10.1162/jocn_a_00325

Lakatos, P., Karmos, G., Mehta, A. D., Ulbert, I., & Schroeder, C. E. (2008). Entrainment of neuronal oscillations as a mechanism of attentional selection. *Science*, *320*(5872), 110–113. doi:10.1126/science.1154735

Lakatos, P., O'Connell, M. N., Barczak, A., Mills, A., Javitt, D. C., & Schroeder, C. E. (2009). The leading sense: Supramodal control of neurophysiological context by attention. *Neuron*, *64*(3), 419–430. doi:10.1016/ j.neuron.2009.10.014

Large, E. W. (2008). Resonating to musical rhythm: Theory and experiment. In S. Grondin (Ed.), *Psychology of time* (pp. 189–232). Bingley: Emerald Group.

Large, E. W., & Jones, M. R. (1999). The dynamics of attending: How people track time-varying events. *Psychological Review*, *106*(1), 119–159. doi:10.1037/0033-295x.106.1.119

Large, E. W., & Kolen, J. F. (1994). Resonance and the perception of musical meter. *Connection Science*, *6*(2–3), 177–208. doi:10.1080/09540099408915723

Large, E. W., & Palmer, C. (2002). Perceiving temporal regularity in music. *Cognitive Science*, *26*(1), 1–37. doi:10.1207/s15516709cog2601_1

Large, E. W., & Snyder, J. S. (2009). Pulse and meter as neural resonance. *Annals of the New York Academy of Sciences*, *1169*, 46–57. doi:10.1111/j.1749- 6632.2009.04550.x

Lewis, P. A., Wing, A. M., Pope, P. A., Praamstra, P., & Miall, R. C. (2004). Brain activity correlates differentially with increasing temporal complexity of rhythms during initialisation, synchronisation, and continuation phases of paced finger tapping. *Neuropsychologia*, *42*(10), 1301–1312. doi:10.1016/j.neuropsychologia.2004 .03.001

Longuet-Higgins, H. C., & Lee, C. S. (1982). The perception of musical rhythms. *Perception*, *11*(2), 115–128. doi:10.1068/p110115

Marchant, J. L., & Driver, J. (2012). Visual and audiovisual effects of isochronous timing on visual perception and brain activity. *Cerebral Cortex, 23*(6), 1290–1298. doi:10.1093/cercor/bhs095

Marie, C., Fujioka, T., Herrington, L., & Trainor, L. J. (2012). The high-voice superiority effect in polyphonic music is influenced by experience: A comparison of musicians who play soprano-range compared with bass-range instruments. *Psychomusicology: Music, Mind, and Brain, 22*(2), 97–104. doi:10.1037/a0030858

Marie, C., & Trainor, L. J. (2012). Development of simultaneous pitch encoding: Infants show a high voice superiority effect. *Cerebral Cortex, 23*(3), 660–669. doi:10.1093/cercor/bhs050

Marie, C., & Trainor, L. J. (2014). Early development of polyphonic sound encoding and the high voice superiority effect. *Neuropsychologia, 57*, 50–58. doi:10.1016/j.neuropsychologia.2014.02.023

Marr, D. (1982). *Vision: A computational investigation in the human representation of visual information.* San Francisco, CA: Freeman.

McAuley, J. D., & Henry, M. J. (2010). Modality effects in rhythm processing: Auditory encoding of visual rhythms is neither obligatory nor automatic. *Attention, Perception, and Psychophysics, 72*(5), 1377–1389. doi:10.3758/app.72.5.1377

Mendez, J. C., Prado, L., Mendoza, G., & Merchant, H. (2011). Temporal and spatial categorization in human and non-human primates. *Frontiers in Integrative Neuroscience, 5*, 1–10. doi:10.3389/fnint.2011.00050

Mendoza, G., & Merchant, H. (2014). Motor system evolution and the emergence of high cognitive functions. *Progress in Neurobiology, 122*, 73–93. doi:10.1016/j.pneurobio.2014.09.001

Mendoza, G., Peyrache, A., Gámez, J., Prado, L., Buzsáki, G., & Merchant, H. (2016). Recording extracellular neural activity in the behaving monkey using a semi-chronic, high-density electrode system. *Journal of Neurophysiology, 116*(2), 563–574. doi:10.1152/jn.00116.2016

Merchant, H., & Averbeck, B. B. (2017). The computational and neural basis of rhythmic timing in medial premotor cortex. *Journal of Neuroscience, 37*(17), 4552–4564. doi:10.1523/JNEUROSCI.0367-17

Merchant, H., and Bartolo, R. (2017). Primate beta oscillations and rhythmic behaviors. *Journal of Neural Transmission.* doi:10.1007/s00702-017-1716-9

Merchant, H., Bartolo, R., Pérez, O., Méndez, J. C., Mendoza, G., Gámez, J., et al. (2014). Neurophysiology of timing in the hundreds of milliseconds: Multiple layers of neuronal clocks in the medial premotor areas. *Neurobiology of Interval Timing, 829*, 143–154. doi:10.1007/978-1-4939-1782-2_8

Merchant, H., Battaglia-Mayer, A., & Georgopoulos, A. P. (2003). Interception of real and apparent motion targets: Psychophysics in humans and monkeys. *Experimental Brain Research, 152*(1), 106–112. doi:10.1007/s00221-003-1514-5

Merchant, H., de Lafuente, V., Peña-Ortega, F., & Larriva-Sahd, J. (2012). Functional impact of interneuronal inhibition in the cerebral cortex of behaving animals. *Progress in Neurobiology, 99*(2), 163–178. doi:10.1016/j.pneurobio.2012.08.005

Merchant, H., Grahn, J., Trainor, L., Rohrmeier, M., & Fitch, W. T. (2015). Finding the beat: A neural perspective across humans and non-human primates. *Philosophical Transactions of the Royal Society B: Biological Sciences, 370*(1664), 1–16. doi:10.1098/rstb.2014.0093

Merchant, H., Harrington, D. L., & Meck, W. H. (2013). Neural basis of the perception and estimation of time. *Annual Review of Neuroscience, 36*, 313–336. doi:10.1146/annurev-neuro-062012-170349

Merchant, H., & Honing, H. (2014). Are non-human primates capable of rhythmic entrainment? Evidence for the gradual audiomotor evolution hypothesis. *Frontiers in Neuroscience, 7*, 1–8. doi:10.3389/fnins.2013.00274

Merchant, H., Pérez, O., Bartolo, R., Méndez, J. C., Mendoza, G., Gámez, J., et al. (2015). Sensorimotor neural dynamics during isochronous tapping in the medial premotor cortex of the macaque. *European Journal of Neuroscience, 41*(5), 586–602. doi:10.1111/ejn.12811

Merchant, H., Pérez, O., Zarco, W., & Gamez, J. (2013). Interval tuning in the primate medial premotor cortex as a general timing mechanism. *Journal of Neuroscience, 33*(21), 9082–9096. doi:10.1523/jneurosci.5513-12.2013

Merchant, H., & Yarrow, K. (2016). How the motor system both encodes and influences our sense of time. *Current Opinion in Behavioral Sciences, 8*, 22–27. doi:10.1016/j.cobeha.2016.01.006

Merchant, H., Zainos, A., Hernadez, A., Salinas, E., & Romo, R. (1997). Functional properties of primate putamen neurons during the categorization of tactile stimuli. *Journal of Neurophysiology, 77*, 1132–1154.

Merchant, H., Zarco, W., Pérez, O., Prado, L., & Bartolo, R. (2011). Measuring time with different neural chronometers during a synchronization-continuation task. *Proceedings of the National Academy of Sciences, 108*(49), 19784–19789. doi:10.1073/pnas.1112933108

Merchant, H., Zarco, W., & Prado, L. (2008). Do we have a common mechanism for measuring time in the hundreds of millisecond range? Evidence from multiple- interval timing tasks. *Journal of Neurophysiology, 99*(2), 939–949. doi:10.1152/jn.01225.2007

Morillon, B., & Schroeder, C. E. (2015). Neuronal oscillations as a mechanistic substrate of auditory temporal prediction. *Annals of the New York Academy of Sciences, 1337*, 26–31. doi:10.1111/nyas.12629

Näätänen, R., Paavilainen, P., Rinne, T., & Alho, K. (2007). The mismatch negativity (MMN) in basic research of central auditory processing: A review. *Clinical Neurophysiology, 118*(12), 2544–2590. doi:10.1016/j.clinph.2007.04.026

Nácher, V., Ledberg, A., Deco, G., & Romo, R. (2013). Coherent delta-band oscillations between cortical areas correlate with decision making. *Proceedings of the National Academy of Sciences, 110*(37), 15085–15090. doi:10.1073/pnas.1314681110

Nagarajan, S. S., Blake, D. T., Wright, B. A., Byl, N., & Merzenich, M. (1998). Practice-related improvements in somatosensory interval discrimination are temporally specific but generalize across skin location, hemisphere, and modality. *Journal of Neuroscience, 18*(4), 1559–1570.

Nagasaka, Y., Chao, Z. C., Hasegawa, N., Notoya, T., & Fujii, N. (2013). Spontaneous synchronization of arm motion between Japanese macaques. *Scientific Reports, 3*, 1–7. doi:10.1038/srep01151

Nelson, A., Schneider, D. M., Takatoh, J., Sakurai, K., Wang, F., & Mooney, R. (2013). A circuit for motor cortical modulation of auditory cortical activity. *Journal of Neuroscience, 33*(36), 14342–14353. doi:10.1523/jneurosci.2275-13.2013

Nozaradan, S., Peretz, I., Missal, M., & Mouraux, A. (2011). Tagging the neuronal entrainment to beat and meter. *Journal of Neuroscience, 31*(28), 10234–10240. doi:10.1523/jneurosci.0411-11.2011

Nozaradan, S., Peretz, I., & Mouraux, A. (2012). Selective neuronal entrainment to the beat and meter embedded in a musical rhythm. *Journal of Neuroscience, 32*(49), 17572–17581. doi:10.1523/jneurosci.3203-12.2012

Patel, A. D. (2014). The evolutionary biology of musical rhythm: Was Darwin wrong? *PLoS Biology, 12*(3), e1001821. doi:10.1371/journal.pbio.1001821

Patel, A. D., & Iversen, J. R. (2014). The evolutionary neuroscience of musical beat perception: The action simulation for auditory prediction (ASAP) hypothesis. *Frontiers in Systems Neuroscience, 8*, 1–14. doi:10.3389/fnsys.2014.00057

Patel, A. D., Iversen, J. R., Chen, Y., & Repp, B. H. (2005). The influence of metricality and modality on synchronization with a beat. *Experimental Brain Research, 163*(2), 226–238. doi:10.1007/s00221-004-2159-8

Pérez, O., Kass, R. E., & Merchant, H. (2013). Trial time warping to discriminate stimulus-related from movement-related neural activity. *Journal of Neuroscience Methods, 212*(2), 203–210. doi:10.1016/j.jneumeth.2012.10.019

Pfurtscheller, G. (1981). Central beta rhythm during sensorimotor activities in man. *Electroencephalography and Clinical Neurophysiology, 51*(3), 253–264. doi:10.1016/0013-4694(81)90139-5

Phillips-Silver, J., & Trainor, L. J. (2005). Feeling the beat: Movement influences infant rhythm perception. *Science, 308*(5727), 1430. doi:10.1126/science.1110922

Phillips-Silver, J., & Trainor, L. J. (2007). Hearing what the body feels: Auditory encoding of rhythmic movement. *Cognition, 105*(3), 533–546. doi:10.1016/j.cognition.2006.11.006

Pollok, B., Südmeyer, M., Gross, J., & Schnitzler, A. (2005). The oscillatory network of simple repetitive bimanual movements. *Cognitive Brain Research, 25*(1), 300–311. doi:10.1016/j.cogbrainres.2005.06.004

Povel, D.-J., & Essens, P. (1985). Perception of temporal patterns. *Music Perception, 2*(4), 411–440. doi:10.2307/40285311

Ramnani, N., & Passingham, R. E. (2001). Changes in the human brain during rhythm learning. *Journal of Cognitive Neuroscience, 13*(7), 952–966. doi:10.1162/089892901753165863

Rao, S. M., Harrington, D. L., Haaland, K. Y., Bobholz, J. A., Cox, R. W., & Binder, J. R. (1997). Distributed neural systems underlying the timing of movements. *Journal of Neuroscience, 17*(4), 5528–5535.

Ravignani, A., Bowling, D. L., & Fitch, W. T. (2014). Chorusing, synchrony, and the evolutionary functions of rhythm. *Frontiers in Psychology, 5*, 1–15. doi:10.3389/fpsyg.2014.01118

Repp, B. H., Iversen, J. R., & Patel, A. D. (2008). Tracking an imposed beat within a metrical grid. *Music Perception, 26*(1), 1–18. doi:10.1525/mp.2008.26.1.1

Repp, B. H., & Penel, A. (2002). Auditory dominance in temporal processing: New evidence from synchronization with simultaneous visual and auditory sequences. *Journal of Experimental Psychology: Human Perception and Performance, 28*(5), 1085–1099. doi:10.1037/0096-1523.28.5.1085

Repp, B. H., & Su, Y.-H. (2013). Sensorimotor synchronization: A review of recent research (2006–2012). *Psychonomic Bulletin and Review, 20*(3), 403–452. doi:10.3758/s13423-012-0371-2

Riecker, A., Kassubek, J., Gröschel, K., Grodd, W., & Ackermann, H. (2006). The cerebral control of speech tempo: Opposite relationship between speaking rate and BOLD signal changes at striatal and cerebellar structures. *NeuroImage, 29*(1), 46–53. doi:10.1016/j.neuroimage.2005.03.046

Riecker, A., Wildgruber, D., Mathiak, K., Grodd, W., & Ackermann, H. (2003). Parametric analysis of rate-dependent hemodynamic response functions of cortical and subcortical brain structures during auditorily cued finger tapping: A fMRI study. *NeuroImage, 18*(3), 731–739. doi:10.1016/s1053-8119(03)00003-x

Rilling, J. K., Glasser, M. F., Preuss, T. M., Ma, X., Zhao, T., Hu, X., et al. (2008). The evolution of the arcuate fasciculus revealed with comparative DTI. *Nature Neuroscience, 11*(4), 426–428. doi:10.1038/nn2072

Romo, R., Merchant, H., Ruiz, S., Crespo, P., & Zainos, A. (1995). Neural activity of primate putamen during categorical perception of somaesthetic stimuli. *Neuroreport, 6*(7), 1013–1017.

Salmelin, R., Hámäaläinen, M., Kajola, M., & Hari, R. (1995). Functional segregation of movement-related rhythmic activity in the human brain. *NeuroImage, 2*(4), 237–243. doi:10.1006/nimg.1995.1031

Schachner, A. (2010). Auditory-motor entrainment in vocal mimicking species. *Communicative and Integrative Biology, 3*(3), 290–293. doi:10.4161/cib.3.3.11708

Schroeder, C. E., & Lakatos, P. (2009). Low-frequency neuronal oscillations as instruments of sensory selection. *Trends in Neurosciences, 32*(1), 9–18. doi:10.1016/j.tins.2008.09.012

Schroeder, C. E., Wilson, D. A., Radman, T., Scharfman, H., & Lakatos, P. (2010). Dynamics of active sensing and perceptual selection. *Current Opinion in Neurobiology, 20*(2), 172–176. doi:10.1016/j.conb.2010.02.010

Schubotz, R. I. (2007). Prediction of external events with our motor system: Towards a new framework. *Trends in Cognitive Sciences, 11*(5), 211–218. doi:10.1016/j.tics.2007.02.006

Schubotz, R. I., Friederici, A. D., & Yves von Cramon, D. (2000). Time perception and motor timing: A common cortical and subcortical basis revealed by fMRI. *NeuroImage, 11*(1), 1–12. doi:10.1006/nimg.1999.0514

Snyder, J., & Krumhansl, C. L. (2001). Tapping to ragtime: Cues to pulse finding. *Music Perception, 18*(4), 455–489. doi:10.1525/mp.2001.18.4.455

Snyder, J. S., & Large, E. W. (2005). Gamma-band activity reflects the metric structure of rhythmic tone sequences. *Cognitive Brain Research, 24*(1), 117–126. doi:10.1016/j.cogbrainres.2004.12.014

Steriade, M., Dossi, R. C., & Nunez, A. (1991). Network modulation of a slow intrinsic oscillation of cat thalamocortical neurons implicated in sleep delta waves: Cortically induced synchronization and brainstem cholinergic suppression. *Journal of Neuroscience, 11*(10), 3200–3217.

Takeya, R., Kameda, M., Patel, A. D., & Tanaka, M. (2017). Predictive and tempo-flexible synchronization to a visual metronome in monkeys. *Scientific Reports, 7*(1), 6127. doi:10.1038/s41598-017-06417-3

Teki, S., Grube, M., Kumar, S., & Griffiths, T. D. (2011). Distinct neural substrates of duration-based and beat-based auditory timing. *Journal of Neuroscience, 31*(10), 3805–3812. doi:10.1523/jneurosci.5561-10.2011

Tian, X., & Poeppel, D. (2010). Mental imagery of speech and movement implicates the dynamics of internal forward models. *Frontiers in Psychology, 1*, 1–23. doi:10.3389/fpsyg.2010.00166

Toma, K., Mima, T., Matsuoka, T., Gerloff, C., Ohnishi, T., Koshy, B., et al. (2002). Movement rate effect on activation and functional coupling of motor cortical areas. *Journal of Neurophysiology, 88*(6), 3377–3385. doi:10.1152/ jn.00281.2002

Trainor, L. J., Marie, C., Bruce, I. C., & Bidelman, G. M. (2014). Explaining the high voice superiority effect in polyphonic music: Evidence from cortical evoked potentials and peripheral auditory models. *Hearing Research, 308*, 60–70. doi:10.1016/j.heares.2013.07.014

van Noorden, L., & Moelants, D. (1999). Resonance in the perception of musical pulse. *Journal of New Music Research, 28*(1), 43–66. doi:10.1076/jnmr.28.1.43.3122

Velasco, M. J., & Large, E. W. (2011). Pulse detection in syncopating rhythms using neural oscillators. In *Proceedings of the 12th Annual Conference of the International Society for Music Information Retrieval* (pp. 185–190). New York: The Printing House.

Vuust, P., Ostergaard, L., Pallesen, K. J., Bailey, C., & Roepstorff, A. (2009). Predictive coding of music—Brain responses to rhythmic incongruity. *Cortex, 45*(1), 80–92. doi:10.1016/j.cortex.2008.05.014

Vuust, P., Roepstorff, A., Wallentin, M., Mouridsen, K., & Ostergaard, L. (2006). It don't mean a thing …: Keeping the rhythm during polyrhythmic tension, activates language areas (BA47). *Neuroimage, 31*(2), 832–841. doi:10.1016/j.neuroimage.2005.12.037

Winkler, I., Háden, G. P., Ladinig, O., Sziller, I., & Honing, H. (2009). Newborn infants detect the beat in music. *Proceedings of the National Academy of Sciences, 106*(7), 2468–2471. doi:10.1073/pnas.0809035106

Yee, W., Holleran, S., & Jones, M. R. (1994). Sensitivity to event timing in regular and irregular sequences: Influences of musical skill. *Perception and Psychophysics, 56*(4), 461–471. doi:10.3758/bf03206737

Zanto, T. P., Large, E. W., Fuchs, A., & Kelso, J. A. S. (2005). Gamma-band responses to perturbed auditory sequences: Evidence for synchronization of perceptual processes. *Music Perception, 22*(3), 531–547. doi:10.1525/mp.2005.22.3.531

Zarco, W., Merchant, H., Prado, L., & Mendez, J. C. (2009). Subsecond timing in primates: Comparison of interval production between human subjects and rhesus monkeys. *Journal of Neurophysiology, 102*(6), 3191–3202. doi:10.1152/jn.00066.2009

Zatorre, R. J., Chen, J. L., & Penhune, V. B. (2007). When the brain plays music: Auditory-motor interactions in music perception and production. *Nature Reviews: Neuroscience, 8*(7), 547–558. doi:10.1038/nrn2152

9

Neural Overlap in Processing Music and Speech

Isabelle Peretz, Dominique T. Vuvan, Marie-Élaine Lagrois, and Jorge L. Armony

Humans are born with the potential to both speak and make music. The relation between language and musicality has been the topic of much debate, and one that is heated because it speaks directly to the nature of evolved human cognition.[1] For some, musicality owes its efficacy to the natural disposition for speech. For example, music may exaggerate particular speech features such as intonation and affective tone that are so effective for bonding (Juslin & Laukka, 2003). In other words, musicality may aim at the language system just as artistic masks target the face recognition system. We can stretch this argument further and envisage that music owes its efficacy to relying on the natural disposition for speech. From this perspective, the language modules are invaded (Sperber & Hirschfeld, 2004). Musicality could have emerged in all cultures because it is so effective at co-opting one or several evolved modules. For others (Jackendoff, 2009), once we take away the tone of voice shared by speech and music, the specialization of language to convey conceptual information and of musicality to express affect is distinct. According to this view, musicality may have preceded language in evolution, and language may build on the natural disposition for musicality.

These divergent perspectives on the origins and functions of musicality are echoed in cognitive neuroscience. On the one hand, an increasing number of neuroimaging studies point to a large and significant neural overlap in the responses to speech and music, taken as evidence of neural sharing (see figure 9.1). On the other hand, a solid body of neuropsychological studies shows that musicality involves multiple processing components that can be selectively impaired without apparent effects on language (or any other cognitive ability; Peretz, 2006, 2009). These conflicting sets of data call for neurocomparative studies.

The neurocomparative study of music and speech processing is obviously limited in animals and must therefore rely on the use of sophisticated and noninvasive methods in humans. The most widely used technique today is functional magnetic resonance imaging (fMRI), which has been recently exploited to compare music and speech perception. As we explain, coactivation of brain regions in fMRI is often interpreted as evidence of the sharing of the underlying neural circuitry. However, activation overlap does not provide sufficient evidence of neural sharing, for which rigorous and direct testing is required. In this chapter, we focus on the use of brain imaging techniques, such as fMRI, to measure neural overlap between music and speech processing. Note that other methodologies, such as brain lesion studies (Peretz, 2006, 2009), transcranial magnetic stimulation, and behavioral interference, can give complementary insight to the question of speech and music overlap (Kunert & Slevc, 2015). However, the findings obtained with these alternative methods have been reviewed elsewhere or are too scarce to be included here.

In this review, we summarize contemporary views of brain organization for music and speech processing with a discussion of the challenges in concluding neural sharing (or separation) from evidence of overlap. We then review some methodological advances that aid in making more definitive conclusions regarding neural overlap. Finally, we discuss the impact of evidence of neural sharing on current views of the neurobiology of musicality.

Brain Specialization: From Regions to Networks

Research in cognitive neuroscience has been guided by the assumption that brain regions are specialized for a function, with each function implemented in a relatively small neural space. For example, the superior temporal sulcus has been associated with voice processing (Belin, Zatorre, Lafaille, Ahad, & Pike, 2000). This voice-preferred region responds, bilaterally, more strongly to human vocalizations, with or without linguistic content, than to nonvocal sounds or vocalizations produced by other animals (Fecteau, Armony, Joanette, & Belin, 2004). A specialized neural system that processes conspecific vocalizations has also been observed in other species. For instance, the superior temporal plane of the macaque monkey responds preferentially to species-specific vocalizations over other vocalizations and sounds (Petkov et al., 2008). Other cortical areas, such as the caudal insular cortex in rhesus monkeys, also appear to be tuned to intraspecies vocalizations over a wide range of auditory stimuli such as environmental sounds and vocalizations from other animals (Remedios, Logothetis, & Kayser, 2009).

Similarly, music processing may rely on a cortical area that is domain specific and neurally separable (Peretz, 2006). For example, the system that maps pitch onto musical keys, termed tonality or tonal encoding of pitch, may be music selective. Research points to the inferior frontal areas as critically involved (Koelsch et al., 2002; Tillmann, Janata, & Bharucha, 2003; Tillmann et al., 2006). However, this localization mostly corresponds to the processing of harmonic structure, a culture-specific elaboration of pitch that is quite recent in music history. Moreover, tonal encoding of pitch is likely to recruit a vast network because it involves multiple processes. For example, Jackendoff and Lerdahl (2006) distinguish three forms of elaboration of pitch hierarchies in a musical context by considering different principles for pitch space, tonal reduction, and tension or relaxation. Thus, it would not be surprising to discover that more than one brain region contributes to the musical interpretation of pitch. A major breakthrough would be to identify one of these brain regions as foundational for tonal encoding of pitch (Peretz, 2006). So far, such an essential component for musicality has not been localized in one specific region. Rather, the evidence points toward the connectivity between the right auditory cortex and inferior frontal gyrus (IFG) as being necessary for normal development of musical encoding of pitch (Albouy et al., 2013; Peretz, 2013).

The key question becomes to what extent parts of the musicality network can be shared both functionally and neurally with language. Logically, music and speech processing could share parts of their respective neural networks, such as the mechanisms for the acoustical analysis of pitch, and still be distinct because musicality and language differ from one another in other respects, notably semantics. The idea that parts of the networks are shared is currently very popular. A Google Scholar search using the keywords "(neural) AND (overlap OR sharing) AND (music) AND (language OR speech)" reveals a linear increase of this notion in published articles over the past decade (figure 9.1). This explosion in research interest in the overlap of music and language is, however, often unsupported by comparative data. To cite a recent example, an fMRI study reporting stronger response to rhythmic musical structure in professional musicians, compared with nonmusicians, in a region typically associated with processing of linguistic syntax (which was not examined in that study), led the authors to conclude that "musical experts seem to rely on the same neural resources during the processing of syntactic violations in the rhythm domain that the brain usually uses for the evaluation of linguistic syntax" (Herdener et al., 2014, p. 841).

It is important to keep in mind that neural overlap does not necessarily entail neural sharing. The neural circuits established for musicality may be intermingled or adjacent to those used for a similar function in language and yet be neurally separable. For example, mirror neurons are interspersed among purely motor-related neurons in premotor regions of the macaque cortex (Rizzolatti & Craighero, 2004). Similarly, the neurons responsible for the computation of some musical feature may be interspersed among neurons involved in similar aspects in speech.

Moreover, most brain structures, such as Broca's area (occupying the left inferior frontal gyrus), which is often the focus of interest in music and language comparisons (see Kunert, Willems, Casasanto, Patel, & Hagoort, 2015, for a recent example), are relatively large and complex and thus can easily accommodate more than one distinct processing network. Common localization of distinct networks or functions can be dictated by neural properties, like dense connectivity with other brain regions, and not dictated by sharing.

Recent network analyses have revealed highly connected network hubs, which may well be shared by music and speech processing. Hubs support efficient control or integration by facilitating the convergence of neuronal signals from different sensory modalities (e.g., auditory and motor) or cognitive domains (e.g., musicality and language; van den Heuvel & Sporns, 2013; Yan & He, 2011). The hubs maintain anatomical and functional connections that span long distances, and they tend to consume metabolic energy at a higher rate

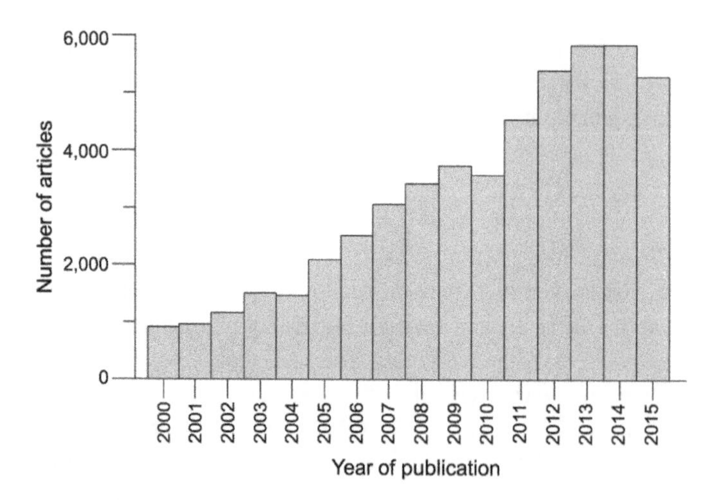

Figure 9.1
Results of a Google Scholar search. The search was performed on January 25, 2016, using the keywords "(neural) AND (overlap OR sharing) AND (music) AND (language OR speech)," with the number of search results plotted per year (noncumulative).

than non-hub regions (Collin, Sporns, Mandl, & van den Heuvel, 2014). Thus, hubs are not only centers of integration but also points of increased hemodynamic responses and vulnerability. Accordingly, the co-activation of brain regions, as well as the sometimes observed co-occurrence of deficits in music and speech processing, may reflect the involvement of these integration centers rather than of distinct parts of their respective networks.

In this network perspective, how can we identify the parts of networks that would constitute evidence of brain specialization, as the subhead of this section suggests? Several fMRI studies have identified a region within the anterior superior temporal gyrus (STG) that responds more strongly to music than to human voice, including speech (Angulo-Perkins et al., 2014; Fedorenko, McDermott, Norman-Haignere, & Kanwisher, 2012; Leaver & Rauschecker, 2010; Rogalsky, Rong, Saberi, & Hickok, 2011; see figure 9.2). Nonetheless, these studies also found large regions within the temporal lobes that responded more to both music and voice, compared to control conditions (e.g., nonvocal sounds or silence), with no significant differences between the two former categories. Moreover, even in regions that were categorized as "music preferred," significant activation in response to speech (compared, for instance, with nonvocal sounds) can be observed (figure 9.2B). Similarly, the so-called language-selective regions in the left posterior temporal lobe show a significant, albeit weaker, response to music (Fedorenko, Behr, & Kanwisher, 2011). Unfortunately, the approach employed in these studies does not allow researchers to

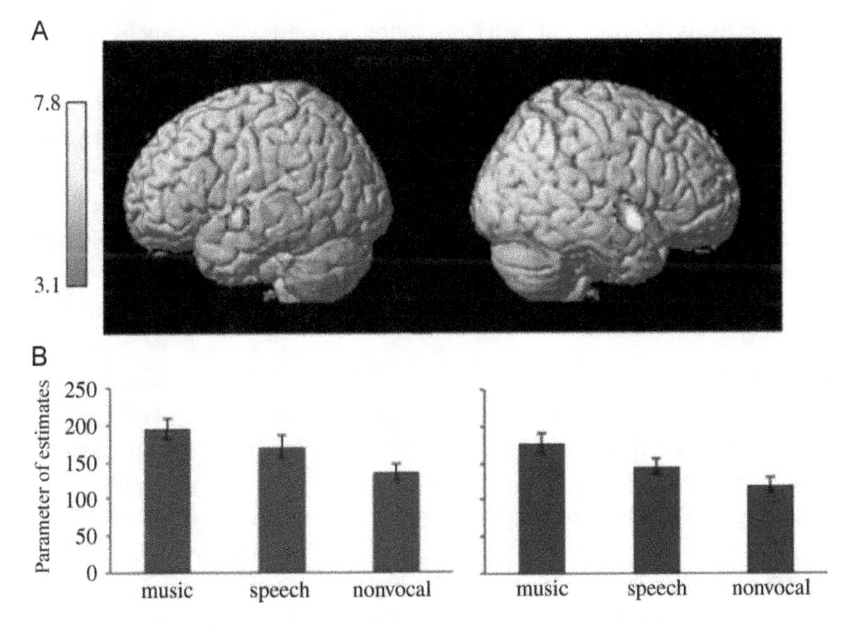

Figure 9.2
(A) Statistical map showing voxels within the temporal lobe with stronger responses to musical stimuli than to human voice (including speech). (B) Left and right cluster-averaged parameter estimates, showing that responses to speech were smaller than to music but were significantly larger compared to human nonvocal stimuli (all pairwise comparisons significant, $p < 0.005$). Data from Angulo-Perkins et al. (2014).

determine whether the same neuronal populations respond to both music and speech (possibly with different strength) or if distinct, neighboring groups of neurons were involved.

Fortunately, new methods have been developed to separate brain responses to different categories of stimuli in the same neural region. These techniques can be exploited to distinguish domain-specific neural activation at a finer-grained level than the standard use of fMRI can offer. We introduce these techniques in the next section.

Evidence of Neural Sharing

Most regions of the brain participate in multiple processes (Anderson, 2010). Moreover, music and speech processing share a large number of properties, from the acoustical analysis of the auditory input to the planning of motor output. Therefore, several brain areas are expected to overlap in the processing of music and speech (Patel, 2013). However, neural overlap does not necessarily mean neural sharing. Here, we review the evidence for and against neural sharing; specifically, we summarize the studies in which the distinct contribution of neural populations to music and speech processing has been examined in overlapping regions. This can be done by a number of neuroimaging techniques, such as multivoxel pattern analysis (MVPA) and fMRI adaptation, and more invasively by intracerebral recordings. We present the main results obtained with each technique.

We chose not to cover lesion studies here because these have been reviewed relatively recently (Peretz, 2006, 2009) and the use of this approach has declined over recent years. Nevertheless, it is worth mentioning that lesion studies suggest neural segregation between musicality and language networks. Brain damage can affect just musical abilities while sparing language and other aspects of cognition (Peretz, 2006, 2009; Peretz & Coltheart, 2003). The possibility of disrupting the operations of musicality without having any impact on language implies that some parts of its neural substrate are not shared with language.

Multivoxel Pattern Analysis

This technique uses machine learning algorithms to categorize neuroimaging data. It differs from the standard approach that assesses, on a voxel-by-voxel basis, the mean difference in activity between conditions and identifies voxels that respond consistently more strongly to one condition (e.g., music) than to another (e.g., speech; Friston, Tononi, Sporns, & Edelman, 1995). Unlike such standard univariate analyses, multivariate decoding methods consider data from several voxels at once to identify patterns of activity associated with a particular stimulus, task, or mental state. The multivariate pattern analysis methods detect distributed representations (i.e., groups of voxels) whose combined activity discriminates between conditions of interest, even if the individual voxels do not exhibit statistically significant (i.e., in the context of the general linear model) differences between them (Norman, Polyn, Detre, & Haxby, 2006). One key advantage of this multivariate approach, which provides complementary information to that obtained through univariate methods (Davis & Poldrack, 2013; Jimura & Poldrack, 2012), is its sensitivity to neural segregation in overlapping regions (Peelen & Downing, 2007; Todd, Nystrom, & Cohen, 2013). This technique has been used in several studies comparing music to speech.

Rogalsky et al. (2011) found large areas of activation in response to music and speech in overlapping portions of auditory cortex. However, multivariate pattern classification analyses indicated that within the regions of overlap, speech and music elicited distinguishable patterns of activation in the STGs. In addition, speech (jabberwocky sentences) elicited more ventrolateral activation, whereas music (novel melodies) elicited a more dorsomedial pattern extending into the parietal lobe. These findings highlight the existence of overlapping but distinct networks for music and speech within the same cortical areas. Abrams et al. (2011) used multivariate pattern analysis for natural and scrambled music and speech excerpts and also found distinct brain patterns of responses to the two categories of sounds in several regions within the temporal lobe and the inferior frontal cortex. Therefore, the pattern of neural activation was distinct between music and speech, although there was overlap in the areas activated by the two domains. New data-driven approaches are starting to be used as a way to uncover the neural signature of different cognitive processes

(Kaplan, Man, & Greening, 2015; see Norman-Haignere, Kanwisher, & McDermott, 2015, for an application of this technique to the comparison of music and speech).

It is important to point out that even if the stimuli are matched for emotional content, attention, memory, subjective interest, arousal, and familiarity (Abrams et al., 2011), any observed category differences in activation strengths or patterns could be owing to acoustical differences. In order to avoid this confound, at least to some extent, one can use sung melodies and spoken lyrics from songs. These are optimal stimuli because they are relatively complex and extend over several seconds. Furthermore, the coexistence of tunes and lyrics in songs makes them very similar in terms of acoustical structure and familiarity. The main acoustic differences between song and speech are the more regular rhythm and pitch stability in each syllable of songs.

Comparing spoken lyrics, sung tunes, and songs (corrected for rhythmic differences) with both multivariate and univariate analyses, Merrill et al. (2012) observed a large overlap between song and speech in the bilateral STGs. The STGs were found to code for differences between words and pitch patterns, whether these were embedded in a song or in speech. They also found that the left IFG coded for spoken words and showed predominance over the right IFG in prosody processing, whereas the right IFG was more active for processing the pitch pattern in songs. Interestingly, this result was found only when using the multivariate decoding method, demonstrating its higher sensitivity to the differential fine-scale coding of information. Another important result is the finding that the intraparietal sulcus shows sensitivity to discrete pitch relations in songs as opposed to the gliding pitches in speech prosody. Thus, as expected, the processing of lyrics, tunes, and songs shares many features that are reflected in a fundamental similarity of brain areas involved in their perception. However, subtle differences between speech and music can lead to distinct patterns of brain activity.

In sum, music and speech stimuli seem to activate distinct neural populations in overlapping regions. However, most of the reported results could be owing, at least in part, to acoustical differences between categories. Another way to circumvent this potential problem is to parametrically manipulate acoustical structure. For example, musical sounds can be morphed into speech sounds gradually, so that the influence of acoustical changes on the brain responses can be measured systematically. So far, white noise has been morphed into a speech sound or a musical instrument sound separately (Merrill et al., 2012), but the morphing technique can be extended to whole sentences and used with other paradigms, like adaptation, to which we now turn.

Functional Magnetic Resonance Imaging Adaptation

Although the nature of the blood oxygen-level-dependent (BOLD) signal and its spatial resolution do not allow direct identification of neurons that are active in response to a given stimulus, we can take advantage of the nonlinear dynamics of neural activity to indirectly address this question by using the so-called fMRI adaptation paradigm (Grill-Spector &

Malach, 2001). This approach, based on the principle of neuronal adaptation/habituation, relies on the fact that the observed BOLD signal to successive stimuli depends on whether they stimulate the same or different neurons. That is, the activity associated with two stimuli will be smaller if they activate the same neuronal pool than if they stimulate different neurons. Although adaptation is strongest when repeating the same stimulus, it can also be observed when different exemplars from the same category are presented and can thus be used to identify those brain regions in which different types of stimuli share a common neural representation.

Thus, in a region that responds to both speech and music, we would expect within-domain (i.e., speech-speech and music-music) adaptation. If the attenuation remains when a switch in domain is introduced (i.e., music-speech or speech-music), it would suggest that the same neuronal population was responding to both categories. In contrast, if the attenuation disappears in the overlapping region after the switch in domains, this would be evidence that fresh neurons from a distinct neural population were responsible for the processing of each of the two domains.

The paradigm of fMRI adaptation has been exploited twice in the neural comparison of music and speech. Sammler et al. (2010) used this technique to induce neural adaptation to listening to lyrics, melodies, and songs. Reductions of the BOLD response were observed along the superior temporal sulcus and gyrus (STS/STG) bilaterally. Within these regions, the left mid-STS showed an interaction of the adaptation effects for lyrics and tunes, suggesting shared processing of the two components. The degree of integration decayed toward more anterior regions of the left STS in which the stronger adaptation for lyrics than for tunes was suggestive of independent processing of lyrics. Evidence for an integrated representation of lyrics and tunes was also found in the left premotor cortex, possibly related to the buildup of a vocal code for singing.

However, in a subsidiary analysis of a recent study (Aubé, Angulo-Perkins, Peretz, Concha, & Armony, 2015), we showed music-music but no speech-music adaptation in the "music-preferred" area in the anterior STG (Armony, Aubé, Angulo-Perkins, Peretz, & Concha, 2015). This result suggests that distinct neural populations underlie the activations observed in this region for speech and music.

Altogether, results show both a neural dissociation as well as a high degree of neural sharing between music and speech, which could arise from the involvement of voice-specific areas. Further studies exploiting this paradigm represent an opportunity for characterizing the nature of the shared mechanisms.

One optimal avenue for adaptation studies is provided by the song illusion that Deutsch, Henthorn, and Lapidis (2011) discovered. This illusion is created by the repetition of a phrase that sounds initially like speech and through repetition, as if it were sung. One possible account of this illusion is that the neurons underlying speech perception get adapted through repetition, whereas the neural population underlying music perception does not.

The robustness of music perception to neural attenuation may originate from the fact that repetition is a characterizing feature of music and not of speech (Margulis, 2013).

Two fMRI studies have tested the neural correlates of the song illusion. In the first one (Tierney, Dick, Deutsch, & Sereno, 2013), different phrases, albeit produced by the same speaker and matched for syllable length, were spoken and sung. Using such stimuli, BOLD response changes were found to be larger in multiple brain regions, including the anterior STG and the right midposterior STG. In a more recent study, Hymers et al. (2015) used the speech-to-song illusion without physically changing the stimulus in an adaptation paradigm. They found that the effect of perceiving a stimulus as song compared with that of speech localized to the right STG and involved the left fronto-temporal region. These findings suggest that illusory song perception recruits a network of brain regions that are predominantly shared with speech perception.

The use of adaptation procedures with fMRI should be pursued. However, like any paradigm, adaptation has its limits. Reduced activity can result from adaptation but also from practice or expectation. Similarly, if adaptation disappears after a change of condition, say from speech to music, this could be because speech-induced neural changes interfered with music processing (see Grill-Spector, Henson, & Martin, 2006, for a discussion of various ways to interpret brain response adaptation).

Intracranial Recordings

The implantation of electrodes for presurgical evaluation of temporal lobe epilepsy represents a rare chance to distinguish neural responses to music and speech with excellent temporal and spatial resolution that largely exceeds that achieved with noninvasive methods. This high spatiotemporal resolution can address the question of shared versus distinct neural populations by revealing the degree to which the time course of neural responses differs in regions in which music and speech processing overlap. So far, depth electrodes have not been used in the comparison of music and speech.

Electrical activity has been recorded intracranially through subdural electrodes located above the left or right perisylvian regions (Sammler et al., 2013). This method presents the advantage of recording temporal activity with high precision. However, the method still faces the difficulty of inferring the location of the sources on the basis of brain surface recordings, especially in overlapping regions.

Using this method to examine the contribution of the STGs to linguistic and musical syntax, Sammler et al. (2013) compared the early negativities evoked by violations of structure in sentences and chord sequences in five patients. The results showed considerable overlap in the bilateral STG but also differences in the hemispheric timing and relative involvement of the frontal and temporal brain structures. Although the combined data lend support for a co-localization of early musical and linguistic syntax processing in the temporal lobe, the mechanisms involved seem to depend on the (music or language) domain considered.

Future Directions

The evidence points toward substantial neural overlap between music and speech processing. A many-to-one mapping between cognitive functions and brain structures seems to characterize the human brain (Anderson, 2010). Therefore, it is more likely to find evidence of overlap than segregation. Nevertheless, there is converging evidence for music-specific responses along the neural pathways. The evidence is still scarce but strengthened by the diversity of the neuroimaging approaches used so far. Therefore, the question of overlap between music and speech processing must still be considered an open question for the field. In this chapter, we have reviewed technological advances that will allow us to tackle this issue more rigorously than in the past.

While using the novel fMRI techniques for cross-domain comparison, it may be useful to consider the following recommendations:

1. *Make comparisons in native (subject-specific) brain space rather than in a common (normalized) one.* Averaging across individually variable anatomies blurs brain activations and can create an artificial overlap of closely neighboring but non-overlapping responses. For example, location of pitch maps in Heschl's gyri and planum temporale varies widely across individuals (Schönwiesner & Zatorre, 2008; Westbury, Zatorre, & Evans, 1999). Such variable organization and localization of cortical maps call for individually determined regions of interest.
2. *Consider connectivity.* The human brain is a highly connected and interactive structure. Domain specificity at high levels can have an impact on low-level processing. That is, lower-level brain areas, such as the brain stem and the primary auditory cortex, are influenced by higher-level areas via efferent connections (Skoe & Kraus, 2012).
3. *Manipulate stimulus and task parameters.* Cognitive demands made by supposedly analogous operations in processing music and speech may widely differ (Asano & Boeckx, 2015; Kunert & Slevc, 2015; LaCroix, Diaz, & Rogalsky, 2015; Tillmann & Bigand, 2015). In order to avoid differences in brain responses driven by task difficulty rather than domain specificity, one can manipulate a common factor (e.g., speed) or use an interference paradigm as is often the case in behavioral studies (Kunert & Slevc, 2015). In doing so, the perceptual demands should always be carefully controlled in order to avoid differences in brain responses driven by purely acoustic differences. As we have noted, Deutsch's song illusion presents an ideal example of this type of control, because the input remains the same, whereas the percept changes through repetition.

Implications and Conclusions

Neural sharing is a key concept for explaining transfer effects between music and language. Patel (2011, 2013) has introduced the OPERA framework to explain why musical training may lead to enhanced speech processing. An essential condition of the OPERA hypothesis is neural overlap, a term Patel used to refer to as "neural sharing." That is, in order for musical training to influence the neural processing of speech, a shared characteristic in both domains must be processed by a population of neurons shared by the musicality and language brain networks.

The original OPERA hypothesis (Patel, 2011) focuses on acoustic features (e.g., waveform periodicity) rather than on cognitive demands (e.g., auditory working memory). This early focus on acoustical features converges with the evidence reviewed here, suggesting that the processing of music and speech overlaps in posterior auditory cortex (basic acoustic processing) and becomes more differentiated anteriorly (domain-specific representation; Leaver & Rauschecker, 2010; Sammler et al., 2010). However, the expanded OPERA hypothesis (Patel, 2013) incorporates the idea of shared *cognitive* processing into the discussion of neural overlap, based on the proposals that musical training enhances auditory attention and working memory (Besson, Chobert, & Marie, 2011; Strait & Kraus, 2011). Indeed, there is abundant evidence that attention sharpens sensory encoding in primary brain areas (see Alho, Rinne, Herron, & Woods, 2014, for a meta-analysis of attention-related modulations in the auditory cortex). However, little is known about the neural specificity of these top-down effects. Backward propagation along neural pathways may be diffuse and affect several distinct sensory systems, not just speech. In that case, neural sharing is not necessary for transfer between musicality and language. Therefore, as research guided by the OPERA hypothesis proceeds, it is important to consider methods that are sufficiently sophisticated to make justifiable claims regarding the neural sharing between music and speech processing.

The systematic search for neural sharing between music and speech is an important avenue not only for clinical and education purposes, but also for understanding the neurobiological origins of musicality. It is common for neural circuits established for one purpose to be reused or recycled during evolution. For example, in the zebra finch, some song nuclei may participate in the learning of nonvocal tasks such as food avoidance. These findings suggest that the specialized forebrain premotor nuclei controlling song evolved from circuits involved in behaviors related to feeding (Tokarev, Tiunova, Scharff, & Anokhin, 2011). In humans, we have examined the possibility that musicality recycles emotion circuits that have evolved for emotional vocalizations (Peretz, Aubé, & Armony, 2013).

Dehaene and Cohen (2007) have proposed an interesting recycling proposal because it entails neural and functional constraints on the newly acquired function. The recycling may occur only if a network of neural structures already has (most of) the structures necessary to support the novel set of cognitive and physical procedures that characterize the new function. As a result, the neural manifestations of novel abilities should have some common characteristics and share some possibilities for learning with nonhuman primates. This theory makes clear some of the limits and costs of neuronal recycling. The greater the distance between the function(s) and the existing cortical structure, the harder the learning process will be, and the more likely that the learning process will disrupt the other functions that the common neural circuitry supports. Therefore, if some core component of musicality can be found to share a brain region or network involved in language, it may reveal a novel pathway by which humans may have achieved their highly sophisticated use of sound.

Note

1. According to the proposal introduced by Honing, ten Cate, Peretz, and Trehub (2015), we are distinguishing between musicality and music, as follows. *Musicality* is defined as a spontaneously developing set of traits based on, and constrained by, our cognitive and biological system. *Music* is defined as a social and cultural construct based on that musicality.

References

Abrams, D. A., Bhatara, A., Ryali, S., Balaban, E., Levitin, D. J., & Menon, V. (2011). Decoding temporal structure in music and speech relies on shared brain resources but elicits different fine-scale spatial patterns. *Cerebral Cortex, 21*(7), 1507–1518. doi:10.1093/cercor/bhq198

Albouy, P., Mattout, J., Bouet, R., Maby, E., Sanchez, G., Aguera, P.-E., et al. (2013). Impaired pitch perception and memory in congenital amusia: The deficit starts in the auditory cortex. *Brain, 136*(5), 1639–1661. doi:10.1093/brain/awt082

Alho, K., Rinne, T., Herron, T. J., & Woods, D. L. (2014). Stimulus-dependent activations and attention-related modulations in the auditory cortex: A meta-analysis of fMRI studies. *Hearing Research, 307*, 29–41. doi:10.1016/j.heares.2013.08.001

Anderson, M. L. (2010). Neural reuse: A fundamental organizational principle of the brain. *Behavioral and Brain Sciences, 33*(4), 245–266. doi:10.1017/S0140525X10000853

Angulo-Perkins, A., Aubé, W., Peretz, I., Barrios, F. A., Armony, J. L., & Concha, L. (2014). Music listening engages specific cortical regions within the temporal lobes: Differences between musicians and non-musicians. *Cortex, 59*, 126–137. doi:10.1016/j.cortex.2014.07.013

Armony, J. L., Aubé, W., Angulo-Perkins, A., Peretz, I., & Concha, L. (2015). The specificity of neural responses to music and their relation to voice processing: An fMRI-adaptation study. *Neuroscience Letters, 593*, 35–39. doi:10.1016/j.neulet.2015.03.011

Asano, R., & Boeckx, C. (2015). Syntax in language and music: What is the right level of comparison? *Frontiers in Psychology, 6*, 942. doi:10.3389/fpsyg.2015.00942

Aubé, W., Angulo-Perkins, A., Peretz, I., Concha, L., & Armony, J. L. (2015). Fear across the senses: Brain responses to music, vocalizations and facial expressions. *Social Cognitive and Affective Neuroscience, 10*(3), 399–407. doi:10.1093/scan/nsu067

Belin, P., Zatorre, R. J., Lafaille, P., Ahad, P., & Pike, B. (2000). Voice-selective areas in human auditory cortex. *Nature, 403*(6767), 309–312. doi:10.1038/35002078

Besson, M., Chobert, J., & Marie, C. (2011). Transfer of training between music and speech: Common processing, attention, and memory. *Frontiers in Psychology, 2*, 94. doi:10.3389/fpsyg.2011.00094

Collin, G., Sporns, O., Mandl, R. C. W., & van den Heuvel, M. P. (2014). Structural and functional aspects relating to cost and benefit of rich club organization in the human cerebral cortex. *Cerebral Cortex, 24*(9), 2258–2267. doi:10.1093/cercor/bht064

Davis, T., & Poldrack, R. A. (2013). Measuring neural representations with fMRI: Practices and pitfalls. *Annals of the New York Academy of Sciences, 1296*(1), 108–134. doi:10.1111/nyas.12156

Dehaene, S., & Cohen, L. (2007). Cultural recycling of cortical maps. *Neuron, 56*(2), 384–398. doi:10.1016/j.neuron.2007.10.004

Deutsch, D., Henthorn, T., & Lapidis, R. (2011). Illusory transformation from speech to song. *Journal of the Acoustical Society of America, 129*(4), 2245–2252. doi:10.1121/1.3562174

Fecteau, S., Armony, J. L., Joanette, Y., & Belin, P. (2004). Is voice processing species-specific in human auditory cortex? An fMRI study. *NeuroImage, 23*(3), 840–848. doi:10.1016/j.neuroimage.2004.09.019

Fedorenko, E., Behr, M. K., & Kanwisher, N. (2011). Functional specificity for high-level linguistic processing in the human brain. *Proceedings of the National Academy of Sciences, 108*(39), 16428–16433. doi:10.1073/pnas.1112937108

Fedorenko, E., McDermott, J. H., Norman-Haignere, S., & Kanwisher, N. (2012). Sensitivity to musical structure in the human brain. *Journal of Neurophysiology, 108*(12), 3289–3300. doi:10.1152/jn.00209.2012

Friston, K. J., Tononi, G., Sporns, O., & Edelman, G. M. (1995). Characterising the complexity of neuronal interactions. *Human Brain Mapping, 3*(4), 302–314. doi:10.1002/hbm.460030405

Grill-Spector, K., Henson, R., & Martin, A. (2006). Repetition and the brain: Neural models of stimulus-specific effects. *Trends in Cognitive Sciences, 10*(1), 14–23. doi:10.1016/j.tics.2005.11.006

Grill-Spector, K., & Malach, R. (2001). fMR-adaptation: A tool for studying the functional properties of human cortical neurons. *Acta Psychologica, 107*(1–3), 293–321. doi:10.1016/S0001-6918(01)00019-1

Herdener, M., Humbel, T., Esposito, F., Habermeyer, B., Cattapan-Ludewig, K., & Seifritz, E. (2014). Jazz drummers recruit language-specific areas for the processing of rhythmic structure. *Cerebral Cortex, 24*(3), 836–843. doi:10.1093/cercor/bhs367

Honing, H., ten Cate, C., Peretz, I., & Trehub, S. E. (2015). Without it no music: Cognition, biology and evolution of musicality. *Philosophical Transactions of the Royal Society of London B: Biological Sciences, 370*(1664), 20140088. doi:10.1098/rstb.2014.0088

Hymers, M., Prendergast, G., Liu, C., Schulze, A., Young, M. L., Wastling, S. J., et al. (2015). Neural mechanisms underlying song and speech perception can be differentiated using an illusory percept. *NeuroImage, 108*, 225–233. doi:10.1016/j.neuroimage.2014.12.010

Jackendoff, R. (2009). Parallels and nonparallels between language and music. *Music Perception, 26*(3), 195–204. doi:10.1525/mp.2009.26.3.195

Jackendoff, R., & Lerdahl, F. (2006). The capacity for music: What is it, and what's special about it? *Cognition, 100*(1), 33–72. doi:10.1016/j.cognition.2005.11.005

Jimura, K., & Poldrack, R. A. (2012). Analyses of regional-average activation and multivoxel pattern information tell complementary stories. *Neuropsychologia, 50*(4), 544–552. doi:10.1016/j.neuropsychologia.2011.11.007

Juslin, P. N., & Laukka, P. (2003). Communication of emotions in vocal expression and music performance: Different channels, same code? *Psychological Bulletin, 129*(5), 770–814. doi:10.1037/0033-2909.129.5.770

Kaplan, J. T., Man, K., & Greening, S. G. (2015). Multivariate cross-classification: Applying machine learning techniques to characterize abstraction in neural representations. *Frontiers in Human Neuroscience, 9*, 1–12. doi:10.3389/fnhum.2015.00151

Koelsch, S., Gunter, T. C., von Cramon, D. Y., Zysset, S., Lohmann, G., & Friederici, A. D. (2002). Bach speaks: A cortical "language-network" serves the processing of music. *NeuroImage, 17*(2), 956–966. doi:10.1016/s1053-8119(02)91154-7

Kunert, R., & Slevc, L. R. (2015). A commentary on: "Neural overlap in processing music and speech." *Frontiers in Human Neuroscience, 9*, 1–3. doi:10.3389/fnhum.2015.00330

Kunert, R., Willems, R. M., Casasanto, D., Patel, A. D., & Hagoort, P. (2015). Music and language syntax interact in Broca's area: An fMRI study. *PLoS One, 10*(11), e0141069. doi:10.1371/journal.pone.0141069

LaCroix, A. N., Diaz, A. F., & Rogalsky, C. (2015). The relationship between the neural computations for speech and music perception is context-dependent: An activation likelihood estimate study. *Frontiers in Psychology, 6*, 1–19. doi:10.3389/fpsyg.2015.01138

Leaver, A. M., & Rauschecker, J. P. (2010). Cortical representation of natural complex sounds: Effects of acoustic features and auditory object category. *Journal of Neuroscience, 30*(22), 7604–7612. doi:10.1523/JNEUROSCI.0296-10.2010

Margulis, E. H. (2013). Repetition and emotive communication in music versus speech. *Frontiers in Psychology, 4,* 1–4. doi:10.3389/fpsyg.2013.00167

Merrill, J., Sammler, D., Bangert, M., Goldhahn, D., Lohmann, G., Turner, R., et al. (2012). Perception of words and pitch patterns in song and speech. *Frontiers in Psychology, 3,* 1–13. doi:10.3389/fpsyg.2012.00076

Norman, K. A., Polyn, S. M., Detre, G. J., & Haxby, J. V. (2006). Beyond mind-reading: Multi voxel pattern analysis of fMRI data. *Trends in Cognitive Sciences, 10*(9), 424–430. doi:10.1016/j.tics.2006.07.005

Norman-Haignere, S., Kanwisher, N. G., & McDermott, J. H. (2015). Distinct cortical pathways for music and speech revealed by hypothesis-free voxel decomposition. *Neuron, 88*(6), 1281–1296. doi:10.1016/j.neuron.2015.11.035

Patel, A. D. (2011). Why would musical training benefit the neural encoding of speech? The OPERA hypothesis. *Frontiers in Psychology, 2,* 1–14. doi:10.3389/fpsyg.2011.00142

Patel, A. D. (2013). Can nonlinguistic musical training change the way the brain processes speech? The expanded OPERA hypothesis. *Hearing Research, 308,* 98–108. doi:10.1016/j.heares.2013.08.011

Peelen, M. V., & Downing, P. E. (2007). Using multi-voxel pattern analysis of fMRI data to interpret overlapping functional activations. *Trends in Cognitive Sciences, 11*(1), 4–5. doi:10.1016/j.tics.2006.10.009

Peretz, I. (2006). The nature of music from a biological perspective. *Cognition, 100*(1), 1–32. doi:10.1016/j.cognition.2005.11.004

Peretz, I. (2009). Music, language and modularity framed in action. *Psychologica Belgica, 49*(2–3), 157–175. doi:10.5334/pb-49-2-3-157

Peretz, I. (2013). The biological foundations of music: Insights from congenital amusia. In D. Deutsch (Ed.), *The psychology of music* (pp. 551–564). London: Academic Press.

Peretz, I., Aubé, W., & Armony, J. L. (2013). Toward a neurobiology of musical emotions. In E. Altenmüller, S. Schmidt, & E. Zimmerman (Eds.), *The evolution of emotional communication: From sounds in nonhuman mammals to speech and music in man* (pp. 277–299). New York: Oxford University Press. doi:10.1093/acprof:oso/9780199583560.003.0017

Peretz, I., & Coltheart, M. (2003). Modularity of music processing. *Nature Neuroscience, 6*(7), 688–691. doi:10.1038/nn1083

Petkov, C. I., Kayser, C., Steudel, T., Whittingstall, K., Augath, M., & Logothetis, N. K. (2008). A voice region in the monkey brain. *Nature Neuroscience, 11*(3), 367–374. doi:10.1038/nn2043

Remedios, R., Logothetis, N. K., & Kayser, C. (2009). An auditory region in the primate insular cortex responding preferentially to vocal communication sounds. *Journal of Neuroscience, 29*(4), 1034–1045. doi:10.1523/jneurosci.4089-08.2009

Rizzolatti, G., & Craighero, L. (2004). The mirror-neuron system. *Annual Review of Neuroscience, 27,* 169–192. doi:10.1146/annurev.neuro.27.070203.144230

Rogalsky, C., Rong, F., Saberi, K., & Hickok, G. (2011). Functional anatomy of language and music perception: Temporal and structural factors investigated using functional magnetic resonance imaging. *Journal of Neuroscience, 31*(10), 3843–3852. doi:10.1523/JNEUROSCI.4515-10.2011

Sammler, D., Baird, A., Valabrègue, R., Clément, S., Dupont, S., Belin, P., et al. (2010). The relationship of lyrics and tunes in the processing of unfamiliar songs: A functional magnetic resonance adaptation study. *Journal of Neuroscience, 30*(10), 3572–3578. doi:10.1523/JNEUROSCI.2751-09.2010

Sammler, D., Koelsch, S., Ball, T., Brandt, A., Grigutsch, M., Huppertz, H.-J., et al. (2013). Co-localizing linguistic and musical syntax with intracranial EEG. *NeuroImage, 64,* 134–146. doi:10.1016/j.neuroimage.2012.09.035

Schönwiesner, M., & Zatorre, R. J. (2008). Depth electrode recordings show double dissociation between pitch processing in lateral Heschl's gyrus and sound onset processing in medial Heschl's gyrus. *Experimental Brain Research, 187*(1), 97–105. doi:10.1007/s00221-008-1286-z

Skoe, E., & Kraus, N. (2012). Human subcortical auditory function provides a new conceptual framework for considering modularity. In P. Rebuschat, M. Rohrmeier, J. Hawkings, & I. Cross (Eds.), *Language and music as cognitive systems* (pp. 269–282). New York: Oxford University Press. doi:10.1093/acprof:oso/9780199553426.001.0001

Sperber, D., & Hirschfeld, L. A. (2004). The cognitive foundations of cultural stability and diversity. *Trends in Cognitive Sciences*, *8*(1), 40–46. doi:10.1016/j.tics.2003.11.002

Strait, D., & Kraus, N. (2011). Playing music for a smarter ear: Cognitive, perceptual and neurobiological evidence. *Music Perception*, *29*(2), 133–146. doi:10.1525/MP.2011.29.2.133

Tierney, A., Dick, F., Deutsch, D., & Sereno, M. (2013). Speech versus song: Multiple pitch sensitive areas revealed by a naturally occurring musical illusion. *Cerebral Cortex*, *23*(2), 249–254. doi:10.1093/cercor/bhs003

Tillmann, B., & Bigand, E. (2015). Response: A commentary on: "Neural overlap in processing music and speech." *Frontiers in Human Neuroscience*, *9*, 1–3. doi:10.3389/fnhum.2015.00491

Tillmann, B., Janata, P., & Bharucha, J. J. (2003). Activation of the inferior frontal cortex in musical priming. *Cognitive Brain Research*, *16*(2), 145–161. doi:10.1016/s0926-6410(02)00245-8

Tillmann, B., Koelsch, S., Escoffier, N., Bigand, E., Lalitte, P., Friederici, A. D., et al. (2006). Cognitive priming in sung and instrumental music: Activation of inferior frontal cortex. *NeuroImage*, *31*(4), 1771–1782. doi:10.1016/j.neuroimage.2006.02.028

Todd, M. T., Nystrom, L. E., & Cohen, J. D. (2013). Confounds in multivariate pattern analysis: Theory and rule representation case study. *NeuroImage*, *77*, 157–165. doi:10.1016/j.neuroimage.2013.03.039

Tokarev, K., Tiunova, A., Scharff, C., & Anokhin, K. (2011). Food for song: Expression of c-Fos and ZENK in the zebra finch song nuclei during food aversion learning. *PLoS One*, *6*(6), e21157. doi:10.1371/journal.pone.0021157

van den Heuvel, M. P., & Sporns, O. (2013). Network hubs in the human brain. *Trends in Cognitive Sciences*, *17*(12), 683–696. doi:10.1016/j.tics.2013.09.012

Westbury, C. F., Zatorre, R. J., & Evans, A. C. (1999). Quantifying variability in the planum temporale: A probability map. *Cerebral Cortex*, *9*(4), 392–405. doi:10.1093/cercor/9.4.392

Yan, C., & He, Y. (2011). Driving and driven architectures of directed small-world human brain functional networks. *PLoS One*, *6*(8), e23460. doi:10.1371/journal.pone.0023460

10

Defining the Biological Bases of Individual Differences in Musicality

Bruno Gingras, Henkjan Honing, Isabelle Peretz, Laurel J. Trainor, and Simon E. Fisher

During the past few decades, our understanding of human biology has been transformed by advances in molecular methods. It has become routine to apply genetic techniques to studies of biomedical disorders, as well as to related traits that show individual variation in the general population. Genetic research has yielded novel mechanistic insights relevant to understanding both disease and normal function. Researchers recently have extended the reach of genetics and genomics beyond standard biomedical traits and have begun to tackle complex human-specific cognitive abilities, such as speech and language, with some success (Deriziotis & Fisher, 2013). Genetic analysis of aspects of musical aptitude is a field that is still in its infancy (Tan, McPherson, Peretz, Berkovic, & Wilson, 2014). In this chapter, we discuss progress thus far and consider the promise that the postgenomic era holds for shedding light on the biological bases of human musicality, broadly defined here as the capacity to perceive (perceptual abilities), reproduce, or create music (production abilities).

As for language, the enormous variability of musical expressions found around the world bears the hallmarks of culture. However, like language, an emerging consensus suggests that musicality may have deep biological foundations and so warrants examination from a genetic perspective (Peretz, 2006). At the same time, if a trait is largely limited to our own species, this poses special challenges for deciphering the underlying biology (see chapters 7 and 11, this volume). When we investigate these kinds of human capacities, it is important that we move beyond questions of species universals and recognize the value of studying variability (Fisher & Vernes, 2015). In particular, major tools of genetics depend on assessing variability in observable aspects of anatomy, physiology, development, cognition, behavior, and so on (*phenotypes*), and then searching for correlations with variations at the genetic level (*genotypes*) (see the glossary below for definitions of italicized technical terms). Variability in musical aptitude is well documented within human populations and is not limited to exceptional cases of virtuoso musicians or, at the other extreme, people who are unable to appreciate or engage with music despite adequate opportunity (Peretz, 2013). Clear evidence has emerged showing considerable individual variation in music-related skills throughout the general population (Müllensiefen, Gingras, Musil, & Stewart,

2014), variation that is likely to have at least some basis in biology (Ullén, Mosing, Holm, Eriksson, & Madison, 2014). Concomitantly, recent efforts to comprehensively catalog the natural variability in modern human genomes have revealed a surprising degree of variation within populations, affecting virtually every genetic locus in some way (Abecasis et al., 2012; Lappalainen et al., 2013). Thus, human populations can be effectively treated as natural experiments for identifying biologically meaningful links between individual variation at different levels (Fisher & Vernes, 2015), allowing researchers to trace causal connections between particular genes and phenotypes of interest—in this case, key features of musicality. Once relevant genes have been pinpointed, they can be used as entry points into the critical neurobiological pathways and potentially complement other approaches to understanding musicality (see chapter 1, this volume).

This should not be taken to imply that there exists a specific "gene for music." Genes cannot directly specify behavioral or cognitive outcomes. They have highly indirect effects at best, encoding molecules (RNAs and proteins) that influence the ways in which neurons proliferate, migrate, differentiate, and connect with each other during brain development or modulate the plasticity of circuits during learning (e.g., Fisher, 2006). Moreover, musicality is a complex multifaceted phenotype, itself comprising many potentially distinct abilities (Levitin, 2012; Müllensiefen et al., 2014), and an array of different genes may be involved. At this point, the genetic architecture underlying music-related skills is largely unknown. While extremes of musical ability might plausibly involve some rare monogenic effects still to be discovered, it is likely that individual differences in the general population involve variants at multiple interacting genetic loci, the number of which has not yet been determined. In addition, environmental influences should not be neglected. Sociocultural variables, exposure to music, and years of music training are well-known environmental factors that have an impact on aptitude (Hannon & Trainor, 2007; Hargreaves & Zimmerman, 1992; Trainor & Unrau, 2012). Indeed, musicality may constitute an ideal system for studying interactions between genes and environment (Baharloo, Johnston, Service, Gitschier, & Freimer, 1998; Hambrick & Tucker-Drob, 2015; Levitin, 2012; Schellenberg, 2015).

People harbor a diverse range of distinct types of genetic variants, which differ in frequency, size, and functional impact.[1] The technology for characterizing genomic variation has been advancing at an astonishing pace as the time and resources needed for genotyping and sequencing have been dramatically reduced. DNA chips allow hundreds of thousands of known genetic variants to be simultaneously genotyped rapidly and at low cost and can easily be scaled up to studies involving thousands of people. The advent of *next-generation DNA sequencing* means that the entire genome of a person can be determined for a few thousand dollars in a matter of days, and the field continues to move forward (Gilad, Pritchard, & Thornton, 2009; Goldstein et al., 2013). Nonetheless, it is important to stress that success in genetic studies of any human trait of interest depends critically on a solid strategy for defining and characterizing the phenotype. Thus, advances in human genomics need to be matched by parallel advances in the area of *phenomics*.

In this chapter, we first review the evidence concerning the links between genes or chromosomal regions that have been associated with "extreme" musical phenotypes—that is, phenotypes that are found in only a small percentage of the general population and correspond to congenital impairments in musical ability on the one hand or rare faculties (such as absolute pitch) on the other hand. We then move on to variability within the normal range of musical aptitudes of the general population, considering traits such as relative pitch, music perception skills, and musical production and creativity. Finally, we outline future research directions for the field and propose concrete suggestions for the development of comprehensive operational tools for the analysis of musical phenotypes.

Musicality at the Extremes

Disorders of Music Perception

Genetic investigations of neurodevelopmental disorders such as speech apraxia, specific language impairment, and dyslexia have been crucial for uncovering the molecular bases of human speech and language skills (Graham & Fisher, 2013). Similar approaches can help to reveal the biological underpinnings of musicality (table 10.1; Peretz, Cummings, & Dube, 2007; Stewart, 2008). About 3 percent of the general population have difficulty detecting notes that are out of key in melodies, against a background of normal hearing, language and intelligence, and adequate environmental exposure (Peretz & Hyde, 2003). The condition, often called tone deafness, is now referred to as congenital amusia to distinguish this lifelong disorder from acquired forms of amusia that occur as the result of brain lesion (Hyde, Zatorre, & Peretz, 2011; Peretz et al., 2002). Congenital amusia is not only characterized by a deficit in detecting mistuning in both melodic and acoustical contexts, but also by an inability to recognize familiar tunes without the help of the lyrics and difficulties singing in tune. In both perception and production, rhythm is relatively spared (Hyde & Peretz, 2004). The biological basis of the condition is further supported by the identification of brain abnormalities affecting gray and white matter in the right auditory and inferior frontal cortex (Hyde et al., 2007), as well as reduced connectivity between these two regions (Hyde et al., 2011).

Congenital amusia tends to show clustering within families (familial aggregation). That is, the condition is present at higher rates in relatives of affected people than expected on the basis of prevalence in the general population. In 2007, Peretz and colleagues studied seventy-one members of nine large families with an amusic *proband* and seventy-five members of ten control families, assessing amusia with an online battery that included an anomalous pitch detection task, a control time asynchrony detection task, and a detailed questionnaire (Peretz et al., 2007). The results confirmed that congenital amusia

Table 10.1
Investigating the biological bases of musicality through extreme phenotypes and known genetic syndromes

Focus	Type of study	Main findings	Citations
Congenital amusia	Familial aggregation	In 9 large families ($n = 71$) with an amusic proband, 39 percent of first-degree relatives were affected, while in 10 control families ($n = 75$), prevalence was only 3 percent. Sibling recurrence risk ratio was estimated at about 10.8.	Peretz et al., 2007
Absolute pitch	Familial aggregation	Different studies estimated sibling recurrence risk ratios of approximately 7.5 to 15.1. Prevalence was higher in people with early musical training and also in families of East Asian ethnicity; direction of causation unknown.	Baharloo et al., 1998, 2000; Gregersen et al., 1999
	Twin study	Concordance in identical twins (78.6 percent, 14 pairs) was significantly higher than that seen in nonidentical twins (45.2 percent, 31 pairs).	Theusch & Gitschier, 2011
	Pharmacology	Adult males taking valproate (a drug hypothesized to affect critical periods) learned to identify pitch better than those taking placebos.	Gervain et al., 2013
	Linkage analysis	A study of 45 European and 19 East Asian families with multiple AP cases found suggestive linkage for multiple chromosomal regions, with inconsistent patterns in the two data sets. Strongest linkage for chromosome 8q24 in European families.	Theusch et al., 2009
	Linkage analysis	Investigation of 53 families (49 European, 4 Asian) failed to replicate top linkage peaks from prior AP work. High rates (20.1 percent) of self-reported synesthesia in AP led the authors to run combined linkage of 53 AP families with 36 synesthesia families. Strongest joint linkage on chromosomes 6q14–q16 and 2q22-q24.	Gregersen et al., 2013
Musicality in known genetic syndromes	Phenotyping	It has been suggested that children with William-Beurens syndrome (due to 7q11.23 microdeletion) have increased auditory sensitivity, musical interest, creativity, and expressivity. Other studies argue that these children show a wide range of musicality profiles, and some may even have elevated risk of amusia.	Levitin, 2005; Lense & Dykens, 2013; Lense et al., 2013
	Phenotyping	Rare mutations of the *FOXP2* transcription factor gene cause a severe speech and language disorder. One study of musical ability in a particularly large family with a *FOXP2* disruption suggested that mutation carriers had selective problems with perception and production of rhythms, while pitch-related abilities were normal.	Alcock et al., 2000; Lai et al., 2001

Note: Examples are given of the different types of approaches discussed in this chapter, along with key results from the relevant studies.

involves deficits in processing musical pitch but not musical time and also showed strong evidence of familial aggregation. In amusic families, 39 percent of first-degree relatives were affected, as compared to only 3 percent in control families (Peretz et al., 2007). The sibling *recurrence risk* ratio was estimated as about 10.8, meaning that siblings of someone with congenital amusia have an almost eleven-fold increased risk of being amusic themselves.

Observations of familial aggregation are supportive of genetic involvement, but might also be (partly or wholly) explained by shared family environment. As explained in box 10.1, twin studies can be used to pull apart these effects and obtain a robust estimate of *heritability*. To our knowledge, no formal twin study of congenital amusia has yet been reported, but a broader study has shown strong heritability for pitch perception (Drayna, Manichaikul, de Lange, Snieder, & Spector, 2001), as we discuss later in this chapter. Nonetheless, by collecting families in which multiple relatives show congenital amusia (Peretz et al., 2007), it becomes possible to try mapping the locations of potential susceptibility genes. Such work is currently under way and will benefit from the recent advances in genomic technologies.

Other forms of congenital amusia that affect rhythm but not pitch have been discovered (Launay, Grube, & Stewart, 2014; Phillips-Silver et al., 2011). So far, the number of cases that have been described is very small. Little is known about the prevalence of such disorders and whether they show familial aggregation. These are potentially interesting areas for future investigation.

Rare Faculties

Absolute pitch (AP), the ability to identify or produce a musical tone (e.g., middle C or concert A) without reference to an external standard (Deutsch, 2013), is an unusual skill found in only a small percentage of people. AP involves at least two separate cognitive skills: memory for pitch, which seems to be widespread among humans (Schellenberg & Trehub, 2003) and nonhuman animals (Weisman, Mewhort, Hoeschele, & Sturdy, 2012), and the ability to attach labels to stimuli (e.g., classifying tones with different spectral characteristics, such as piano or voice, and consequently labeling their pitch class), which appears to be rarer (Deutsch, 2013). In early reports, the prevalence of AP in the general population was estimated to be 1 in 10,000 (Bachem, 1955), but more recent studies suggest that it may be found in as many as 1 in 1,500 people (Profita & Bidder, 1988). It has been proposed that this is a dichotomous trait, with a clear phenotypic separation between AP possessors and non-AP possessors (Athos et al., 2007). Recent structural neuroimaging studies have suggested that AP is associated with altered cortical thickness and connectivity in a number of brain regions (Dohn et al., 2015). As a discrete and easily quantifiable cognitive phenotype, AP may be particularly suited for genetic studies (Athos et al., 2007; Gregersen, 1998; but see Vitouch, 2003). However, its relevance to musicality

Box 10.1
Do Genes Contribute?

Even without molecular data, it is possible to investigate contributions of genetic factors to phenotypes of interest. For a qualitatively defined trait, such as the presence or absence of a particular disorder, researchers can ask whether cases tend to cluster within families and assess whether inheritance is consistent with simple single-gene patterns of transmission or more likely to involve multiple factors. Increased incidence of a trait in relatives of a proband is often taken as evidence of genetic involvement, but it could also be due to environmental factors shared by family members. Twin studies allow these types of contributing factors to be teased apart.

In its simplest form, this approach assesses concordance of a phenotype in pairs of identical twins (who have almost identical genomes) and compares it to the concordance seen for pairs of nonidentical twins (who share around 50 percent of their genetic variations, just like nontwin siblings). Elevated concordance in the identical twins provides evidence of genetic involvement. In fact, twin designs typically go further by directly incorporating quantitative trait data and using the twin-twin correlation structure to partition the phenotypic variation into that due to additive genetic factors, common environment (shared by twins), and unique environment (unshared by twins). The proportion of phenotypic variance that is accounted for by genetics gives a formal estimate of heritability. Statistical tools have become more sophisticated over the years, and it is now routine to apply structural equation modeling and maximum-likelihood methods to large twin data sets, asking questions that extend far beyond heritability estimation. What is the contribution of genetic factors at different ages, and is this due to the same or different sets of genes? How much of the covariance between two correlated traits involves common genetic or environmental contributions? Are sex differences likely to play a role? Is there evidence of gene-environment interaction or correlation underlying a trait?

Quantitative methods can also be used in multigenerational families for partitioning the observed phenotypic variance and estimating heritability (variance component models). Quantitative genetic methods depend on certain assumptions (outside the scope of this chapter), some of which have been challenged. More important, the concept of heritability itself is often misunderstood by nonspecialists. Heritability is a useful statistic that describes variance in a given population at a specific time with a particular set of genetic variations and environmental factors. It is not an intrinsic fixed property of a phenotype, and it does not reveal anything about the biology of an individual or of how malleable a trait might be. For example, heritability estimates of certain features (including general intelligence) are well known to increase with age. Changes in environment (such as many of the developments of modern medicine) can radically alter the heritability of a trait, either diminishing or exaggerating the relative contributions of genetic variation.

remains questionable, especially given that most professional musicians do not possess AP (Gregersen, 1998).

Profita and Bidder (1988) were among the first to explore the hypothesis of a genetic basis for the condition in a study of thirty-five people with AP across nineteen families. Subsequent familial aggregation studies reported sibling recurrence risk ratios between 7.5 and 15.1 (Baharloo et al., 1998; Baharloo, Service, Risch, Gitschier, & Freimer, 2000; Gregersen, Kowalsky, Kohn, & Marvin, 1999), consistent with a role for genetic factors. Further evidence of a significant genetic contribution has been found in studies of twins with AP; the concordance of the condition in fourteen pairs of identical twins was 78.6 percent, compared to a concordance of 45.2 percent in thirty-one pairs of nonidentical twins (Theusch & Gitschier, 2011).

Environmental factors are also strongly implicated in AP, albeit in a complex manner. A robust link between AP and early music training has been uncovered (Baharloo et al., 1998; Gregersen et al., 1999), with a significantly higher prevalence of the condition in people who began their musical training at a very young age. Thus, early music training could potentially be a crucial environmental factor contributing to AP. On the other hand, this same pattern of data could be explained by assuming that a genetic predisposition to AP increases the likelihood that a child receives early music training. Hence, the direction of causation is difficult to establish (Baharloo et al., 1998). In any case, it seems likely that early musical training and genetic predisposition contribute together to the development of AP. Another unexplained observation concerns the higher rates of AP for people of East Asian ethnicity (Gregersen et al., 1999). Again, several alternative hypotheses could account for this well-documented effect; certain cultural groups may respond to early signs of AP with more intensive parental efforts at music education, the increased AP prevalence may be a consequence of culture-specific educational systems that are more effective at fostering this ability, or the findings may have a genetic explanation, reflecting ethnic differences in frequencies of susceptibility *alleles* (Gregersen et al., 1999).

A recent intriguing observation comes from studies of valproate, an inhibitor of the histone-deacetylase enzyme, which can act to put a brake on critical-period learning (Morishita & Hensch, 2008). Administration of this enzyme to adult males apparently reopens the critical-period learning of absolute pitch (Gervain et al., 2013). Neuroimaging studies have also been revealing. Relative to non-AP possessors, AP possessors exhibit anatomical differences in the temporal lobe and other areas (Bermudez, Lerch, Evans, & Zatorre, 2009; Loui, Li, Hohmann, & Schlaug, 2011), as well as differences in the cortical processing of pitch information (Loui, Zamm, & Schlaug, 2012; Zatorre, Perry, Beckett, Westbury, & Evans, 1998).

Researchers studying AP have used linkage analyses in families (see box 10.2) to search for chromosomal regions that may harbor genes involved in the condition (Gregersen et al., 2013; Theusch, Basu, & Gitschier, 2009). In a 2009 study, Theusch and colleagues investigated seventy-three families with multiple AP possessors, including forty-five families of

Box 10.2
Tracing Connections between Genotypes and Phenotypes

Familial clustering and twin studies may provide support for genetic involvement in a human trait. How do we pinpoint the critical genes? In the early days of gene mapping, linkage analysis came to the fore. In this approach, researchers treat polymorphic genetic markers like signposts marking different chromosomal regions. They track how such genetic markers are transmitted to different members of a family, asking whether any particular chromosomal interval is linked to inheritance of the trait of interest. Robust statistical methods are used to ensure that an observed co-segregation between a genetic marker and the phenotype is not a chance finding. Linkage analysis is equally applicable to qualitative (i.e., dichotomous or yes/no) and quantitative traits, and can involve a prespecified genetic model or be model free. Data from different families may be combined; if the same genetic factors influence the phenotype (even if the precise mutation differs in each family), this may help localize the gene(s) responsible. Nonetheless, linkage has low resolution, implicating large regions (loci) containing multiple genes, and it is not well suited for detecting genetic effects that account for only a small proportion of phenotypic variance.

Association analysis, a complementary method with different strengths and weaknesses, tests for correlations between particular gene variants and a trait at the population level. It has greater power than linkage to uncover small effect sizes and allows higher resolution mapping. Still, due to linkage disequilibrium, a polymorphism that shows significant association is often not causal, but could be indexing a causal variant (as yet undiscovered) nearby. The first association studies typically focused on testing small numbers of polymorphisms from selected candidate genes, either based on hypotheses about the biology of the trait or targeting regions highlighted by linkage. In recent years, it became quick and inexpensive to carry out systematic genome-wide genotyping capturing much of the polymorphic content of a phenotyped sample, allowing researchers to perform hypothesis-free association screening at high density across the genome. These screens involve an enormous amount of multiple testing (hundreds of thousands of polymorphisms in each individual), so rigorously adjusted thresholds for statistical significance are required to avoid false positives. Together with the fact that most complex traits are likely to involve many genes with very small effect sizes, such studies require sample sizes of thousands of individuals to achieve adequate power. In the postgenomic era, scientists also now make use of CNV data and rare variations emerging from next-generation sequencing studies. Again, the key to success is the use of robust statistics, and replication in independent samples, to discount spurious genotype-phenotype relationships. Ultimately studies of gene function in model systems are needed to demonstrate true causal connections.

European descent and nineteen families with East Asian ancestry. They found suggestive evidence for linkage to several different chromosomal regions, with strongest evidence on *chromosomal band* 8q24.21 in the European families (see figure 10.1A). There was little consistency between the pattern of findings in the European and East Asian data sets; this genetic heterogeneity is interesting in light of the documented population differences in AP prevalence.

In 2013, Gregersen and colleagues studied an independent set of fifty-three AP families (forty-nine European, four Asian) and identified modest evidence of linkage implicating different chromosomal regions from the prior work. More intriguing, this later study also uncovered evidence suggesting phenotypic and genetic overlaps between AP and synesthesia, another rare condition. For people with synesthesia, a stimulus in one sensory modality automatically evokes a perceptual experience in another modality; for instance, particular pitches, keys, or timbres may evoke specific sensations of color. Like AP, synesthesia is thought to involve genetic contributions, with some clues as to chromosomal regions of interest but no definitive genes yet identified (Asher et al., 2009; Bosley & Eagleman, 2015). Gregersen et al. (2013) uncovered unusually high rates (20.1 percent) of self-reported synesthesia in people with AP, which motivated them to do a joint linkage study, combining their set of fifty-three AP families together with thirty-six families from a prior screen of synesthesia (Asher et al., 2009). Joint evidence of linkage was seen on chromosomes 6 and 2, but since the regions implicated are large and contain many genes (see box 10.2), further studies are needed to pinpoint potential causal variants.

A possible drawback of most AP studies published so far is that they rely on the explicit labeling of pitches and therefore are limited to people with musical training. However, methods have been developed for detecting AP without requiring explicit labeling (Plantinga & Trainor, 2008; Ross & Marks, 2009; Ross, Olson, Marks, & Gore, 2004). Thus, musical training may not be necessary for AP (Ross, Olson, & Gore, 2003), underlining the need to test for the presence of this condition in nonmusicians.

Altered Musicality in Known Genetic Syndromes

The discussions so far concern identification of rare music-specific conditions, followed by a search for genetic correlates. A complementary approach is to target existing syndromes, where the causative gene or genes are already known, and investigate whether there are any consequences for the musicality of affected people (table 10.1). This is an area that has been little explored but could prove fruitful. Here, we briefly mention two examples from the literature, both of which (by coincidence) involve genes on chromosome 7 (figure 10.1B). Williams-Beuren syndrome (WBS) is a well-characterized microdeletion syndrome with a prevalence of about 1 in 7,500 people, in which as many as twenty-eight neighboring genes in 7q11.23 may be deleted (Martens, Wilson, & Reutens, 2008). People with WBS often show a distinctive cognitive/behavioral profile, which has

Figure 10.1
Connecting genes to musicality: Some selected examples from the literature. Ideograms of chromosomes are shown with the cytogenetic bands of interest indicated. Each chromosome has a short (p) arm and a long (q) arm, separated by a structure called a centromere. When treated with certain stains, chromosomes display consistent banding patterns that are used to denote specific locations with respect to the centromere. (A) Linkage analysis of extreme phenotypes. The first linkage screen of families with AP highlighted a peak on chromosome 8q24.2 (Theusch et al., 2009). Subsequent AP studies have pointed instead to other regions elsewhere in the genome, some of which overlap with linkages to synesthesia (Gregersen et al., 2013). No specific AP-related genes have yet been identified. Linkage analysis has also been used to investigate musical aptitudes using quantitative phenotypes, as detailed in the main text. (B) Studies of musicality in known genetic disorders. Williams-Beuren Syndrome (WBS; Martens et al., 2008) and *FOXP2*-associated speech/language disorder (Fisher & Scharff, 2009), both involving chromosome 7, have been investigated in relation to musicality (Lense & Dykens, 2013; Lense et al., 2013; Levitin, 2005; Levitin et al., 2004) and rhythm (Alcock et al., 2000). (C) Candidate genes. In some cases, particular candidate genes have been targeted based on hypotheses about their biological effects, and polymorphisms have been tested for association with music-related phenotypes. The *AVPR1A* gene is one well-studied example (Bachner-Melman et al., 2005; Morley et al., 2012; Ukkola-Vuoti et al., 2011; Ukkola et al., 2009). However, recent genome-wide screens failed to find significant effects for any prior-studied candidates, including *AVPR1A* (Oikkonen et al., 2014; Park et al., 2012). (D) Copy number variants (CNVs). A recent study searched for CNVs in people with low or high musical aptitude or musical creativity (Ukkola-Vuoti et al., 2013). A number of interesting regions were reported, such as the PCDHα cluster on chromosome 5, found in some individuals with low musical aptitude. Nonetheless, for rare CNVs observed in only a few individuals, it can be difficult to show causality, and so these findings await confirmation in independent samples. (E) Combined approach, using linkage, association, CNV analyses, and sequencing. Park and colleagues (2012) studied pitch production accuracy in a multistage approach. They began with a linkage screen, identifying a broad linkage peak on chromosome 4q23, and followed up with association analyses of the surrounding region, eventually zooming in on the UGT8 gene in 4q26 as a candidate. Further independent evidence to support UGT8 came from identification of a CNV spanning that region as well as variants identified by large-scale sequencing.

been much studied by researchers interested in tracing connections between genes and brain functions. The typical WBS phenotype includes mild to moderate cognitive impairments, disparity between verbal and spatial skills, with receptive language being a relative strength as compared to other abilities, as well as hypersociability, increased empathy, anxiety, and attention deficits (Levitin et al., 2004; Martens et al., 2008; Ng, Lai, Levitin, & Bellugi, 2013). It has been argued that people with WBS show increased auditory sensitivity, heightened emotional responses to music, and relative strengths in musical interest, creativity, and expressivity, in contrast with other neurodevelopmental disorders (Levitin, 2005). Close examinations of particular music perception, production, and learning skills associated with WBS have revealed a more complex story, however, with considerable phenotypic variability from one affected person to another (Lense & Dykens, 2013). These issues are beyond the scope of this chapter but have been discussed in detail recently by Lense, Shivers, and Dykens (2013) and colleagues, who found that the incidence of amusia in WBS is probably higher than that seen in the general population and that the neural correlates of amusia appear to be similar in people with WBS and typically developing individuals (Lense, Dankner, Pryweller, Thornton-Wells, & Dykens, 2014).

Elsewhere on chromosome 7 lies *FOXP2*, a regulatory gene that modulates the expression of other genes (Fisher & Scharff, 2009; see figure 10.1B). Rare mutations that disrupt

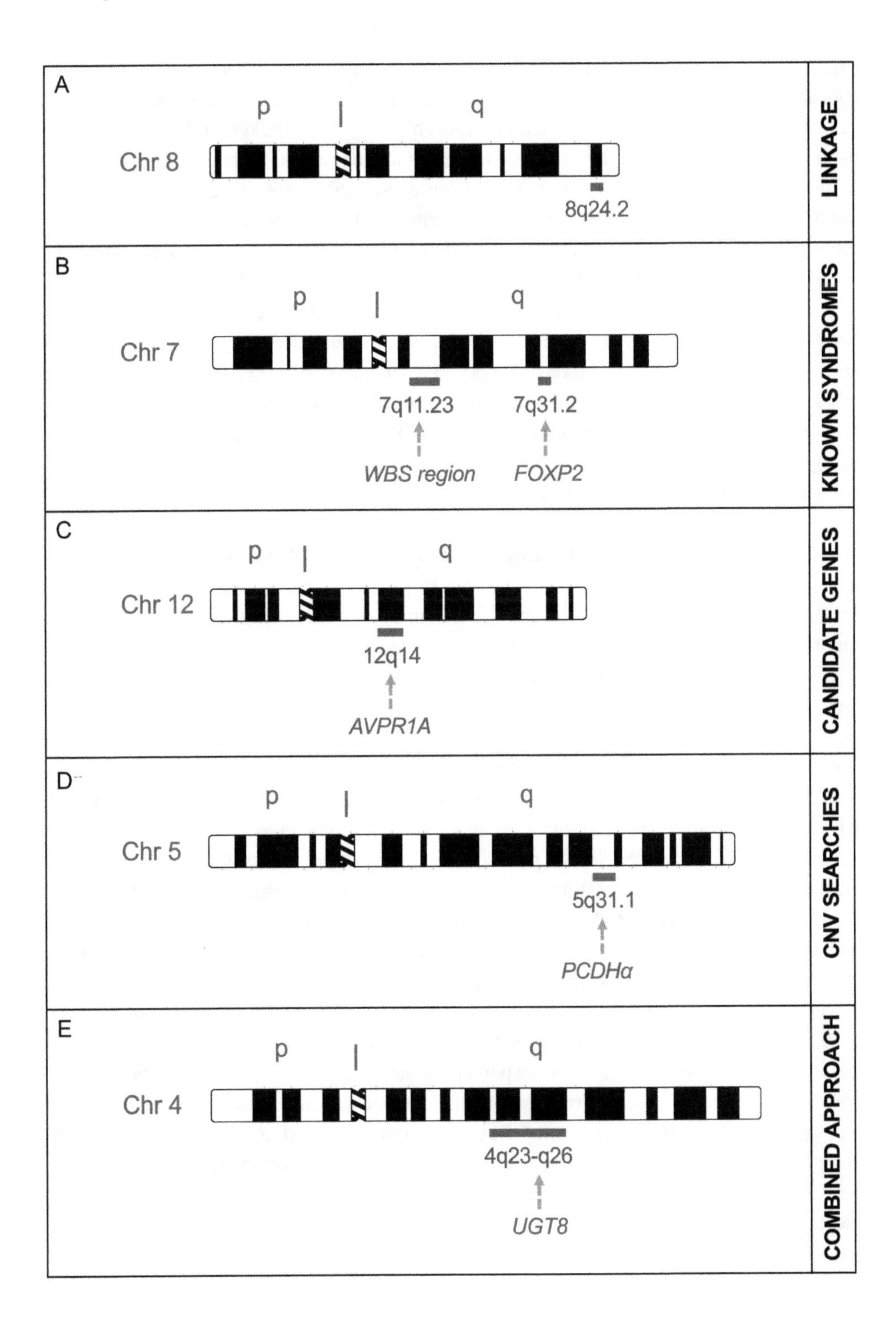
A
p
q
Chr 8
8q24.2
LINKAGE
B
p
q
Chr 7
7q11.23
7q31.2
WBS region
FOXP2
KNOWN SYNDROMES
C
p
q
Chr 12
12q14
AVPR1A
CANDIDATE GENES
D
p
q
Chr 5
5q31.1
PCDHα
CNV SEARCHES
E
p
q
Chr 4
4q23-q26
UGT8
COMBINED APPROACH

FOXP2 cause a severe speech and language disorder. Affected people have problems coordinating sequences of orofacial movements during speech (known as developmental verbal dyspraxia or childhood apraxia of speech), as well as spoken and written impairments in many aspects of expressive and receptive language. A number of different *FOXP2* mutations have been identified thus far (Fisher & Scharff, 2009). One of these has been particularly intensively studied, since it was found in fifteen affected relatives of a large multigenerational pedigree, known as the KE family (Fisher, Vargha-Khadem, Watkins, Monaco, & Pembrey, 1998; Lai, Fisher, Hurst, Vargha-Khadem, & Monaco, 2001). A study of musical ability in affected members of this family reported reduced performance in tasks involving perception and production of vocal and manual rhythms, while pitch-related abilities appeared to be preserved (Alcock, Passingham, Watkins, & Vargha-Khadem, 2000). These findings are interesting in light of functional evidence implicating *FOXP2* in sensorimotor integration and motor skill learning (Fisher & Scharff, 2009). Further studies of rhythmic abilities in the KE family and other independent cases of *FOXP2* disruption are needed to shed more light on this area.

Genetic Contributions to Individual Differences in the General Population

We now turn our attention to the normal spectrum of musical abilities and survey findings linking individual phenotypic differences to genetic variation (summarized in table 10.2).

Given that pitch perception is a central component of musicality, and perhaps one of the most amenable for large-scale testing, it is not surprising that this facet has been examined more thoroughly than others. In one of the earliest twin studies conducted on music perception abilities in the general population, 136 identical and 148 nonidentical twin pairs were administered the Distorted Tunes Test (Kalmus & Fry, 1980) in which they judged whether familiar melodies contained "wrong notes" (Drayna et al., 2001). The scores on this test, considered a proxy for the participants' ability to judge successive pitch intervals, were estimated to have a heritability of 71 to 80 percent, with no significant effect of shared environment. In a recent study of young adult twins from Finland, 69 identical twin pairs and 44 nonidentical twin pairs (as well as 158 individual twins without their co-twin), were given online tests to assess their melody perception abilities (Seesjarvi et al., 2016). For detection of pitch changes in a task comparing two melodies, about 58 percent of the variance appeared to be due to additive genetic effects. When it came to detection of incongruities in key or rhythm in single-melody perception tasks, a larger proportion of the variance could be explained by the environment, with shared environmental effects accounting for 61 percent of the variance in an out-of-key detection task, and nonshared environmental effects explaining 82 percent of the variance in an "off-beat" task (Seesjarvi et al., 2016). As is standard for twin studies, this report did not identify which specific genes or environmental factors might be involved.

Table 10.2
Investigating the biological bases of musical perception and production through individual differences in the general population

Focus	Type of study	Key findings	Citation
Pitch perception	Twin study	Performance on the Distorted Tunes Test in 136 identical and 148 nonidentical twin pairs from general population yielded heritability estimates of approximately 71 to 80 percent.	Drayna et al., 2001
Musical memory	Candidate genes	Targeted study in 82 students reported provisional association of musical memory with an epistatic interaction between common promoter variants of the genes *AVPR1A* and *SLC6A4*.	Granot et al., 2007
	Pharmacology	Arginine vasopressin was administered to 25 males, yielding effects on musical memory, mood, and attentiveness without affecting digit span.	Granot et al., 2013
Battery of music perception tasks (Karma Test, Seashore Pitch and Rhythm Subtests)	Genome-wide linkage scan	Phenotypic scores in 15 families ($n = 234$) had heritabilities of 42 percent (Karma), 57 percent (Seashore Pitch), 21 percent (Seashore Rhythm), and 48 percent (composite score). Linkage screening revealed a significant peak on chromosome 4q22 and suggestive evidence at 8q13–21. A linkage region on 18q overlapped with one seen in prior studies of dyslexia.	Pulli et al., 2008
	Candidate genes	19 families ($n = 343$) were genotyped for polymorphisms of *AVPR1A*, *SLC6A4*, *COMT*, *DRD2*, and *TPH1*. Some haplotypes of *AVPR1A* were associated with aptitude on the music perception tasks. The other candidate genes showed no significant associations after multiple-testing correction.	Ukkola et al., 2009
	Screen for copy number variants	Study of 5 families ($n = 170$) and 172 unrelated subjects. Nine people with low perception scores carried a 5q31 deletion spanning *PCDHα*; deletion frequency did not significantly differ in people with high scores. Duplication of 8q24 (cf. AP) seen in one person with low scores, but absent in low-scoring relatives. Genome-wide CNV burden did not differ between people with high/low scores.	Ukkola-Vuoti et al., 2013
	Genome-wide linkage and association scan	Linkage scan in 76 families ($n = 767$) identified strongest evidence at 4p14–13 and 4p12–q12, other peaks at 16q21–22, 18q12–21, 22q11. Locations of most linkages differed from prior music-related studies. Genome-wide association scan in same data set found top evidence at 3q21 near *GATA2*. Some association seen for *PCDH7* in 4p15. No association for usual candidate genes (e.g., *AVPR1A*).	Oikkonen et al., 2014
Pitch production	Genome-wide linkage scan and targeted association	Seventy-three families ($n = 1008$) completed pitch production task. Linkage screen in 70 families ($n = 862$) found a significant peak on 4q23. Genotyping of single-nucleotide polymorphisms (SNPs) from the region in 53 families ($n = 630$) revealed significant association near *UGT8*. Authors subsequently identified a nonsynonymous SNP in *UGT8*, and a CNV deletion in the region, each showing association.	Park et al., 2012

Note: Key example studies are shown. For investigations of self-reported musical creativity, see main text.

Genetic contributions to AP have been studied more extensively, most likely because it can be treated as a dichotomous trait, but relative pitch (RP) abilities are probably more relevant to everyday music listening (Miyazaki, 2004). Indeed, RP allows a listener to identify a familiar tune by means of its interval structure (or contour) instead of its constituent pitches (or absolute frequencies), and it allows the detection of "wrong notes." Importantly, AP and RP appear to correspond to two different pitch-processing abilities (Ziv & Radin, 2014), and the RP performance of AP possessors is fairly variable (Miyazaki, 1993; Miyazaki & Rakowski, 2002; Renninger, Granot, & Donchin, 2003). A study by Hove, Sutherland, and Krumhansl (2010) shows that as with AP, individuals of East Asian ethnicity tend to display better RP abilities than Caucasian subjects. Interestingly, this East Asian advantage did not extend to a rhythm perception task and was not modulated by tone language experience.

Few studies have examined the genetic correlates of musical memory, and so far these have focused on testing particular candidate genes for association (see box 10.2 and figure 10.1D). The choice of candidate genes has been motivated by prior studies outside the music domain; for example, some studies of musical memory have targeted arginine vasopressin receptor 1a (*AVPR1A*) and serotonin transporter (*SLC6A4*) genes because common *polymorphisms* of those genes had been previously reported to be associated with creative dance performance (Bachner-Melman et al., 2005). A study of musical and phonological memory in eighty-two students found provisional evidence that these skills were associated with a gene × gene *epistatic* interaction between *promoter region* polymorphisms of the two candidate genes (Granot et al., 2007). In a follow-up to this work, intranasal administration of the arginine vasopressin hormone in twenty-five males was reported to affect musical working memory as well as mood and attentiveness levels, without influencing digit span test scores, suggesting a complex interaction between this hormone, musical memory, and affective states (Granot, Uzefovsky, Bogopolsky, & Ebstein, 2013). Arginine vasopressin and its receptor have been broadly implicated in social behaviors in rodents and humans (Insel, 2010).

A series of studies investigated genetic contributions to musical aptitudes (at multiple levels from heritability to linkage mapping and *association analyses*) in an expanding sample of extended Finnish families (table 10.2; Oikkonen et al., 2014; Pulli et al., 2008; Ukkola-Vuoti et al., 2011, 2013; Ukkola, Onkamo, Raijas, Karma, & Jarvela, 2009). In the first of these studies (Pulli et al., 2008), 15 families (234 people) were tested on a battery of music perception tests comprising the Karma Music Test (Karma, 2007), which measures participants' ability to detect structural changes in abstract sound patterns, and Seashore's pitch and rhythm subtests, which are based on pairwise comparisons (see Ukkola et al., 2009). By analyzing the quantitative phenotype data using a variance component model (box 10.1), the authors obtained heritability estimates of 42 percent for the Karma Music Test, 57 percent for Seashore's pitch subtest, and 21 percent for Seashore's rhythm subtest. Linkage mapping using the quantitative traits revealed significant linkage on chromosome 4q22, as well as suggestive evidence on chromosome 8q13–21 (Pulli et al., 2008). The latter shows some overlap with a region of suggestive linkage identified in one of the AP studies (Theusch et al., 2009), thus implying a potential link between general music perception aptitudes and rare faculties. Interestingly, there was also some evidence of linkage to a region on 18q that had previously been implicated in developmental dyslexia (Fisher et al., 2002). As we have noted in this chapter, linkage regions are typically large and contain many different genes, so findings of overlapping linkages with distinct phenotypes need further investigation to establish whether there is indeed a shared genetic basis.

Later work by the Finnish group (Ukkola et al., 2009) tested for association of selected candidate genes (based on biological hypotheses from previous literature) with musical aptitudes, as measured by the Karma and Seashore tests, in an expanded data set of 19 Finnish families (343 individuals). Participants were also probed about their musical creativity using a questionnaire; the resulting scores were highly heritable and correlated with performance on the music perception tests. The authors reported that certain *haplotypes* of *AVPR1A* (figure 10.1C) were associated with music perception aptitudes, while there was little support for the variants of the other candidate genes that they tested (serotonin transporter *SLC6A4*, catecol-O-methyltranferase *COMT*, dopamine receptor D2 *DRD2*, and tyrosine hydroxylase 1 *TPH1*). In a follow-up study involving *AVPR1A* and *SCL6A4* polymorphisms (Ukkola-Vuoti et al., 2011), the music listening activities of 31 Finnish families (437 members) were surveyed, suggesting associations between *AVPR1A* haplotypes, but not *SCL6A4* haplotypes, and active music listening.

The same research team also performed a preliminary investigation of genome-wide *copy number variations* (CNVs) in 5 extended families and in 172 unrelated participants (Ukkola-Vuoti et al., 2013). They used the quantitative scores on the Karma and Seashore tests to define cases of low musical aptitude in their sample. A deletion at 5q31.1 (figure 10.1D) was found in 54 percent of "low" cases in two of the extended families, although the frequencies in the other members of these families were not reported, so the strength of the genotype-phenotype correlation remains unclear. In the set of unrelated participants, deletion of 5q31.1 was observed in two of twenty-eight "low" cases (7 percent), as compared to zero of forty cases of "high" musical aptitude, but this difference in frequency is not statistically significant. Nonetheless, since the deletion spans the protocadherin alpha (*PCDHα*) gene cluster (figure 10.1D), which encodes cell adhesion proteins that are important for brain development, the observations warrant further investigation in samples with adequate power. One case of low musical aptitude in one of the large families carried a duplication of 8q24.22, overlapping with the top linkage region from an early study of AP (Theusch et al., 2009), but this CNV did not segregate with the phenotype in the family, making the finding difficult to interpret. The authors also performed CNV analyses in relation to self-reports of musical creativity (Ukkola-Vuoti et al., 2013). For example, they highlighted a duplication of 2p22.1 found in 27 percent of "creative" relatives within two families; this CNV spanned galactose mutarotase (*GALM*), a gene that is linked to serotonin metabolism. There was no evidence that high or low musical aptitude or musical creativity was associated with an overall increase in CNVs or with an excess of large CNVs (Ukkola-Vuoti et al., 2013).

The Finnish group went on to conduct a genome-wide study of 767 individuals from 76 families, phenotyped with the music perception tests already described (Oikkonen et al., 2014). They screened hundreds of thousands of single nucleotide polymorphisms (SNPs) across the genome, using these data to test not only for linkage but also association (see box 10.2). The best evidence for linkage was found on chromosome 4, with strongest peaks at 4p14–13 and 4p12–q12. In this study, there were also weaker regions of linkage at other genomic locations on chromosome 4, including one that showed some overlap with the 4q22 interval implicated in the prior linkage screen on a smaller subset of the families (Pulli et al., 2008). Additional regions elsewhere in the genome showed evidence of linkage in the set of 76 families, including 16q21–22.1, 18q12.3–21.1, and 22q11.1–.21, but they did not replicate any findings from prior studies of music-related phenotypes. Moreover, none of the top linkage regions contained SNPs that showed robust evidence of association with the traits. Although linkage and association are different types of tests (box 10.2), it is unusual that there were no genetic markers showing convergent evidence from both methods (Oikkonen et al., 2014). Neighboring the 4p14 linkage peak but outside the region of linkage evidence, the authors identified association with SNPs that were next to protocadherin 7 (*PCDH7*), a gene known to be expressed in the cochlea and the amygdala. The strongest associations in the genome were observed for SNPs in 3q21.3, in the vicinity of the *GATA2* (GATA binding protein 2) gene. This gene encodes a transcription factor that determines the identity of GABAergic neurons in the midbrain and has been implicated in the development of several organs, including cochlear hair cells and the inferior colliculus. Overall, the study suggested interesting connections to known molecular pathways implicated in auditory processing, but did not support the findings from prior targeted studies on candidate genes such as *AVPR1A* (Oikkonen et al., 2014).

Most recently, the Finnish team followed up their genome-wide study of music perception (Oikkonen et al., 2014) by studying self-reports of musical creativity in this same data set (Oikkonen et al., 2016). For each of two music-related traits—arranging and composing—they used the available self-reported data to produce a dichotomous yes/no classification, that is, designating people who reported such creativity as "cases," and then searched for regions of the genome that were linked to this phenotypic status. No statistically significant linkage was found for musical creativity, but suggestive evidence was observed in a number of regions (Oikkonen et al., 2016). Suggestive linkage for arranging (based on 120 cases) was found for 16p12.1-q12.1. This linkage maps on the same chromosome as the 16q21–22.1 linkage from the authors' prior study of musical aptitude (Oikkonen et al., 2014), although the peaks of the signals lie a large distance from each other, so it is not clear whether shared genes are involved. More intriguing, a suggestive linkage for composing (based on 103 cases) was observed for 4q22.1, a region that has been highlighted in a number of other studies of musical phenotypes. The authors also studied nonmusical creativity, finding linkage to Xp11.23, based on 259 cases. In fact, the strongest findings of linkage were found for a fourth trait: when people with musical experience but no creative activity (i.e., neither composing nor arranging) were designated as cases, this yielded linkage to 18q21 (based on 149 cases; Oikkonen et al., 2016). In assessing biological pathways highlighted by the suggestive findings of their study, the authors found overrepresentation of genes involved in cerebellar long-term depression, a form of synaptic plasticity that has been well studied over the years. Overall, as with the molecular studies of musical perception reported thus far, the available results point to interesting candidate genes and potential pathways, but lacking statistically significant findings or robust associations for individual genes, it is still difficult to draw firm conclusions concerning the underlying biology.

Few large-scale twin studies have focused specifically on music production abilities. One of the first such studies was conducted by Coon and Carey (1989), who analyzed music-related data obtained from an earlier survey containing a battery of personality and interest questionnaires. Heritability estimates were higher for participation in singing activities than for self-reported music abilities, and heritability was higher for males than for females. A later study used self-reported data from 1,685 twin pairs (twelve to twenty-four years old) to estimate the heritability of aptitude and exceptional talent across different domains such as language, mathematics, and sports, as well as music (Vinkhuyzen, van der Sluis, Posthuma, & Boomsma, 2009). Heritability estimates for music aptitudes were again higher for males (66 percent) than for females (30 percent). However, in both studies, no objective assessment of musical abilities was obtained. More recently, a large-scale study on 10,975 Swedish twins found that the heritability of music achievement was substantially higher for males (57 percent) than for females (9 percent, a nonsignificant influence). The heritability estimates for objectively assessed musical aptitude, however, were 38 percent for males and 51 percent for females (Mosing et al., 2015).

Other studies have explored the interaction between genes, the amount of musical practice, and musical achievement. Using the same large set of Swedish twins, Mosing and colleagues found that music practice was substantially heritable (40 to 70 percent) and that associations between music practice and objectively assessed musical aptitude were predominantly genetic (Mosing, Madison, Pedersen, Kuja-Halkola, & Ullen, 2014). Another study on the Swedish twin sample showed that shared genetic influences explain the association between openness, music flow, and music practice (Butkovic, Ullén, & Mosing, 2015). Working with a sample of 850 twin pairs, Hambrick and Tucker-Drob (2015) reported moderate heritability estimates of music practice and music achievement, and found that genetic influences on musical achievement were strongest among people who engage in music practice, suggesting the presence of a gene-environment interaction.

Research exploring genetic contributions to music production abilities has largely focused on singing abilities, probably the most widespread such behavior in the general population. Morley and colleagues investigated the *AVPR1A* and *SLC6A4* polymorphisms that were previously associated with musical abilities (Granot et al., 2007; Ukkola et al., 2009; but see Oikkonen et al., 2014, for nonreplication) and creative dancing (Bachner-Melman et al., 2005), testing for their association with choir participation in 523 subjects (Morley et al., 2012). Significant association was detected for a *SLC6A4* polymorphism but not for *AVPR1A* haplotypes proposed to be connected with musical skills in other studies.

Park and colleagues (2012) invited 1,008 individuals from 73 Mongolian families to participate in a pitch-production accuracy test. Family-based linkage analyses using over a thousand genetic markers across the genome identified a peak on 4q23 (figure 10.1E), in an interval that shows some overlap with regions of interest in studies of music perception (Oikkonen et al., 2014; Pulli et al., 2008; note, however, that the genomic positions of the peak regions of chromosome 4 linkage in the most recent Finnish study [Oikkonen et al., 2014] were somewhat different from earlier work on smaller samples [Pulli et al., 2008]). The authors went on to investigate the linked region in detail, using data obtained from SNP genotyping in 53 of the families, and testing for association. They were eventually able to zoom in on a SNP near to the gene *UGT8* (figure 10.1E) showing a highly significant association with performance on the production task. Further analyses uncovered a *nonsynonymous* SNP as well as a CNV in this region that provided further support for a relationship between *UGT8* variations and musical phenotypes (Park et al., 2012). UDP glycosyltransferase 8 catalyzes the transfer of galactose to ceramide, a key step in the biosynthesis of galactocerebrosides, which are important components of myelin membranes in the nervous system.

Phenomics of Musicality in the Postgenomic Era

Dramatic advances in molecular technologies, particularly the development of next-generation DNA sequencing, are set to make a major impact on gene mapping studies of families with music-related disorders or exceptional skills. As for other cognitive traits, the road ahead will still be challenging, since it remains difficult to pinpoint etiological gene variants against a genomic background containing many potential candidates, but developments in analyses of gene function will help to resolve this (Deriziotis & Fisher, 2013). Moreover, the advent of high-throughput large-scale genotyping and sequencing now raises the potential to reliably detect complex genetic effects on musical abilities in the general population. Crucially, investigations of other complex human phenotypes indicate that thousands of participants are needed to achieve adequate power for *genome-wide association scans* (GWAS; see box 10.2). The largest genetic association studies of musical skills reported thus far (e.g., Oikkonen et al., 2014; Park et al., 2012; table 10.2) have involved sample sizes that are small when compared to GWAS studies in other complex genetic traits, and so have been relatively underpowered. Studies with low power may fail to detect effects that are biologically real. Underpowered studies are also more susceptible to false-positive results in which spurious genotype-phenotype correlations are observed (Button et al., 2013). The lack of replication of linkage and association findings in music-related studies thus far may stem in part from this issue of low power, especially given that the underlying genetic architecture (e.g., number of genes involved, effect sizes) is still unknown. Indeed, this is a problem that has broadly affected studies across human genetics as a whole, including investigations of many standard biomedical traits. These difficulties are now being overcome by improved study designs with high power to accommodate small genetic effect sizes or substantial degrees of heterogeneity.

The success of genetic studies of musical ability also depends critically on a robust, objective, and reliable measure of the phonotype. Yet many of the studies discussed so far have used self-reports (e.g., musical creativity studies [Ukkola-Vuoti et al., 2013; Ukkola et al., 2009], twin studies on music production aptitudes [Coon & Carey, 1989; Vinkhuyzen et al., 2009]). Furthermore, as Levitin (2012) pointed out, scores obtained on traditional assessments of musical aptitude, like the Seashore test, are not highly correlated with real-world musical achievement. The great majority of earlier tests were designed for specific music education purposes (Boyle & Radocy, 1987), and consequently they tend to overlook more general musical skills such as the abilities to verbally communicate about music and use music to efficiently modulate emotional states (Müllensiefen et al., 2014; Murphy, 1999).

Thus, there is a need for objective, validated measures that correlate with expressed musicality and can be used to systematically assess large numbers of people. Ideally, a test battery would have the following characteristics:

1. Capture a broad array of musical abilities including the perception, memory, and production of pitch and rhythm
2. Be designed to be administered to individuals with limited or no formal musical training in order to obtain measures that are widely applicable to the general population
3. Have a version appropriate for preschool children to investigate phenotypic differences before formal musical training
4. Cover a wide range of difficulty so that there is power to detect differences at both the low and high ends of ability, which may be most informative
5. Show only a weak or moderate correlation with broader cognitive measures such as general intelligence or working memory

6. Be culture independent, or at least have culture-independent components, thus allowing comparisons between people from different cultures and reducing confounding factors when assessing potential genetic predispositions associated with specific phenotypes
7. Include covariates such as amount of musical training
8. Be designed to be administered robustly online to enable rapid large-scale phenotyping
9. Be of sufficiently short duration that large numbers of people will agree to participate

A test battery that met these criteria could be administered to population cohorts that have already received genome-wide genotyping for studies unrelated to musical abilities. This kind of phenotype-driven approach could potentially be applied across multiple cohorts, and meta-analyses of the resulting GWAS data sets would yield suitably large sample sizes to achieve high power for detecting subtle genotype-phenotype connections. Other potential practical applications include fractionating musical ability by examining which aspects of musical ability correlate specifically with other cognitive traits or genetic characteristics (Levitin, 2012).

While there have been critiques on fundamental issues of method and control in web-delivered experiments (McGraw, Tew, & Williams, 2000; Mehler, 1999), this type of data collection has great potential for music perception and cognition research, especially in domains where versatility and ecological validity are at stake (Honing & Ladinig, 2008; Honing & Reips, 2008). Probing music perceptual skills can now be done reliably due to recent technological advances in presenting audio over the Internet, for example by using file formats like MPEG4 that guarantee optimal sound quality on different computer platforms at different transmission rates. However, when it comes to collecting and uploading individual sound files, there remains a lack of standardization, most notably with respect to timing. Therefore, music production experiments (such as tapping or singing along with a stimulus) are still unreliable. Hence, it is still most realistic to focus on phenotypes related to music perception abilities while also collecting information on other aspects of the phenotype through survey-style questionnaires.

There are several candidate components of musicality suggested in the literature (see chapter 1, this volume). With regard to perceptual abilities, relative pitch (Justus & Hutsler, 2005; Trehub, 2003), tonal encoding of pitch (Peretz & Coltheart, 2003), beat or pulse perception (Fitch, 2013; Honing, 2012), and metrical encoding of rhythm (Fitch, 2013) are a good starting point for a phenomics of musicality. For example, the following specific tests could, in principle, be implemented in an Internet-based survey that could be administered to a broad population in less than thirty minutes:

- Relative pitch ability (Müllensiefen & Halpern, 2014)
- Melodic memory (Müllensiefen et al., 2014)

- Beat perception: identifying the tempo of a musical excerpt by comparing two excerpts in different tempi and judging whether they are different, or judging whether an isochronous rhythm is on or off the beat with respect to the underlying music (task based on Iversen & Patel, 2008; Iversen, Patel, & Ohgushi, 2008; cf. Honing, 2013)
- Meter perception: judging whether two excerpts are rhythmically (dis)similar using classes of rhythms in simple and compound meters (classification task based on Hannon & Trehub, 2005; cf. Cao, Lotstein, & Johnson-Laird, 2014).

Two test batteries covering most of these aspects, the Goldsmiths Musical Sophistication Index (Gold-MSI) questionnaire and test battery (Müllensiefen et al., 2014) and the Swedish Musical Discrimination Test (SMDT; Ullén et al., 2014) have been validated on large populations. The Gold-MSI, which can be completed in twenty minutes, includes a melodic memory task based on a comparison paradigm (Bartlett & Dowling, 1980), a beat tracking task (based on Iversen & Patel, 2008), and a self-report questionnaire covering a broad spectrum of musical behaviors. Furthermore, data from the Gold-MSI have been correlated and validated with other test batteries and with personality measures such as the TIPI inventory (Gosling, Rentfrow, & Swann, 2003). The SMDT, which takes approximately fifteen minutes to complete, includes tasks related to melody, pitch, and rhythm.

Of course, for a fuller understanding of genetic contributions to musicality, there are many aspects of phenotypic variation beyond what is proposed above that could prove to be important. Certain of these aspects could potentially be probed in a less objective manner in questionnaires, or some of them might be administered to subsets of the thousands participating in the core thirty-minute test. For example, sensitivity to expressive timing nuances (Honing & Ladinig, 2009) or musical timbre (Law & Zentner, 2012) might be connected with consistent genetic variation. Psychophysical tasks measuring auditory streaming abilities (Huron, 1989) or the sensitivity to acoustical features such as roughness and harmonicity (Cousineau, McDermott, & Peretz, 2012; McDermott, Lehr, & Oxenham, 2010) could also prove informative, although the sound fidelity they require could be difficult to ensure in an online setting. Despite these difficulties with administration, such tasks are relatively culture free and could form the basis for a test of musicality that could be administered across cultures. Musical production abilities such as pitch reproduction accuracy (e.g., Park et al., 2012) and meter tapping accuracy are undoubtedly critical components of the musical phenotype, but can be evaluated more reliably in the laboratory than using Internet-based experiments. Finally, it would be of great benefit to obtain indexes of social and emotional responses to music, as well as musical behavior in the sense of attendance at and participation in musical events. It would be possible to get at least crude estimates of these attributes through online questionnaires (cf. Mas-Herrero, Zatorre, Rodriguez-Fornells, & Marco-Pallares, 2014).

Broader Perspectives

A primary focus of this chapter has concerned the potential biological bases of individual differences in musical abilities. We note that the phenomics of musicality can also be investigated at the level of populations, although such studies typically involve comparing musical cultures and genetic relationships rather than assessing musical aptitudes. For example, one study has described a relationship between genetic distance and similarity in the folk music styles across thirty-one Eurasian nations (Pamjav, Juhasz, Zalan, Nemeth, & Damdin, 2012). A more recent report obtained significant correlations between folk song structure and mitochondrial DNA variation among nine indigenous Taiwanese populations (Brown et al., 2014). The magnitude of these correlations was similar to that of the correlations between linguistic distance (based on lexical cognates) and genetic distance for the same populations. Interestingly, although musical and linguistic distances were both correlated with genetic distance, musical and linguistic distances were not significantly correlated with one another.

Another complementary approach for deciphering biological pathways implicated in diverse human traits involves analysis of levels of expression of different genes. For example, one might screen a tissue sample to determine how many copies of messenger RNA transcripts are made from a gene of interest. With advances in molecular technology, it is now possible to perform simultaneous systematic expression profiling of all the many thousands of different transcripts in a sample—known as the transcriptome (Lappalainen et al., 2013). By comparing transcriptomes from different cell types in the same person from the same cell type in different people or under different conditions, particular genes or sets of genes that are important for a certain biological pathway can be identified. Two recent studies have sought to determine how transcriptomes might be affected when a person listens to music (Kanduri et al., 2015b) or gives a musical performance (Kanduri et al., 2015a). Detailed discussion of these investigations is beyond the scope of this chapter, but it is worth noting that for practical reasons, such analyses are currently limited to studying changes in gene expression levels in peripheral blood rather than assessing them directly in brain tissue. In such studies, any transcriptome changes that are observed in blood necessarily reflect systemic effects (for example on stress levels) rather than specific changes that are occurring in brain circuits underpinning music processing, so their relevance to learning and memory seems unclear. Although it is established that engaging in musical activities has effects on structural and functional aspects of certain neural circuits and that such effects may well be mediated by altered gene expression in the relevant parts of the brain, most of these are localized transcriptomic changes that do not get transmitted to peripheral blood and are unlikely to be captured by screening nonneural tissue. As an analogy, screening only an arm or a hand in an MRI scanner does not provide a readout of potential changes in the connectivity of the brain tissues involved in practicing or playing a musical instrument. With further advances in gene expression profiling, it might eventually be possible to assess how transcriptomes change within the circuits of a living human brain, a development that could revolutionize the understanding of molecular mechanisms underlying key cognitive traits such as language and music. In the meantime, it is worth being cautious about the reach of transcriptomics for this field.

Crucially, molecular genetic studies of humans should be seen as one part of a broader framework for identifying the underpinnings of musicality. This might include comparative work assessing relevant skills in nonhuman animals (see chapter 7, this volume). Moreover, new possibilities are opened up once key genes have been identified; their evolutionary history can be traced, including searching for signs of Darwinian selection (for a recent example, see Liu et al., 2016), molecular networks can be teased apart in human neurons, and ancestral functions can be studied in animal models. At the same time, the evolutionary history of cultural markers, including music, can be informed by phylogenetic studies comparing human populations and, possibly, nonhuman animals. New technologies offer promising prospects in both respects. First, the fields of molecular and developmental neurobiology provide an ever-growing tool kit of sophisticated methods that can be used to decipher how particular genes of interest contribute to the development and plasticity of neural circuits in model systems and humans themselves. Second, the implementation of online-based testing procedures enables a systematic assessment of musical aptitudes on an unprecedented scale. Together, these developments will likely result in a paradigmatic shift in this research field, ushering in a new era for the exploration of the biological bases of musicality.

Glossary

Allele Alternative forms of the same gene or genomic position.

Association analysis Statistical test of whether there are nonrandom correlations between a certain phenotypic trait (either a qualitatively defined affection status or a quantitative measure) and specific allelic variants.

Chromosomal band Each human chromosome has a short arm (p) and long arm (q), separated by a centromere. Each chromosome arm is divided into regions, or cytogenetic bands, that can be seen using a microscope and special stains. These bands are labeled p1, p2, p3, q1, q2, q3, … counting from the centromere outward. At higher resolutions, subbands can be seen within the bands, also numbered from centromere outwards.

Copy number variation (CNV) Structural alteration of a chromosome giving an abnormal number of copies of a particular section of DNA due to a region of the genome being deleted or duplicated. Copy number variants may occur in an array of sizes, from hundreds to several million nucleotides.

Epistasis When a single phenotype involves interactions between two or more genes.

Genome-wide association scans (GWAS) Systematic hypothesis-free testing of association at hundreds of thousands (perhaps millions) of different genetic markers across the entire genome. Involves a huge amount of multiple testing, requiring appropriate adjustments when evaluating significance of results, in order to avoid false-positive findings.

Genotype The genetic constitution of an individual. Can refer to the entire complement of genetic material, a specific gene, or a set of genes.

Haplotype A cluster of several neighboring polymorphisms on a chromosome that tend to be inherited together.

Heritability The proportion of variability in a characteristic that can be attributed to genetic influences. A statistical description that applies to a specific population, it can vary if the environment changes.

Linkage analysis Enables mapping of the rough genomic location of a gene implicated in a given trait. This method tracks the inheritance of polymorphic genetic markers as they are transmitted in families, assessing whether they cosegregate with a trait of interest in a way that is unlikely to be due to chance.

Next-generation DNA sequencing Newly emerged high-throughput technologies that allow DNA sequences to be determined at dramatically lower cost and much more rapidly than the standard approaches that were previously available.

Nonsynonymous A nucleotide change in a gene that alters the amino acid sequence of the encoded protein. Contrasts with synonymous substitutions that preserve the usual amino acid sequence.

Phenomics The robust measurement of physical, biochemical, physiological, and behavioral traits of organisms and how they alter due to changes in genes and environment.

Phenotype The appearance of an individual in terms of a particular characteristic (e.g., physical, biochemical, physiological) resulting from interactions between genotype, environment and random factors.

Polymorphism A position in the genome that contains variation in the population and therefore has more than one possible allele. At present, the most commonly studied of these are single nucleotide polymorphisms (SNPs) involving a single nucleotide at a specific point in the genome.

Proband The index case who triggers investigation of a particular family to isolate the potential genetic factors involved in a given trait.

Promoter region A region at the beginning of each gene that is responsible for its regulation, allowing it to be switched on or off in distinct cell types and developmental stages.

Recurrence risk The likelihood that a trait will be observed elsewhere in a pedigree, given that at least one family member exhibits the trait. Can be defined for specific types of relationships, such as siblings.

Acknowledgments

This chapter is based on ideas exchanged during the "What Makes Us Musical Animals? Cognition, Biology and the Origins of Musicality" workshop that took place at the Lorentz Center (Leiden, Netherlands) in April 2014. It represents an updated version of an article that was published in *Philosophical Transactions of the Royal Society B*, *370* (2015), in the special issue "Biology, Cognition, and Origins of Musicality." We thank David Huron for his suggestions and comments on an earlier version of the manuscript.

Note

1. A primer on genes and genomes is available online at https://mitpress.mit.edu/books/origins-musicality.

References

Abecasis, G. R., Auton, A., Brooks, L. D., DePristo, M. A., Durbin, R. M., Handsaker, R. E., et al. (2012). An integrated map of genetic variation from 1,092 human genomes. *Nature*, *491*(7422), 56–65. doi:10.1038/nature11632

Alcock, K. J., Passingham, R. E., Watkins, K., & Vargha-Khadem, F. (2000). Pitch and timing abilities in inherited speech and language impairment. *Brain and Language*, *75*(1), 34–46. doi:10.1006/brln.2000.2323

Asher, J. E., Lamb, J. A., Brocklebank, D., Cazier, J. B., Maestrini, E., Addis, L., et al. (2009). A whole-genome scan and fine-mapping linkage study of auditory visual synesthesia reveals evidence of linkage to chromosomes 2q24, 5q33, 6p12, and 12p12. *American Journal of Human Genetics*, *84*(2), 279–285. doi:10.1016/j.ajhg.2009.01.012

Athos, E. A., Levinson, B., Kistler, A., Zemansky, J., Bostrom, A., Freimer, N., et al. (2007). Dichotomy and perceptual distortions in absolute pitch ability. *Proceedings of the National Academy of Sciences of the United States of America, 104*(37), 14795–14800. doi:10.1073/pnas.0703868104

Bachem, A. (1955). Absolute pitch. *Journal of the Acoustical Society of America, 27*, 1180–1185.

Bachner-Melman, R., Dina, C., Zohar, A. H., Constantini, N., Lerer, E., Hoch, S., et al. (2005). AVPR1a and SLC6A4 gene polymorphisms are associated with creative dance performance. *PLOS Genetics, 1*(3), e42. doi:10.1371/journal.pgen.0010042

Baharloo, S., Johnston, P. A., Service, S. K., Gitschier, J., & Freimer, N. B. (1998). Absolute pitch: An approach for identification of genetic and nongenetic components. *American Journal of Human Genetics, 62*(2), 224–231. doi:10.1086/301704

Baharloo, S., Service, S. K., Risch, N., Gitschier, J., & Freimer, N. B. (2000). Familial aggregation of absolute pitch. *American Journal of Human Genetics, 67*(3), 755–758. doi:10.1086/303057

Bartlett, J. C., & Dowling, W. J. (1980). Recognition of transposed melodies: A key-distance effect in developmental perspective. *Journal of Experimental Psychology: Human Perception and Performance, 6*(3), 501–515. doi:10.1037/0096-1523.6.3.501

Bermudez, P., Lerch, J. P., Evans, A. C., & Zatorre, R. J. (2009). Neuroanatomical correlates of musicianship as revealed by cortical thickness and voxel-based morphometry. *Cerebral Cortex, 19*(7), 1583–1596. doi:10.1093/cercor/bhn196

Bosley, H. G., & Eagleman, D. M. (2015). Synesthesia in twins: Incomplete concordance in monozygotes suggests extragenic factors. *Behavioural Brain Research, 286*, 93–96. doi:10.1016/j.bbr.2015.02.024

Boyle, J. D., & Radocy, R. E. (1987). *Measurement and evaluation of musical experiences*. New York: Schirmer Books.

Brown, S., Savage, P. E., Ko, A. M., Stoneking, M., Ko, Y. C., Loo, J. H., et al. (2014). Correlations in the population structure of music, genes and language. *Proceedings of the Royal Society of London B: Biological Sciences, 281*(1774), 20132072. doi:10.1098/rspb.2013.2072

Butkovic, A., Ullén, F., & Mosing, M. A. (2015). Personality related traits as predictors of music practice: Underlying environmental and genetic influences. *Personality and Individual Differences, 74*, 133–138. doi:10.1016/j.paid.2014.10.006

Button, K. S., Ioannidis, J. P., Mokrysz, C., Nosek, B. A., Flint, J., Robinson, E. S., et al. (2013). Power failure: Why small sample size undermines the reliability of neuroscience. *Nature Reviews: Neuroscience, 14*(5), 365–376. doi:10.1038/nrn3475

Cao, E., Lotstein, M., & Johnson-Laird, P. (2014). Similarity and families of musical rhythms. *Music Perception, 31*(5), 444–469. doi:10.1525/mp.2014.31.5.444

Coon, H., & Carey, G. (1989). Genetic and environmental determinants of musical ability in twins. *Behavior Genetics, 19*(2), 183–193. doi:10.1007/bf01065903

Cousineau, M., McDermott, J. H., & Peretz, I. (2012). The basis of musical consonance as revealed by congenital amusia. *Proceedings of the National Academy of Sciences, 109*(48), 19858–19863. doi:10.1073/pnas.1207989109

Deriziotis, P., & Fisher, S. E. (2013). Neurogenomics of speech and language disorders: The road ahead. *Genome Biology, 14*(4), 204. doi:10.1186/gb-2013-14-4-204

Deutsch, D. (2013). *Absolute pitch psychology of music* (3rd ed.). San Diego, CA: Elsevier.

Dohn, A., Garza-Villarreal, E. A., Chakravarty, M. M., Hansen, M., Lerch, J. P., & Vuust, P. (2015). Gray- and white-matter anatomy of absolute pitch possessors. *Cerebral Cortex, 25*(5), 1379–1388. doi:10.1093/cercor/bht334

Drayna, D., Manichaikul, A., de Lange, M., Snieder, H., & Spector, T. (2001). Genetic correlates of musical pitch recognition in humans. *Science, 291*(5510), 1969–1972. doi:10.1126/science.291.5510.1969

Fisher, S. E. (2006). Tangled webs: Tracing the connections between genes and cognition. *Cognition, 101*(2), 270–297. doi:10.1016/j.cognition.2006.04.004

Fisher, S. E., Francks, C., Marlow, A. J., MacPhie, I. L., Newbury, D. F., Cardon, L. R., et al. (2002). Independent genome-wide scans identify a chromosome 18 quantitative-trait locus influencing dyslexia. *Nature Genetics, 30*(1), 86–91. doi:10.1038/ng792

Fisher, S. E., & Scharff, C. (2009). FOXP2 as a molecular window into speech and language. *Trends in Genetics, 25*(4), 166–177. doi:10.1016/j.tig.2009.03.002

Fisher, S. E., Vargha-Khadem, F., Watkins, K. E., Monaco, A. P., & Pembrey, M. E. (1998). Localisation of a gene implicated in a severe speech and language disorder. *Nature Genetics, 18*(2), 168–170. doi:10.1038/ng0298-168

Fisher, S. E., & Vernes, S. C. (2015). Genetics and the language sciences. *Annual Review of Linguistics, 1*, 289–310. doi:10.1146/annurevlinguist-030514-125024

Fitch, W. T. (2013). Rhythmic cognition in humans and animals: Distinguishing meter and pulse perception. *Frontiers in Systems Neuroscience, 7*, 68. doi:10.3389/fnsys.2013.00068

Gervain, J., Vines, B. W., Chen, L. M., Seo, R. J., Hensch, T. K., Werker, J. F., et al. (2013). Valproate reopens critical-period learning of absolute pitch. *Frontiers in Systems Neuroscience, 7*, 102. doi:10.3389/fnsys.2013.00102

Gilad, Y., Pritchard, J. K., & Thornton, K. (2009). Characterizing natural variation using next-generation sequencing technologies. *Trends in Genetics, 25*(10), 463–471. doi:10.1016/j.tig.2009.09.003

Goldstein, D. B., Allen, A., Keebler, J., Margulies, E. H., Petrou, S., Petrovski, S., et al. (2013). Sequencing studies in human genetics: Design and interpretation. *Nature Reviews: Genetics, 14*(7), 460–470. doi:10.1038/nrg3455

Gosling, S. D., Rentfrow, P. J., & Swann, W. B., Jr. (2003). A very brief measure of the Big-Five personality domains. *Journal of Research in Personality, 37*, 504–528. doi:10.1016/s0092-6566(03)00046-1

Graham, S. A., & Fisher, S. E. (2013). Decoding the genetics of speech and language. *Current Opinion in Neurobiology, 23*(1), 43–51. doi:10.1016/j.conb.2012.11.006

Granot, R. Y., Frankel, Y., Gritsenko, V., Lerer, E., Gritsenko, I., Bachner-Melman, R., et al. (2007). Provisional evidence that the arginine vasopressin 1a receptor gene is associated with musical memory. *Evolution and Human Behavior, 28*, 313–318. doi:10.1016/j.evolhumbehav.2007.05.003

Granot, R. Y., Uzefovsky, F., Bogopolsky, H., & Ebstein, R. P. (2013). Effects of arginine vasopressin on musical working memory. *Frontiers in Psychology, 4*, 712. doi:10.3389/fpsyg.2013.00712

Gregersen, P. K. (1998). Instant recognition: The genetics of pitch perception. *American Journal of Human Genetics, 62*(2), 221–223. doi:10.1086/301734

Gregersen, P. K., Kowalsky, E., Kohn, N., & Marvin, E. W. (1999). Absolute pitch: Prevalence, ethnic variation, and estimation of the genetic component. *American Journal of Human Genetics, 65*(3), 911–913. doi:10.1086/302541

Gregersen, P. K., Kowalsky, E., Lee, A., Baron-Cohen, S., Fisher, S. E., Asher, J. E., et al. (2013). Absolute pitch exhibits phenotypic and genetic overlap with synesthesia. *Human Molecular Genetics, 22*(10), 2097–2104. doi:10.1093/hmg/ddt059

Hambrick, D. Z., & Tucker-Drob, E. M. (2015). The genetics of music accomplishment: Evidence for gene-environment correlation and interaction. *Psychonomic Bulletin and Review, 22*(1), 112–120. doi:10.3758/s13423-014-0671-9

Hannon, E. E., & Trainor, L. J. (2007). Music acquisition: Effects of enculturation and formal training on development. *Trends in Cognitive Sciences, 11*(11), 466–472. doi:10.1016/J.Tics.2007.08.008

Hannon, E. E., & Trehub, S. E. (2005). Metrical categories in infancy and adulthood. *Psychological Science, 16*(1), 48–55. doi:10.1111/j.0956-7976.2005.00779.x

Hargreaves, D. J., & Zimmerman, M. P. (1992). Developmental theories of music learning. In R. Colwell (Ed.), *Handbook of research on music teaching and learning: A Project of the Music Educators National Conference* (pp. 377–391). New York: Schirmer.

Honing, H. (2012). Without it no music: Beat induction as a fundamental musical trait. *Annals of the New York Academy of Sciences, 1252*, 85–91. doi:10.1111/J.1749-6632.2011.06402.X

Honing, H. (2013). Structure and interpretation of rhythm in music. In D. Deutsch (Ed.), *Psychology of music* (3rd ed., pp. 369–404). London: Academic Press.

Honing, H., & Ladinig, O. (2008). The potential of the Internet for music perception research: A comment on lab-based versus web-based studies. *Empirical Musicology Review, 3*(1), 4–7.

Honing, H., & Ladinig, O. (2009). Exposure influences expressive timing judgments in music. *Journal of Experimental Psychology: Human Perception and Performance*, *35*(1), 281–288. doi:10.1037/a0012732

Honing, H., & Reips, U. (2008). Web-based versus lab-based studies: A response to Kendall. *Empirical Musicology Review*, *3*(2), 73–77.

Hove, M. J., Sutherland, M. E., & Krumhansl, C. L. (2010). Ethnicity effects in relative pitch. *Psychonomic Bulletin and Review*, *17*(3), 310–316. doi:10.3758/PBR.17.3.310

Huron, D. (1989). Voice denumerability in polyphonic music of homogeneous timbres. *Music Perception*, *6*(4), 361–382. doi:10.2307/40285438

Hyde, K. L., Lerch, J. P., Zatorre, R. J., Griffiths, T. D., Evans, A. C., & Peretz, I. (2007). Cortical thickness in congenital amusia: When less is better than more. *Journal of Neuroscience*, *27*(47), 13028–13032. doi:10.1523/JNEUROSCI.3039-07.2007

Hyde, K. L., & Peretz, I. (2004). Brains that are out of tune but in time. *Psychological Science*, *15*(5), 356–360. doi:10.1111/j.0956-7976.2004.00683.x

Hyde, K. L., Zatorre, R. J., & Peretz, I. (2011). Functional MRI evidence of an abnormal neural network for pitch processing in congenital amusia. *Cerebral Cortex*, *21*(2), 292–299. doi:10.1093/cercor/bhq094

Insel, T. R. (2010). The challenge of translation in social neuroscience: A review of oxytocin, vasopressin, and affiliative behavior. *Neuron*, *65*(6), 768–779. doi:10.1016/j.neuron.2010.03.005

Iversen, J. R., & Patel, A. D. (2008). The Beat Alignment Test (BAT): Surveying beat processing abilities in the general population. In K. Miyazaki, M. Adachi, Y. Hiraga, Y. Nakajima, & M. Tsuzaki (Eds.), *Proceedings of the 10th International Conference on Music Perception and Cognition* (CD-ROM; pp. 465–468). Adelaide: Causal Productions.

Iversen, J. R., Patel, A. D., & Ohgushi, K. (2008). Perception of rhythmic grouping depends on auditory experience. *Journal of the Acoustical Society of America*, *124*(4), 2263–2271. doi:10.1121/1.2973189

Justus, T., & Hutsler, J. J. (2005). Fundamental issues in the evolutionary psychology of music: Assessing innateness and domain specificity. *Music Perception*, *23*(1), 1–27. doi:10.1525/Mp.2005.23.1.1

Kalmus, H., & Fry, D. B. (1980). On tune deafness (dysmelodia): Frequency, development, genetics and musical background. *Annals of Human Genetics*, *43*(4), 369–382. doi:10.1111/j.1469-1809.1980.tb01571.x

Kanduri, C., Kuusi, T., Ahvenainen, M., Philips, A. K., Lahdesmaki, H., & Jarvela, I. (2015a). The effect of music performance on the transcriptome of professional musicians. *Scientific Reports*, *5*, 9506. doi:10.1038/srep09506

Kanduri, C., Raijas, P., Ahvenainen, M., Philips, A. K., Ukkola-Vuoti, L., Lahdesmaki, H., et al. (2015b). The effect of listening to music on human transcriptome. *PeerJ*, *3*, e830. doi:10.7717/peerj.830

Karma, K. (2007). Musical aptitude definition and measure validation: Ecological validity can endanger the construct validity of musical aptitude tests. *Psychomusicology: Music, Mind, and Brain*, *19*, 79–90. doi:10.1037/h0094033

Lai, C. S., Fisher, S. E., Hurst, J. A., Vargha-Khadem, F., & Monaco, A. P. (2001). A forkhead-domain gene is mutated in a severe speech and language disorder. *Nature*, *413*(6855), 519–523. doi:10.1038/35097076

Lappalainen, T., Sammeth, M., Friedlander, M. R., t Hoen, P. A., Monlong, J., Rivas, M. A., … Dermitzakis, E. T. (2013). Transcriptome and genome sequencing uncovers functional variation in humans. *Nature*, *501*(7468), 506–511. doi:10.1038/nature12531

Launay, J., Grube, M., & Stewart, L. (2014). Dysrhythmia: A specific congenital rhythm perception deficit. *Frontiers in Psychology*, *5*, 18. doi:10.3389/fpsyg.2014.00018

Law, L. N., & Zentner, M. (2012). Assessing musical abilities objectively: Construction and validation of the profile of music perception skills. *PLoS One*, *7*(12), e52508. doi:10.1371/journal.pone.0052508

Lense, M. D., Dankner, N., Pryweller, J. R., Thornton-Wells, T. A., & Dykens, E. M. (2014). Neural correlates of amusia in Williams syndrome. *Brain Sciences*, *4*(4), 594–612. doi:10.3390/brainsci4040594

Lense, M. D., & Dykens, E. (2013). Musical learning in children and adults with Williams syndrome. *Journal of Intellectual Disability Research*, *57*(9), 850–860. doi:10.1111/j.1365-2788.2012.01611.x

Lense, M. D., Shivers, C. M., & Dykens, E. M. (2013). (A)musicality in Williams syndrome: Examining relationships among auditory perception, musical skill, and emotional responsiveness to music. *Frontiers in Psychology*, *4*, 525. doi:10.3389/fpsyg.2013.00525

Levitin, D. J. (2005). Musical behavior in a neurogenetic developmental disorder: Evidence from Williams syndrome. *Annals of the New York Academy of Sciences, 1060*, 325–334. doi:10.1196/annals.1360.027

Levitin, D. J. (2012). What does it mean to be musical? *Neuron, 73*(4), 633–637. doi:10.1016/j.neuron.2012.01.017

Levitin, D. J., Cole, K., Chiles, M., Lai, Z., Lincoln, A., & Bellugi, U. (2004). Characterizing the musical phenotype in individuals with Williams syndrome. *Child Neuropsychology: A Journal on Normal and Abnormal Development in Childhood and Adolescence, 10*(4), 223–247. doi:10.1080/09297040490909288

Liu, X., Kanduri, C., Oikkonen, J., Karma, K., Raijas, P., Ukkola-Vuoti, L., et al. (2016). Detecting signatures of positive selection associated with musical aptitude in the human genome. *Scientific Reports, 6*, 21198. doi:10.1038/srep21198

Loui, P., Li, H. C., Hohmann, A., & Schlaug, G. (2011). Enhanced cortical connectivity in absolute pitch musicians: A model for local hyperconnectivity. *Journal of Cognitive Neuroscience, 23*(4), 1015–1026. doi:10.1162/jocn.2010.21500

Loui, P., Zamm, A., & Schlaug, G. (2012). Enhanced functional networks in absolute pitch. *NeuroImage, 63*(2), 632–640. doi:10.1016/j.neuroimage.2012.07.030

Martens, M. A., Wilson, S. J., & Reutens, D. C. (2008). Research review: Williams syndrome: A critical review of the cognitive, behavioral, and neuroanatomical phenotype. *Journal of Child Psychology and Psychiatry, and Allied Disciplines, 49*(6), 576–608. doi:10.1111/j.1469-7610.2008.01887.x

Mas-Herrero, E., Zatorre, R. J., Rodriguez-Fornells, A., & Marco-Pallares, J. (2014). Dissociation between musical and monetary reward responses in specific musical anhedonia. *Current Biology, 24*(6), 699–704. doi:10.1016/J.Cub.2014.01.068

McDermott, J. H., Lehr, A. J., & Oxenham, A. J. (2010). Individual differences reveal the basis of consonance. *Current Biology, 20*(11), 1035–1041. doi:10.1016/j.cub.2010.04.019

McGraw, K. O., Tew, M. D., & Williams, J. E. (2000). The integrity of web-delivered experiments: Can you trust the data? *Psychological Science, 11*(6), 502–506. doi:10.1111/1467-9280.00296

Mehler, J. (1999). Experiments carried out over the web. *Cognition, 71*(3), 187–189.

Miyazaki, K. (1993). Absolute pitch as an inability: Identification of musical intervals in a tonal context. *Music Perception, 11*(1), 55–72. doi:10.2307/40285599

Miyazaki, K. (2004). How well do we understand absolute pitch? *Acoustical Science and Technology, 25*, 426–432. doi:10.1250/ast.25.426

Miyazaki, K., & Rakowski, A. (2002). Recognition of notated melodies by possessors and nonpossessors of absolute pitch. *Perception and Psychophysics, 64*(8), 1337–1345. doi:10.3758/bf03194776

Morishita, H., & Hensch, T. K. (2008). Critical period revisited: Impact on vision. *Current Opinion in Neurobiology, 18*(1), 101–107. doi:10.1016/j.conb.2008.05.009

Morley, A. P., Narayanan, M., Mines, R., Molokhia, A., Baxter, S., Craig, G., et al. (2012). AVPR1A and SLC6A4 polymorphisms in choral singers and non-musicians: A gene association study. *PLoS One, 7*(2), e31763. doi:10.1371/journal.pone.0031763

Mosing, M. A., Madison, G., Pedersen, N. L., Kuja-Halkola, R., & Ullen, F. (2014). Practice does not make perfect: No causal effect of music practice on music ability. *Psychological Science, 25*(9), 1795–1803. doi:10.1177/0956797614541990

Mosing, M. A., Verweij, K. J., Madison, G., Pedersen, N. L., Zietsch, B. P., & Ullén, F. (2015). Did sexual selection shape human music? Testing predictions from the sexual selection hypothesis of music evolution using a large genetically informative sample of over 10,000 twins. *Evolution and Human Behavior, 36*(5), 359–366. doi:10.1016/j.evolhumbehav.2015.02.004

Müllensiefen, D., Gingras, B., Musil, J., & Stewart, L. (2014). The musicality of non-musicians: An index for assessing musical sophistication in the general population. *PLoS One, 9*(2), e89642. doi:10.1371/journal.pone.0089642

Müllensiefen, D., & Halpern, A. R. (2014). The role of features and context in recognition of novel melodies. *Music Perception, 31*(5), 418–435. doi:10.1525/Mp.2014.31.5.418

Murphy, C. (1999). How far do tests of musical ability shed light on the nature of musical intelligence? *British Journal of Music Education, 16*(1), 39–50. doi:10.1017/s0265051799000133

Ng, R., Lai, P., Levitin, D. J., & Bellugi, U. (2013). Musicality correlates with sociability and emotionality in Williams syndrome. *Journal of Mental Health Research in Intellectual Disabilities*, *6*(4), 268–279. doi:10.1080/19315864.2012.683932

Oikkonen, J., Huang, Y., Onkamo, P., Ukkola-Vuoti, L., Raijas, P., Karma, K., et al. (2014). A genome-wide linkage and association study of musical aptitude identifies loci containing genes related to inner ear development and neurocognitive functions. *Molecular Psychiatry*, *20*(2), 275–282. doi:10.1038/mp.2014.8

Oikkonen, J., Kuusi, T., Peltonen, P., Raijas, P., Ukkola-Vuoti, L., Karma, K., et al. (2016). Creative activities in music: A genome-wide linkage analysis. *PLoS One*, *11*(2), e0148679. doi:10.1371/journal.pone.0148679

Pamjav, H., Juhasz, Z., Zalan, A., Nemeth, E., & Damdin, B. (2012). A comparative phylogenetic study of genetics and folk music. *Molecular Genetics and Genomics*, *287*(4), 337–349. doi:10.1007/s00438-012-0683-y

Park, H., Lee, S., Kim, H. J., Ju, Y. S., Shin, J. Y., Hong, D., et al. (2012). Comprehensive genomic analyses associate UGT8 variants with musical ability in a Mongolian population. *Journal of Medical Genetics*, *49*(12), 747–752. doi:10.1136/jmedgenet-2012-101209

Peretz, I. (2006). The nature of music from a biological perspective. *Cognition*, *100*(1), 1–32. doi:10.1016/j.cognition.2005.11.004

Peretz, I. (2013). The biological foundations of music: Insights from congenital amusia. In D. Deutsch (Ed.), *Psychology of music*. San Diego, CA: Elsevier.

Peretz, I., Ayotte, J., Zatorre, R. J., Mehler, J., Ahad, P., Penhune, V. B., et al. (2002). Congenital amusia: A disorder of fine-grained pitch discrimination. *Neuron*, *33*(2), 185–191

Peretz, I., & Coltheart, M. (2003). Modularity of music processing. *Nature Neuroscience*, *6*(7), 688–691. doi:10.1038/Nn1083

Peretz, I., Cummings, S., & Dube, M. P. (2007). The genetics of congenital amusia (tone deafness): A family-aggregation study. *American Journal of Human Genetics*, *81*(3), 582–588. doi:10.1086/521337

Peretz, I., & Hyde, K. L. (2003). What is specific to music processing? Insights from congenital amusia. *Trends in Cognitive Sciences*, *7*(8), 362–367. doi:10.1016/s1364-6613(03)00150-5

Phillips-Silver, J., Toiviainen, P., Gosselin, N., Piche, O., Nozaradan, S., Palmer, C., et al. (2011). Born to dance but beat deaf: A new form of congenital amusia. *Neuropsychologia*, *49*(5), 961–969. doi:10.1016/j.neuropsychologia.2011.02.002

Plantinga, J., & Trainor, L. J. (2008). Infants' memory for isolated tones and the effects of interference. *Music Perception*, *26*(2), 121–127. doi:10.1525/Mp.2008.26.2.121

Profita, J., & Bidder, T. G. (1988). Perfect pitch. *American Journal of Medical Genetics*, *29*(4), 763–771. doi:10.1002/ajmg.1320290405

Pulli, K., Karma, K., Norio, R., Sistonen, P., Goring, H. H., & Jarvela, I. (2008). Genome-wide linkage scan for loci of musical aptitude in Finnish families: Evidence for a major locus at 4q22. *Journal of Medical Genetics*, *45*(7), 451–456. doi:10.1136/jmg.2007.056366

Renninger, L. B., Granot, R. I., & Donchin, E. (2003). Absolute pitch and the P300 component of the event-related potential: An exploration of variables that may account for individual differences. *Music Perception*, *20*(4), 357–382. doi:10.1525/mp.2003.20.4.357

Ross, D. A., & Marks, L. E. (2009). Absolute pitch in children prior to the beginning of musical training. *Annals of the New York Academy of Sciences*, *1169*, 199–204. doi:10.1111/j.1749-6632.2009.04847.x

Ross, D. A., Olson, I. R., & Gore, J. C. (2003). Absolute pitch does not depend on early musical training. *Annals of the New York Academy of Sciences*, *999*, 522–526. doi:10.1196/annals.1284.065

Ross, D. A., Olson, I. R., Marks, L. E., & Gore, J. C. (2004). A nonmusical paradigm for identifying absolute pitch possessors. *Journal of the Acoustical Society of America*, *116*(3), 1793–1799. doi:10.1121/1.1758973

Schellenberg, E. G. (2015). Music training and speech perception: A gene-environment interaction. *Annals of the New York Academy of Sciences*, *1337*, 170–177. doi:10.1111/nyas.12627

Schellenberg, E. G., & Trehub, S. E. (2003). Good pitch memory is widespread. *Psychological Science*, *14*(3), 262–266. doi:10.1111/1467-9280.03432

Seesjarvi, E., Sarkamo, T., Vuoksimaa, E., Tervaniemi, M., Peretz, I., & Kaprio, J. (2016). The nature and nurture of melody: A twin study of musical pitch and rhythm perception. *Behavior Genetics*, *46*(4), 506–515. doi:10.1007/s10519-015-9774-y

Stewart, L. (2008). Fractionating the musical mind: Insights from congenital amusia. *Current Opinion in Neurobiology, 18*(2), 127–130. doi:10.1016/j.conb.2008.07.008

Tan, Y. T., McPherson, G. E., Peretz, I., Berkovic, S. F., & Wilson, S. J. (2014). The genetic basis of music ability. *Frontiers in Psychology, 5*, 658. doi:10.3389/fpsyg.2014.00658

Theusch, E., Basu, A., & Gitschier, J. (2009). Genome-wide study of families with absolute pitch reveals linkage to 8q24.21 and locus heterogeneity. *American Journal of Human Genetics, 85*(1), 112–119. doi:10.1016/j.ajhg.2009.06.010

Theusch, E., & Gitschier, J. (2011). Absolute pitch twin study and segregation analysis. *Twin Research and Human Genetics, 14*(2), 173–178. doi:10.1375/twin.14.2.173

Trainor, L. J., & Unrau, A. (2012). Development of pitch and music perception. *Human Auditory Development, 42*, 223–254. doi:10.1007/978-1-4614-1421-6_8

Trehub, S. E. (2003). The developmental origins of musicality. *Nature Neuroscience, 6*(7), 669–673. doi:10.1038/Nn1084

Ukkola-Vuoti, L., Kanduri, C., Oikkonen, J., Buck, G., Blancher, C., Raijas, P., et al. (2013). Genome-wide copy number variation analysis in extended families and unrelated individuals characterized for musical aptitude and creativity in music. *PLoS One, 8*(2), e56356. doi:10.1371/journal.pone.0056356

Ukkola-Vuoti, L., Oikkonen, J., Onkamo, P., Karma, K., Raijas, P., & Jarvela, I. (2011). Association of the arginine vasopressin receptor 1A (AVPR1A) haplotypes with listening to music. *Journal of Human Genetics, 56*(4), 324–329. doi:10.1038/jhg.2011.13

Ukkola, L. T., Onkamo, P., Raijas, P., Karma, K., & Jarvela, I. (2009). Musical aptitude is associated with AVPR1A-haplotypes. *PLoS One, 4*(5), e5534. doi:10.1371/journal.pone.0005534

Ullén, F., Mosing, M. A., Holm, L., Eriksson, H., & Madison, G. (2014). Psychometric properties and heritability of a new online test for musicality, the Swedish Musical Discrimination Test. *Personality and Individual Differences, 63*, 87–93. doi:10.1016/j.paid.2014.01.057

Vinkhuyzen, A. A., van der Sluis, S., Posthuma, D., & Boomsma, D. I. (2009). The heritability of aptitude and exceptional talent across different domains in adolescents and young adults. *Behavior Genetics, 39*(4), 380–392. doi:10.1007/s10519-009-9260-5

Vitouch, O. (2003). Absolutist models of absolute pitch are absolutely misleading. *Music Perception, 21*, 111–117. doi:10.1525/mp.2003.21.1.111

Weisman, R. G., Mewhort, D. J. K., Hoeschele, M., & Sturdy, C. B. (2012). New perspectives on absolute pitch in birds and mammals. In E. A. Wasserman & T. R. Zentall (Eds.), *Handbook of comparative cognition* (pp. 67–82). New York: Oxford University Press.

Zatorre, R. J., Perry, D. W., Beckett, C. A., Westbury, C. F., & Evans, A. C. (1998). Functional anatomy of musical processing in listeners with absolute pitch and relative pitch. *Proceedings of the National Academy of Sciences, 95*(6), 3172–3177. doi:10.1073/pnas.95.6.3172

Ziv, N., & Radin, S. (2014). Absolute and relative pitch: Global versus local processing of chords. *Advances in Cognitive Psychology/University of Finance and Management in Warsaw, 10*(1), 15–25. doi:10.2478/v10053-008-0152-7

IV

STRUCTURE, AFFECT, AND HISTORY

11

Formal Models of Structure Building in Music, Language, and Animal Song

Willem Zuidema, Dieuwke Hupkes, Geraint A. Wiggins, Constance Scharff, and Martin Rohrmeirer

Human language, music, and the complex vocal sequences of animal songs constitute ways of sonic communication that evolved a remarkable degree of structural complexity, for which—extensive research notwithstanding—completely satisfactory explanatory and descriptive models have yet to be found. Formal models of structure have been most commonly proposed in the field of natural language, often building on the foundational work of Shannon and Chomsky in the 1940s and 1950s (Chomsky, 1956; Shannon, 1948). Research in mathematical and computational linguistics has resulted in extensive knowledge of the formal properties of such models, as well as of their fit to phenomena in natural languages. Such formal methods have been much less prevalent in modeling music and animal song. Research using formal models of sequential structure has often focused on comparing the structure of human language to that of learned animal songs, focusing particularly on songbirds, but also whales and bats (e.g., Bolhuis, Okanoya, & Scharff, 2010; Doupe & Kuhl, 1999; Hurford, 2007; Knoernschild, 2014). Such comparisons addressed aspects of phonology (e.g., Spierings & ten Cate, 2014; Yip, 2006) and syntax (e.g., Berwick, Okanoya, Beckers, & Bolhuis, 2011; Markowitz, Ivie, Kligler, & Gardner, 2013; Sasahara, Cody, Cohen, & Taylor, 2012; ten Cate, Lachlan, & Zuidema, 2013), aiming to identify both species-specific principles of structure building and cross-species principles underlying the sequential organization of complex communication. In such comparisons, there has been a big focus on the role of recursion as a core mechanism of the language faculty in the narrow sense (Hauser, Chomsky, & Fitch, 2002) and its uniqueness to both humans and human language.

However, although recursion and the potential uniqueness of other features of language are important topics, it is certainly not the only relevant topic for comparative studies (see also Fitch, 2006; Rothenberg, Roeske, Voss, Naguib, & Tchernichovski, 2014). Structurally and functionally, music, language, and animal song not only share certain aspects but also have important differences (Asano & Boeckx, 2015). A three-way comparison of language, music, and animal songs and the techniques that are used to model and explain them has the potential to benefit research in all three domains by highlighting shared and unique mechanisms, as well as hidden assumptions in current research paradigms.

In this chapter, we present an overview of research considering structure building and sequence generation in language, music, and animal song. Our starting point is the work of Shannon and Chomsky from the 1940s and 1950s, which has been prominent in establishing a tradition of research in formal models of structure in natural language. We discuss issues concerning building blocks, Shannon's *n*-gram models, and the Chomsky hierarchy (CH), as well as the limitations of both frameworks in relation to empirical observations from the biological and cognitive sciences. We then proceed with discussing ways of addressing these limitations, including extending the CH with more fine-grained classes, the addition of probabilities and meaning representations to symbolic grammars, and replacing abstract symbols with numerical vectors. At the end of the chapter, we reflect on what type of conclusions can be drawn from comparing and using these models and what impact this may have for further research.

Building Blocks and Sequential Structure

Models for sequence generation highly depend on the choice of atomic units of the sequence. Before considering models of structure building, we may first want to identify what the elementary building blocks are that sequences—be it in language, music, or animal vocalizations—are built up from. This, however, turns out to be much more complicated than we might naively expect.

Elementary Units of Models of Language

One of the classical universal design features of human language is duality of patterning (Hockett, 1960), which refers to the fact that all languages show evidence of at least two combinatorial systems: one where meaningless units of sounds are combined into words and morphemes and one where those meaningful morphemes and words are further combined into words, phrases, sentences, and discourse. Although the two systems are not independent, in this chapter we pragmatically focus mainly on the second combinatorial system, which combines already meaningful units into larger pieces, because this is the target of the most heavily studied models of structure building in natural language. Later in the chapter, we briefly consider the interplay between the two systems.

But even when restricting ourselves to meaning-carrying units, it turns out to be far from trivial to identify phoneme, morpheme, syllable, or word boundaries based on cues in the observable signal (i.e., a spectrogram) alone (Liberman, Cooper, Shankweiler, & Studdert-Kennedy, 1967). The choice of elementary units of models for structure in language is therefore usually not based on features of the acoustic information but, rather, on semantic information accessible through introspection. Most commonly, models considering structure in language are defined over words.

Building Blocks of Animal Song

Like language, animal songs combine units of sound into larger units in a hierarchical way, but the comparability of the building blocks and the nature of the hierarchical structure in language, music, and animal song is not at all straightforward. In particular, there are no clear analogues for words, phrases, or even sentences in animal song (Besson, Frey, & Aramaki, 2011; Scharff & Petri, 2011), and regardless of the approach taken to establish the smallest unit of the sequence (in birdsong commonly referred to with the term *note* or *element*), making decisions that are somewhat arbitrary seems unavoidable. A common way of identifying units in animal songs is to study their spectrogram and delineate units based on acoustic properties such as silent gaps (e.g., Adam et al., 2013; Fehér, Wang, Saar, Mitra, & Tchernichovski, 2009; Isaac & Marler, 1963; Marler & Pickert, 1984) or changes in the acoustic signal (e.g., Clark & Feo, 2008; Payne & McVay, 1971). In addition, evidence about perception and production of different acoustic structures is often used to motivate a particular choice of building blocks (e.g., Amador & Margoliash, 2013; Cynx, 1990; Franz & Goller, 2002; Tierney, Russo, & Patel, 2011). Importantly, choices regarding building blocks might also be made by studying patterns of recombination and co-occurrence (Podos, Peters, Rudnicky, Marler, & Nowicki, 1992; ten Cate et al., 2013), an observation that illustrates the interdependence of the choice of building blocks and models for structure building, an issue that we revisit in the section "Function, Motivation, and Context"). For a detailed review of the different methods used to identify units in animal vocalizations, we refer to Kershenbaum et al. (2014).

Basic Elements in Music

In music—aside from the lack of a (compositional) semantic interpretation—the complexity of the musical surface (i.e., an interplay of different features like rhythm, meter, melody, and harmony)[1] leaves an even larger spectrum of possible choices for building blocks. Models can be defined not only over notes or chords but also intervals and durations of notes, or other more complex features could be used as elementary units of the sequence. Traditionally, much of the discussion of structure in music has focused on Western classical music and has built on building blocks of melody, voice leading (e.g., Callender, Quinn, & Tymoczko, 2008; Quinn & Mavromatis, 2011; Tymoczko, 2006), outer voices (e.g., Aldwell, Schachter, & Cadwallader, 2010), harmony (Rohrmeier, 2007, 2011; Winograd, 1968), combinations of harmony and voice leading (Aldwell, Schachter, & Cadwallader., 2010; Lerdahl & Jackendoff, 1983; Neuwirth & Rohrmeier, 2015), or complex feature combinations derived from monophonic melody (Conklin & Witten, 1995; Pearce, 2005), and harmony (Rohrmeier & Graepel, 2012; Whorley, Wiggins, Rhodes, & Pearce, 2013).

The choice of building blocks is thus a difficult issue in all three domains we consider, and any choice will have important consequences for the models of structure that can be

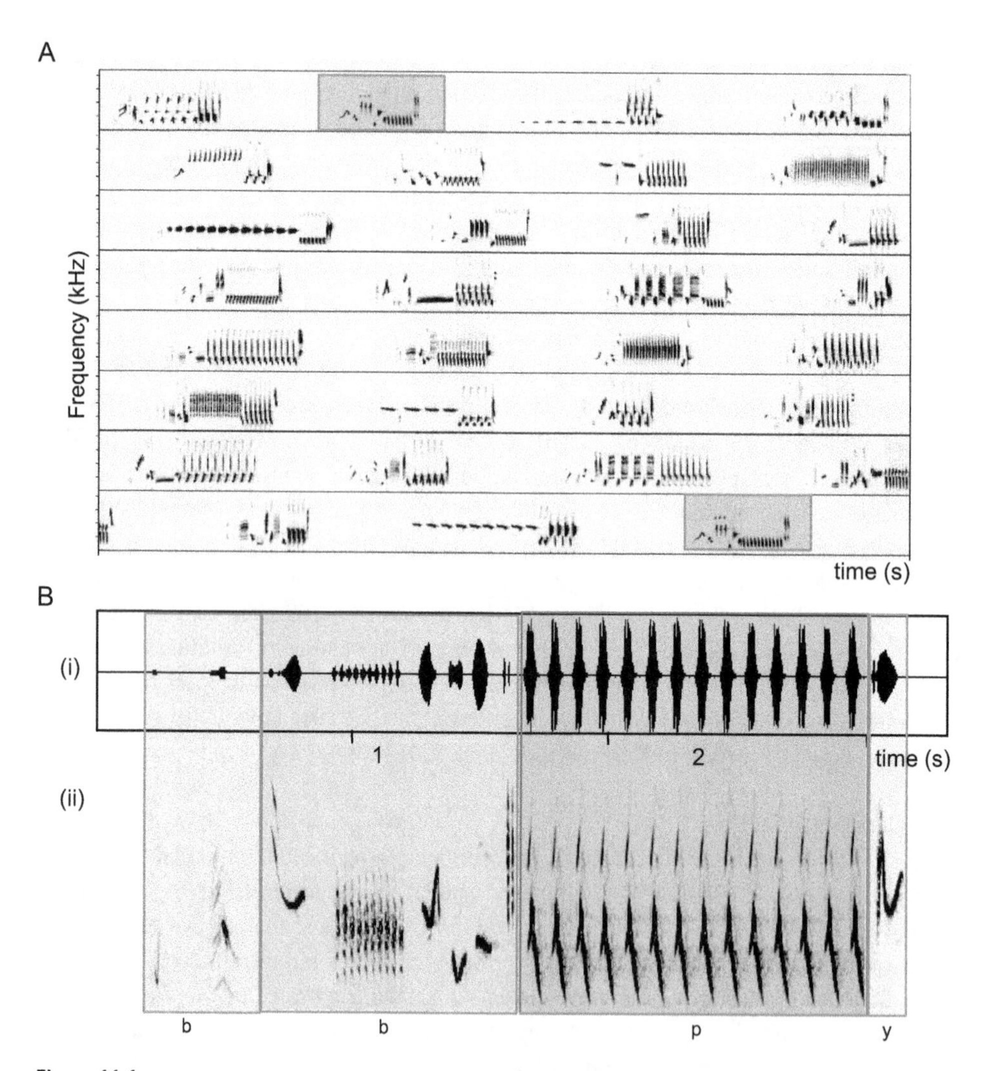

Figure 11.1
Hierarchical organization of nightingale song. (A) Spectrogram of about two minutes of continuous nocturnal singing of a male nightingale. Shown are thirty sequentially delivered unique songs. The thirty-first song is the same song type as the second one (highlighted by a box). On average, a male has a repertoire of 180 distinct song types, which can be delivered in variable but nonrandom order for hours continuously. (B) The structural components of one song type. Individual sound elements are sung at different loudnesses, represented by different heights of the amplitude envelope (i) and are acoustically distinct in the frequency range, modulation, emphasis, and temporal characteristics, represented in (ii). (C) The structural similarities in three different song types (i, ii, iii): Song types begin usually with one or more very softly sung elements, followed by a sequence of distinct individual elements of variable loudness. All song types contain one or more sequences of loud note repetitions (third frame in each of the spectrograms in (i), (ii) and (iii)) and usually end by a single, acoustically distinct element. (D) The same song type can vary in the number of element repetitions in the repeated section. For details, see Hultsch and Todt (1998). Spectrograms modified from Henrike Hultsch (unpublished). (Online version in color.)

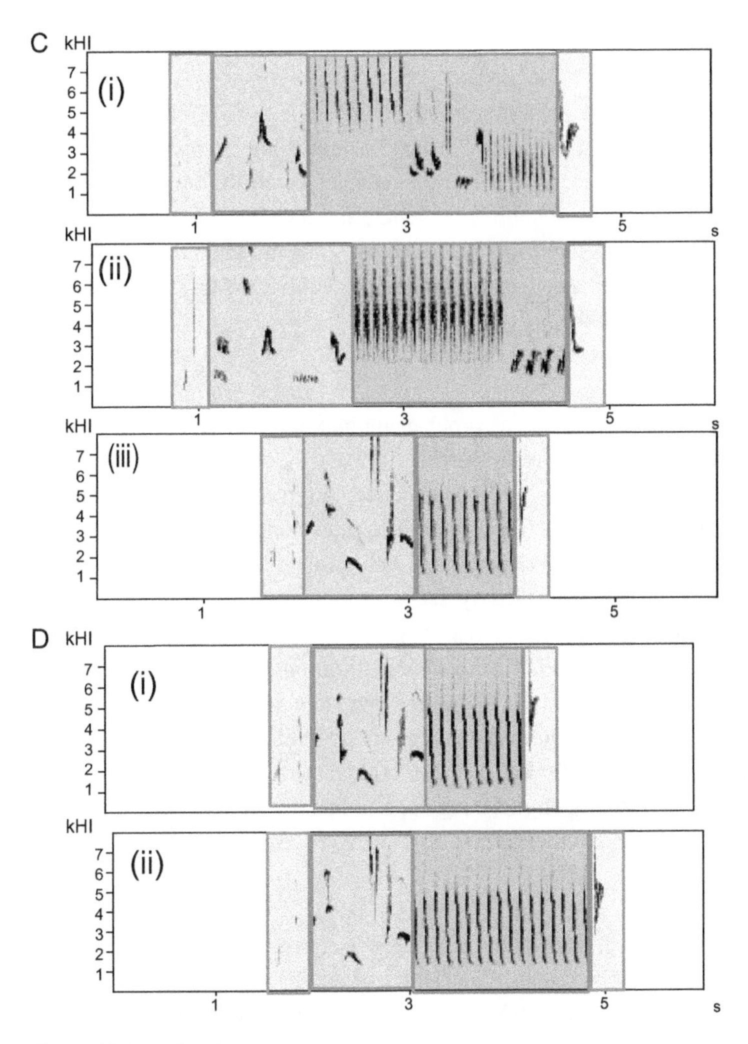

Figure 11.1 continued

defined over these building blocks. The fact that choices regarding the units of comparison may strongly affect the conclusions that can be drawn is frequently overlooked in the literature comparing birdsong, music, and language. Nevertheless, it is often best to make pragmatic decisions about the building blocks in order to move on. As it turns out, some of the questions about building blocks can be addressed only after having considered models of structure (at which point applying model selection, a topic we revisit later in the chapter, can help to revisit choices about the building blocks).

Shannon's *n*-grams

In the slipstream of his major work on information theory, Shannon (1948) introduced *n*-gram models as a simple model of sequential structure in language. An *n*-gram defines the probability of generating the next symbol in a sequence given the previous $(n - 1)$ symbols generated. When $n = 2$, the probability of generating the next word depends only on what the current word is. The *n*-gram model—called a bigram model in this case—then simply models transition probabilities. *n*-gram models are equivalent to $(n - 1)$th-order Markov models over the same alphabet.

Probability Estimation

n-gram probabilities can be estimated from a corpus using maximum likelihood estimation (or relative frequency estimation: Jurafsky & Martin, 2000). In theory, the bigger the value of *n*, the better one can predict the next word in a sentence, but in practice, no natural language corpus is large enough to estimate the probabilities of events in the long Zipfian tail of the probability distribution with relative frequency estimation.[2] When human language is modeled, this problem is usually addressed by decreasing the probability of the counted *n*-grams and reassigning the resulting probability mass to unseen events, a process called smoothing or discounting (Weikum, 2002). Smoothed *n*-gram models have long been the state of the art for assigning probabilities to natural language sentences, and although better-performing language models (in terms of modeling the likelihood of corpora of sentences, e.g., Mikolov, 2012; Schwenk & Gauvain, 2005) have been developed now, *n*-gram models are still heavily used in many engineering applications in speech recognition and machine translation due to their convenience and efficiency.

n-grams Models of Birdsong

n-gram models (often simple bigrams) have also been frequently applied to birdsong (Briefer, Osiejuk, Rybak, & Aubin, 2009; Chatfield & Lemon, 1970; Isaac & Marler, 1963; Markowitz et al., 2013; Okanoya, 2004; Samotskaya, Opaev, Ivanitskii, Marova, & Kvartalnov, 2016; Slater, 1983b) and music (Ames, 1989; Pearce & Wiggins, 2012). For many bird species, bigrams in fact seem to give an adequate description of the sequential structure. Chatfield and Lemon (1970) studied the song of the cardinal and reported that a 3-gram (trigram) model modeled song data only marginally better than a bigram model, measured by the likelihood of the data under each of these models. There is a single, small data set used for extracting *n*-grams and measuring likelihood, which makes drawing firm conclusions from this classic analysis difficult. More recent work with birds that were exposed to artificially constructed songs as they were raised suggests that transitional probabilities between adjacent elements are the most important factor in the organization of the songs also in zebra finches and Bengalese finches (Lipkind et al., 2013), although many other examples of birdsong also require richer models (Katahira, Suzuki, Kagawa,

& Okanoya, 2013; Katahira, Suzuki, Okanoya, & Okada, 2011; ten Cate et al., 2013; ten Cate & Okanoya, 2012).

n-gram Models for Music

In music, numerous variants of *n*-gram models have been used to model musical expectancy (Eerola, 2003; Krumhansl, 1995; Narmour, 1992; Schellenberg, 1996, 1997), but also to account for the perception of tonality and key (which has been argued to be governed by pitch distributions that correspond to the unigram model; Krumhansl, 2004; Krumhansl & Kessler, 1982) and to describe melody and harmony (Conklin & Witten, 1995; Pearce, 2005; Ponsford, Wiggins, & Mellish, 1999; Reis, 1999; Rohrmeier, 2006; Rohrmeier & Graepel, 2012; Whorley et al., 2013). In particular, in the domain of harmony, Piston's (1948) table of common root progressions and Rameau's (1971) theory (of the *basse fondamentale*; may be argued to have the structure of a first-order Markov model (a bigram model) of the root notes of chords (Hedges & Rohrmeier, 2011; Temperley, 2001). In analogy with the findings in birdsong research, several music modeling studies find trigrams optimal with respect to modeling melodic structure (Pearce & Wiggins, 2004) or harmonic structure (Rohrmeier & Graepel, 2012), although here too, the size of the data sets used is too small to draw firm conclusions.

In choosing the optimal value of *n*, some additional aspects that play a role are usually not considered in language and animal song. For instance, because of the interaction of melody with metrical structure, not all surface symbols have the same salience when forming a sequence. This could be an argument to—in the face of data sparsity—prefer a 4-gram model over a 3-gram model to model music with a three-beat metrical structure, as a 3-gram necessarily cannot capture the fact that the first beat of a bar is, in harmonic terms, more musically salient than the other two (Ponsford et al., 1999). More generally, the interaction between different single-stream features in music forms a challenge for *n*-gram models, an aspect that is not as inescapable when modeling language and animal song (but see Ullrich, Norton, & Scharff, 2016, for an interesting, multistream pattern in zebra finch vocalizations and dance). One model that addresses this problem by combining *n*-gram models over different features and combined feature spaces was proposed in Conklin and Witten (1995).

The Classical Chomsky Hierarchy

Shannon's *n*-grams are simple and useful descriptions of some aspects of local sequential structure in animal communication, music, and language. It is, however, often argued that they are unable to model certain key structural aspects of natural language. In theoretical linguistics, *n*-grams, no matter how large their *n*, were famously dismissed as useful models of syntactic structure in natural language in the foundational work of Noam Chomsky from the mid-1950s (Chomsky, 1956). Chomsky argued against incorporating probabilities into language models; in his view, the core issues for linguists concern the symbolic, syntactic structure of language. He proposed an idealization of natural language where a language is conceived of as a potentially infinite set of sentences and a sentence is simply a sequence of words (or morphemes). By systematically analyzing the ways in which such sets of sequences of words could be generated, Chomsky discovered a hierarchy of increasingly powerful grammars, relevant for both linguistics and computer science, that has been named the Chomsky hierarchy (CH).

Four Classes of Grammars and Languages

In its classical formulation, the CH distinguishes four classes of grammars and their corresponding languages: regular languages, context-free languages, context-sensitive languages, and recursively enumerable languages. Each class contains an infinite number of sets and is strictly contained in all classes that are higher up in the hierarchy: every regular language is also context free, every context-free language is also context sensitive, and every context sensitive language is recursively enumerable. When probabilities are stripped off, *n*-grams correspond to a proper subset of the regular languages.

The Chomsky Hierarchy and Cognitive Science

For cognitive science, the relevance of the hierarchy comes from the fact that the four classes can be defined by the kinds of rules that generate structures as well as by the kind of computations needed to parse the sets of sequences in the class (the corresponding formal automaton). Informally, regular languages are the sets of sequences that can be characterized by a flowchart description, which corresponds to a finite-state automaton. Regular languages can be straightforwardly processed (and generated) from left to right in an incremental fashion. Crucially, when generating or parsing the next word in a sentence of a regular language, we need only to know where we currently are on the flowchart, not how we got there (for an example, see figure 11.5).

At all higher levels of the CH, some sort of memory is needed by the corresponding formal automaton that recognizes or generates the language. The next level up in the classical CH are context-free languages (CFLs), generated by context-free grammars (CFGs), equivalent to so-called push-down automata, that employ a simple memory in the form of a stack. CFGs consist of (context-free) rewrite rules that specify which symbols (representing a category of words or other building blocks, or categories of phrases) can be rewritten to which lists of symbols. Chomsky observed that natural language syntax allows for nesting of clauses (center embedding) and argued that finite-state automata are incapable of accounting for such phenomena. In contrast, context-free grammars can express multiple forms of nesting as well as forms of counting elements in a sequence. An example of such nesting, and a context-free grammar that can describe it, is given in figure 11.2.

The sentences *The song the bird sang was beautiful* and *The song the bird the linguists observed sang was beautiful* are examples of sentences with center embedding in English. The latter sentence can be derived from the start symbol S by subsequently applying rules 1, 2a, 2b, 3, 2a, 2c, 2a, 2d, 4c, 4b, 4a. (Note that traditionally, the analysis of the sentence contains a so-called trace connecting the VP to its subject; left out here for clarity.)

(1) S → NP VP
(2a) NP → NP SBAR
(2b) NP → the song
(2c) NP → the bird
(2d) NP → the linguists
(3) SBAR → NP VP
(4a) VP → was beautiful
(4b) VP → sang
(4c) VP → observed

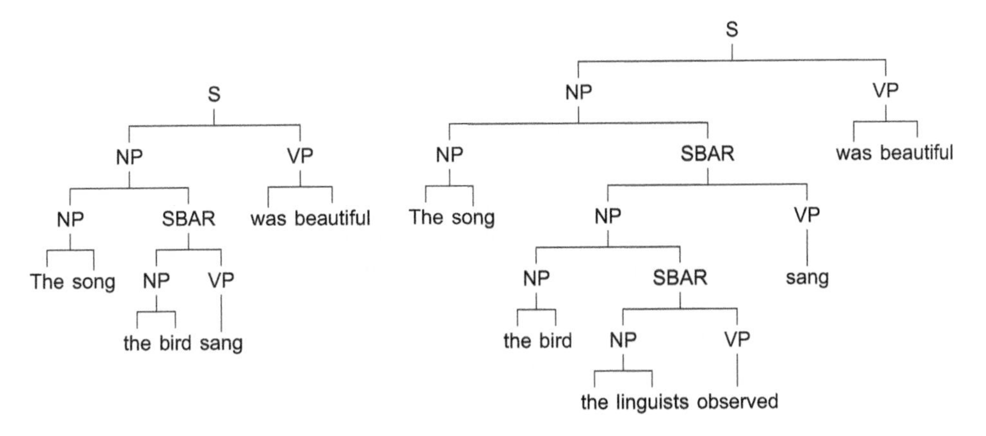

Figure 11.2
Center embedding in English. NP and VP denote noun and verb phrases, respectively, S a declarative clause, and SBAR a clause introduced by a (possibly empty) subordinating conjunction.

Using the Chomsky Hierarchy to Model Music

The success of the CH in linguistics and computer science and Chomsky's demonstration that natural language syntax is beyond the power of finite-state automata has influenced many researchers to examine the formal structures underlying animal song and music (though there is no comprehensive comparison of models in either domain in terms of the CH yet). In music, there appears to be evidence for a number of nontrivial structure-building operations at work that invite an analysis in terms of the CH or related frameworks. While more cross-cultural research is necessary (Cross, 2012a), key structural operations that we can already identify include repetition and variation (Margulis, 2014), element-to-element implication (e.g., note-note, chord-chord; Huron, 2007; Narmour, 1992), hierarchical organization and tree structure, and nested dependencies and insertions (e.g., Jackendoff & Lerdahl, 2006; Widdess & Wolpert, 1981). Most of these operations are more naturally expressed using CFGs than with finite-state-automata, and indeed a rich tradition that emphasizes hierarchical structure, categories, and, particularly, recursive insertion and embedding exists to characterize Western tonal music (Granroth-Wilding & Steedman, 2014; Haas, Rohrmeier, & Wiering, 2009; Keiler, 1983; Keller, 1978; Lerdahl &

Jackendoff, 1983; Narmour, 1992; Neuwirth & Rohrmeier, 2015; Rohrmeier, 2007, 2011; Steedman, 1984, 1996; Winograd, 1968).

However, unlike language, music does not convey propositional, denotational semantics. The function of the proposed hierarchical structures therefore cannot be the communication of a hierarchical, compositional semantics (Slevc & Patel, 2011), and one cannot appeal to semantics or binary grammaticality judgments used (for better or for worse) to make the formal argument that language is trans-finite-state. Rather, a common thread in research about structure in music is that at any point in a musical sequence, listeners are computing expectations about how the sequence will continue regarding timing details and classes of pitches or other building blocks (Huron, 2007; Rohrmeier & Koelsch, 2012). Composers can play with these expectations: meet expectations, violate them, or even put them on hold. In this play with expectations lie both the explanation for the existence of nested context-free structure in music and the way to make a more or less formal argument to place music on the CH. This is because the fact that an event may be prolonged (i.e., extended through another event, an idea originating with Schenker, 1935) and events may be prepared or implied by other events creates the possibility of having multiple and recursive preparations. Employing an event as a new tonal center (musical modulation) could be formally interpreted as an instance of recursive context-free embedding of a new diatonic space into an overarching one (somewhat analogous to a relative clause in language; Hofstadter, 1980; Rohrmeier, 2007, 2011), which provides a motivation of the context-freeness of music through complex patterns of musical tension (Lehne, Rohrmeier, & Koelsch, 2013; Lerdahl & Krumhansl, 2007). Figure 11.3 shows an example of a syntactic analysis of the harmonic structure of a Bach chorale that illustrates an instance of recursive center embedding in the context of modulation.

Although the fact that composers include higher-level structure in their pieces is uncontroversial, whether listeners are actually sensitive to such structures in day-to-day listening is a debated topic (Farbood, Heeger, Marcus, Hasson, & Lerner, 2015; Heffner & Slevc, 2015; Koelsch, Rohrmeier, Torrecuso, & Jentschke, 2013). An alternative potential explanation for the existence of hierarchical structure in music could be found in the notation system and tradition of formal teaching and writing, a factor that may be relevant even for complexity differences in written and spoken languages in communities that may differ with respect to their formal education (Zengel, 1962). However, there are also analytical findings that suggest that principles of hierarchical organization may be found in classical North Indian music (Widdess & Wolpert, 1981) that is based on a tradition of extensive oral teaching. More cross-cultural research on other cultures and structures in more informal and improvised music is required before conclusions may be drawn concerning structural complexity and cross-cultural comparisons.

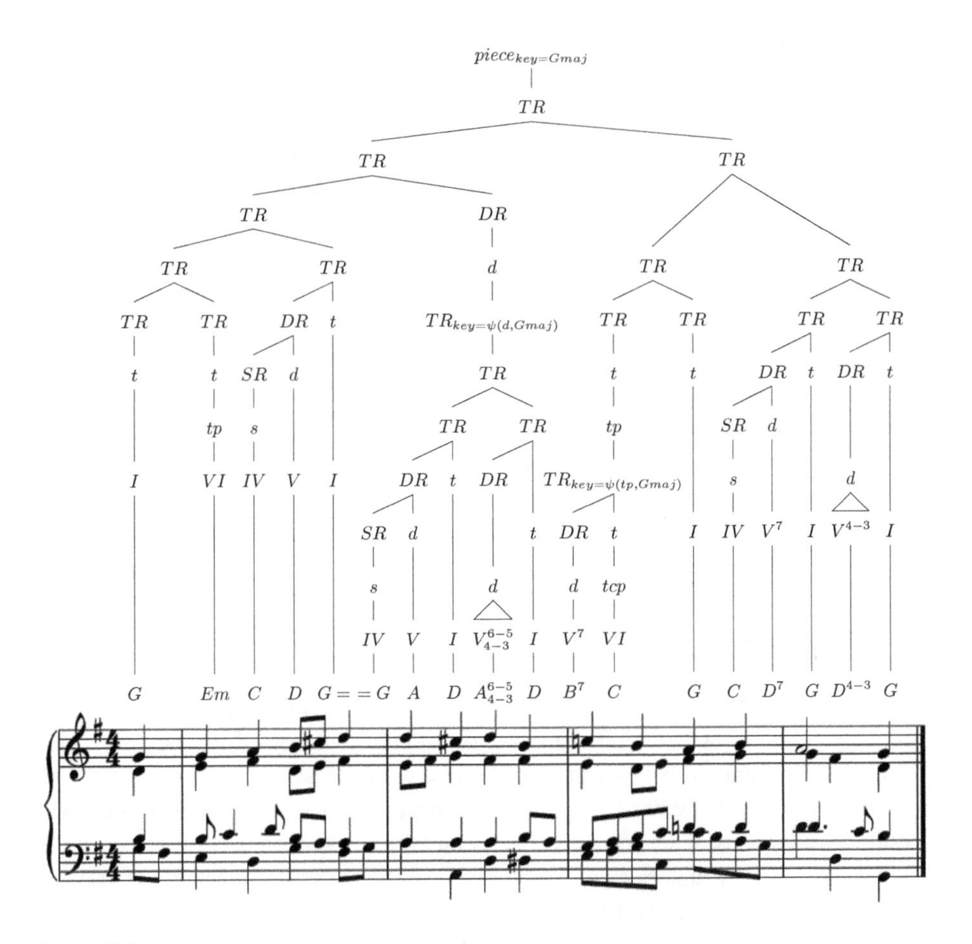

Figure 11.3
Analysis of Bach's chorale "Ermuntre Dich, mein schwacher Geist" according to the GSM proposed by Rohrmeier (2011). The analysis illustrates hierarchical organization of tonal harmony in terms of piece (*piece*), functional regions (*TR, DR, SR*), scale-degree (roman numerals), and surface representations (chord symbols). The analysis further exhibits an instance of recursive center embedding in the context of modulation in tonal harmony. The transitions involving *TRkey* = *ψ*(*x,ykey*) denote a change of key such that a new tonic region (TR) is instantiated from an overarching tonal context of the tonal function *x* in the key *ykey*. One remarkable detail is that the chorale sets up the expectation toward an e minor modulation, as would be standard at this formal position, with the B7 chord. However, the e minor modulation is effectively not realized but continued in a deceptive way with the C chord. Accordingly the subtree expresses an analysis from the alluded e minor key, which derives the B7 chord, with a V–I prototype that then is transformed in a deceptive way by the unary tcp modification of the tonic, following a Riemannian analytical framework.

The Complexity of Animal Vocalizations

In animal vocalizations, there is little evidence that nonhuman structures or communicative abilities (in either production or reception) exceed finite-state complexity. However, a number of studies have examined abilities to learn trans-finite-state structures (e.g., Chen, Jansen, & ten Cate, 2016; Chen, Van Rossum, & Weiss, 2015; Fitch & Hauser, 2004; Lipkind et al., 2013). Claims have been made—and refuted—that songbirds are able to learn such instances of context-free-structures (see Abe & Watanabe, 2011; Gentner, Fenn, Margoliash, & Nusbaum, 2006; and respective responses Beckers, Bolhuis, Okanoya, & Berwick, 2012; Corballis, 2007; ten Cate, 2014; van Heijningen, de Visser, Zuidema, & ten Cate, 2009; Zuidema, 2013a). Hence further targeted research with respect to trans-finite-stateness of animal song is required to shed light on this question. By contrast, a number of studies argue for implicit acquisition of context-free structure (and even [mildly] context-sensitive structure) in humans from abstract stimulus materials from language and music (Jiang et al., 2012; Kuhn & Dienes, 2005; Li, Jiang, Guo, Yang, & Dienes, 2013; Rohrmeier & Cross, 2008; Rohrmeier, Fu, & Dienes, 2012; Rohrmeier & Rebuschat, 2012; Uddén, Ingvar, Hagoort, & Petersson, 2012).

Practical Limitations of the Chomsky Hierarchy

It has turned out to be difficult to decide empirically where to place language, music, and animal song on the CH due to a number of different but related issues. One of the more easily addressable problems concerns the fine-grainedness of the levels of the CH. It was observed in many studies that (plain) *n*-grams are inadequate models of the structure of the vocalizations of several bird species on both the syllable and phrase (e.g., Jin & Kozhevnikov, 2011; Katahira et al., 2011; Markowitz et al., 2013; Okanoya, 2004) and song (Slater, 1983a,b; Todt & Hultsch, 1996) level. Although *n*-gram models seem to suffice for modeling songs of, for instance, mistle thrushes (Chatfield & Lemon, 1970; Isaac & Marler, 1963) and zebra finches (Zann, 1993), richer models are needed to characterize the vocalizations of Bengalese finches (e.g., Katahira et al., 2011, 2013), blackbirds (ten Cate et al., 2013; Todt, 1975), and other birds singing complex songs (see Kershenbaum et al., 2014; ten Cate & Okanoya, 2012, for a review). However, this difference in complexity is not captured by the CH, as the richer models proposed (e.g., hidden Markov models; Rabiner & Juang, 1986) are still finite-state models that fall into the lowest complexity class of the CH: regular languages.

A similar issue occurs on higher levels of the hierarchy, when one tries to establish the formal complexity of natural languages such as English. It was noticed already in the 1980s that some natural languages seem to display structures that are not adequately modeled by CFGs (Culy, 1985; Huybregts, 1984; Shieber, 1985). However, the class of context-sensitive languages—one level up in the hierarchy—subsumes a much larger set of complex generalizations, many of which are never observed in natural language.

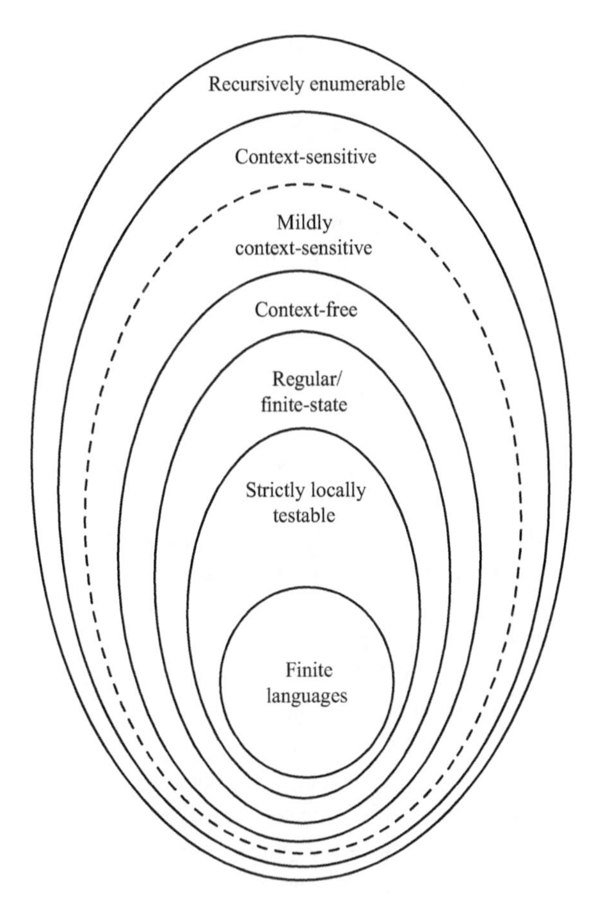

Figure 11.4
The extended Chomsky hierarchy. Note that area sizes in this diagrams do not reflect the actual sizes of the classes they represent; for example, the class of recursively enumerable languages is much larger than the class of context-sensitive languages.

Adding Extra Classes

Both issues can be addressed by extending the CH with more classes. At the lowest level of the hierarchy, Rogers and colleagues (Jäger & Rogers, 2012) describe a hierarchy of subregular languages that contains the set of strictly local (SL) languages, which constitute the nonprobabilistic counterpart of *n*-gram models. At the high end of the hierarchy, Joshi, Vijay-Shanker, and Weir (1991) looked at a number of formalisms that were proposed to address the inadequacy of context-free grammars to model natural language (e.g., tree-adjoining grammars [Joshi et al., 1975] and combinatorial categorical grammar [Steedman, 2000]),[3] and pointed out that a number of linguistic formalisms are formally equivalent with respect to the class of languages they are describing. These languages, collectively referred to as mildly context-sensitive languages (MCSLs: Joshi, 1985), can be roughly characterized by the fact that they are a proper superset of context-free languages,[4] can be parsed in polynomial time, capture only certain kinds of dependencies, and have constant growth property (Joshi et al., 1991).

Empirically Establishing the Complexity of Different Languages

Orthogonal to this granularity problem is the more difficult problem of empirically evaluating membership of a class on the (extended) CH. The mere fact that a set of sequences can be built by a grammar from a certain class does not constitute a valid form of argument to place the system in question at that level on the (extended) CH. This is because a system lower in the hierarchy can approximate a system higher in the hierarchy with arbitrary precision by simply listing instances. For instance, Markov models are successfully used to describe some statistical features of corpora of music (e.g., De Clercq & Temperley, 2011; Huron, 2007; Rohrmeier & Cross, 2008; Tymoczko & Meeùs, 2003), but crucially, this fact does not imply that a Markov model is also the best model. The best model might be a context-free model that involves a single rule that captures a generalization not captured by many specific nodes in the HMM.

To drive home this point further, consider again the example of center embedding (described in the "The Classical Chomsky Hierarchy"). Note that arbitrarily deep center embeddings do not occur in practice (and even center embeddings of very limited depth are rarely observed and are shown to be incomprehensible for most humans; see, e.g., Miller & Isard, 1964; Stolz, 1967). In any real-world data, there is thus always only a finite (and, in fact, relative small) number of center embeddings; this finite set is easily modeled with a finite-state automaton that contains a different state for each depth. A finite-state account, however, loses a generalization: different states lead to the same types of sequences, and we suddenly have a strict upper bound on the depth of possible embeddings.

A similar issue arises when long-distance dependencies are used to prove the inadequacy of finite-state models. For instance, when a bird sings songs of the structure AB^nC and DB^nE, a long-distance dependency between A and C and between D and E can be observed, but the songs can be easily modeled with finite-state automata (see figure 11.5) by just assuming two different (hidden) states from which the Bs are generated: one for the condition starting with A and ending with C, and one for the other. This explains why some efforts to empirically demonstrate the context freeness of birdsong or music may not be convincing from a formal language theory perspective if they are based on just demonstrating a long-distance dependency. However, a long-distance dependency does have consequences for the underlying model that can be assumed in terms of its strong generative capacity (i.e., the set of structures it can generate) and compressive power: in the example shown in figure 11.5, we were forced to duplicate the state responsible for generating B; in fact we require $2m$ states (where m is the number of nonlocal dependency pairs, such as $A ... C$ or $D ... E$, that need to be encoded). Therefore, if there are multiple (finite),

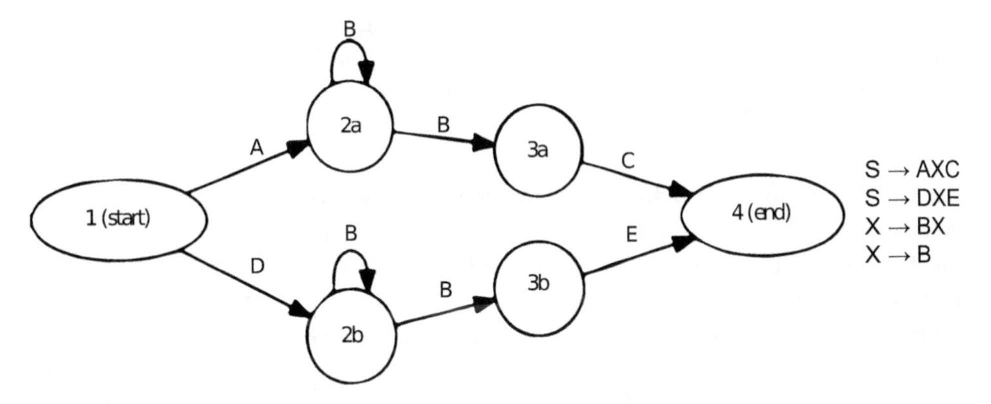

Figure 11.5
A finite-state automaton and a context-free grammar generating a repertoire consisting of two sequences: AB^nC and DB^nE (with $n > 0$). Note that the finite-state automaton is redundant in comparison with the CFG in the way that it contains multiple instances of the same structure B^n.

potentially nested nonlocal dependencies, the number of required states grows exponentially, which is arguably unsatisfactory when considering strong generative capacity arguments (see also the comparable argument regarding the implicit acquisition of such structures in Rohrmeier, Dienes, Guo, & Fu, 2014; Rohrmeier et al., 2012). If the intervening material in a long-distance dependency is very variable, even if not technically unbounded, considerations of parsimony and efficiency provide strong reasons to prefer a model other than the minimally required class in the CH, or a different type of model altogether.

Learnability

A third type of problem concerns the learnability of certain types of structures from examples and the complexity of the inference process that this requires. A common paradigm to probe the type of generalizations made by humans and other species is to generate a sequence of sentences from an underlying grammar and study how well test subjects learn them in an artificial grammar learning (AGL) experiment (Pothos, 2007; Reber, 1967). Such experiments show robustly, for instance, that humans are able to distinguish grammatical from ungrammatical sentences generated by a CFG without having very explicit knowledge of the rules they are using to make this distinction (Fitch, Friederici, & Hagoort, 2012). Many examples of AGL experiments with birds (not even necessarily songbirds; see, e.g., Herbranson & Shimp, 2008, for an AGL study with pigeons) and other animals (such as rats or monkeys; Murphy, Mondragón, & Murphy, 2008; Wilson et al., 2013, respectively) can be found in the literature.

The results of AGL experiments remain difficult to interpret, as the inference procedures (and their complexity) to learn even relatively simple structures from examples are not well understood. Reviews of the ample number of AGL studies with both humans and nonhuman animals, as well as more formal accounts of their interpretability, are given by Fitch et al. (2012), ten Cate (2016), and Fitch and Friederici (2012) and Pothos (2007), respectively.

Relating the Chomsky Hierarchy to Cognitive and Neural Mechanisms

Another class of arguments to move beyond the confines of the CH comes from considering its relation with cognition. Historically, the CH is a theoretical construct that organizes types of structures according to different forms of rewrite rules and has little immediate connection with cognitive motivations or constraints (such as limited memory). And although it has been extensively used in recent debates on human and animal cognitive capacities, the CH may be quite fundamentally unsuitable for informing cognitive or structural models that capture frequent structures in language, music, and animal song. This may be a surprising claim given the long and proud history of the Chomsky hierarchy, the fact that its classes are organized in mutual superset relations and the fact that the top level contains all recursively enumerable languages: everything has its place on the Chomsky hierarchy. However, all the mathematical sophistication of the classes on the CH does not motivate their reification in terms of mental processes, cognitive constraints, or neural correlates. A metaphor might help drive this point home. All squares on a chessboard can be reached by a knight in a finite number of steps. We can therefore compute the distance between two fields in terms of the number of moves a knight needs. This metric is universal in some sense (it applies to any two squares), but in general it is unhelpful because its primitive operation (the knight's jump) is not representative for other chess pieces. Similarly, the CH's metric of complexity is universal, but its usefulness is restricted by the primitive operations (rewrite operations) it assumes.

A well-known issue that illustrates this point is the fact that repetition, repetition under a modification (such as musical transposition) and cross-serial dependencies constitute types of structures that require complex rewrite rules (see also the example of context-sensitive rewrite rules expressing cross-serial dependencies in Rohrmeier et al., 2014), while such phenomena, in contrast, are frequent forms of form building in music and animal song and much more naturally modeled in systems that have repeat as a primitive operation. Mechanisms that can recognize and generate context-free languages are not limited to rewrite rules or even phrasal constituents (consider, e.g., dependency grammars in Tesnière, 1966, that describe the structure of a string in terms of binary connections between its elements), and the mismatch between the simplicity of repetitive structures and the high CH class it is mapped onto might be one of many motivations to consider different types of models.

Other motivations to move beyond the confinements of the CH lie in the modeling of real-world structures that undermine some of the assumptions of the CH. Generally the observation that music involves not only multiple parallel streams of voices but also correlated streams of different features and complex timing constitutes a theme that receives considerable attention in the domain of music cognition, but it does not easily match with the principles that underlie the CH, which is based on modeling a single sequence of words. Similarly, one can argue that the CH is incapable of dealing with several essential features and characteristics of language, such as the fact that language is primarily used to convey messages with a complex semantic structure and the gradedness of syntactic acceptability (Aarts, 2004; Sorace & Keller, 2005).

In summary, the CH does not constitute an inescapable a priori point of reference for all kinds of models of structure building or processing, but it has inspired research in terms of a framework that allowed the comparison of different models and formal negative arguments against the plausibility of certain formal languages or corresponding computational mechanisms. Such formal comparison and proofs should inspire future modeling endeavors, yet better forms of structural or cognitive models may involve distinctions orthogonal to the CH and may be designed and evaluated in the light of modeling data and its inherent structure as well as possible.

Moving Toward Different Types of Models

Considering the challenges we have mentioned, what are some different aspects that new models of structure building and corresponding cognitive models should take into account? In the slew of possible desiderata for new models, we observe two categories of requirements that such models should address.

Modeling Observed Data

The first category regards the suitability of models to deal with the complexity of actual real-world structures, which includes being able to deal with graded syntactic acceptability but also handling semantics and form-meaning interactions. One main aspect that is particularly relevant is the notion of grammaticality or well-formedness. The CH relies quite strongly on this notion for establishing and testing symbolic rules, but the idea that grammaticality is a strictly binary concept is problematic in the light of real-world data. Even if the underlying system would prescribe so in theory, models should be able to account for the fact that in practice, grammaticality is graded rather than binary (e.g., Abney, 1996). In the case of music, it is not clear whether ungrammatical or irregular structures are clear-cut or distinguished in agreement by nonexpert or expert subjects. This problem is even more prominent in the domain of animal songs, where introspection cannot be used to assess the grammaticality of sequences or the salience of proposed structures and research can typically be based only on so-called positive data—examples conforming with the proposed rules. It is significantly more difficult to establish the validity and extension of rules in the absence of negative data—where humans or animals explicitly reject a malformed sequence—which is hard to obtain in case of animal research.

Evaluation and Comparison

A second category of requirements concerns the evaluation and comparison of different models. As we have pointed out, the CH is not particularly useful for selecting or even distinguishing models based on empirical data, as it provides no means to quantify the fit of a certain model with observed data. To overcome this problem, new models should include some mechanism that allows the modeler to evaluate which model better describes experimental data, for instance, by evaluating the agreement of their complexity judgments with empirical findings from the sentence processing literature (e.g., Gibson & Thomas, 1999), their assessment of the likelihood of observed or made-up sequences, or by their predictive power. The last method of evaluating models seems particularly suitable for music, where empirical data are often focused around the expectations listeners are computing about how the sequence will continue. Further considerations to prefer one model over another could be grounded in descriptive parsimony or minimum description length (Mavromatis, 2009).

In the remainder of this chapter, we discuss three important extensions of the CH that address some of the previously mentioned issues.

Dealing with Noisy Data: Adding Probabilities

An important way to build better models of cognition and deal with issues from both categories comes from reintroducing the probabilities that Chomsky abandoned along with his rejection of finite-state models. A hierarchy of probabilistic grammars can be defined that is analogous to the classical (and extended) CH and exhibits the same expressive power. We already mentioned that augmenting the automata generating SL languages yields *n*-gram models, whereas the probabilistic counterpart of a finite state automatonnis a hidden Markov model (HMM). Similarly, CFGs and CSGs can be straightforwardly extended to probabilistic CFGs (PCFGs) and probabilistic CSGs (PCSGs), respectively.

Adding probabilities to the grammars defined in the CH addresses many of the issues we have mentioned. Probabilistic models can deal with syntactic gradience by comparing the likelihood of observing particular sentences, songs, or musical structures (although accounting for human graded grammaticality judgments is not easy; see, e.g., Lau, Clark, & Lappin, 2015). Furthermore, they lend themselves well to information-theoretic methodologies such as model comparison, compression, or minimum description length (Grünwald, 2007; Mackay, 2003). Probabilities allow us to quantify degrees of fit and thus select models in a Bayesian model comparison paradigm by selecting the model with the posterior probability given the data and prior beliefs or requirements. In addition, probabilistic models permit defining a probability distribution over possible next words, notes, or chords in a sequence, which matches well with many experimental data about sentence and music processing.

The use of probabilistic models is widespread in music, language, and animal song. Aside from *n*-gram models, frequently applied in all three domains, more expressive probabilistic models have also been widely used. Pearce's IDyOM model, an extension of the multiple feature *n*-gram models proposed by Conklin and Witten (1995), has been shown to be successful in the domains of both music and language (Pearce & Wiggins, 2012; Wiggins, 2012). Recent modeling approaches generalized the notion of modeling parallel feature streams into dynamic Bayesian networks that combine the advantages of HMMs with modeling feature streams (Murphy, 2002; Paiement, 2008; Raczynski, Vincent, & Sagayama, 2013; Rohrmeier & Graepel, 2012).

In general, HMMs—which assume that the observed state is generated by a sequence of underlying (hidden) states that emit surface symbols according to a given probability distribution (for a comprehensive tutorial, see Rabiner, 1989)—have been used extensively to model sequences in music (e.g., Mavromatis, 2005; Raphael & Stoddard, 2004; Rohrmeier & Graepel, 2012) and animal song (e.g., Jin & Kozhevnikov, 2011; Katahira et al., 2011).

HMMs are also frequently practiced in modeling human language, although their application is usually limited to tasks regarding more shallow aspects of structure, such as part-of-speech tagging (e.g., Brants, 2000) or speech recognition (e.g., Juang & Rabiner, 1991; Rabiner & Juang, 1993). For modeling structural aspects of natural language, researchers usually resort to probabilistic models higher up the hierarchy, such as PCFGs (e.g., Petrov & Klein, 2007), lexicalized tree-adjoining grammars (Joshi et al., 1991), or combinatory categorial grammars (Steedman, 2000).

Dealing with Meaning: Adding Semantics

One crucial aspect of human language of language that does not play a role in the CH is semantics. Chomsky's original work stressed the independence of syntax from semantics, but that does not mean that semantics is unimportant for claims about human uniqueness or structure building operations in language, even for linguists working within a Chomskyan paradigm. Berwick et al. (2011), for instance, use the point that birdsong crucially lacks underlying semantic representations to argue against the usefulness of birdsong as a comparable model system for human language. Their argument is that in natural language the trans-finite-state structure is not some idiosyncratic feature of the word streams we produce, but something that plays a key role in mediating between thought (the conceptual-intentional system in Chomsky's terms) and sound (the articulatory-perceptual system). Note that while the relevance of the interlinkedness of thought and sound in language is an important point, we are not sure on which evidence Berwick et al. (2011) ground their statement that birdsong lacks semantic representations.

Transducers

Crucially, the conceptual-intentional system is also a hierarchical, combinatorial system (most often modeled using some variety of symbolic logic, most famously the system of Montague, 1970). From that perspective, grammars from the (extended) CH describe only half of the system; a full description of natural language would involve a transducer that maps meanings to forms and vice versa (e.g., Jurafsky & Martin, 2000; Zuidema, 2013b). For instance, finite-state grammars can be turned into finite-state transducers and context-free grammars into synchronous context-free grammars. All of the classes of grammars in the CH have a corresponding class of transducers (see Knight & Graehl, 2005, for an overview). Depending on the type of interaction we allow between syntax and semantics, there might or might not be consequences for the set of grammatical sentences that a grammar allows if we extend the grammar with semantics. In any case, the extension is relevant for assessing the adequacy of the combined model—for example, we can ask whether a particular grammar supports the required semantic analysis—as well as for determining the likelihood of sentences and alternative analyses of a sentence.

Semantics in Music

Whether we need transducers to model structure building in animal songs and music is a question that remains to be answered. There have been debates about forms of musical meaning and its neurocognitive correlates. A large number of researchers in the field agree that music may feature simple forms of associative meaning and connotations, as well as illocutionary forms of expression, but it lacks kinds of more complex forms of combinatorial semantics (see the discussion of Davies, 2011; Fitch & Gingras, 2011; Koelsch, 2011; Reich, 2011; Slevc & Patel, 2011). However, it is possible to conceive of complex forms of musical tension that involve nested patterns of expectancy and prolongation as an abstract secondary structure and motivate syntactic structures at least in Western tonal music, and in analogy would require characterizing a transducer mapping syntactic structure and corresponding structures of musical tension in future research.

Semantics in Animal Song

Similarly, there have been debates about the semantic content of animal communication. There are a few reported cases of potential compositional semantics in animal communication (e.g., Arnold & Zuberbühler, 2012), but these concern sequences of only two elements and thus do not come close to needing the expressiveness of finite-state or more complex transducers. For all animal vocalizations that have nontrivial structure, such as the songs of nightingales (Weiss, Hultsch, Adam, Scharff, & Kipper, 2014), blackbirds (ten Cate et al,, 2013; Todt, 1975), pied butcher-birds (Taylor & Lestel, 2011), or humpback whales (Payne & Payne, 1985; Payne & McVay, 1971), it is commonly assumed that no combinatorial semantics underlies it. However, it is important to note that the ubiquitous claim that animal songs do not have combinatorial, semantic content is actually based on few to no experimental data. As long as the necessary experiments are not designed and performed, the absence of evidence of semantic content should not be taken as evidence of absence.

If animal songs do indeed lack semanticity, they would be more analogous to human music than to human language. The analogy to music would then not primarily be based on the surface similarity to music on the level of the communicative medium (use of pitch, timbre, rhythm, or dynamics), but on functional considerations such as that they do not constitute a medium to convey types of (propositional) semantics or simpler forms of meaning, but are instances of comparably free play with form and displays of creativity (Wiggins, Tyack, Scharff, & Rohrmeier, 2015).

A Music-Language Continuum?

Does this view on music-animal song analogies have any relevance for the study of language? There are reasons to argue it does, because music and human language may be regarded as constituting a continuum of forms of communication that is distinguished in terms of specificity of meaning (Brown, 2001; Cross, 2012b; Cross & Woodruff, 2010). Consider, for instance, several forms of language that may be considered closer to a "musical use" in terms of their use of pitch, rhythm, meter, and semantics, such as motherese, prayers, mantras, poetry, and nursery rhymes, as well as perhaps forms of the utterance "huh" (see Dingemanse, Torreira, & Enfield, 2014).

Animal vocalizations may be motivated by forms of meaning (that are not necessarily comparable with combinatorial semantics), such as expressing aggression or submission, warning of predators, group cohesion, or social contagion, or they may constitute free play of form for display of creativity—for instance (but not necessarily), in the context of reproduction. Given that structure and structure building moving from the language end to the music end is less constrained by semantic forms, more richness of structural play and creativity is expected to occur on the musical side (see chapter 12, this volume).

Dealing with Gradations: Adding Continuous-Valued Variables

An entirely different approach to modeling natural language, parallel to the symbolic one employed by the Chomsky hierarchy, is one where the symbols and categories of the CH are replaced by vectors and the rules are projections in a vector space (implicitly) defined in matrix vector algebra. Thus, instead of having a rule X → Y Z, where X, Y, and Z are symbolic objects (such as a prepositional phrase [PP] in linguistics, or a motif in a zebra finch song), we treat X, Y, and Z as *n*-dimensional vectors of numbers (which can be binary, integer, rational, or real numbers; for example [0, 1, 0 …] or [0.453, 0.3333, –0.211, …]) and → becomes an operation on vectors that describes how the vector for *Z* can be computed given the vectors for *X* and *Y*. Vector grammars offer a natural way to model similarity between words and phrases, which can be defined as their distance in the vector space. Consequently, as one can compute how close a vector is to its prototypical version, vector grammars can straightforwardly deal with noisy data and exhibit a gradual decrease of performance when inputs become longer or noisier, properties that are attractive for cognitive models of language.

Vector Grammars and Connectionism

Vector grammars bear a close relation to connectionist neural network models of linguistic structure that were introduced in the 1990s (Elman, 1990; Pollack, 1990). After being practically abandoned as models of linguistic structure for over a decade, neural networks are experiencing a new wave of excitement in computational linguistics, following some successes with learning such grammars from data for practical natural language processing tasks, such as next word prediction (Mikolov, Karafiát, Burget, Černocký, & Khudanpur, 2010), sentiment analysis (Le & Zuidema, 2014; Socher, Bauer, Manning, & Ng, 2013; Socher, Manning, & Ng, 2010), generating paraphrases (Iyyer et al., 2014; Le & Mikolov, 2014), and machine translation (Bahdanau, Cho, & Bengio, 2014). As they can straightforwardly deal with phenomena that are continuous in nature (such as loudness, pitch variation, or beat), as well as conveniently handle multiple streams at the same time, vector grammars or neural networks—although not frequently applied in this field—also seem very suitable to model music (see, e.g., Cherla, Tran, Weyde, & d'Avila Garcez, 2015; Spiliopoulou & Storkey, 2011, for some examples of recent work in which neural network models are used to model aspects of music).

Whether neural network models are fundamentally up to the task of modeling structure in language and what they can teach us about the nature of mental processes and representations (if anything at all) has been the topic of a longstanding (heated) debate (some influential papers are Fodor & Pylyshyn, 1988; Pinker & Mehler, 1990; Pollack, 1988; Rumelhart & McClelland, 1986). Whichever side one favors in this debate, it seems undoubtedly true that the successes of neural networks in performing natural language processing tasks are difficult to interpret and that the underlying mechanisms are difficult to characterize in terms of the structure building operations familiar from the CH tradition. Part of this difficulty comes from the fact that neural network models are typically trained on approximating a particular input–output relation (end-to-end) and do not explicitly model structure. Although one might argue that many end-to-end tasks require (implicit) knowledge about the underlying structure of language, it is not obvious what this structural knowledge actually entails. Analyzing the internal dynamics to interpret how solutions are encoded in the vector space is notoriously hard for networks that have more than a couple of nodes and the resulting systems—a few exceptions aside (e.g., Karpathy, Johnson, & Fei-Fei, 2015)—often remain black boxes. Furthermore, finding the right vectors and operations that encode a certain task is a complicated task; the research focus is therefore typically more on finding optimization techniques to more effectively search through the tremendous space of possibilities than on interpretation (e.g., Hochreiter & Schmidhuber, 1997; Kingma & Ba, 2015; Zeiler, 2012).

Expressivity of Vector Grammars

The focus or difficulties in the field aside, however, one can observe that the expressivity of connectionist models reduces the need for more complex architectures (such as MCSGs), as vector grammars are computationally much more expressive than symbolic systems with similar architectures. For instance, Rodriguez (2001) demonstrated that a simple recurrent network (or SRN, on an architectural level similar to a finite-state automaton) can implement the counter language a^nb^n, a prime example of a context-free language (see figure 11.6). Theoretically, one can prove that an SRN with a nonlinear activation function is a Turing complete system that can implement any arbitrary input–output mapping (Siegelmann & Sontag, 1992). Although it is not well understood what this means in practice—we lack methods to find the parameters to do so, appropriate techniques to understand potential solutions, and, arguably, even suitable input–output pairs—the theoretical possibility nevertheless calls into question the a priori plausibility of the hypothesis that context freeness is uniquely human. Rodriguez's results demonstrate a continuum between finite-state and (at least some) context-free languages, which casts doubt on the validity of the focus on architectural constraints on structure building operations that dominate the CH. As such, while much theoretical work exploring their expressive power is still

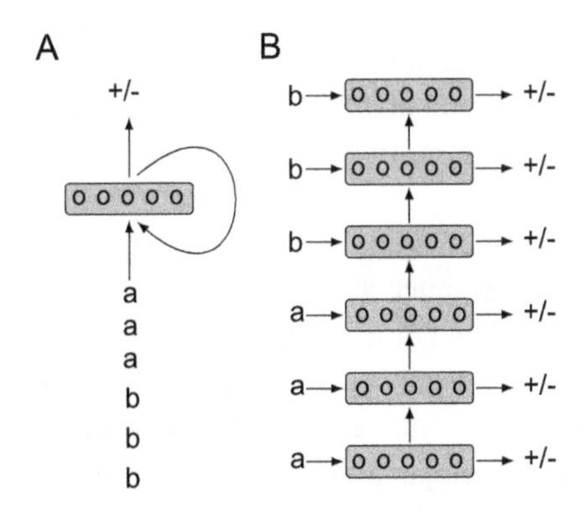

Figure 11.6
(A) A simple recurrent network (SRN), equivalent to the one proposed by Elman (1990). The network receives a sequence of inputs (in this case, a a a b b b) and outputs whether this is a grammatical sequence (+/-). The arrows represent so-called weight matrices that define the projections in the vector space used to compute the new vector activations from the previous ones. Traditionally the SRN does not classify whether sequences belong to a certain language but predicts at every point in the sequence its next element (including the end-of-string marker. Whether a sequence is grammatical can then be evaluated by checking if the network predicted the right symbol at every point in the sequence where this was actually possible (thus, in the case of anbn, predicting correctly the end of the string, as well as all the b's but the first one). (B) The same network but unfolded over time. On an architectural level, the SRN is similar to a finite-state automaton.

necessary, vector grammars provide another motivation to move on to probabilistic, non-symbolic models that go beyond the constraints of the CH.

Discussion

We have discussed different formal models of syntactic structure building, building blocks, and functional motivations of structure in language, music, and birdsong. We aimed to lay a common ground for future formal and empirical research addressing questions about the cognitive mechanisms underlying structure in each of these domains and about commonalities as well as differences between music and language and between species. This chapter can thus be seen as a long-overdue effort to bring theoretical approaches in music and animal vocalization into common terms that can be compared with approaches established in formal and computational linguistics, complementing the literature responding to Hauser, Chomsky, and Fitch's (2002) provocative hypothesis concerning the "exceptional" role of the human cognitive/communicative abilities.

Our journey through the computational models of structure building—from Shannon's *n*-grams via the CH to vector grammars and models beyond the CH—has uncovered many useful models for how sequences of sound might be generated and processed. We arrived at a discussion of recent models that add probabilities, semantics, and graded categories to classical formal grammars. Graded category models, which we called vector grammars, link formal grammar and neural network approaches and add the power to deal with structures that are inherently continuous. An important finding using such vector grammars is that one relatively simple architecture can predict sequences of different complexity in the CH and therefore have the potential to undermine assumptions concerning categorically different cognitive capacities between human and animal forms of communication (Hauser et al., 2002).

Perhaps the most important lesson we can draw from comparing models for structure building is that the chosen level of description determines much about the type of conclusions that can be drawn and that there is no single true level of description; the choice of model should therefore depend strongly on the question under investigation. This is true for choices within a certain paradigm (such as the choice of basic building blocks, or the level of comparison) and between paradigms. Most comparative research to structure building compares words in language with notes in music and animal song, and sentences to songs. But this ignores the potential structure in bouts of songs. If songs were in fact to be compared with words rather than sentences, such models would be comparing the bird's phonology with human syntax (Yip, 2006).

Moreover, the choice of model determines which aspects of structure building we are comparing across domains. What should be considered in this case is not which model is better in itself but which model is better for a certain purpose. For instance, fully symbolic models from the CH provide a useful perspective for the comparison of different theoretical approaches and predictions concerning properties of sets of sequences, but at the same time, they might not necessarily tell us much about the cognitive mechanisms that underlie learning, processing, or generating such sets of sequences. This means that even if we can show that animal song and human language are of a different complexity class in the CH, we cannot automatically assume that there is a qualitative difference in the cognitive capacities of humans and nonhuman animals, as demonstrated by the relatively simple neural network architectures that can predict sequences of different complexity in the CH.

Acknowledgments

This chapter is a thoroughly revised version of Rohrmeier et al. (2015). We thank the reviewers and editors who offered advice and comments on the different versions of the manuscript. Special thanks go to the late Remko Scha, whose skepticism and encouragement have indirectly had much influence on our discussion of the cognitive relevance of the Chomsky hierarchy.

Notes

1. Although many of these aspects also occur in speech (cf. prosody), the structural aspect of human language appears to be easier to isolate.
2. This is an even bigger problem when trying to model birdsong, where data sets are generally small, and in many cases, the number of possible transition probabilities—despite the comparably small number of elementary units—vastly exceeds the number of examples in the entire set of empirical data.
3. This was also proposed in the music domain to describe harmony in jazz (Steedman, 1996).
4. As an anonymous reviewer points out, there is no mathematically precise characterization of the mildly context-sensitive languages in terms of a formal automaton, as can be given for the context-free and context-sensitive languages. The status of MCSLs as a level on the CH can therefore be debated.

References

Aarts, B. (2004). Modelling linguistic gradience. *Studies in Language*, *28*(1), 1–49. doi:10.1075/sl.28.1.02aar

Abe, K., & Watanabe, D. (2011). Songbirds possess the spontaneous ability to discriminate syntactic rules. *Nature Neuroscience*, *14*(8), 1067–1074. doi:10.1038/nn.2869

Abney, S. (1996). Statistical methods and linguistics. In J. Klavans & P. Resnik (Eds.), *The balancing act: Combining symbolic and statistical approaches to language* (pp. 1–23). Cambridge, MA: MIT Press.

Adam, O., Cazau, D., Gandilhon, N., Fabre, B., Laitman, J. T., & Reidenberg, J. S. (2013). New acoustic model for humpback whale sound production. *Applied Acoustics*, *74*(10), 1182–1190. doi:10.1016/j.apacoust.2013.04.007

Aldwell, E., Schachter, C., & Cadwallader, A. (2011). *Harmony and voice leading* (4th ed.). Boston: Schirmer.

Amador, A., & Margoliash, D. (2013). A mechanism for frequency modulation in songbirds shared with humans. *Journal of Neuroscience*, *33*(27), 11136–11144. doi:10.1523/JNEUROSCI.5906-12.2013

Ames, C. (1989). The Markov process as a compositional model: A survey and tutorial. *Leonardo*, *22*(2), 175–187. doi:10.2307/1575226

Arnold, K., & Zuberbühler, K. (2012). Call combinations in monkeys: Compositional or idiomatic expressions? *Brain and Language*, *120*(3), 303–309. doi:10.1016/j.bandl.2011.10.001

Asano, R., & Boeckx, C. (2015). Syntax in language and music: What is the right level of comparison? *Frontiers in Psychology*, *6*, 942. doi:10.3389/fpsyg.2015.00942

Bahdanau, D., Cho, K., & Bengio, Y. (2014). Neural machine translation by jointly learning to align and translate. *Proceedings of the ICLR 2015*, *26*(1), 1–15.

Beckers, G. J. L., Bolhuis, J. J., Okanoya, K., & Berwick, R. C. (2012). Birdsong neurolinguistics. *Neuroreport*, *23*(3), 139–145. doi:10.1097/WNR.0b013e32834f1765

Berwick, R. C., Okanoya, K., Beckers, G. J. L., & Bolhuis, J. J. (2011). Songs to syntax: The linguistics of birdsong. *Trends in Cognitive Sciences*, *15*(3), 113–121. doi:10.1016/j.tics.2011.01.002

Besson, M., Frey, A., & Aramaki, M. (2011). Is the distinction between intra- and extra-musical meaning implemented in the brain? Comment on "Towards a neural basis of processing musical semantics" by Stefan Koelsch. *Physics of Life Reviews*, *8*(2), 112–113. doi:10.1016/j.plrev.2011.05.006

Bolhuis, J. J., Okanoya, K., & Scharff, C. (2010). Twitter evolution: Converging mechanisms in birdsong and human speech. *Nature Reviews: Neuroscience*, *11*(11), 747–759. doi:10.1038/nrn2931

Brants, T. (2000). TnT: A statistical part-of-speech tagger. In *Proceedings of the Sixth Conference on Applied Natural Language Processing* (pp. 224–231). Stroudsburg, PA: Association for Computational Linguistics. doi:10.3115/974147

Briefer, E., Osiejuk, T. S., Rybak, F., & Aubin, T. (2009). Are bird song complexity and song sharing shaped by habitat structure? An information theory and statistical approach. *Journal of Theoretical Biology*, *262*(1), 151–164. doi:10.1016/j.jtbi.2009.09.020

Brown, S. (2001). The "musilanguage" model of music evolution. In N. Wallin, B. Merker, & S. Brown (Eds.), *The origins of music* (pp. 271–300). Cambridge, MA: MIT Press.

Callender, C., Quinn, I., & Tymoczko, D. (2008). Generalized voice-leading spaces. *Science*, *320*(5874), 346–348. doi:10.1126/science.1153021.

Chatfield, C., & Lemon, R. E. (1970). Analysing sequences of behavioural events. *Journal of Theoretical Biology*, *29*(3), 427–445. doi:10.1016/0022-5193(70)90107-4

Chen, J., Jansen, N., & ten Cate, C. (2016). Zebra finches are able to learn affixation-like patterns. *Animal Cognition*, *19*(1), 65–73. doi:10.1007/s10071-015-0913-x

Chen, J., Van Rossum, D., & Weiss, D. (2015). Artificial grammar learning in zebra finches and human adults: XYX vs XXY. *Animal Cognition*, *18*(1), 151–164. doi:10.1007/s10071-014-0786-4

Cherla, S., Tran, S. N., Weyde, T., & d'Avila Garcez, A. (2015). Hybrid long- and short-term models of folk melodies. In M. Müller & F. Wiering (Eds.), *Proceedings of the 16th International Society for Music Information Retrieval Conference* (pp. 584–590). Victoria, Canada: International Society for Music Information Retrieval.

Chomsky, N. (1956). Three models for the description of language. *I.R.E. Transactions on Information Theory*, *2*(3), 113–124. doi:10.1109/TIT.1956.1056813

Clark, C. J., & Feo, T. J. (2008). The Anna's hummingbird chirps with its tail: A new mechanism of sonation in birds. *Proceedings of the Royal Society of London B: Biological Sciences*, *275*(1637), 955–962. doi:10.1098/rspb.2007.1619

Conklin, D., & Witten, I. H. (1995). Multiple viewpoint systems for music prediction. *Journal of New Music Research*, *24*(1), 51–73. doi:10.1080/09298219508570672

Corballis, M. C. (2007). Recursion, language, and starlings. *Cognitive Science*, *31*(4), 697–704. doi:10.1080/15326900701399947

Cross, I. (2012a). Cognitive science and the cultural nature of music. *Topics in Cognitive Science*, *4*(4), 668–677. doi:10.1111/j.1756-8765.2012.01216.x

Cross, I. (2012b). Music and biocultural evolution. In R. M. Clayton & T. Herbert (Eds.), *The cultural study of music: A critical introduction* (pp. 19–30). New York: Routledge. 10.1080/0749446032000150906

Cross, I., & Woodruff, G. E. (2010). Music as a communicative medium. In *The prehistory of language* (Vol. 11, p. 77). Oxford: Oxford University Press. doi:10.1093/acprof:oso/9780199545872.003.0005

Culy, C. (1985). The complexity of the vocabulary of Bambara. *Linguistics and Philosophy*, *8*, 345–351. doi:10.1007/BF00630918

Cynx, J. (1990). Experimental determination of a unit of song production in the zebra finch (*Taeniopygia guttata*). *Journal of Comparative Psychology, 104*(1), 3. doi:10.1037/0735-7036.104.1.3

Davies, S. (2011). Questioning the distinction between intra-and extra-musical meaning: Comment on "Towards a neural basis for processing musical semantics" by Stefan Koelsch. *Physics of Life Reviews, 8*(2), 114–115. doi:10.1016/j.plrev.2011.05.005

De Clercq, T., & Temperley, D. (2011). A corpus analysis of rock harmony. *Popular Music, 30*(1), 47–70. doi:10.1017/S026114301000067X

Dingemanse, M., Torreira, F., & Enfield, N. J. (2014). Correction: Is "Huh?" a universal word? Conversational infrastructure and the convergent evolution of linguistic items. *PLoS One, 8*(11), e78273. doi:10.1371/journal.pone.0094620

Doupe, A. J., & Kuhl, P. K. (1999). Birdsong and human speech: Common themes and mechanisms. *Annual Review of Neuroscience, 22*(1), 567–631. doi:10.1146/annurev.neuro.22.1.567

Eerola, T. (2003). *The dynamics of musical expectancy: Cross-cultural and statistical approaches to melodic expectations*. Jyväskylä: University of Jyväskylä.

Elman, J. L. (1990). Finding structure in time. *Cognitive Science, 14*(2), 179–211. doi:10.1207/s15516709cog1402_1

Farbood, M. M., Heeger, D. J., Marcus, G. F., Hasson, U., & Lerner, Y. (2015). The neural processing of hierarchical structure in music and speech at different timescales. *Frontiers in Neuroscience, 9*. doi:10.3389/fnins.2015.00157

Fehér, O., Wang, H., Saar, S., Mitra, P. P., & Tchernichovski, O. (2009). De novo establishment of wild-type song culture in the zebra finch. *Nature, 459*(7246), 564–568. doi:10.1038/nature07994

Fitch, W. T. (2006). The biology and evolution of music: A comparative perspective. *Cognition, 100*(1), 173–215. doi:10.1016/j.cognition.2005.11.009

Fitch, W. T., & Friederici, A. D. (2012). Artificial grammar learning meets formal language theory: An overview. *Philosophical Transactions of the Royal Society of London B: Biological Sciences, 367*(1598), 1933–1955. doi:10.1098/rstb.2012.0103

Fitch, W. T., Friederici, A. D., & Hagoort, P. (2012). Pattern perception and computational complexity: Introduction to the special issue. *Philosophical Transactions of the Royal Society of London B: Biological Sciences, 367*(1598), 1925–1932. doi:10.1098/rstb.2012.0099

Fitch, W. T., & Gingras, B. (2011). Multiple varieties of musical meaning: Comment on "Towards a neural basis of processing musical semantics" by Stefan Koelsch. *Physics of Life Reviews, 8*(2), 108–109. doi:10.1016/j.plrev.2011.05.004

Fitch, W. T., & Hauser, M. D. (2004). Computational constraints on syntactic processing in a nonhuman primate. *Science, 303*(5656), 377–380. doi:10.1126/science.1089401

Fodor, J. A., & Pylyshyn, Z. W. (1988). Connectionism and cognitive architecture: A critical analysis. *Cognition, 28*, 3–71. doi:10.1016/0010-0277(88)90031-5

Franz, M., & Goller, F. (2002). Respiratory units of motor production and song imitation in the zebra finch. *Journal of Neurobiology, 51*(2), 129–141. doi:10.1002/neu.10043

Gentner, T. Q., Fenn, K. M., Margoliash, D., & Nusbaum, H. C. (2006). Recursive syntactic pattern learning by songbirds. *Nature, 440*(7088), 1204–1207. doi:10.1038/nature04675

Gibson, E., & Thomas, J. (1999). Memory limitations and structural forgetting: The perception of complex ungrammatical sentences as grammatical. *Language and Cognitive Processes, 14*(3), 225–248. doi:10.1080/016909699386293

Granroth-Wilding, M., & Steedman, M. (2014). A robust parser-interpreter for jazz chord sequences. *Journal of New Music Research, 43*(4), 355–374. doi:10.1080/09298215.2014.910532

Grünwald, P. (2007). *The minimum description length principle*. Cambridge, MA: MIT Press.

Haas, W. B., Rohrmeier, M., & Wiering, F. (2009). Modeling harmonic similarity using a generative grammar of tonal harmony. In *Proceedings of the 10th International Society for Music Information Retrieval Conference* (pp. 549–554).

Hauser, M. D., Chomsky, N., & Fitch, W. T. (2002). The faculty of language: What is it, who has it, and how did it evolve? *Science, 298*(5598), 1569–1579. doi:10.1126/science.298.5598.1569

Hedges, T., & Rohrmeier, M. (2011). Exploring Rameau and beyond: A corpus study of root progression theories. In C. Agon, E. Amiot, M. Andreatta, G. Assayag, J. Bresson, & J. Mandereau (Eds.), *Mathematics and computation in music: Lecture notes in artificial intelligence* (Vol. 6726, pp. 334–337). Berlin: Springer; doi:10.1007/978-3-642-21590-2_27

Heffner, C. C., & Slevc, L. R. (2015). Prosodic structure as a parallel to musical structure. *Frontiers in Psychology, 6*. doi:10.3389/fpsyg.2015.01962

Herbranson, W. T., & Shimp, C. P. (2008). Artificial grammar learning in pigeons. *Learning and Behavior, 36*(2), 116–137. doi:10.3758/LB.36.2.116

Hochreiter, S., & Schmidhuber, J. (1997). Long short-term memory. *Neural Computation, 9*(8), 1–32. doi:10.1162/neco.1997.9.8.1735

Hofstadter, D. R. (1980). *Gödel, Escher, Bach: An eternal golden braid*. New York: Basic Books.

Hockett, C. F. (1960). The origin of speech. *Scientific American, 203*, 88–111.

Horner, V., Whiten, A., Flynn, E., & de Waal, F. B. M. M. (2006). Faithful replication of foraging techniques along cultural transmission chains by chimpanzees and children. *Proceedings of the National Academy of Sciences, 103*(37), 13878–13883. doi:10.1073/pnas.0606015103

Hultsch, H., & Todt, D. (1998). How songbirds deal with large amounts of serial information: Retrieval rules suggest a hierarchical song memory. *Biological Cybernetics, 79*(6), 487–500. doi:10.1007/s004220050498

Hurford, J. R. (2007). *The origins of meaning*. Oxford: Oxford University Press.

Huron, D. (2007). Sweet anticipation: Music and the psychology of expectation. *Music Perception, 24*(5), 511–514. doi:10.1525/mp.2007.24.5.511

Huybregts, R. (1984). The weak inadequacy of context-free phrase structure grammars. In G. Haan, M. Trommelen, & W. Zonneveld (Eds.), *Van Periferie naar Kern* (pp. 81–99). Dordrecht: Foris.

Isaac, D., & Marler, P. (1963). Ordering of sequences of singing behaviour of mistle thrushes in relationship to timing. *Animal Behaviour, 11*(1), 179–188. doi:10.1016/0003-3472(63)90027-7

Iyyer, M., Boyd-Graber, J., Claudino, L., Socher, R., & Daumé III, H. (2014). A neural network for factoid question answering over paragraphs. In *Proceedings of the 2014 Conference on Empirical Methods in Natural Language Processing* (pp. 633–644). Stroudsburg, PA: Association for Computational Linguistics. doi:10.3115/v1/d14-1070

Jackendoff, R., & Lerdahl, F. (2006). The capacity for music: What is it, and what's special about it? *Cognition, 100*(1), 33–72. doi:10.1016/j.cognition.2005.11.005

Jäger, G., & Rogers, J. (2012). Formal language theory: Refining the Chomsky hierarchy. *Philosophical Transactions of the Royal Society of London B: Biological Sciences, 367*(1598), 1956–1970. doi:10.1098/rstb.2012.0077

Jiang, S., Zhu, L., Guo, X., Ma, W., Yang, Z., & Dienes, Z. (2012). Unconscious structural knowledge of tonal symmetry: Tang poetry redefines limits of implicit learning. *Consciousness and Cognition, 21*(1), 476–486. doi:10.1016/j.concog.2011.12.009

Jin, D. Z., & Kozhevnikov, A. A. (2011). A compact statistical model of the song syntax in Bengalese finch. *PLoS Computational Biology, 7*(3), e1001108. doi:10.1371/journal.pcbi.1001108

Joshi, A. K. (1985). Tree adjoining grammars: How much context-sensitivity is required to provide reasonable structural descriptions? In D. Dowty, L. Karttunen, & A. Zwicky (Eds.), *Natural language parsing: Psychological, computational, and theoretical perspectives* (pp. 206–250). New York: Cambridge University Press; 10.1017/cbo9780511597855.007

Joshi, A. K., Levy, L. S., & Takahashi, M. (1975). Tree adjunct grammars. *Journal of Computer and System Sciences, 10*(1), 136–163. doi:10.1016/s0022-0000(75)80019-5

Joshi, A. K., Vijay-Shanker, K., & Weir, D. (1991). The convergence of mildly context-sensitive grammar formalisms. In P. Sells, S. M. Shieber, & T. Wasow (Eds.), *Foundational issues in natural language processing* (pp. 31–81). Cambridge, MA: MIT Press.

Juang, B.-H., & Rabiner, L. (1991). Hidden Markov models for speech recognition. *Technometrics, 33*(3), 251–272. doi:10.1080/00401706.1991.10484833

Jurafsky, D., & Martin, J. H. (2000). *Speech and language processing*. Noida: Pearson Education India.

Karpathy, A., Johnson, J., & Fei-Fei, L. (2015). Visualizing and understanding recurrent networks. In *Proceedings of the International Conference on Learning Representations* (pp. 1–13).

Katahira, K., Suzuki, K., Kagawa, H., & Okanoya, K. (2013). A simple explanation for the evolution of complex song syntax in Bengalese finches. *Biology Letters*, *9*(6), 20130842. doi:10.1098/rsbl.2013.0842

Katahira, K., Suzuki, K., Okanoya, K., & Okada, M. (2011). Complex sequencing rules of birdsong can be explained by simple hidden Markov processes. *PLoS One*, *6*(9), e24516. doi:10.1371/journal.pone.0024516

Keller, A. (1978). Bernstein's "The unanswered question" and the problem of musical competence. *Musical Quarterly*, *64*(2), 195–223. doi:10.1093/mq/LXIV.2.195

Keiler, A. (1983). On some properties of Schenker's pitch derivations. *Music Perception*, *1*(2), 200–228. doi:10.2307/40285256

Kershenbaum, A., Bowles, A. E., Freeberg, T. M., Jin, D. Z., Lameira, A. R., & Bohn, K. (2014). Animal vocal sequences: Not the Markov chains we thought they were. *Proceedings of the Royal Society of London B: Biological Sciences*, *281*(1792), 20141370. doi:10.1098/rspb.2014.1370

Kingma, D. P., & Ba, L. J. (2015). Adam: A method for stochastic optimization. In *Proceedings of the International Conference for Learning Representations* (pp. 1–13).

Knight, K., & Graehl, J. (2005). An overview of probabilistic tree transducers for natural language processing. In *Computational linguistics and intelligent text processing* (Vol. 3406, pp. 1–24). Berlin: Springer. doi:10.1007/978-3-540-30586-6_1

Knoernschild, M. (2014). Male courtship displays and vocal communication in the polygynous bat. *Behaviour*, *151*(6), 781–798. doi:10.1163/1568539x-00003171

Koelsch, S. (2011). Towards a neural basis of processing musical semantics. *Physics of Life Reviews*, *8*(2), 89–105. doi:10.1016/j.plrev.2011.04.004

Koelsch, S., Rohrmeier, M., Torrecuso, R., & Jentschke, S. (2013). Processing of hierarchical syntactic structure in music. *Proceedings of the National Academy of Sciences*, *110*(38), 15443–15448. doi:10.1073/pnas.1300272110

Krumhansl, C. L. (1995). Music psychology and music theory: Problems and prospects. *Music Theory Spectrum*, *17*(1), 53–80. doi:10.1525/mts.1995.17.1.02a00030

Krumhansl, C. L. (2004). The cognition of tonality—as we know it today. *Journal of New Music Research*, *33*(3), 253–268. doi:10.1080/0929821042000317831

Krumhansl, C. L., & Kessler, E. J. (1982). Tracing the dynamic changes in perceived tonal organization in a spatial representation of musical keys. *Psychological Review*, *89*(4), 334–368. doi:10.1037/0033-295X.89.4.334

Kuhn, G., & Dienes, Z. (2005). Implicit learning of nonlocal musical rules: Implicitly learning more than chunks. *Journal of Experimental Psychology: Learning, Memory, and Cognition*, *31*(6), 1417–1432. doi:10.1037/0278-7393.31.6.1417

Lau, J. H., Clark, A., & Lappin, S. (2015). Unsupervised prediction of acceptability judgements. In *Proceedings of the 53rd Annual Meeting of the Association for Computational Linguistics and the 7th International Joint Conference on Natural Language Processing* (vol. 1, pp. 1618–1628). Stroudsburg, PA: Association for Computational Linguistics. doi:10.3115/v1/P15-1156

Le, P., & Zuidema, W. (2014). The inside-outside recursive neural network model for dependency parsing. In *Proceedings of the 2014 Conference on Empirical Methods in Natural Language Processing* (pp. 729–739). Stroudsburg, PA: Association for Computational Linguistics. doi:10.3115/v1/D14-1081

Le, Q., & Mikolov, T. (2014). Distributed representations of sentences and documents. In *Proceedings of the 31st International Conference on Machine Learning, 32*, 1188–1196.

Lehne, M., Rohrmeier, M., & Koelsch, S. (2014). Tension-related activity in the orbitofrontal cortex and amygdala: An fMRI study with music. *Social Cognitive and Affective Neuroscience*, *9*(10), 1515–1523. doi:10.1093/scan/nst141.

Lerdahl, F., & Jackendoff, R. (1983). *A general theory of tonal music*. Cambridge, MA: MIT Press.

Lerdahl, F., & Krumhansl, C. L. (2007). Modeling tonal tension. *Music Perception*, *24*(4), 329–366. doi:10.1525/mp.2007.24.4.329

Li, F., Jiang, S., Guo, X., Yang, Z., & Dienes, Z. (2013). The nature of the memory buffer in implicit learning: Learning Chinese tonal symmetries. *Consciousness and Cognition*, *22*(3), 920–930. doi:10.1016/j.concog.2013.06.004

Liberman, A. M., Cooper, F. S., Shankweiler, D. P., & Studdert-Kennedy, M. (1967). Perception of the speech code. *Psychological Review*, *74*(6), 431–461. doi:10.1037/h0020279

Lipkind, D., Marcus, G. F., Bemis, D. K., Sasahara, K., Jacoby, N., Takahasi, M., et al. (2013). Stepwise acquisition of vocal combinatorial capacity in songbirds and human infants. *Nature*, *498*(7452), 104–108. doi:10.1038/nature12173

Mackay, D. J. C. (2003). *Information theory, inference, and learning algorithms*. Cambridge: Cambridge University Press.

Margulis, E. H. (2014). *On repeat: How music plays the mind*. Oxford: Oxford University Press.

Markowitz, J. E., Ivie, E., Kligler, L., & Gardner, T. J. (2013). Long-range order in canary song. *PLoS Computational Biology*, *9*(5), e1003052. doi:10.1371/journal.pcbi.1003052

Marler, P., & Pickert, R. (1984). Species-universal microstructure in the learned song of the swamp sparrow (*Melospiza georgiana*). *Animal Behaviour*, *32*(3), 673–689. doi:10.1016/S0003-3472(84)80143-8

Mavromatis, P. (2005). A hidden Markov model of melody production in Greek church chant. *Computing in Musicology*, *14*, 93–112.

Mavromatis, P. (2009). Minimum description length modelling of musical structure. *Journal of Mathematics and Music*, *3*(3), 117–136. doi:10.1080/17459730903313122

Mikolov, T. (2012). *Statistical language models based on neural networks* (Unpublished doctoral dissertation). Brno University of Technology, Brno, Czech Republic.

Mikolov, T., Karafiát, M., Burget, L., Černocký, J., & Khudanpur, S. (2010). Recurrent neural network based language model. In *Proceedings of the 11th Annual Conference of the International Speech Communication Association* (pp. 1045–1048). Makuhari: International Speech Communication Association.

Miller, G. A., & Isard, S. (1964). Free recall of self-embedded English sentences. *Information and Control*, *7*(3), 292–303. doi:10.1016/S0019-9958(64)90310-9

Montague, R. (1970). Universal grammar. *Theoria*, *36*(3), 373–398. doi:10.1111/j.1755-2567.1970.tb00434.x

Murphy, K. P. (2002). *Dynamic Bayesian networks: Representation, inference and learning*. (Unpublished doctoral dissertation). University of California, Berkeley.

Murphy, R. A., Mondragón, E., & Murphy, V. A. (2008). Rule learning by rats. *Science*, *319*(5871), 1849–1851. doi:10.1126/science.1151564

Narmour, E. (1992). *The analysis and cognition of melodic complexity: The implication-realization model*. Chicago: University of Chicago Press.

Neuwirth, M., & Rohrmeier, M. (2015). Towards a syntax of the classical cadence. In M. Neuwirth & P. Bergé (Eds.), *What is a cadence?* (pp. 287–338). Leuven: Leuven University Press.

Okanoya, K. (2004). The Bengalese finch: A window on the behavioral neurobiology of birdsong syntax. *Annals of the New York Academy of Sciences*, *1016*(1), 724–735. doi:10.1196/annals.1298.026

Paiement, J.-F. (2008). *Probabilistic models for music* (Unpublished doctoral dissertation). École Polytechnique Fédérale de Lausanne, Lausanne, Switzerland.

Payne, K., & Payne, R. S. (1985). Large scale changes over 19 years of songs of Humpback whales in Bermuda. *Zeitschrift für Tierpsychologie*, *68*(2), 89–114. doi:10.1111/j.1439-0310.1985.tb00118.x

Payne, R. S., & McVay, S. (1971). Songs of humpback whales. *Science*, *173*(3997), 585–597. doi:10.1126/science.173.3997.585

Pearce, M. T. (2005). *The construction and evaluation of statistical models of melodic structure in music perception and composition structure in music perception and composition* (Unpublished doctoral dissertation). City University, London, UK.

Pearce, M., & Wiggins, G. A. (2004). Improved methods for statistical modelling of monophonic music. *Journal of New Music Research*, *33*(4), 367–385. doi:10.1080/0929821052000343840

Pearce, M. T., & Wiggins, G. A. (2012). Auditory expectation: The information dynamics of music perception and cognition. *Topics in Cognitive Science*, *4*(4), 625–652. doi:10.1111/j.1756-8765.2012.01214.x

Petrov, S., & Klein, D. (2007). Improved inference for unlexicalized parsing. In *Human Language Technologies 2007: The Conference of the North American Chapter of the Association for Computational Linguistics; Proceedings of the Main Conference* (Vol. 7, pp. 404–411). Rochester, NY: Association for Computational Linguistics.

Pinker, S., & Mehler, J. (1990). *Connections and symbols: Artificial Intelligence*. Cambridge, MA: MIT Press.

Piston, W. (1948). *Harmony*. New York: Norton.

Podos, J., Peters, S., Rudnicky, T., Marler, P., & Nowicki, S. (1992). The organization of song repertoires in song sparrows: Themes and variations. *Ethology, 90*(2), 89–106. doi:10.1111/j.1439-0310.1992.tb00824.x

Pollack, J. B. (1988). *Recursive auto-associative memory: Devising compositional distributed representations*. Las Cruces: Computing Research Laboratory, New Mexico State University.

Pollack, J. B. (1990). Recursive distributed representations. *Artificial Intelligence, 46*(1), 10. doi:10.1016/0004-3702(90)90005-K

Ponsford, D., Wiggins, G. A., & Mellish, C. (1999). Statistical learning of harmonic movement. *Journal of New Music Research, 28*(2), 150–177. doi:10.1076/jnmr.28.2.150.3115

Pothos, E. M. (2007). Theories of artificial grammar learning. *Psychological Bulletin, 133*(2), 227–244. doi:10.1037/0033-2909.133.2.227

Quinn, I., & Mavromatis, P. (2011). Voice-leading prototypes and harmonic function in two chorale corpora. In C. Agon, M. Andreatta, G. Assayag, E. Amiot, J. Bresson, & J. Mandereau (Eds.), *Mathematics and computation in music* (pp. 230–240). Berlin: Springer. doi:10.1007/978-3-642-21590-2_18

Rabiner, L. R. (1989). A tutorial on hidden Markov models and selected applications in speech recognition. *Proceedings of the IEEE, 77*(2), 257–286. doi:10.1109/5.18626

Rabiner, L., & Juang, B.-H. (1993). *Fundamentals of speech recognition*. Upper Saddle River, NJ: Prentice Hall.

Rabiner, L., & Juang, B.-H. (1986). An introduction to hidden Markov models. *IEEE Signal Processing Magazine, 3*(1), 4–16.

Raczynski, S. A., Vincent, E., & Sagayama, S. (2013). Dynamic Bayesian networks for symbolic polyphonic pitch modeling. *IEEE Transactions on Audio, Speech, and Language Processing, 21*(9), 1830–1840. doi:10.1109/TASL.2013.2258012

Rameau, J. P. (1971). *Treatise on harmony*. (Philip Gossett, Trans.) New York: Dover.

Raphael, C., & Stoddard, J. (2004). Functional harmonic analysis using probabilistic models. *Computer Music Journal, 28*(3), 45–52. doi:10.1162/0148926041790676

Reber, A. S. (1967). Implicit learning of artificial grammars. *Journal of Verbal Learning and Verbal Behavior, 6*(6), 855–863. doi:10.1016/S0022-5371(67)80149-X

Reich, U. (2011). The meanings of semantics. *Physics of Life Reviews, 8*(2), 120–121. doi:10.1016/j.plrev.2011.05.012

Reis, B. Y. (1999). *Simulating music learning with autonomous listening agents: Entropy, Ambiguity and Context* (Unpublished doctoral dissertation). University of Cambridge, Cambridge, UK.

Rodriguez, P. (2001). Simple recurrent networks learn context-sensitive languages by counting. *Neural Computation, 13*(9), 2093–2118. doi:10.1162/089976601750399326

Rohrmeier, M. (2006). *Towards modelling harmonic movement in music: Analysing properties and dynamic aspects of PC set sequences in Bach's chorales*. Technical Report DCRR-004 Darwin College. Cambridge, MA.

Rohrmeier, M. (2007). A generative grammar approach to diatonic harmonic structure. In C. Spyridis, A. Georgaki, G. Kouroupetroglou, & C. Anagnostopoulou (Eds.), *Proceedings of the Fourth Sound and Music Computing Conference* (pp. 97–100). Athens: National and Kapodistrian University of Athens.

Rohrmeier, M. (2011). Towards a generative syntax of tonal harmony. *Journal of Mathematics and Music, 5*(1), 35–53. doi:10.1080/17459737.2011.573676.

Rohrmeier, M., & Cross, I. (2008). Statistical properties of tonal harmony in Bach's chorales. In K. Miyazaki, Y. Hiraga, M. Adachi, Y. Nakajima, & M. Tsuzaki (Eds.), *Proceedings of the 10th International Conference on Music Perception and Cognition* (pp. 619–627).

Rohrmeier, M., Dienes, Z., Guo, X., & Fu, Q. (2014). Implicit learning and recursion. In F. Lowenthal & L. Lefebvre (Eds.), *Language and recursion* (pp. 67–85). Berlin: Springer. 10.1007/978-1-4614-9414-0_6

Rohrmeier, M., Fu, Q., & Dienes, Z. (2012). Implicit learning of recursive context-free grammars. *PLoS One*, *7*(10), e45885. doi:10.1371/journal.pone.0045885

Rohrmeier, M., & Graepel, T. (2012). Comparing feature-based models of harmony. In R. Kronland-Martinet, S. Ystad, M. Aramaki, M. Barthet, & S. Dixon (Eds.), *Proceedings of the Ninth International Symposium on Computer Music Modeling and Retrieval* (pp. 357–370). London: Springer.

Rohrmeier, M., & Koelsch, S. (2012). Predictive information processing in music cognition. A critical review. *International Journal of Psychophysiology*, *83*(2), 164–175. doi:10.1016/j.ijpsycho.2011.12.010

Rohrmeier, M., & Rebuschat, P. (2012). Implicit learning and acquisition of music. *Topics in Cognitive Science*, *4*(4), 525–553. doi:10.1111/j.1756-8765.2012.01223.x

Rohrmeier, M., Zuidema, W., Wiggins, G. A., & Scharff, C. (2015). Principles of structure building in music, language and animal song. *Philosophical Transactions of the Royal Society of London B: Biological Sciences*, *370*(1664), 20140097. doi:10.1098/rstb.2014.0097

Rothenberg, D., Roeske, T. C., Voss, H. U., Naguib, M., & Tchernichovski, O. (2014). Investigation of musicality in birdsong. *Hearing Research*, *308*, 71–83. doi:10.1016/j.heares.2013.08.016

Rumelhart, D. E., & McClelland, J. L. (1986). PDP models and general issues in cognitive science. In *Parallel distributed processing: Explorations in the microstructure of cognition* (vol. 1, pp. 110–149). Cambridge, MA: MIT Press.

Samotskaya, V. V., Opaev, A. S., Ivanitskii, V. V., Marova, I. M., & Kvartalnov, P. V. (2016). Syntax of complex bird song in the large-billed reed warbler (*Acrocephalus orinus*). *Bioacoustics*, *25*(2), 127–143. doi:10.1080/09524622.2015.1130648

Sasahara, K., Cody, M. L., Cohen, D., & Taylor, C. E. (2012). Structural design principles of complex bird songs: A network-based approach. *PLoS One*, *7*(9), e44436. doi:10.1371/journal.pone.0044436

Scharff, C., & Petri, J. (2011). Evo-devo, deep homology and FoxP2: Implications for the evolution of speech and language. *Philosophical Transactions of the Royal Society of London B: Biological Sciences*, *366*(1574), 2124–2140. doi:10.1098/rstb.2011.0001

Schellenberg, E. G. (1996). Expectancy in melody: Tests of the implication realization model. *Cognition*, *58*(1), 75–125. doi:10.1016/0010-0277(95)00665-6

Schellenberg, E. G. (1997). Simplifying the implication-realization model of melodic expectancy. *Music Perception*, *14*(3), 295–318. doi:10.2307/40285723

Schenker, H. (1935). *Der freie Satz. Neue Musikalische Theorien und Phantasien*. Liège, Belgium: Margada.

Schwenk, H., & Gauvain, J.-L. (2005). Training neural network language models on very large corpora. In *Proceedings of the Conference on Human Language Technology and Empirical Methods in Natural Language Processing* (pp. 201–208). Vancouver, CA. doi:10.3115/1220575.1220601

Shannon, C. E. (1948). A mathematical theory of communication. *Bell System Technical Journal*, *27*(3), 379–423.

Shieber, S. M. (1985). Evidence against the non-context-freeness of natural language. *Studies in Linguistics and Philosophy*, *8*, 333–343. doi:10.1007/BF00630917

Siegelmann, H. T., & Sontag, E. D. (1992). On the computational power of neural nets. In *Proceedings of the Fifth Annual workshop on Computational Learning Theory 2* (vol. 50, pp. 440–449). Pittsburgh, PA: Association for Computing Machinery. doi:10.1145/130385.130432

Slater, P. J. B. (1983a). Bird song learning: Theme and variations. In A. H. Brush & A. C. J. George (Eds.), *Perspectives in ornithology: Essays presented for the centennial of the American Ornithologists' Union* (pp. 475–499). Cambridge: Cambridge University Press; 10.1017/cbo9780511754791.009

Slater, P. J. B. (1983b). Sequences of song in chaffinches. *Animal Behaviour*, *31*(1), 272–281. doi:10.1016/S0003-3472(83)80197-3

Slevc, L. R., & Patel, A. D. (2011). Meaning in music and language: Three key differences. *Physics of Life Reviews*, *8*(2), 110–111. doi:10.1016/j.plrev.2011.05.003

Socher, R., Bauer, J., Manning, C. D., & Ng, A. Y. (2013). Parsing with compositional vector grammars. In *Proceedings of the 51st Annual Meeting of the Association for Computational Linguistics* (pp. 455–465). Seattle, WA: Association for Computational Linguistics.

Socher, R., Manning, C. D., & Ng, A. Y. (2010). Learning continuous phrase representations and syntactic parsing with recursive neural networks. In *Proceedings of the NIPS-2010 Deep Learning and Unsupervised Feature Learning Workshop* (pp. 1–9).

Sorace, A., & Keller, F. (2005). Gradience in linguistic data. *Lingua, 115*(11), 1497–1524. doi:10.1016/j.lingua.2004.07.002

Spierings, M. J., & ten Cate, C. (2014). Zebra finches are sensitive to prosodic features of human speech. *Proceedings of the Royal Society of London B: Biological Sciences, 281*(1787), 20140480. doi:10.1098/rspb.2014.0480

Spiliopoulou, A., & Storkey, A. (2011). Comparing probabilistic models for melodic sequences. In *Lecture notes in computer science* (pp. 289–304). Berlin: Springer. doi:10.1007/978-3-642-23808-6_19

Steedman, M. (1984). A generative grammar for jazz chord sequences. *Music Perception, 2*(1), 52–77. doi:10.2307/40285282.

Steedman, M. (1996). The blues and the abstract truth: Music and mental models. In J. Oakhill & A. Garnham (Eds.), *Mental models in cognitive science: Essays in honour of Phil Johnson-Laird* (pp. 305–318). Mahwah, NJ: Erlbaum.

Steedman, M. (2000). *The syntactic process: Computational linguistics*. Cambridge, MA: MIT Press.

Stolz, W. S. (1967). A study of the ability to decode grammatically novel sentences. *Journal of Verbal Learning and Verbal Behavior, 6*(6), 867–873. doi:10.1016/S0022-5371(67)80151-8

Taylor, H., & Lestel, D. (2011). The Australian pied butcherbird and the natureculture continuum. *Journal of Interdisciplinary Music Studies, 5*(1), 57–83. doi:10.4407/jims.2011.07.004

Temperley, D. (2001). *The cognition of basic musical structures*. Cambridge, MA: MIT Press

ten Cate, C. (2014). On the phonetic and syntactic processing abilities of birds: From songs to speech and artificial grammars. *Current Opinion in Neurobiology, 28*, 157–164. doi:10.1016/j.conb.2014.07.019

ten Cate, C. (2016). Assessing the uniqueness of language: Animal grammatical abilities take center stage. *Psychonomic Bulletin and Review, 24*(1), 91–96. doi:10.3758/s13423-016-1091-9

ten Cate, C., Lachlan, R. F., & Zuidema, W. (2013). Analyzing the structure of bird vocalizations and language: Finding common ground. In J. Bolhuis & M. Everaert (Eds.), *Birdsong, speech, and language: Exploring the evolution of mind and brain* (pp. 243–260). Cambridge, MA: MIT Press. doi:40022179514.

ten Cate, C., & Okanoya, K. (2012). Revisiting the syntactic abilities of non-human animals: Natural vocalizations and artificial grammar learning. *Philosophical Transactions of the Royal Society of London B: Biological Sciences, 367*(1598), 1984–1994. doi:10.1098/rstb.2012.0055

Tesnière, L. (1966). *Eléments de syntaxe structurale*. Librairie C. Klincksieck.

Tierney, A. T., Russo, F. A., & Patel, A. D. (2011). The motor origins of human and avian song structure. *Proceedings of the National Academy of Sciences, 108*(37), 3–8. doi:10.1073/pnas.1103882108

Todt, D. (1975). Social learning of vocal patterns and modes of their application in grey parrots (*Psittacus erithacus*). *Zeitschrift für Tierpsychologie, 39*(1–5), 178–188. doi:10.1111/j.1439-0310.1975.tb00907.x.

Todt, D., & Hultsch, H. (1996). Acquisition and performance of song repertoire: Ways of coping with diversity and verstatility. In D. E. Kroodsma & E. H. Miller (Eds.), *Ecology and evolution of acoustic communication in birds* (pp. 79–96). Ithaca, NY: Cornell University Press.

Tymoczko, D. (2006). The geometry of musical chords. *Science, 313*(72), 72–74. doi:10.1126/science.1126287

Tymoczko, D., & Meeùs, N. (2003). Progressions fondamentales, fonctions, degrés: Une grammaire de l'harmonie tonale élémentaire. *Musurgia, 10*(3–5), 35–64.

Uddén, J., Ingvar, M., Hagoort, P., & Petersson, K. M. (2012). Implicit acquisition of grammars with crossed and nested non-adjacent dependencies: Investigating the push-down stack model. *Cognitive Science, 36*(6), 1078–1101. doi:10.1111/j.1551-6709.2012.01235.x

Ullrich, R., Norton, P., & Scharff, C. (2016). Waltzing Taeniopygia: Integration of courtship song and dance in the domesticated Australian zebra finch. *Animal Behaviour, 112*, 285–300. doi:10.1016/j.anbehav.2015.11.012.

van Heijningen, C., de Visser, J., Zuidema, W., & ten Cate, C. (2009). Simple rules can explain discrimination of putative recursive syntactic structures by a songbird species. *Proceedings of the National Academy of Sciences, 106*(48), 20538–20543. doi:10.1073/pnas.0908113106

Weikum, G., & Vossen, G. (2002). *Transactional information systems: Theory, algorithms and the practice of concurrency control and recovery*. San Francisco, CA: Morgan Kaufmann.

Weiss, M., Hultsch, H., Adam, I., Scharff, C., & Kipper, S. (2014). The use of network analysis to study complex animal communication systems: A study on nightingale song. *Proceedings of the Royal Society of London B: Biological Sciences, 281*, 201404. doi:10.1098/rspb.2014.0460

Whorley, R. P., Wiggins, G. A., Rhodes, C., & Pearce, M. T. (2013). Multiple viewpoint systems: Time complexity and the construction of domains for complex musical viewpoints in the harmonization problem. *Journal of New Music Research, 42*(3), 237–266. doi:10.1080/09298215.2013.831457

Widdess, D. R., & Wolpert, R. F. (1981). Aspects of form in North Indian ālāp and dhrupad. In D. R. Widess & R. F. Wolpert (Eds.), *Music and tradition: Essays on Asian and other musics presented to Laurence Picken* (pp. 143–181). Cambridge: Cambridge University Press.

Wiggins, G. A. (2012). I let the music "speak": Cross-domain application of a cognitive model of musical learning. In P. Rebuschat & J. Williams (Eds.), *Statistical learning and language acquisition* (pp. 1–19). Amsterdam: Mouton De Gruyter.

Wiggins, G. A., Tyack, P., Scharff, C., & Rohrmeier, M. (2015). The evolutionary roots of creativity: mechanisms and motivations. *Philosophical Transactions of the Royal Society of London B: Biological Sciences, 370*(1664), 20140099. doi:10.1098/rstb.2014.0099

Wilson, B., Slater, H., Kikuchi, Y., Milne, A. E., Marslen-Wilson, W. D., Smith, K., & Petkov, C. I. (2013). Auditory artificial grammar learning in macaque and marmoset monkeys. *Journal of Neuroscience, 33*(48), 18825–18835. doi:10.1523/jneurosci.2414-13.2013.

Winograd, T. (1968). Linguistics and the computer analysis of tonal harmony. *Journal of Music Therapy, 12*(1), 2–49. doi:10.2307/842885.

Yip, M. J. (2006). The search for phonology in other species. *Trends in Cognitive Sciences, 10*(10), 442–446. doi:10.1016/j.tics.2006.08.001.

Zann, R. (1993). Structure, sequence and evolution of song elements in wild Australian zebra finches. *Auk, 110*, 702–715.

Zeiler, M. D. (2012). *ADADELTA: An adaptive learning rate method* (Technical Report). arXiv:1212.5701.

Zengel, M. S. (1962). Literacy as a factor in language change. *Readings in the Sociology of Language, 64*(1), 132–139. doi:10.1525/aa.1962.64.1.02a00120

Zuidema, W. (2013a). Context-freeness revisited. In M. Knauff, M. Pauen, N. Sebanz, & I. Wachsmuth (Eds.), *Proceedings of the 35th Annual Meeting of the Cognitive Science Society* (pp. 1664–1669). Austin, TX: Cognitive Science Society

Zuidema, W. (2013b). Language in nature: On the evolutionary roots of a cultural phenomenon. In P. M. Binder & K. Smith (Eds.), *The language phenomenon* (pp. 163–189). Berlin: Springer. doi:10.1007/978-3-642-36086-2_8

12

The Evolutionary Roots of Creativity: Mechanisms and Motivations

Geraint A. Wiggins, Peter Tyack, Constance Scharff, and Martin Rohrmeier

One of the defining features of humanity is the ability to be creative. This ability is exhibited throughout human society and is a fundamental force in the development of humankind. However, the concept of creativity itself is shrouded in imprecision and subjectivity, making it difficult to address from a scientific perspective. One approach to the study of creativity in humans is to consider it from an evolutionary perspective, aiming to identify related behaviors in other species that carry aspects of creativity and may be evolutionary precursors to what falls under the broad umbrella of human creativity. In this chapter, we consider the origins of human creativity by comparison with other species, particularly in relation to vocal communication, thus relating to language and music. We contribute to the debate on musicality, since all musicality is dependent a priori on creativity.

For the purpose of our argument, *evolution* refers to the process of the gradual change of form and behavior as a result of differential advantages of some forms or behavior over others. In the case of biological evolution, we talk of fitness and mean the number of offspring produced and surviving.

In the next section, we decompose the idea of creativity into tractable components to allow us to examine whether music and other forms of vocal communication (including language in humans) share similar functional roots and may have evolved out of similar cognitive precursors. Whether a society values or eschews creativity, whether we agree on what constitutes good or bad music, or where one stands in the balance between humans as cognitive individuals and humans as cultural components, is secondary to understanding the essence of the concepts.

Components of Creativity

Valuing Creativity and Creating Value

We begin our decomposition of creativity with the relationship between perceived creativity and attributed value. In Western society, *creativity* is most commonly used to refer to the embodied cognitive process that gives rise to pieces of music, sculpture, paintings, poems, and other things that are taken or presented as art. We include science and engineering in our list of creative endeavors.

Creativity is intensely context dependent: reproducing the style of Monteverdi in the twentieth century would be regarded negatively as pastiche or plagiarism or an exercise of style replication. Creativity is heavily dependent on the nature of the creator. For example, Harold Cohen's AARON painter program (McCorduck, 1991; http://www.aaronshome.com/) has made paintings that have hung in galleries and sold for thousands of dollars. His daughter was also an artist, producing the kind of drawings one might expect from a three-year-old, for which most people would not be inclined to pay. Cohen, however, rates his daughter's creativity as much greater than that of his program (Cohen, 1999). Accordingly, he makes a distinction between Big-C creativity and Little-C creativity, also seen elsewhere in the literature (Cohen, 1999), where Big-C is Picasso level and Little-C is what AARON can manage. Boden (1990) makes another, perhaps more tractable, distinction between *psychological creativity*—the act of generating an artifact that is novel and of value to an individual—and *historical creativity*—that of generating an artifact that is novel and valued in historical terms. However, this notion must be generalized: rather than two discrete kinds of creativity, value and novelty should not be thought of as simple quantities but as relations between observers and the created artifact. Thus, for example, we can account for cycles of fashion: retro styles may be valued by both teenagers and their parents, the former enjoying their (relative) novelty and the latter doing exactly the opposite. We return to the matter of novelty below.

Value is dependent not only on the observer but also on the context in which the observation is made. It is present in many more pursuits than the artistic ones already mentioned and in manifold ways. A prime example is mathematics, where the creation of the proof of a theorem is more highly valued if it is "elegant," according to the principles of the particular branch of mathematics to which it applies; mathematics has its own aesthetics, as does engineering. Often the aesthetic of one context is utterly incomprehensible, and even offensive, to observers comfortable in another. Consider, for example, the riot that followed the premiere of Stravinsky's *The Rite of Spring* in the 1920s. Thus, the value relation is between not just the observer and the artifact but between the observer and the artifact in a given context. Finally, value is also a function of the creator. Expectations are based on experience. We are disappointed when our favorite author, admired musician, or best-loved car company turns out a product that underperforms.

In summary, we treat value as a relation between an artifact, its creator, and its observers, and the context in which creation and observation take place.

Exploration, Transformation, and the Paradigm Shift

Boden (1990) also introduces an important philosophical distinction, between exploratory creativity, where the conceptual space being explored is fixed (though possibly not all visible, and possibly infinite) and exploration occurs within that space (e.g., different songs in a particular style) and transformational creativity in which the space itself is subject to change (developing from one style to another). Coupled with successful persuasion, transformational creativity is what leads to a paradigm shift in Kuhn's philosophy of science (Kuhn, 1962). Boden proposes that Little-C creativity is exploratory and Big-C creativity is transformational, but history is littered with exceptions to this. Mozart, for example, perfected a style that Haydn introduced, but Mozart is universally regarded as the greater creator. Wiggins (2006) shows that in any case, transformational creativity is formally exploratory creativity at the metalevel, where the conceptual space of artifacts is replaced by the conceptual space of conceptual spaces. This way of thinking, where the conceptual space can be taken to define the class of artifacts at which a creator is aiming, yields some elegant ways of discussing what happens when a creator pushes the boundaries of the expected, in a process taxonomized as different kinds of aberration by Wiggins (2006). This concept allows further objective, mechanistic description and prediction of creative behavior.

Boden's (1990) concept of exploratory creativity within a style may be relevant for some forms of animal communication. The songs of humpback whales within a population progressively change from year to year. Payne and Payne (1985) analyzed the structure of humpback songs recorded near Bermuda over nineteen years; figure 12.1 illustrates a typical structure. All of the whales recorded within a year produced similar songs, but songs separated by several years were made up of very different sounds. However, all humpback songs share basic structural similarities. For example, all humpback songs are made up of themes that occur in a regular sequence. Theme transitions are so regular that Frumhoff (1983) called the few backward transitions therein "aberrant" (though the

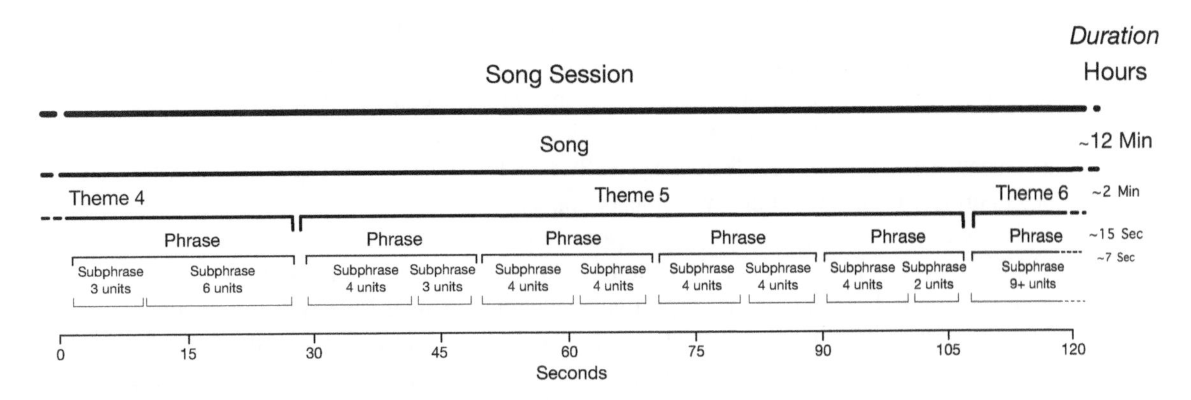

Figure 12.1
The hierarchical structure of all humpback whale songs, using a traced spectrogram. Timings are approximate. Vertical lines indicate phrase boundaries. After a diagram by Payne, Tyack, and Payne (1983). Progressive changes in the songs of humpback whales (*Megaptera novaeangliae*): A detailed analysis of two seasons in Hawaii.

common terminology here is coincidental). Other elements of the song are more variable. Each theme is made up of a varying number of phrases, and phrases are made up of a varying number of units. Payne and Payne (1985) report three kinds of theme in the humpback songs recorded in their sample from Bermuda: static, shifting, and unpatterned. The phrases in static themes are nearly identical in every repetition. Unpatterned themes lack the structure of repeated phrases. Successive phrases in shifting themes change gradually from one form to another. The variation in structure is somewhat like a theme and variations, but the same shifting theme recurs over and over within a year, so while they fit Boden's exploratory framework, their generation is not as free as that term might suggest. A striking feature about the songs of humpback whales is the contrast between how much the song changes from year to year, and the convergence of whales within a population using the same set of sounds and following several styles of theme to construct the song typical of any one time.

A restricted exploration of a language in this way can be modeled as a Boden conceptual space accompanied by a value measure, which filters out unvalued artifacts; we return to this below.

Creativity: Process or Property?

Boden's (1990) approach raises some interesting questions concerning the conceptual space and the attribution of value to artifacts in it. These things are separable, and the conceptual space is neutral with respect to both value and novelty, and it inherently captures cognitive generation, not the subsequent value or novelty of that which is generated. Thus, the paintings of Cohen (for he was a successful computer-less artist before AARON), and of AARON, and of Cohen's daughter, all coexist equally in the conceptual space of paintings; it is only when they are evaluated by an observer (possibly the artist) that issues of novelty and value arise. In a less Western-centric perspective, we might conflate these two and argue that novelty is a kind of value, since in some cultures, it does not have the high status accorded in the West, and in some, it is actively eschewed in favor of the strict maintenance of tradition. This feature of creativity in the social context does not decrease the importance of novelty in the evolutionary context.

Thus, we see that the production of the painting per se is not what guarantees its value. While, of course, the artifact must exist to be valued, it is interaction between production and (probably, at least initially, introspective) evaluation by an artist, and then by a social community, that identifies relative value and relative novelty of both the artifact and the way it was made. Thus, we can decompose creativity into a series of steps and tests within a process, of which a "creative" agent is capable, and can begin to study it. This is altogether more scientifically tractable than the philosophical debate about the ineffable nature of creativity itself.

Size Does Not Matter

Given the nature of the conceptual space as distinct from the novelty and value of the concepts in the space, a natural question to ask is: Need there be a difference in kind between big-C and little-C creativity? For some authors, the answer is clearly no. Plotkin (1998) describes creativity as the sine qua non of everyday language generation. For others, the word should be applied only to the great creators of great historical import.

From the perspective of this chapter, this latter view is destructively problematic. We aim here to understand what evolutionary advantage may have been given to humans and animals by the ability to be creative. At the extreme level, it is hard to argue for evolutionary advantage in the authorship of very large-scale created constructs such as symphonies. However, it has been argued that sexual selection may be a factor in smaller creativity (Bown & Wiggins, 2009; van den Broek & Todd, 2009). Thus, if we were to restrict our definition to great human creators, ruling out minor creative acts, we would also rule out a priori the possibility of the incremental development of creative faculties over evolutionary time. Instead, it is necessary to look for the roots of that ability in both humans and nonhumans, with a view to understanding how the extreme ("great creativity" in the terms of the relevant culture) emerged from the ordinary (everyday creative activity). One unbiased way of approaching the question of how creativity evolved is to deconstruct the components and explore which ones exist in nonhuman animals and to what degree.

Equally, there is no scientific evidence to support the position that the ability to create did not evolve, step by step, as opposed to merely appearing fully formed in humans, and there is evidence of creativity or protocreativity in other species, in both animals belonging to the same direct evolutionary lineage (Miller, 1997) and those more distantly related (Noad, Cato, Bryden, Jenner, & Jenner, 2000). Therefore, when studying the development of creativity in our own and other species, it is necessary to admit and value the creation of less-than-amazing artifacts (as we do in our children) in order to encompass the overall development of the faculty over evolutionary and ontogenetic time.

Novelty and Its Perception

Another key dimension of creativity is novelty, and the ability to perceive it. In Western culture, the attribution of creativity entails the attribution of novelty; various authors have argued that the human creative drive is the search for novelty (Martindale, 1990) or, differently termed, curiosity (Saunders, 2002). While this is not the case in all cultures, the fact remains that novelty detection is a feature of creative behavior—whether it is a feature to be valued or (in some contexts) suppressed. Regardless of one's response to novelty, the fact that one can respond to it means that it can be detected, and we propose that this is a fundamental component of creative behavior. However, too much novelty prevents recognition, a fact embodied in the famous Wundt curve of hedonic response to novelty (Margulis & Beatty, 2008; Wundt, 1874). The inverted-U shape captures the notion that

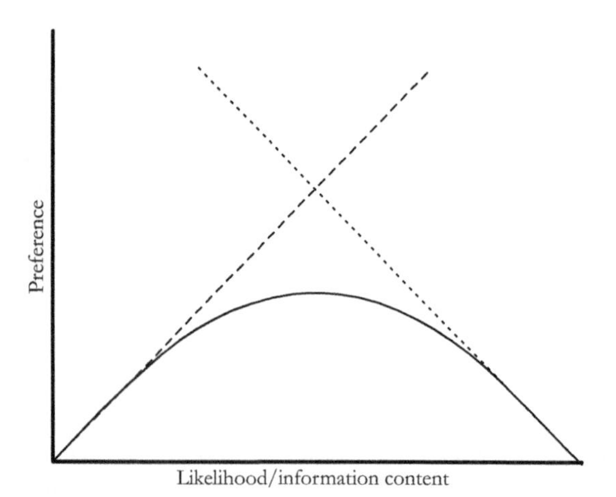

Figure 12.2
The Wundt curve illustrates the rise and fall of preference (y-axis) in perceivers for complexity of stimulus (x-axis). Very simple stimuli are uninteresting, while extremely complex ones are inaccessible, either case producing dissatisfaction. Intermediate levels of complexity are preferred.

not enough variation is boring, while too much is unpleasantly incomprehensible, yielding a sweet spot in between. This is illustrated in figure 12.2.

Novelty detection is a requirement for noticing changes in the environment, a feature all animals need for survival. A pattern that deviates from the known is novel and can signal good things (a new food source) or bad things (a new type of predator). Thus, animals carefully need to balance the exploration of novelty because it can open up new niches that enhance evolutionary fitness or are detrimental to it. In this context, novelty detection as a complex behavioral trait seems to depend partly on particular gene variants in both humans and songbirds. A number of studies have found gene polymorphisms in the dopamine receptor 4 associated with variation in novelty seeking and exploration behavior (Korsten et al., 2010; Schinka, Letsch, & Crawford, 2002). The ability to detect novelty in the environment also allows animals to detect novelty in behavior of conspecifics. Famous examples are the cultural transmission of novel behavior through a population as observed for sweet potato washing in Japanese macaques and opening the aluminum foil covers of milk bottles by British tits (Shettleworth, 2010).

Huron (2006) extends this argument to affective response, exapted to music. Because the outcome of a novel experience is sometimes dangerous, it is appropriate for an animal to be alert and prepared for fight or flight in the face of novel circumstances. Thus, there is evolutionary incentive to perceive not just danger but uncertainty and novelty in their own right. In humans, this situation is experienced as tension, leading to arousal and, in extremis, to fear; simple observation suggests that other species share the same affective response. The experience of tension entails its subsequent release, which seems to be accompanied by positive affective states. Huron (2006), following Meyer (1956), suggests that tension stimulated by expectation, and its denial or fulfillment, is in large part responsible for affect stimulated by Western music, whose emotive content is frequently theoretically conceived as an ebb and flow of tension of various kinds. This affective experience is highly valued and is altogether more subtle and dynamic than the common labeling of emotional analysis of music as, for example, tender or sad (Juslin & Sloboda, 2010; Skowronek, McKinney, & van de Par, 2006).

It is possible to mathematically model expectations over a well-defined symbol system (musical melodies constructed from a known range of pitches and durations, or bird- or whale-song categorized into appropriate symbol sets) using uncomplicated statistical techniques (Manning & Schütze, 1999). Zuidema et al. discuss the various ways of learning these kinds of sequential structures in chapter 11 in this book. From these models, human melodic expectations can be estimated (Conklin & Witten, 1995; Pearce & Wiggins, 2012) and birdsong can be modeled (ten Cate & Okanoya, 2012). Pearce's model of Western tonal musical melody, IDyOM (Information Dynamics of Music: Pearce, 2005; Pearce & Wiggins, 2012), predicts human expectations very well ($r = 0.91$ in four studies; Pearce, Herrojo Ruiz, Kapasi, Wiggins, & Bhattacharya, 2010; Pearce & Wiggins, 2006). Expectations are expressed as probability distributions over the set of symbols allowed (musical pitches, here). Given such a probability distribution, we can estimate the unexpectedness of an event drawn from it, using Shannon's (1948) information theory. It is important to understand that this property is relative. It is computed in terms of the statistical model, so unexpectedness is relative to the information that the model contains about the set of sequences being modeled and to the immediately precedent sequence. Thus, we can model an individual's memory and predict the unexpectedness of perceived events. Two quantities, entropy and information content, model uncertainty and unexpectedness, respectively (Pearce & Wiggins, 2012). More recent work on physiological and behavioral measures of human response to live music suggests that the unexpectedness value of pitch, calculated as above, explains a significant part of the variance in physiological measures (heart rate, skin conductivity) that correspond to arousal (Egermann, Pearce, Wiggins, & McAdams, 2013). This constitutes evidence that unexpectedness in music correlates with arousal in listeners, supporting Huron's (2006) hypothesis concerning the source of enjoyment in musical listening and that both correlated with the predictions of the model. These model-driven empirical methods can be applied to any form of vocal communication given enough examples.

Ikebuchi, Futamatsu, and Okanoya (2003) showed that female Bengalese finch hearts respond with tachycardia to more complex male song with higher information content. This is a result comparable to the human musical response we have already outlined (Egermann et al., 2013). Further investigation of these phenomena using the models we introduce in the "Modeling the Process of Creativity" section may yield understanding of the relationships between birds' reaction to song and humans' reaction to music.

A recent study by Weiss, Hultsch, Adam, Scharff, and Kipper (2014) investigated the song of nightingales, who are famous for their quite variable delivery of hundreds of different song types during European summer nights (figure 12.3A). Looking at the sequential organization of nightingale song, scientists discovered two patterns.

Particular song types could act as branch points; for example, they could be followed by many different song types. Other song types acted as bottlenecks; for example, many different song types converged onto a particular bottleneck song type (figure 12.3B). Importantly, Weiss and his team found that when nightingales heard a playback containing many song types with branch transition patterns, they responded with song types with bottleneck transition patterns. Conversely, when they heard song types with bottleneck transition patterns, they responded with song types that tended to be branching transitions in their population—that is, they responded with the unexpected (figure 12.3C). While it remains unclear why this behavior would arise, the fact that it does so entails the ability to detect high and low entropy distributions or high and low information content, as in Pearce et al.'s (2010) human studies and Huron's (2006) evolutionary argument.

Here, then, is a scientific question that the study of creativity can ask: Do species other than humans exhibit similar responses to novelty or complexity, and if so, how does their behavior inform our understanding of our own? Given a sufficient amount of song

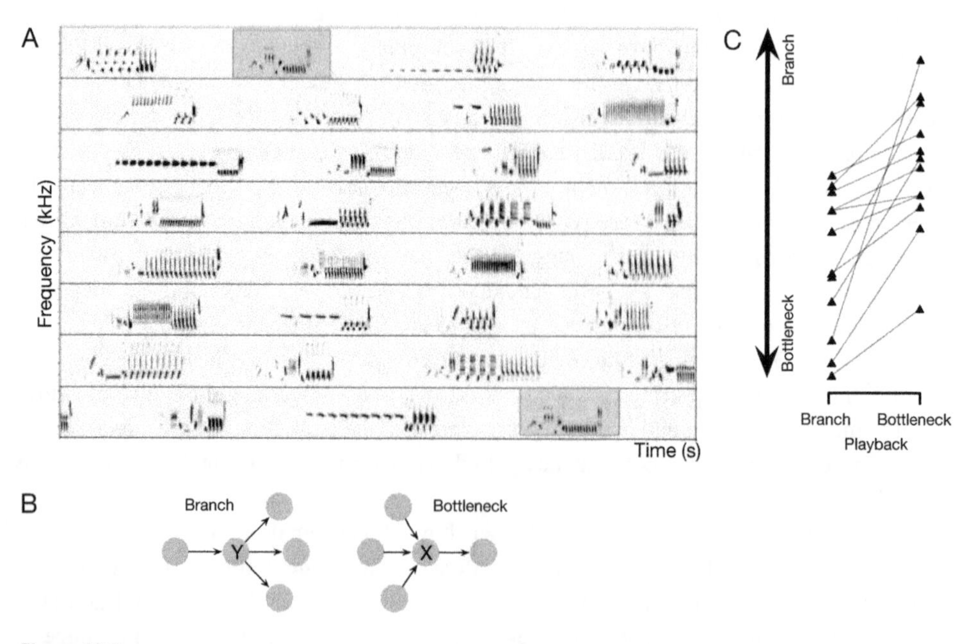

Figure 12.3
Patterns in nightingale song. (A) Time-domain audio waveforms show the bursts of song. (B) Illustration of the branch/bottleneck distinction (figure 1c from Weiss, Hultsch, Adam, Scharff, & Kipper, 2014). (C) Alternating correspondence between song types (figure 3, Weiss et al., 2014).

produced by a particular species or even an individual, we can construct a model of the sequences using the techniques we have set out and generate new sequences from it, with particular information-theoretic properties (e.g., surprising, neutral, or very obvious). We can monitor the response of the relevant animal to the constructed sequence by means of judicious audio editing and thus test hypotheses regarding the value of novelty and complexity in vocal display. This view of song construction raises the possibility that the song is valued by its own species for some of the same reasons that humans value music: the affect of rising and falling tension caused by unexpectedness, information content, or complexity. This is a testable hypothesis.

Modeling the Process of Creativity

To study creativity effectively, we need a rigorous frame of reference, including the ability to simulate perception and creative generation. Historically, there are not many scientific theories of creativity, and those that do exist are fundamentally qualitative. We now survey them, in contrast with a newer, quantitative approach.

Wallas (1926) focuses on the cognitive process of creativity. He identifies four parts of a sequence: preparation, in which the creative goal is identified and considered; incubation, during which conscious attempts at creativity are not made; illumination, the moment of enlightenment when an idea appears in conscious awareness, sometimes called the "aha!" moment; and verification, in which the new idea is applied. These ideas highlight a further distinction that is useful in focusing on creativity: that between conscious, or deliberate, creativity and nonconscious, or spontaneous, creativity (Wiggins, 2012). The former of these is the creativity where, for example, a professional composer must produce a TV theme in too short a time to wait for inspiration; she consciously applies rules of her craft to create what is necessary. The latter is the creativity where an idea or concept appears in one's awareness, apparently without bidding, effort, or intention, in the way Mozart described as the beginning of his mode of creativity (Holmes, 2009, 317–318):

> When I am, as it were, completely myself, entirely alone, and of good cheer say traveling in a carriage, or walking after a good meal, or during the night when I cannot sleep; it is on such occasions that my ideas flow best and most abundantly. Whence and how they come, I know not; nor can I force them. Those ideas that please me I retain in memory, and am accustomed, as I have been told, to hum them to myself.
>
> All this fires my soul, and provided I am not disturbed, my subject enlarges itself, becomes methodized and defined, and the whole, though it be long, stands almost completed and finished in my mind, so that I can survey it, like a fine picture or a beautiful statue, at a glance.

Most human creativity processes, including Mozart's overall description, are probably a cyclic combination of the two. Wallas (1926), however, is considering spontaneous creativity resulting from earlier conscious consideration, and he considers the illumination point to be the arrival of a spontaneously produced concept in consciousness: the "aha!" moment. Wiggins and Bhattacharya (2014) propose that the "noise" encountered in electrophysiological studies during reported cognitive resting states may be the incubation process at work.

Wallas's theory requires created artifacts to undergo validation, where they are examined to make sure they are fit for purpose. This may suggest that the theory is meant to account for larger-scale acts of creativity than, for example, spontaneous sentence production, or maybe successful communication of meaning would fulfill the definition in this example. In any case, the theory does not propose an underlying mechanism but rather describes a series of stages. As such, it at most provides an overarching framework for the study of creativity.

Guilford's (1967) model is more qualitative but does not contradict Wallas. Guilford proposes a phase of divergent thinking, where possibilities are opened, followed by one of convergent thinking, in which the creator homes in on her idea. Both phases could happen consciously or nonconsciously, and one can also imagine repeating cycles of the two phases. The model has less predictive power than Wallas's, however, and we will not refer to it further. More recently Csikszentmihalyi (1996) described the subjective experience of creativity, involving the state of flow; again, this lacks quantitative analysis and predictive power. A final theory worthy of mention is that of Koestler (1964). The cognitive operation of bisociation is proposed, enabling cognitive structures representing two or more ideas to be combined to produce new concepts. This theory, though convincing, is not specified with mathematical precision.

None of the four frameworks we have outlined affords a quantitative means to examine creative processes in detail. A more recent hypothetical mechanism for a cognitive creative process is provided by the information dynamics of thinking (IDyOT) cognitive architecture (Forth, Agres, Purver, & Wiggins, 2016; Wiggins, 2012; Wiggins & Forth, 2015), based on Baars's (1988) global workspace theory and using the same information-theoretic notions as the information dynamics of music (IDyOM) model already cited (Pearce & Wiggins, 2012). The key idea is that cognitive creativity is a result of prediction, which itself is a means for managing information and action in the world. Statistical generators continually predict outcomes from sensory inputs, based on statistical models trained by unsupervised observation. They compete in terms of the information content of their predictions (quantified in terms of Shannon information theory: Shannon, 1948) for access to the global workspace (GW), which equates with conscious awareness. When an item enters the GW, it may be novel, or it may be a predictable part of an ongoing experience. In the former case, creativity has happened, and passage into the GW corresponds with Wallas's moment of illumination, the preceding activity being incubation. What enters the GW is recorded in memory and becomes available for future prediction, and thus the cycle repeats. This theory gives a concrete mechanism for creative production and is applicable directly to discrete and continuous symbolic data represented on a computer. Thus, it can be applied to transcriptions of birdsong and whale song, with a view to comparing their information-theoretic properties. This approach, then, can be used directly on data to make testable predictions about animal behavior, as it has done for humans.

Affording Creative Behavior

Darwin described two primary mechanisms of selection as driving biological evolution: natural selection and sexual selection. The critical elements for evolution by natural selection are variation in traits within a population, differential reproduction of animals with the differing traits, and inheritance of the trait from one generation to the next. Sexual selection can be viewed as a special case of natural selection that acts on an individual's ability to mate. Some traits—for example, ones that increase fighting ability—may improve an individual's ability to compete with members of the same sex for mating, while others—such as ornaments or song—may make a member of the other sex more likely to select an individual for mating.

The topic of mate choice is important for our discussion of selection for creative behavior, especially for creativity in communication. Biologists have investigated a variety of modes of sexual selection for mate choice. The simplest selection would be for a characteristic that provides a direct benefit, such as when a female bird chooses a male whose genes produced a tail of the optimal size for flight. But suppose males also use the tail in a display to impress females. Females might have a sensory bias to choose males with even larger tails than optimal for flight because the display is more visible (Endler & Basolo, 1998). Here sexual selection might drive the evolution of tails that are longer than optimal under natural selection. And if a population of females tends to have a preference for longer tails, this could lead to a runaway process of evolution of longer and longer tails until the benefit from sexual selection is outweighed by other natural selection pressures (Fisher, 1930). The evolution of large complex ornaments in males raises the question of why a female should choose a male with a trait that may make it more visible to predators and less able to escape. Zahavi (1975) argued that males with such a handicap might have to be better quality, thus suggesting that handicaps help a female choose a better-quality male.

How does creative behavior fit onto these categories? Creative behavior could result in biological selective advantage in all the above cases. Perhaps the most celebrated case of animal innovation involves a young female Japanese macaque that invented the idea of washing the sand off potatoes in the ocean and then, three years later, the idea of separating grain from sand by throwing the mixture in water and scooping out the floating grain. Both of these innovations would be selected because they improve foraging. This kind of innovation is particularly important in species capable of social learning so that beneficial innovations diffuse through the population. Creative behavior may also be the substrate for sexual selection: mimicking the sounds of other birds and adding them to his own repertoire of song may signal to the female lyre bird that her mate has particularly good cognitive skills that will also help to raise their young and pass on his intelligence as well. This logic has been applied to a more specific issue for the songs of birds. Nowicki, Peters, and Podos (1998) pointed out that the nuclei in the brain that control song develop during critical periods of development. If a young bird does not have adequate nutrition at this time, it may suffer broader developmental problems. They reasoned that large and complex repertoires of song may indicate a history of good nutrition, and they suggest that females might select males with large and complex song repertoires for this reason. Or creative behavior could evolve as by-product of something else: the need to explore to find new food sources or new territories might have selected animals that are less neophobic and more curious, leading to more novel behaviors—not all of them necessarily beneficial to survival and reproduction. Thus, the expense of creative behavior in terms of time, energy, and risk, which might at first seem problematic, can be motivated in biological terms in terms of introducing beneficial behaviors, creating a particularly attractive display, or as a demonstration of a valuable capacity that underlies creativity itself.

However, while the substrate of the variability required for biological evolution is genetic, the behaviors we are considering are complex, learned, and cultural, involving not just the generation of short sequences grounded in action but substantial long-term abstract sequence production. The larger question is therefore, "Why and when is there selection for innovation—forming new combinations of behaviors, versus reliance on unlearned behaviors or social learning of successful behaviors?" Laland (2004) discusses strategies animals might use for selecting when to rely on unlearned behaviors, when and who to copy in social learning, and when to innovate. However, he is primarily considering instrumental behaviors for solving nonsocial problems rather than learning about signaling for communication.

We next consider cases of animal communication that appear to be examples of creative behavior in the terms proposed here and then discuss how the dynamics of the communication might be quantitatively studied using a computational framework such that we propose.

Creativity in Animal Communication

The very attribution of the word *song* to the vocal communication behaviors of birds and whales is based on the problematic romanticization of that phenomenon, akin to the romanticization of creativity. First, then, we must dissociate ourselves from the metaphorical notion of theatrical or concert hall performance and focus instead on the functional, communicative aspects of the behaviors (Trehub, Becker, & Morley, 2015). Here we follow Fitch (2006) in defining songs as learned complex vocalizations. A comparable danger is the naive assumption that the behaviors we will describe are due to the same mechanisms as superficially similar behaviors in humans. Indeed, this claim is one we would like to test. One means of doing so might be through the observation-based model of Wallas

(1926). However, it is hard to know whether the preparation and incubation phases exist in animals; they cannot be asked, and current understanding of the human mechanisms at the neural level is not detailed enough to make a search for comparable effects in animals possible. We are currently limited to measures such as EEG-frequency bandpower that do not explain mechanism (e.g., Schwab, Benedek, Papousek, Weiss, & Fink, 2014). Illumination and verification may be more accessible because they may manifest behaviorally, as when an animal immediately repeats material once it has been internalized. More work is needed in this area.

However, there is evidence of the effect of information content and entropy with respect to a context on humans, measurable directly from physiological responses (Egermann et al., 2013) and information content on birds (Okanoya, 2012). A more direct comparison of these two phenomena can be made using Pearce et al.'s (2010) tripartite empirical approach: a computer program is used to embody the proposed mechanism, and its predictions are then tested empirically with both behavioral responses and electrophysiological measures. Here, the idea is extended to a comparison between species.

The setting in animal communication where the concept of creativity seems most relevant concerns reproductive advertisement displays called songs, which are a product of sexual selection. In some animal species, the songs of each individual singer are learned through listening to the songs of other individuals. When one individual learns the song of another, the copy will probably not be perfect; there may be errors in the stored memory and differences in the vocal production apparatus between individuals. This process of vocal copying within a community of animal singers leads to vocal traditions, which may be formalized as conceptual spaces, that often map onto habitats as geographical dialects in song; similar effects of vocal tradition, coupled with migratory patterns, arise in human folk music (Pamjav, Juhász, Zalán, Neméth, & Damdin, 2012). However, there are also situations when an animal actively appears to innovate, producing sounds that are more novel than would be expected to arise from copy errors alone in a step akin to transformational creativity. Work in this area focuses on what is copied, at the expense of studying the "unrecognizable" new material: the corollary, invention of new song types, seems not to have been studied formally.

Kroodsma (1983) suggests a relationship between site fidelity and mode of vocal learning in birds, with high site fidelity being correlated with imitation and low site fidelity correlated with improvisation. One possible explanation for this would be the need to associate a clear signal with territory so as to mark it, but also for the signal to vary as fledglings leave the nest and lay out their own territorial boundaries: it is necessary first to innovate and then to fix so as to identify a difference between the territories, while maintaining recognizability to members of one's own species for the purposes of sexual attraction.

Trehub et al. (2015) and Savage, Brown, Sakai, and Currie (2015) analyze cross-cultural similarities in music performed by humans. They emphasize that many shared traits relate to roles that music plays in maintaining group coordination and cohesion. Some animal species such as killer whales, *Orcinus orca*, learn to form group-distinctive repertoires of calls that are also used to maintain group coordination and cohesion (Ford, 1991; Yurk et al., 2002). New groups are thought to be formed by fission of a larger group. The correct amount of innovation is required for each group to develop a group-distinctive dialect while at the same time changing slowly enough that closely related groups, called vocal clans, retain enough shared calls to provide information about relatedness between groups. If there was not the correct balance between innovation and stability as young individuals learn the calls of their group, then groups and clans could not distinguish one another.

In both of these cases of song learning in songbirds and call learning in killer whales, the ability to recognize and value just enough variation is paramount. This notion of just enough corresponds with the maximum of the Wundt curve; it also corresponds with a middling, moderate value of information content, as measured by the models outlined in the "Modeling the Process of Creativity" section.

The development of song through vocal learning is common among songbirds but rare among nonhuman mammals (Janik & Slater, 1997). Some of the best evidence comes from the songs of bats (Morell, 2014) and humpback whales. At any one time, the songs of different individuals within a population of humpback whales are quite similar (Guinee, Chu, & Dorsey, 1983), but many acoustic features of the songs change rapidly enough that they can be tracked from month to month (Payne, Tyack, & Payne, 1983). The changes are progressive over time in the sense that if a sound is increasing in frequency or decreasing in duration, that trend is likely to continue for some time rather than vary randomly. The rapidity of the song change coupled with the similarity among whales at one time makes it difficult to identify whether some individuals are innovators that are copied, especially since it is so difficult to make repeated recordings from the same individual at different times given such a large and mobile population. However, it is clear that a strong pressure for conformity must drive each whale to copy the song of the moment, while at the same time there must be a selection for specific innovations that are picked up by the population to change the vocal tradition. Until now, there has been no way of studying this process or identifying the benefits and process of innovation. Computational modeling based on creativity theory may help.

Following the Boden analysis of creativity and our subsequent suggestion that value and novelty should be thought of as relations between observers and the created artifact, we can evaluate differences in the value of novel sounds produced by different singers in terms of whether they are copied by others. We know little about the psychological process by which an individual animal generates a novel sound or about what value the sound may offer to that individual. But in parallel with our explicitly relativistic version of Boden's "historical" creativity, we can study what novel sounds are incorporated into the vocal tradition of the population, modeling the whale song as a conceptual space.

A striking case of adoption of novel songs involves the song of the humpback whale. Two populations of humpback whales winter off the coast of Australia: one on the east coast and one on the west. Males sing on their winter breeding grounds and as they migrate to and from the breeding grounds. These two populations are separated by thousands of kilometers. With little interchange, each population is characterized by one song at any one time, and the songs of the two populations are usually very different. However, in 1996, Noad et al. (2000) noticed an unusual pattern when two of eighty-two singers recorded off the east coast produced a song that was completely different from the rest of this population but matched the 1996 song of the west coast population. This movement is illustrated in figure 12.4.

During 1997, some songs mixed features of both west and east coast whales, but by the end of the year, nearly all of the east coast whales had switched to singing the west coast song. By 1998, no whales were left singing the old east coast song, and all had switched to the west coast song. The rarity of west coast songs recorded in the east during 1996, coupled with the following independent evolution of the west coast song on both coasts, led Noad et al. (2000) to conclude that only a few singers transferred from west to east during 1996, bringing the new vocal tradition with them. This rapid and complete replacement of one vocal tradition with another suggests recognition of a value for very specific kinds of novelty is what drives the change in the song, even when this is usually a less radical process driven from within the population.

Analysis of songs recorded from 1998 to 2008 from eastern Australia and the other populations of the South Pacific show a remarkable pattern. Garland et al. (2011) report that

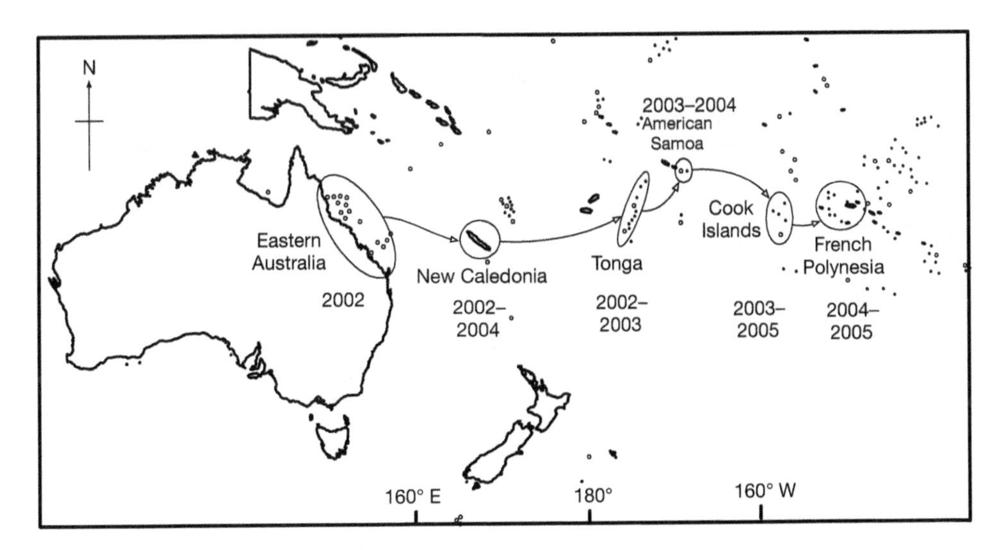

Figure 12.4
Map showing the trajectory of whale song change across the ocean basin. The marked trajectory follows the song first detected in 2002 in Eastern Australia, song 5 in table 12.1. Note that the arrow indicates general direction and not specific whale movement. (After supplementary material from Garland et al., 2011.)

over this time period, eight different song types originated in the eastern Australia population and spread over several years across six humpback populations from west to east, all the way to French Polynesia, 5,000 kilometers away. They suggest that as with the uptake of a new song as reported by Noad et al. (2000), diffusion of a vocal tradition occurs when individual males from adjacent populations spend enough time together for one to learn the others' song. However, this does not explain the directionality of information transfer. Data on the movement of individuals from one population to adjacent ones suggest that this is bidirectional with no bias to the eastern direction in which song information flows. Garland et al.'s (2011) suggestion for the remarkable directionality of the change is that the eastern Australia population is much larger than the others. While this may account for a more likely flow of animals from eastern Australia to the adjacent population to the east, it fails to account for the broader eastward pattern of information flow. These patterns are summarized in table 12.1.

One way to think about this pattern from the current perspective would be to consider the value of particular innovations within the context of a particular vocal tradition at a particular time. The 1998 song in eastern Australia was the song originally from western Australia that was valued so highly that it swept through the population in 1997. At this point, this song started to evolve within the eastern Australia population, and at the same time, its high value made it likely to spread to populations to the east. Given the time this took and the speed at which song evolves within a population, the large eastern Australia population

Table 12.1
Summary of pattern transmission in South Pacific whale song

Year	East Australia	New Caledonia	Tonga	American Samoa	Cook Islands	French Polynesia
1998	1	1	1		2	2
1999	1	1	1			1
2000	3	1	1/4/4			
2001	3	3	4/3/3		1/4/4	1/1/4
2002	5	5	5			
2003	6	5	5	5	5	
2004	6	5/6/6		5	5	5
2005	7	6		6	5	5/6/5
2006	8	7	7	7	7	5/6/5
2007	9	8	8		8/8/10	10/7/10
2008	11	9	9		10	10

Source: After Garland et al. (2011).
Note: Song types identified in the South Pacific region from 1998 to 2008. Populations are listed from west to east across the region. Each number represents a distinct song type. Empty cells represent no data available. Three numbers within a year/location indicate different song types were present, early/middle/late season. The patterns of eastward transmission are clearly visible.

had an advantage in being more likely to offer high-value changes within the shared vocal tradition, and these high-value changes would maintain the directionality as they spread to other populations to the east. Once this dynamic was set up, if the easternmost populations were several years behind in the process of innovation and selection for value, then it was less likely for any innovations in this setting to spread west.

There has been growing interest in studying the strategies animals might adopt when they learn from others. Laland (2004) points out that the costs and benefits of social learning depend on the context, and he suggests more attention be paid to strategies of when to copy and whom to copy. One strategy as to when suggests copying another singer when the copier's current behavior is unproductive. From the perspective of a singer, this would suggest copying if you are not attracting females or are failing in competition with males. One strategy as to whom might be to copy the majority, which would lead to conformist behavior. Another is to copy the most successful. If singing whales can monitor the success of others and if successful whales have variations in their song, this could drive a process of change, although it is difficult to see how it would lead to the progressive evolution observed most of the time in humpback song. McLoughlin et al. (2016) used a spatially explicit multiagent model to test what different song patterns are generated by different movement, grouping, and learning patterns. Preliminary results generated different songs from different patterns but convergence within populations. We still do not understand what drives the conformity in humpback song, what drives the selection of specific novelties, and how or whether this is driven by sexual selection.

The movement and variation of whale song bear comparison with the movement of human music during migration. Pamjav et al. (2012) conducted a large study of musical melody styles for thirty-one Eurasian nations. They found that close musical relations indicate close genetic relations ($F_{ST} < 0.05$;[1] Wright, 1969) with probability 82 percent. This is one of the largest studies ever done of folk music and almost certainly the largest computational study: they used databases of 1,000 to 2,500 melodies for each of the thirty-one cultures. The notion of musical similarity here revolves around a Euclidean distance metric derived from a self-organizing map (SOM; Kohonen, 1998), and this is an area requiring further validation: musical similarity is strongly context dependent, and the workings of a SOM are somewhat inscrutable. Nevertheless, this work presents an interesting opportunity, given the models of music that we propose, to compare the whale song behavior with the human musical behavior in detailed and explicable ways.

A Research Program on Creativity in Vocal Communication in Humans and Nonhumans

In this chapter, we have identified parallels between human and animal vocal communication behaviors at the immediate phenomenological level and suggested that they are worthy of further investigation in the context of creativity research.

We decomposed the notion of creativity into an objective process of generation, coupled with a combination of relative value judgments, some of which, notably novelty, can be objectively modeled. This added objectivity allows us to ask questions that were not previously scientifically formulable regarding the nature of vocal communication, its effect on humans, and other species, and the mechanisms that underlie it.

We have deployed Boden's philosophical approach to human creativity (Boden, 1990) to hypothesize a possible explanation for new song construction in migrating whales and identified evidence of music migration in humans. We have presented evidence from normally separate research fields of comparable physiological responses to aural sequence perception in birds and humans, which might suggest similar processes at deeper levels, suggesting a computational method by which these empirical studies can be implemented.

We propose that when we examine the evolution of vocal communication in animals and humans from the perspective of creativity, we can shed new light on processes that seem to be common (though probably not commonly derived) between very distantly related species. Therefore, we suggest that the philosophical framework we have outlined is a potentially fruitful means of addressing the communicative behavior of animals that improvise (individually or collectively) and perhaps understanding better the mechanisms that underlie human communication and human creativity.

More specifically, we can propose:

- Comparative studies on heart rate and other physiological and electrophysiological measures in birds and in humans in response to complex aural stimulation, relative to a known vocal communication form; subsequent neural studies to seek neural correlates of information content (Pearce et al., 2010)
- Comparative studies on the dynamics of whale migration and song variation as compared with the dynamics of human migration and song variation; subsequent modeling to compare the processes, novelty, and complexity involved
- The development of new measurement techniques to allow the physiological and neural analysis of birds, whales, and other improvising animals to be compared with human analysis, and then modeled as we have described in this chapter
- Studies of species stimulated with utterances generated by computational models of their own communication, enabling direct measurement of the information-theoretic properties of the interaction and relating them to its observed effects.

We believe that these approaches and others entailed by questioning the relationship between creative behavior in humans and the superficially similar behaviors in other species offer a new and exciting approach to understanding the cognitive mechanisms involved in both vocal communication and creativity.

Acknowledgments

We gratefully acknowledge the advice and support of Björn Merker and two anonymous reviewers in assembling this chapter. We gratefully acknowledge the important work of Henkjan Honing in organizing the "What Makes Us Musical animals" workshop, kindly hosted by the Lorentz Center. G.A.W. is funded by the Lrn2Cre8 and ConCreTe projects, which acknowledge the financial support of the Future and Emerging Technologies (FET) program within the Seventh Framework program for Research of the European Commission, under FET grant numbers 610859 and 611733. P.L.T. acknowledges the support of the MASTS (Marine Alliance for Science and Technology for Scotland) pooling initiative. MASTS is funded by the Scottish Funding Council (grant reference HR09011) and contributing institutions.

Note

1. F_{ST} is the correlation between two randomly sampled gametes from the same sub-population when the correlation of two randomly sampled gametes from the total population is set to zero.

References

Baars, B. J. (1988). *A cognitive theory of consciousness.* Cambridge: Cambridge University Press.

Boden, M. A. (1990). *The creative mind: Myths and mechanisms.* London: Weidenfield and Nicholson.

Bown, O., & Wiggins, G. A. (2009). From maladaptation to competition to cooperation in the evolution of musical behaviour. *Musicae Scientiae, 13*(2), 387–411. doi:10.1177/1029864909013002171

Cohen, H. (1999). Colouring without seeing: A problem in machine creativity. *AISB Quarterly, 102,* 26–35.

Conklin, D., & Witten, I. H. (1995). Multiple viewpoint systems for music prediction. *Journal of New Music Research, 24*(1), 51–73. doi:10.1080/09298219508570672

Csikszentmihalyi, M. (1996). *Creativity: Flow and the psychology of discovery and invention.* New York: HarperCollins.

Egermann, H., Pearce, M., Wiggins, G., & McAdams, S. (2013). Probabilistic models of expectation violation predict psychophysiological emotional responses to live concert music. *Cognitive, Affective and Behavioral Neuroscience, 13*(3), 533–553. doi:10.3758/s13415-013-0161-y

Endler, J. A., & Basolo, A. L. (1998). Sensory ecology, receiver biases and sexual selection. *Trends in Ecology and Evolution, 13*(10), 415–420. doi:10.1016/s0169-5347(98)01471-2

Fisher, R. A. (1930). *The genetical theory of natural selection.* Oxford: Clarendon Press. doi:10.5962/bhl.title.27468

Fitch, W. T. (2006). The biology and evolution of music: A comparative perspective. *Cognition, 100*(1), 173–215. doi:10.1016/j.cognition.2005.11.009

Ford, J. K. B. (1991). Vocal traditions among resident killer whales (*Orcinus orca*) in coastal waters of British Columbia. *Canadian Journal of Zoology, 69*(6), 1454–1483. doi:10.1139/z91-206

Forth, J., Agres, K., Purver, M., & Wiggins, G. A. (2016). Entraining IDyOT: Timing in the information dynamics of thinking. *Frontiers in Psychology, 7,* 1–19. doi:10.3389/fpsyg.2016.01575

Frumhoff, P. (1983). Aberrant songs of humpback whales (*Megaptera novaeangliae*): Clues to the structure of humpback songs. In R. Payne (Ed.), *Communication and behavior of whales* (pp. 81–127). Boulder, CO: Westview Press.

Garland, E. C., Goldizen, A. W., Rekdahl, M. L., Constantine, R., Garrigue, C., Hauser, N. D., et al. (2011). Dynamic horizontal cultural transmission of humpback whale song at the ocean basin scale. *Current Biology, 21*(8), 687–691. doi:10.1016/j.cub.2011.03.019

Guilford, J. (1967). *The nature of human intelligence*. New York: McGraw-Hill.

Guinee, L. N., Chu, K., & Dorsey, E. M. (1983). Changes over time in the songs of known individual humpback whales (*Megaptera novaeangliae*). In R. Payne (Ed.), *Communication and behavior of whales* (pp. 59–80). Boulder, CO: Westview Press.

Holmes, E. (2009). *The life of Mozart: Including his correspondence*. Cambridge: Cambridge University Press.

Huron, D. (2006). *Sweet anticipation: Music and the psychology of expectation*. Cambridge, MA: MIT Press.

Ikebuchi, M., Futamatsu, M., & Okanoya, K. (2003). Sex differences in song perception in Bengalese finches as measured by cardiac response. *Animal Behaviour*, *65*(1), 123–130. doi:10.1006/anbe.2002.2012

Janik, V., & Slater, P. J. B. (1997). Vocal learning in mammals. *Advances in the Study of Behavior*, *26*, 59–99. doi:10.1016/s0065-3454(08)60377-0

Juslin, P., & Sloboda, J. (2010). *Handbook of music and emotion: Theory, research, applications*. Oxford: Oxford University Press.

Koestler, A. (1964). *The act of creation*. London: Hutchinson.

Kohonen, T. (1998). The self-organizing map. *Neurocomputing*, *21*, 1–6. doi:10.1016/S0925-2312(98)00030-7

Korsten, P., Mueller, J. C., Hermannstädter, C., Bouwman, K. M., Dingemanse, N. J., Drent, P. J., et al. (2010). Association between DRD4 gene polymorphism and personality variation in great tits: A test across four wild populations. *Molecular Ecology*, *19*(4), 832–843. doi:10.1111/j.1365-294x.2009.04518.x

Kroodsma, D. E. (1983). *Acoustic communication in birds*. New York, NY: Academic Press. doi:10.1016/c2009-0-03022-0

Kuhn, T. S. (1962). *The structure of scientific revolutions*. Chicago: University of Chicago Press.

Laland, K. N. (2004). Social learning strategies. *Animal Learning and Behavior*, *32*(1), 4–14. doi:10.3758/bf03196002

Manning, C. D., & Schtze, H. (1999). *Foundations of statistical natural language processing*. Cambridge, MA: MIT Press.

Margulis, E. H., & Beatty, A. P. (2008). Musical style, psychoaesthetics, and prospects for entropy as an analytic tool. *Computer Music Journal*, *32*(4), 64–78. doi:10.1162/comj.2008.32.4.64

Martindale, C. (1990). *The clockwork muse: The predictability of artistic change*. New York: Basic Books

McCorduck, P. (1991). *Aaron's code: Meta-art, artificial intelligence and the work of Harold Cohen*. New York: Freeman.

McLoughlin, M., Lamoni, L., Garland, E., Ingram, S., Kirke, A., Noad, M., et al. (2016). Preliminary results from a computational multi agent modelling approach to study humpback whale song cultural transmission. In S. Roberts, C. Cuskley, L. McCrohon, L. Barcelóe-Coblijn, O. Feher, & T. Verhoef (Eds.), *The evolution of language: Proceedings of the 11th International Conference*. http://www.evolang.org.

Meyer, L. B. (1956). *Emotion and meaning in music*. Chicago: University of Chicago Press.

Miller, G. F. (1997). Protean primates: The evolution of adaptive unpredictability in competition and courtship. In A. Whiten & R. W. Byrne (Eds.), *Machiavellian intelligence II: Extensions and evaluations* (pp. 312–340). Cambridge: Cambridge University Press.

Morell, V. (2014). When the bat sings. *Science*, *344*(6190), 1334–1337. doi:10.1126/science.344.6190.1334

Noad, M., Cato, D., Bryden, M. M., Jenner, M.-N., & Jenner, K. (2000). Cultural revolution in whale songs. *Nature*, *408*, 537. doi:10.1038/35046199

Nowicki, S., Peters, S., & Podos, J. (1998). Song learning, early nutrition, and sexual selection in songbirds. *American Zoologist*, *38*(1), 179–190. doi:10.1093/icb/38.1.179

Okanoya, K. (2012). Behavioural factors governing song complexity in Bengalese finches. *International Journal of Comparative Psychology*, *25*, 44–59.

Pamjav, H., Juhász, Z., Zalán, A., Neméth, E., & Damdin, B. (2012). A comparative phylogenetic study of genetics and folk music. *Molecular Genetics and Genomics*, *287*(4), 337–349. doi:10.1007/s00438-012-0683-y

Payne, K., & Payne, R. (1985). Large scale changes over 19 years in songs of humpback whales in Bermuda. *Zeitschrift für Tierpsychologie*, *68*(2), 89–114. doi:10.1111/j.1439-0310.1985.tb00118.x

Payne, K., Tyack, P., & Payne, R. (1983). Progressive changes in the songs of humpback whales (*Megaptera novaeangliae*): A detailed analysis of two seasons in Hawaii. In R. Payne (Ed.), *Communication and behavior of whales* (pp. 9–57). Boulder, CO: Westview Press.

Pearce, M. T. (2005). *The construction and evaluation of statistical models of melodic structure in music perception and composition* (Unpublished doctoral dissertation). City University, London, UK.

Pearce, M. T., Herrojo Ruiz, M., Kapasi, S., Wiggins, G. A., & Bhattacharya, J. (2010). Unsupervised statistical learning underpins computational, behavioural and neural manifestations of musical expectation. *NeuroImage*, *50*(1), 303–314. doi:10.1016/j.neuroimage.2009.12.019

Pearce, M. T., & Wiggins, G. A. (2006). Expectation in melody: The influence of context and learning. *Music Perception*, *23*(5), 377–405. doi:10.1525/mp.2006.23.5.377

Pearce, M. T., & Wiggins, G. A. (2012). Auditory expectation: The information dynamics of music perception and cognition. *Topics in Cognitive Science*, *4*(4), 625–652. doi:10.1111/j.1756-8765.2012.01214.x

Plotkin, H. (1998). *Evolution in mind: An introduction to evolutionary psychology*. Cambridge, MA: Harvard University Press.

Saunders, R. (2002). *Curious design agents and artificial creativity: A synthetic approach to the study of creative behaviour* (Unpublished doctoral dissertation). University of Sydney, Sydney, Australia.

Savage, P. E., Brown, S., Sakai, E., & Currie, T. E. (2015). Statistical universals reveal the structures and functions of human music. *Proceedings of the National Academy of Sciences*, *112*(29), 8987–8992. doi:10.1073/pnas.1414495112

Schinka, J. A., Letsch, E. A., & Crawford, F. C. (2002). DRD4 and novelty seeking: Results of meta-analyses. *American Journal of Medical Genetics*, *114*(6), 643–648. doi:10.1002/ajmg.10649

Schwab, D., Benedek, M., Papousek, I., Weiss, E. M., & Fink, A. (2014). The time-course of EEG alpha power changes in creative ideation. *Frontiers in Human Neuroscience*, *8*, 1–8. doi:10.3389/fnhum.2014.00310

Shannon, C. (1948). A mathematical theory of communication. *Bell System Technical Journal*, *27*, 379–423. doi:10.1002/j.1538-7305.1948.tb01338.x

Shettleworth, S. J. (2010). *Cognition, evolution, and behavior*. New York: Oxford University Press.

Skowronek, J., McKinney, M. F., & van de Par, S. (2006). Ground truth for automatic music mood classification. In R. Dannenberg & K. Lemström (Eds.), *Proceedings of ISMIR 2006* (pp. 395–396).

ten Cate, C., & Okanoya, K. (2012). Revisiting the syntactic abilities of non-human animals: Natural vocalizations and artificial grammar learning. *Philosophical Transactions of the Royal Society of London B: Biological Sciences*, *367*(1598), 1984–1994. doi:10.1098/rstb.2012.0055

Trehub, S. E., Becker, J., & Morley, I. (2015). Cross-cultural perspectives on music and musicality. *Philosophical Transactions of the Royal Society of London B: Biological Sciences*, *370*(1664), 20140096–20140096. doi:10.1098/rstb.2014.0096

van den Broek, E. M. F., & Todd, P. M. (2009). Evolution of rhythm as an indicator of mate quality. *Musicae Scientiae*, *13*(2), 369–386. doi:10.1177/1029864909013002161

Wallas, G. (1926). *The art of thought*. New York: Harcourt.

Weiss, M., Hultsch, H., Adam, I., Scharff, C., & Kipper, S. (2014). The use of network analysis to study complex animal communication systems: A study on nightingale song. *Proceedings of the Royal Society of London B: Biological Sciences*, *281*(1785), 1–9. doi:10.1098/rspb.2014.0460

Wiggins, G. A. (2006). A preliminary framework for description, analysis and comparison of creative systems. *Journal of Knowledge Based Systems*, *19*(7), 449–458. doi:10.1016/j.knosys.2006.04.009

Wiggins, G. A. (2012). The mind's chorus: Creativity before consciousness. *Cognitive Computation*, *4*(3), 306–319. doi:10.1007/s12559-012-9151-6

Wiggins, G., & Bhattacharya, J. (2014). Mind the gap: An attempt to bridge computational and neuroscientific approaches to study creativity. *Frontiers in Human Neuroscience*, *8*(540), 1–15. doi:10.3389/fnhum.2014.00540

Wiggins, G. A., & Forth, J. (2015). IDyOT: A computational theory of creativity as everyday reasoning from learned information. *Computational Creativity Research: Towards Creative Machines*, *8*, 127–148. doi:10.2991/978-94-6239-085-0_7

Wright, S. (1969). *Evolution and the genetics of populations* (Vol. 2). Chicago: University of Chicago Press.

Wundt, W. M. (1874). *Grundzüge der physiologischen Psychologie*. Leipzig: Engelmann.

Yurk, H., Barrett-Lennard, L., Ford, J. K. B., & Matkin, C. O. (2002). Cultural transmission within maternal lineages: Vocal clans in resident killer whales in southern Alaska. *Animal Behaviour*, *63*(6), 1103–1119. doi:10.1006/anbe.2002.3012

Zahavi, A. (1975). Mate selection—a selection for a handicap. *Journal of Theoretical Biology*, *53*(1), 205–213. doi:10.1016/0022-5193(75)90111-3

13

Affect Induction through Musical Sounds: An Ethological Perspective

David Huron

When asked why they listen to music, people commonly allude to the pleasure of how music makes them feel. Music is capable of evoking a wide range of feeling states from the pedestrian to the sublime (Gabrielsson, 2011; Gabrielsson & Lindström Wik, 2003). Not all emotions are equally easy to evoke, however. Music might evoke feelings of exuberance, compassion, or tenderness, but it is less common that music evokes or portrays emotions such as jealousy or guilt (Zentner, Grandjean, & Scherer, 2008).

Conceptually, a distinction can be made between emotional representations or portrayals in music and the emotions that are evoked or induced by music. Theoretically a musical passage might be regarded as portraying, say, fear but evoking, say, jocularity instead. Considerable research has focused on the associations between various acoustical features and the representation or portrayal of various affects through music (cf. Juslin & Laukka, 2003, for summary). It is generally acknowledged, however, that the more difficult problem is identifying the means by which affective states are evoked in listeners.

Several possible mechanisms have been proposed to account for musically evoked emotions. The many possible sources include innate auditory responses, learned associations, and mirror neuron processes (e.g., Juslin, 2013; Molnar-Szakacs & Overy, 2006). Efforts to identify possible sources have been partly confounded by the variability of responses both between different cultural communities and between individuals within a given culture. Even with regard to the perception or recognition of affective portrayals in music, research has documented considerable variability. For example, whether a particular musical passage is perceived as expressing or representing a given emotion is known to be influenced by personality: listeners who score high on neuroticism are more likely to characterize a passage as sad sounding (Ladinig & Schellenberg, 2012).

The variability found between individuals and cultures has led some researchers to conclude that music-induced affect must arise principally or exclusively due to learning. This variability further implies that any presumed biological foundations for music are likely to be shallow, and so calls into question several proposals for possible evolutionary bases for the origins of music.

Both conditioned responses and learned associations may have an arbitrary relationship to the musical stimulus: a song that typically evokes happiness might evoke sadness for some listeners due to an association with a failed romantic relationship. While some learned associations are widespread across a population (such as the use of a tune in a popular television show), many associations are idiosyncratic to specific listeners. Even if it were the case that learned associations provoke the strongest and most reliable emotions in listeners, unless these associations are broadly shared across listeners, there is little incentive for musicians to use these associations as compositional tools for evoking particular responses. The existence of a musical culture depends on shared experiences.

From a musical perspective, purported mirror neuron processes appear to offer a more attractive method for inducing listener emotions since mirror neurons imply a close echoing of expressed and evoked emotion (at least when a mirrored action is involved). Gallese, Keysers, and Rizzolatti (2004) have suggested that mirror neurons provide a plausible neurological mechanism for empathetic responses. Singer et al. (2004) carried out brain scans while they inflicted obvious pain on a romantic partner of the person being scanned. They found similar patterns of activation in brain regions associated with pain in the observing romantic partner—suggesting that empathetic responses rely on emulating the emotional experiences of others. Although Singer's work does not specifically link empathetic responses to mirror neuron processes, several studies are consistent with the idea that mirror neuron systems lead to empathetic feeling states (Carr, Iacoboni, Dubeau, Mazziotta, & Lenzi, 2003; Gallese, 2003). Evidence consistent with auditory-evoked mirror responses exists (Aziz-Zadeh, Iacoboni, Zaidel, Wilson, & Mazziotta, 2004; Buccino et al., 2005), and Molnar-Szakacs and Overy (2006) have explicitly proposed that music evokes empathetic affective states in listeners through a mirror neuron process.

The promise of mirror neurons notwithstanding, in this chapter, we propose and explore other ways by which music might induce emotions. Drawing on concepts in ethology, we suggest that ethological *signals*, *cues*, and *indexes* provide additional useful analytical tools for approaching auditory-induced affect. In particular, we suggest that these concepts can help to decipher whether a given musical or acoustical feature is likely to induce similar affective responses in listeners across cultures.

Size Matters

One of the best cross-species generalizations one can make about behavior is that appearing to be large is associated with threat, whereas appearing to be small is associated with deference or submission. This principle is not restricted to vision. One of the best generalizations that can be made regarding acoustics is that large masses or volumes vibrate at a lower frequency. Morton (1977) observed this principle in a broad sample of vocalizations from fifty-six species. In general, high-pitched calls are associated with fear, affinity, or submission, whereas low-pitched calls are associated with threat or aggression. This principle has also been observed in human speech. Across cultures, high pitch is associated with friendliness or deference, whereas low pitch is associated with seriousness or aggression (Bolinger, 1978).

The same relationship has been observed in music. When melodies are transposed lower in pitch, they are perceived as less polite, less submissive, and more threatening (Huron, Kinney, & Precoda, 2006). This effect is clearly apparent in opera, where heroic roles tend to be assigned to high-pitched tenors and sopranos, whereas villainous roles tend to be assigned to basses and contraltos (Shanahan & Huron, 2014). In speech, an important acoustical distinction is made between the source and filter components (Fant, 1960). The vocal folds provide the sound source whose vibrations determine the fundamental frequency (F_0) or pitch of the voice. The vocal tract (pharynx, oral cavity, and nasal passages) provides the filter component.

The volume and length of the oral cavity can be manipulated, changing the resonance of the ensuing sound. Low-resonance vowels are associated with a dropped chin and protruding lips that both lengthen the vocal tract and increase the oral cavity volume. Conversely, high-resonance vowels are associated with a raised chin and retracted lips, resulting in higher-frequency resonances. The independence of the source and filter in speech means that there are two distinct frequency components: the fundamental frequency and the vocal tract resonance.

Ohala (1980) proposed a theory of the human smile that deserves to be better known. Why would a display that commonly reveals a person's teeth be construed as anything other than a threat display? Tartter (1980; Tartter & Braun, 1994) showed that listeners easily recognize the sound of a smiling voice. Pulling the lips taut against the teeth shortens the vocal tract length (VTL)—shifting the main resonance higher in frequency. Ohala suggested that the smile originated as an auditory rather than as a visual display. Specifically, the upward shift of the smiling-voice resonance is consistent with the association of small size with affiliative or deferential behaviors.

Ohala's (1980) conjecture suggests that the opposite of the smile may not be the frown but the pout, where the lips are thrust outward. The association of a pout or pucker display with threat has been observed in many animal species (Epple, 1967; Fitch, 1994; Fox & Cohen, 1977; Kaufman, 1967; Miller, 1975; Pruitt & Burghardt, 1977). Among humans, the aggressiveness of the pout is evident in film portrayals of the hooligan where the speaker's lips are pushed forward—producing a bellicose or hostile-sounding timbre. Although Ohala has drawn attention to the auditory properties of smiling, we will see that the combination of visual and auditory features likely represents the more seminal observation.

Signals and Cues

Consider the distinction ethologists make between signals and cues (Lorenz, 1939; Lorenz, 1970). A signal is an evolved purposeful communicative behavior, as evident in a rattlesnake's rattle, for example. The rattle is a special anatomical organ that has evolved explicitly for communicative purposes. When the snake is threatened, the rattling behavior alerts the observer to the possibility of a preemptive or retaliatory attack.

By contrast, an ethological cue is a nonfunctional behavior that is nevertheless informative. An example is the buzzing sound produced by a mosquito, suggesting an imminent attack. Experience suggests that when we hear the buzzing of a mosquito, there is a good likelihood we will be bitten. However, in this case, the buzzing sound is simply an unintended consequence of the insect flapping its wings, not a functional communicative act.

Both the rattlesnake's rattle and the mosquito's buzzing presage the possibility of an attack. But one is a functional signal, whereas the other is an artifactual cue. Both sounds are informative, but only one sound is overtly communicative.

Signals are important because their purpose is to change the behavior of the observer, to the mutual benefit of signaler and observer (Maynard Smith & Harper, 2003; Silk, Kaldor, & Boyd, 2000). Among ethologists, the ensuing changes in behavior are thought to arise because the signal activates coevolved circuits intended to influence motivation. Among humans, these motivational circuits are more commonly deemed emotions (Tomkins, 1980). Unlike cues, signals are evolved behaviors that induce stereotypical feeling states in an observer, as when crying evokes feelings of compassion in a bystander (Gelstein et al., 2011; Kottler, 1996; Lane, 2006). In music, acoustical features that imitate true signals (in the ethological sense) are most likely to induce affective states in listeners that are similar cross-culturally. That is, signal-like musical features are likely to be interpreted similarly.

Moreover, with regard to affect induction, the concept of an ethological signal provides a commonly overlooked mechanism for generating affect in observers. If ethologists are correct, these behavioral changes are largely automatic and species-wide.

Multimodal Signals

Determining whether a particular behavior represents a cue or signal is nontrivial. Ethologists have proposed a number of characteristics that help in making this distinction. Most important, signals tend to be conspicuous (Johnstone, 1997; Wiley, 1983). If the purpose is to communicate, the display behavior is unlikely to be timid or subtle. One of the best ways to increase conspicuousness is to make the signal multimodal. In the case of the rattlesnake, for example, the raised shaking tail is visually distinctive, even if the rattling sound is not heard. Cues may also be multimodal, but because the behaviors are artifacts, they are less likely to exhibit multimodal features. In short, cues tend to be unimodal, whereas signals tend to be multimodal (Maynard Smith & Harper, 2003).

Whether or not Ohala's theory (1980) of the acoustical origin of the smile is correct, the important point is that smiling exhibits both characteristic visual and auditory features, consistent with a multimodal tendency for ethological signals. That is, the smile display is consistent with a coevolved communicative function.

Note that the smile and pout relate only to the filter component of the source-filter foundations of vocalization. This raises the question of the signaling status of the source (F_0) component of vocalization. Unlike the smile, it is not immediately apparent whether changes in F_0 are accompanied by a distinctive visual element. Does a low voice signal aggression (in the ethological sense), or is it merely an artifactual cue? If F_0 is a signal, we might expect to see an accompanying visual correlate consistent with the multimodal tendency of ethological signals.

In chimpanzees, F_0 is positively correlated with teeth and lip opening distances (Bauer, 1987). However, among *Homo sapiens*, the key visual element appears to reside in the eyebrows. Huron, Dahl, and Johnson (2009) showed that the height of the eyebrows tracks vocal pitch. When asked to raise or lower the pitch of the voice, participants' eyebrows tend to rise and fall in tandem with the pitch. Moreover, assessments of the resulting facial expressions are consistent with affective assessments of F_0: photographs of faces when producing a high-pitch vocalization were deemed much friendlier than matched facial photographs of low-pitch vocalization. In the production of a low pitch, the head tends to tilt down, the chin drops, and the eyebrows are lowered; conversely, in the production of a high pitch, the head tends to tilt up, the mouth tends to form a smile, and the eyebrows are raised.

In order to control for a possible confounding motor relationship between the larynx and chin position, a second control experiment was carried out in which the photographs were cropped so that subjects could view only the region above the nose. Nevertheless, faces were clearly deemed friendlier or more aggressive when producing high and low pitches, respectively (Huron et al., 2009). In a subsequent study (Huron & Shanahan, 2013), the reverse causation was also demonstrated. When instructed to move their eyebrows up or down, F_0 was found to move in tandem. The observations that moving pitch causes the eyebrows to move and that moving the eyebrows causes the pitch to move implies the existence of a central motor process that controls both the eyebrows and the vocal folds. This multimodal relationship is consistent with an ethological signal—suggesting that eyebrow position and vocal pitch may be components of a single communicative display.

In light of these apparently successful reinterpretations of the smile and pitch/eyebrow displays, we might consider applying a similar ethological analysis to other facial expressions and consider their possible repercussions for understanding music-induced affect.

Sneer as Signal

Ekman and Friesen (1986) published a seminal description of the facial expression of contempt (cross-culturally replicated in Ekman & Heider, 1988). Ekman (Ekman & Friesen, 1986) proposed that contempt is a variant of the disgust display. In an expression of disgust, the upper lip is elevated and the nostrils are flexed. Contempt can be viewed as an asymmetrical version of disgust, in which the elevation of the upper lip and nostril flexion occurs on only one side of the face. According to Ekman, disgust is a general expression of revulsion (such as when exposed to repugnant smells or tastes), whereas contempt is a social expression, used to communicate hostility toward another person or social situation (see also Kelly, 2011). Applying the ethological question, we might once again ask whether contempt is a signal. More specifically, is there a characteristic sound that accompanies a characteristic visual display?

Contemptuousness and sarcasm are readily conveyed through speech prosody. Even if one does not understand the language, contempt is often apparent. Sarcastic or contemptuous speech appears to be linked with high nasalization. The nasal quality of contempt is exemplified by playground taunts like *nya, nya.*

In the vocal tract, the nasal passages normally behave as static acoustical cavities with a fixed formant or resonance. However, the overall acoustical effect is influenced by the amount of airflow through the nose. Nasalization arises when the airflow is high (Ladefoged, 1962). So-called *nasals*—phonemes such as *m* and *n*—are accompanied by high nasal airflow. The acoustical effect of nasalization is that the first formant becomes broader. As a consequence, more energy is shifted to the upper partials.

The volume of nasal airflow is determined by the size of the velopharyngeal opening. In general, the bigger the opening, the greater the nasality. Sundberg (personal communication, 2014) has suggested that nasalization is not directly caused by flexing nose muscles. Instead, there is some motor connection between the nose and the velum/pharynx that causes one to influence the other.

With regard to music, pertinent work has been carried out by Plazak (2011), who asked instrumentalists to perform various passages (including single tones) in happy, sad, angry, and sarcastic manners. In a forced-choice paradigm, Plazak found that listeners had little difficulty distinguishing these different represented or expressed affects. Plazak then analyzed the various sounds in an effort to identify acoustical correlates with musical sarcasm. For the instruments most capable of conveying sarcasm, Plazak found that the sarcastic renditions exhibited elevated levels of nasality as measured using standard speech analysis methods. Although further research is warranted, these observations suggest an association between a facial expression (sneer) and a recognizable change in vocal sound (nasalization), consistent with multimodal tendencies of ethological signals. Moreover, we see evidence that these features generalize beyond speech and can be observed in musical timbres as well.

Grief as Signal: Sadness as Cue

Darwin (1872) distinguished sadness (a low-arousal affect) from grief (a high-arousal affect). Compared with normal speech, sad speech is slower, quieter, lower in pitch, more monotone, more mumbled, and exhibits a dark timbre (lower spectral centroid). Grief-related vocalizations, by comparison, exhibit high pitch, pharyngealization, breaking voice, and ingressive phonation (Huron, 2012).

Huron (2012) offered a detailed argument that sadness bears the hallmarks of an ethological cue, whereas grief bears the hallmarks of an ethological signal. Space limitations preclude reviewing the argument here. By way of summary, all of the features of sad speech can be attributed to low physiological arousal and cannot be easily distinguished from similar low-arousal states such as relaxed speech or sleepy speech. Similarly, a nominally sad facial expression is not easily distinguished from a relaxed or sleepy facial expression. Moreover, sad individuals tend to remain mute, which is not consistent with a communicative function. By contrast, grief exhibits highly distinctive visual and acoustical features. Grief is accompanied by a compulsion to vocalize, exemplified by the rare phenomenon of phonating while inhaling (gasping sound). In addition, Gelstein et al. (2011) have shown that psychic (emotionally evoked) tears contain a pheromone that induces observers to behave in a more compassionate way. In short, grief exhibits multimodal characteristics (visual, acoustical, and olfactory) that result in notable changes in the feeling states of observers, consistent with an ethological signal. Sadness, by contrast, is easily confused with relaxation, and these confusions are echoed in musical confusions, such as disagreements among listeners about whether New Age music sounds sad or relaxing. As noted in the introduction to this chapter, these judgments are known to be confounded by personality (Ladinig & Schellenberg, 2012), suggesting that any presumed communicative function is not robust.

An especially distinctive acoustical feature of grief is the sound of breaking voice, caused by an abrupt transition between falsetto and modal phonation. The breaking sounds are evident in many musical styles, including Western opera and country music. Paul and Huron (2010) carried out a study of breaking voice in commercially recorded music and found that moments of breaking voice are positively correlated with grief-related content in the lyrics.

Recall that ethological signals are produced in order to change the behavior of the observer to the benefit of the signaler. In the case of grief, crying behavior has a tendency to evoke strong feelings of compassion in observers. This suggests that listening to musical passages that emulate grief-related acoustical features should tend to evoke, not feelings of sadness or grief, but feelings of compassion and tenderness in listeners.

Cuteness as Index

In an exploratory study of auditory cuteness, we found that listeners exhibit high intersubjective agreement when characterizing the degree of cuteness for various sounds (Huron, 2005). Listeners heard a wide range of sounds, including those produced by assorted toys and musical instruments. Listeners describing sounds scoring high in cuteness (such as squeeze toys, ocarina, sopranino recorder, or music box) used terms such as *fragile*, *delicate*, *innocent*, and *vulnerable*. The sounds appear to induce nurturing and protective attitudes. In short, the responses are consistent with the evoking of parenting behaviors appropriate for interacting with infants.

An analysis of the sound-producing mechanisms established that cute sounds tend to be associated with a small resonant cavity activated by a small amount of energy. Although no formal tests were carried out, the optimum size of a cute cavity appears to resemble the vocal tract of infants. Vorperian et al. (2005) measured VTLs across various ages, from neonates to adults. A typical newborn infant has a vocal tract near 8 centimeters in length, whereas adult VTLs are typically 14 to 17 centimeters.

Fitch (1997) has noted that VTL is consistent with the ethological concept of an *index*, defined as a type of signal whose variability is causally related to the quality being signaled and cannot be faked (Maynard Smith & Harper, 1995). An example of an index is the marking of trees by tigers. Tigers mark trees by reaching as high as they can along the trunk. Consequently, the position of the smell provides a reliable indicator of the size of the animal.

In the acoustical domain, formant dispersion has been identified as a reliable indicator of size (Fitch, 1994, 1997). Specifically, formant dispersion is directly proportional to VTL, and VTL is strongly correlated with body size. In this regard, formant dispersion contrasts notably with vocal pitch. Although the sex, age, and size of an individual establish a general tessitura or range for vocal pitch, pitch variability can be voluntarily controlled within this range. In short, pitch provides a useful signal reflecting affective states such as aggressive or submissive attitudes, whereas formant dispersion provides a useful index of the individual's size, largely independent of affective state or intention. As noted, the VTL is susceptible to some modification (due to retracting or extending the lips). However, in contrast to pitch variability, VTL is much more constrained—much less susceptible to voluntary control, and therefore a reliable or honest indicator of body size.

In music, examples of cuteness are readily found. A good example is the innocent "little girl" voice of Helen Kane, a diminutive popular American singer from the 1930s. Kane became the model for the subsequent well-known Betty Boop cartoon character. In light of the idea that VTL represents an (honest) ethological index of size, it is understandable how these sounds might provoke descriptions suggestive of parenting behaviors—specifically, that the sounds might be described as innocent and vulnerable, while evoking feelings of care, tolerance, or nurturing in listeners.

Signals, cues, and indexes are classes of adaptations. Apart from the adaptive concepts, Tinbergen (1951) would encourage consideration of some of the causal mechanisms that may be involved. To this end, we might contrast the ethological concepts with two plausible mechanisms for music-induced affect mentioned earlier, notably mirror processes and associations.

Tempo as Mirror

Mirror neuron systems are closely linked to motor behaviors. The stimuli most likely to evoke mirror responses are those implying a particular motor action rather than an emotion. When we hear someone speaking with a "frog" in his or her throat, we may feel an unconscious compulsion to want to clear our own throat. Any evoked emotions are presumed to arise from a Jamesian process of self-perception (Laird, 2007). Mirror systems may be the basis for emotional contagion (Hatfield, Cacioppo, & Rapson, 1994).

The close relationship to motor action suggests that mirror responses are most likely to be evoked by musical features related to action. A large body of research has examined the relationship between music and movement. The foremost behavior examined has been locomotion, notably walking.

Recording acceleration data from different body parts over sustained periods of daily activity, MacDougall and Moore (2005) showed that the distribution of preferred musical tempos is very closely related not to the movement of legs or arms, but to the motion of the head. (When walking, head movements are twice the frequency of individual leg or arm movements.) Consistent with other music research, the key to tempo perception appears to be vestibular activation. It appears that what makes 40 beats per minute sound sluggish and lethargic (whereas 120 beats per minute sounds spirited and zestful) is that these frequencies match head movements for lethargic and spirited walking, respectively. Moreover, it is not simply that these slower and faster beat rates represent or convey lethargic and spirited activity; as beat rate is echoed in the motor cortex, mirror processes might be expected to evoke feelings of lethargy or energy (respectively) in an engaged listener. In short, a mirror process may explain how fast and slow music succeeds in inducing languid or exuberant feeling states. Although further research is warranted, the induced affects arising from different tempos appear consistent with a mirror process linked to the vestibular system.

Vibrato as Association

As noted, many associations are idiosyncratic to individual listeners. However, some associations are widespread across a culture, such as the association of "Taps" with funeral or memorial events. Other associations might be deemed universal due to shared experiences. For example, virtually all humans have broadly similar experiences with gravity, breathing, walking, thirst, happiness, and other phenomena. One of these stock phenomena is the experience of fear.

When a person is in a state of fear, trembling may result. The trembling is attributable to high levels of adrenaline and associated peripheral acetylcholine—which increases muscle reactivity. All skeletal muscles are influenced, including the muscles involved in vocal production. Fear or nervousness is often evident in the trembling sound of the voice, with a distinctive modulation of pitch. Although trembling may also be visible in hand motions, it is most likely to become evident first in the voice.

The stylized vocal trembling musicians call *vibrato* can be observed in many musical cultures around the world. Moreover, the same frequency modulation is commonly found in purely instrumental music, where special performance techniques are required in order to emulate the acoustical effects of trembling voice. The effect of vibrato is commonly described as rendering the sound more emotional, consistent with the etiology of trembling voice (Seashore, 1932).

It is possible that vocal vibrato represents an ethological signal for which a learned association provides the proximal affect-inducing mechanism. However, the trembling is a straightforward consequence of a general physiological response, and so has the appearance of being artifactual. This suggests that vocal trembling is more likely to be a cue or a learned association.

At this point, it is appropriate to consider how cues might differ from the psychological concept of a learned association. In the first instance, learned associations are mechanisms that may or may not be adaptive, whereas cues (by definition) are necessarily adaptive. Also, cues may or may not be innate. On the one hand, mosquitoes have an innate disposition to fly in the direction of an increasing carbon dioxide (CO_2) gradient. That is, they are attracted to the exhaled breath of animals. Because the production of CO_2 is an artifact of respiration, it is not likely to represent a functional communicative behavior. At the same time, mosquitoes do not need to learn to follow a CO_2 trail. In many other cases, simple associative or Pavlovian learning is sufficient to account for adaptive behaviors without appealing to an innate biological mechanism. That is, for ethologists, cues can entail responses that can be either wholly learned or innate. When wholly learned, associations may be regarded as a causal mechanism that enables cues.

At least in Western culture, vibrato is typically slower than fear-induced trembling. Presumably this connotes a more controlled effect, suggesting high emotionality without necessarily evoking fear or terror in a listener.

Conclusion

In attempting to understand the emotional dimensions of music, how sounds induce emotions in listeners is one of the more difficult questions. This chapter has explored several potential emotion-inducing approaches, including signals, cues, indexes, mirrors, and associations. Learned associations may be emotionally compelling, but they are of limited compositional utility unless the associations are widely shared among listeners. Because associations can be entirely arbitrary (with no resemblance between the sounds and the evoked memory), they may be useful when accounting for idiosyncratic or paradoxical responses to particular stimuli. Mirrors appear to be important mechanisms that provide a ready explanation for how expressed emotion may be echoed as evoked emotion, with the important caveat that mirrors are restricted to action-oriented behaviors. Signals, by contrast, offer a unique opportunity to induce feeling states that are not simply reflections of the displayed affect. Signals warrant particular interest since, as evolved communicative behaviors, both the displaying behavior and the observer's response are innate, and so presumably shared across all human listeners. In humans at least, the responses are not the involuntary innate releasing mechanisms envisioned by early ethologists. Nevertheless, the feeling states evoked by observing a signal do tend to be stereotypic, making them especially attractive tools when composing music intended to induce emotions in listeners. Finally, indexes are notably honest signals whose connotations ought to be self-evident to listeners.

The affect-inducing concepts discussed here are not intended to be exhaustive or exclusive. Juslin (Juslin & Västfjäll, 2008, revised in Juslin, 2013), for example, has outlined a framework for music-induced emotion that posits eight mechanisms, referred to using the acronym BRECVEMA: brain stem reflexes, rhythmic entrainment, evaluative conditioning, contagion, visual imagery, episodic memory, musical expectancy, and aesthetic judgment. As with signals and cues, the specific physiological and cognitive pathways for each mechanism discussed in this chapter await detailed exposition. It may turn out that some of the mechanisms distinguished here share identical neural origins that warrant a single description.

In addition, none of the mechanisms for auditory-evoked affect discussed here should be considered mutually exclusive. For example, learned associations, mirror processes, and cognitive appraisals can conceivably coexist, and any given acoustical stimulus might well activate each of these mechanisms concurrently. Parallel systems suggest the possibility of evoking mixed emotions, a state that is commonly reported among listeners (Hunter, Schellenberg, & Schimmack, 2010). Crying provides an illustrative example. Mirror mechanisms suggest that observing someone crying might lead to an empathetic response in which the observer also feels disposed to cry. However, crying also bears the hallmarks of an ethological signal. Here, the research suggests that crying commonly evokes sympathetic feelings of compassion that motivate the observer to offer assistance (or terminate aggressive behaviors). As a result, the combination of mirror and signaling processes implies that for many observers, a crying display may evoke simultaneously both empathetic feelings of grief or sadness, mixed with sympathetic feelings of compassion and tenderness. To the extent that a musical passage emulates crying-related acoustical features, such as breaking voice, we might expect many listeners to experience a similar mixture of grief, sadness, compassion, or tenderness.

The different mechanisms for auditory-induced emotion distinguished in this chapter provide possible starting points for addressing five puzzles in music-related affect: Why is music able to induce only certain emotions (e.g., exuberance, compassion and tenderness) but not others (e.g., jealousy and guilt)? Why are some induced emotions similar to the displayed emotion, whereas in other cases, the induced emotion differs from the displayed emotion? Why do listeners often report feeling mixed emotions? Why are some emotional connotations similar across musical cultures, whereas others are not? Why do musicians appear to rely on some emotion-inducing mechanisms more than others?

Particular stress has been placed here on the potential role of ethological signals in music-related emotion. Identifying signals provides a potentially useful strategy for discerning those aspects of musical expression that are likely to be shared across cultures. One useful heuristic in this effort is attending to possible multimodal aspects of purported displays. Because signals are intended to be communicative, they tend to be conspicuous, and one way of ensuring conspicuousness is using multimodal displays. For decades, emotion researchers have tended to focus on facial expressions without attending to acoustical concomitants of emotion. Researchers interested in acoustical aspects of emotion would be wise to avoid a similar myopia by attending to multimodal correlates.

References

Aziz-Zadeh, L., Iacoboni, M., Zaidel, E., Wilson, S., & Mazziotta, J. (2004). Left hemisphere motor facilitation in response to manual action sounds. *European Journal of Neuroscience*, *19*(9), 2609–2612. doi:10.1111/j.0953-816X.2004.03348.x

Bauer, H. R. (1987). Frequency code: Orofacial correlates of fundamental frequency. *Phonetica*, *44*(3), 173–191. doi:10.1159/000261793

Bolinger, D. L. (1978). Intonation across languages. In J. H. Greenberg, C. A. Ferguson, & E. A. Moravcsik (Eds.), *Universals of human language: Phonology* (Vol. 2, pp. 471–524). Stanford, CA: Stanford University Press.

Buccino, G., Riggio, L., Melli, G., Binkofski, F., Gallese, V., & Rizzolatti, G. (2005). Listening to action-related sentences modulates the activity of the motor system: A combined TMS and behavioral study. *Cognitive Brain Research*, *24*(3), 355–363. doi:10.1016/j.cogbrainres.2005.02.020

Carr, L., Iacoboni, M., Dubeau, M. C., Mazziotta, J. C., & Lenzi, G. L. (2003). Neural mechanisms of empathy in humans: A relay from neural systems for imitation to limbic areas. *Proceedings of the National Academy of Sciences*, *100*(9), 5497–5502. doi:10.1073/pnas.0935845100

Darwin, C. (1872). *The expression of the emotions in man and animals*. London: John Marry.

Ekman, P., & Friesen, W. V. (1986). A new pan-cultural facial expression of emotion. *Motivation and Emotion*, *10*(2), 159–168. doi:10.1007/BF00992253

Ekman, P., & Heider, K. G. (1988). The universality of a contempt expression: A replication. *Motivation and Emotion*, *12*(3), 303–308. doi:10.1007/BF00993116

Epple, G. (1967). Vergleichende untersuchungen über sexual-und sozialverhalten der krallenaffen (Hapalidae). *Folia Primatologica*, *7*(1), 37–65. doi:10.1159/000155095

Fant, G. (1960). *Acoustic theory of speech production*. The Hague: Mouton.

Fitch, W. T. (1994). *Vocal tract length perception and the evolution of language* (Unpublished doctoral dissertation). Brown University, Providence, RI.

Fitch, W. T. (1997). Vocal tract length and formant frequency dispersion correlate with body size in rhesus macaques. *Journal of the Acoustical Society of America*, *102*(2), 1213–1222. doi:10.1121/1.421048

Fox, M. W., & Cohen, J. A. (1977). Canid communication. In T. A. Sebeok (Ed.), *How animals communicate* (pp. 728–748). Bloomington: Indiana University Press.

Gabrielsson, A. (2011). *Strong experiences with music: Music is much more than just music*. Oxford: Oxford University Press.

Gabrielsson, A., & Lindström Wik, S. (2003). Strong experiences related to music: A descriptive system. *Musicae Scientiae*, *7*(2), 157–217. doi:10.1177/102986490300700201

Gallese, V. (2003). The roots of empathy: The shared manifold hypothesis and the neural basis of intersubjectivity. *Psychopathology*, *36*(4), 171–180. doi:10.1159/000072786

Gallese, V., Keysers, C., & Rizzolatti, G. (2004). A unifying view of the basis of social cognition. *Trends in Cognitive Sciences*, *8*(9), 396–403. doi:10.1016/j.tics.2004.07.002

Gelstein, S., Yeshurun, Y., Rozenkrantz, L., Shushan, S., Frumin, I., Roth, Y., et al. (2011). Human tears contain a chemosignal. *Science*, *331*(6014), 226–230. doi:10.1126/science.1198331

Hatfield, E., Cacioppo, J., & Rapson, R. L. (1994). *Emotional contagion*. Cambridge: Cambridge University Press.

Hunter, P. G., Schellenberg, E. G., & Schimmack, U. (2010). Feelings and perceptions of happiness and sadness induced by music: Similarities, differences, and mixed emotions. *Psychology of Aesthetics, Creativity, and the Arts*, *4*(1), 47–56. doi:10.1037/a0016873

Huron, D. (2005). The plural pleasures of music. In *Proceedings of the Music and Music Science Conference* (pp. 1–13). Stockholm: Kungliga Musikhögskolan & KTH (Royal Institute of Technology).

Huron, D. (2012). Understanding music-related emotion: Lessons from ethology. In E. Cambouropoulos, C. Tsougras, P. Mavromatis, & K. Pastiadis (Eds.), *Proceedings of the 12th International Conference on Music Perception and Cognition/8th Conference of the European Society for the Cognitive Sciences of Music* (pp. 473–481). Thessaloniki: Aristotle University of Thessaloniki/ESCOM.

Huron, D., Dahl, S., & Johnson, R. (2009). Facial expression and vocal pitch height: Evidence of an intermodal association. *Empirical Musicology Review*, *4*(3), 93–100.

Huron, D., Kinney, D., & Precoda, K. (2006). Influence of pitch height on the perception of submissiveness and threat in musical passages. *Empirical Musicology Review*, *1*(3), 170–177.

Huron, D., & Shanahan, D. (2013). Eyebrow movements and vocal pitch height: Evidence consistent with an ethological signal. *Journal of the Acoustical Society of America*, *133*(5), 2947–2952. doi:10.1121/1.4798801

Johnstone, R. A. (1997). The evolution of animal signals. In J. R. Krebs & N. B. Davies (Eds.), *Behavioural ecology: An evolutionary approach* (pp. 155–178). Oxford: Blackwell Science.

Juslin, P. N. (2013). From everyday emotions to aesthetic emotions: Towards a unified theory of musical emotions. *Physics of Life Reviews*, *10*(3), 235–266. doi:10.1016/j.plrev.2013.05.008

Juslin, P. N., & Laukka, P. (2003). Communication of emotions in vocal expression and music performance: Different channels, same code? *Psychological Bulletin*, *129*(5), 770–814. doi:10.1037/0033-2909.129.5.770

Juslin, P. N., & Västfjäll, D. (2008). Emotional responses to music: The need to consider underlying mechanisms. *Behavioral and Brain Sciences*, *31*(5), 559–575. doi:10.1017/S0140525X08005293

Kaufman, J. H. (1967). Social relations of adult males in a free-ranging band of rhesus monkeys. In S. Altmann (Ed.), *Social communication among primates* (pp. 73–98). Chicago: University of Chicago Press.

Kelly, D. (2011). *Yuck! The nature and moral significance of disgust*. Cambridge, MA: MIT Press.

Kottler, J. A. (1996). *The language of tears*. San Francisco, CA: Jossey-Bass.

Ladefoged, P. N. (1962). *Elements of acoustic phonetics*. Chicago: University of Chicago Press.

Ladinig, O., & Schellenberg, E. G. (2012). Liking unfamiliar music: Effects of felt emotion and individual differences. *Psychology of Aesthetics, Creativity, and the Arts*, *6*(2), 146–154. doi:10.1037/a0024671

Laird, J. D. (2007). *Feelings: The perception of the self*. Oxford: Oxford University Press.

Lane, C. (2006). *Evolution of gender differences in adult crying* (Unpublished doctoral dissertation). University of Texas at Arlington.

Lorenz, K. (1939). Vergleichende Verhaltensforschung. *Zoologischer Anzeiger Supplementband*, *12*, 69–102.

Lorenz, K. (1970). *Studies in animal and human behaviour* (Vol. 1). London: Methuen.

MacDougall, H. G., & Moore, S. T. (2005). Marching to the beat of the same drummer: The spontaneous tempo of human locomotion. *Journal of Applied Physiology*, *99*(3), 1164–1173. doi:10.1152/japplphysiol.00138.2005

Maynard Smith, J., & Harper, D. G. C. (1995). Animal signals: Models and terminology. *Journal of Theoretical Biology*, *177*(3), 305–311. doi:10.1006/jtbi.1995.0248

Maynard Smith, J., & Harper, D. G. C. (2003). *Animal signals*. Oxford: Oxford University Press.

Miller, E. H. (1975). A comparative study of facial expressions of two species of pinnipeds. *Behaviour, 53*(3), 268–284. doi:10.1163/156853975X00227

Molnar-Szakacs, I., & Overy, K. (2006). Music and mirror neurons: From motion to 'e'motion. *Social Cognitive and Affective Neuroscience, 1*(3), 235–241. doi:10.1093/scan/nsl029

Morton, E. S. (1977). On the occurrence and significance of motivation-structural rules in some bird and mammal sounds. *American Naturalist, 111*(981), 855–869. doi:10.1086/283219

Ohala, J. J. (1980). The acoustic origin of the smile. *Journal of the Acoustical Society of America, 68*(1), 33. doi:10.1121/1.2004679

Paul, B., & Huron, D. (2010). An association between breaking voice and grief-related lyrics in country music. *Empirical Musicology Review, 5*(2), 27–35.

Plazak, J. (2011). *Encoding and decoding sarcasm in instrumental music: A comparative study* (Unpublished doctoral dissertation). Ohio State University, Columbus.

Pruitt, C. H., & Burghardt, G. M. (1977). Communication in terrestrial carnivores: Mustelidae, Procyonidae, and Ursidae. In T. A. Sebeok (Ed.), *How animals communicate* (pp. 767–793). Bloomington: Indiana University Press.

Seashore, C. E. (1932). *The vibrato.* Iowa City: University of Iowa.

Shanahan, D., & Huron, D. (2014). Heroes and villains: The relationship between pitch tessitura and sociability of operatic characters. *Empirical Musicology Review, 9*(2), 141–153. doi:10.18061/emr.v9i2.4441

Silk, J. B., Kaldor, E., & Boyd, R. (2000). Cheap talk when interests conflict. *Animal Behaviour, 59*(2), 423–432. doi:10.1006/anbe.1999.1312

Singer, T., Seymour, B., O'Doherty, J., Kaube, H., Dolan, R. J., & Frith, C. D. (2004). Empathy for pain involves the affective but not sensory components of pain. *Science, 303*(5661), 1157–1162. doi:10.1126/science.1093535

Tartter, V. C. (1980). Happy talk: Perceptual and acoustic effects of smiling on speech. *Perception and Psychophysics, 27*(1), 24–27. doi:10.3758/BF03199901

Tartter, V. C., & Braun, D. (1994). Hearing smiles and frowns in normal and whisper registers. *Journal of the Acoustical Society of America, 96*(4), 2101–2107. doi:10.1121/1.410151

Tinbergen, N. (1951). *The study of instinct.* New York: Clarendon Press.

Tomkins, S. S. (1980). Affect as amplification: Some modifications in theory. In R. Plutchik & H. Kellerman (Eds.), *Emotion: Theory, research, and experience* (pp. 141–187). New York: Academic Press.

Vorperian, H. K., Kent, R. D., Lindstrom, M. J., Kalina, C. M., Gentry, L. R., & Yandell, B. S. (2005). Development of vocal tract length during early childhood: A magnetic resonance imaging study. *Journal of the Acoustical Society of America, 117*(1), 338–350. doi:10.1121/1.1835958

Wiley, R. H. (1983). The evolution of communication: Information and manipulation. In T. R. Halliday & P. J. B. Slater (Eds.), *Animal behaviour: Communication* (Vol. 2, pp. 156–189). Oxford: Blackwell.

Zentner, M., Grandjean, D., & Scherer, K. R. (2008). Emotions evoked by the sound of music: Characterization, classification, and measurement. *Emotion, 8*(4), 494–521. doi:10.1037/1528-3542.8.4.494

14

Carl Stumpf and the Beginnings of Research in Musicality

Julia Kursell

In 2012, a century after its first publication, Carl Stumpf's *The Origins of Music* appeared in an English translation. As systematic musicologist de la Motte-Haber (2012) pointed out in her introductory essay to the edition, this coincided with a renewed interest in the "evolutionary origins of music" (p. 3). Music's origins had come center stage in publications such as the collective volume with the same title as Stumpf's book (Wallin, Merker, & Brown, 2000),[1] special issues of journals, and chapters in monographs (Patel, 2008, especially chap. 7; Peretz, 2006a; Vitouch & Ladinig, 2009–2010). Many of the authors emphasized that, for bridging between the study of music as a cultural phenomenon and the investigation of the human biological condition more generally, it might help to state a human "capacity" for music (Brown, Merker, & Wallin, 2000; Jackendoff & Lerdahl, 2006; Peretz, 2006b).

Most recently, the *Philosophical Transactions of the Royal Society B* issue, "Biology, Cognition and Origins of Musicality" (Honing, ten Cate, Peretz, & Trehub, 2015a), on which this book is based, proposed to assemble the relevant capacities under the umbrella term *musicality*. As Honing, ten Cate, Peretz, and Trehub (2015b) suggested in the introductory article to that issue, "musicality can be defined as a natural, spontaneously developing trait based on and constrained by biology and cognition" (p. 1). This enabled defining music and musicality independently from one another, admitting that music can be "a social and cultural construct" that nevertheless builds on biological conditions and constraints. They thereby induced a terminological shift that may free the biological investigation of human traits from the need to explain cultural history, while at the same time opening up a rich array of research topics that can use music's phenotype without declaring it to be a natural given.

This chapter takes advantage of this shift for looking once more at Stumpf's work. As a setting for an account of Stumpf's views on music's origins, I first address the role of the concept of musicality in his work more generally. Stumpf was the first to undertake systematic research with nonmusical subjects. A philosopher by training and profession, he did not infer from their condition to the physiological foundations of music cognition. Nevertheless, the subjects' lack of ability to assess certain aspects of music provided him with insights into music cognition that seem pertinent to this book. More specifically, this chapter reviews the analytical power of his concepts at a moment in history when systematic experimental research in cognition was only emerging. An experimental psychologist *avant la lettre*, Stumpf grounded his work on a use of philosophical concepts that aimed at the utmost precision, which in turn earned him the respect of his contemporaries. William James (1891, p. 282), for one, enumerated Stumpf among the "most philosophical and profound" writers in *The Principles of Psychology*.[2]

I discuss three phases in Stumpf's work that span roughly half a century in detail. The first part introduces the elementary methodological operation that Stumpf developed in the two volumes of his *Psychology of Tone* (1883, 1890) and compared judgments of musical and nonmusical subjects. The second part discusses Stumpf's encounter with non-European music, in which he found himself in the position of an incapable listener and the effect this encounter had on his views on music's origins. The third part traces how he hoped to find an answer to the questions opened by his reflections of music's origins in comparing music and language.

In presenting these three case studies from Stumpf's work, the chapter seeks to understand how Stumpf worked toward an empirical study of music cognition within his own philosophical, framework. As I will argue, Stumpf maintained an acute awareness of the conditions and constraints that shaped his research. Among these, three aspects stand out: his embedding in a specific culture, the use of tools and media, and, most important, the careful use of terminology and language more generally. While these can be considered standards of any scientific endeavor today, Stumpf could not take them for granted when creating new research domains from scratch. This chapter, in short, suggests taking inspiration from this situation for setting up the study of musicality as a new research domain.

Psychology of Tone

Historical Background

The idea that music might yield valuable insights into human cognition (e.g., Zatorre, 2005) has a long history. Stumpf was probably the first to stress that music as an object of investigation provided a privileged access to cognition (Klotz, 2008). He singled out music in the introduction to his *Psychology of Tone* as the "sensory content" (Stumpf, 1883, p. vi)[3] that posed particularly fascinating problems to the study of the mind, such as the mechanism of attention and the problem of distinguishing between simultaneous sensations. Music as an object of study involved a great diversity of psychological methods, such as introspection, observation, and the collection of statistical data, as well as comparisons of cultures and peoples, historical times, and individual biographies, and therefore, in Stumpf's view, offered as many opportunities for philosophical research. Many had shied away from the study of so complex an object. Perhaps for the same reason, Stumpf's mentor, Hermann Lotze, had stated that music was not among the favorite subjects of German philosophers.[4] What made matters even more complicated was the amount of detail, which "only a fancier would have the patience to consider" (Stumpf, 1883, p. vi).

At a moment when neither psychology nor musicology was established as a university discipline, the study of music as an object of psychology was thus doubly hindered by the lack of previous research and institutional support and by the complexity and intricacy of music itself. To cut this Gordian knot, Stumpf confronted the complexity of the phenomena with judgments by those who were unable to grasp this very complexity. Western music of the nineteenth century, which one could describe as a system with a code of its own that comprises certain phenomena and rules, enabled him to research "false judgments":

> The physicist looks for the motivation of false judgments only to eliminate them. They may provoke the physiologist as such to speculate about unknown processes in the brain. For the psychologist, however, they are essential for better understanding the genesis and conditions of judgments in general. For him, deceptions are perhaps more important than true judgments, with non-trained observers, whom the physicist would reject at the outset, he studies the influence of practice, and with a-musical ones the conditions of the musical sense (*Musikgefühl*). (Stumpf, 1883, p. vii)

The decisive step Stumpf proposed was to switch from the question, "How does music work?" to, "What happens when it doesn't?" To give insight into cognition through their failure in applying the rules of music was the role Stumpf foresaw for the *amusoi*, as he called the nonmusical experimental subjects he invited to his study in a Greek-style euphemism. For understanding the preconditions of judging more generally, he tested the limitations of judging in music. Stumpf invited colleagues, students, and staff who asserted that they were utterly amusical and checked whether their ideas about their musical abilities withstood the test of experimentation. He invited them to sit down next to his piano and asked them to tell which of two notes he played was higher (1883).[5] Simple as it was, this test served as a first experiment for investigating the reliability of tone judgments in the most general sense. Whatever the subjects actually perceived and however they had come to perceive it, their perception did not allow them to distinguish the pitch of the two notes.[6] They were not able to handle pitch and thus likely to fail in carrying out the basic operations of Western musical harmony.

Stumpf used this for a methodological shift. With one stroke, the focus on false judgments did away with problems previous researchers had encountered in several domains of research. The opposition of correct and false judgments freed him first from the material aspects of hearing, sound, and music. His method was independent of anatomy, physics, and physiology—areas that had struggled with a lack of knowledge that had become apparent since the beginning of the century. Second, Stumpf replaced a synthetic with an analytic approach. To make his achievement palpable, this history of the psychological study of music and auditory perception will be briefly outlined here.

In the standard account by Boring (1929, 1942), the history of psychology is rooted in psychophysics. While this resulted in a meaningful account for the state of the field in the mid-twentieth century, it goes against a more recent account of physiology's important role in introducing experimentation into the study of life. Ernst Wilhelm Weber, who is held to be one of the early protagonists of an experimental approach to sensory perception, still adhered to the paradigm of anatomical physiology. He nevertheless worked toward an alternative to postmortem neuroanatomy (Hoffmann, 2001, 2006). His experiments on distinguishing the number and direction of the two needles of a compass he applied to the skin were meant to obtain a better understanding of the shape and extent of the sensory "apparatus" in the first instance (Weber, 1835, 1846).

Fechner, a founding figure of experimental psychology, is known for having coined the term *psychophysics* (Fechner, 1860). He was the first to systematically relate physically measurable magnitudes to sensations. But, although he was certainly steeped in the musical culture of his time (Hui, 2013), he refrained from a systematic study of hearing. As a trained physicist and former editor of the *Repertorium der Physik* [Repertory of physics], a journal that regularly published updates on the state of the art in the various subdisciplines of physics, Fechner was well aware of the difficulties in acoustic measurement. The physics of sound were not far enough advanced to allow reliable measurements that would enable consistent comparison with mental data. No measurement for intensity was available, let alone the intensities of wave components.

Hermann von Helmholtz was the first to bring the mathematics of periodic functions and the phenomenal appearance of tones together. He also was first to claim that the components of a sound and their respective intensities were responsible for the sound's quality. In his influential study of human hearing, *On the Sensations of Tone as a Physiological Basis for the Theory of Music* (1863, first English translation, 1875), however, Helmholtz carefully avoided speaking of "the tone" as the counterpart in perception to the written note. Instead, he introduced a terminological distinction between simple tones of a single sine wave frequency (*Ton*) and compounds containing many sine wave components (*Klang*). He also explained that the tones heard in music were in fact compounds, (physically speaking, *Klänge*), and he never commented on the seemingly obvious capacity of music listeners to form musical "notes" from compounds.

Sensation and Analysis

Stumpf had served as an experimental subject for Fechner in whose house he had met the Weber brothers. He was thus thoroughly familiar with the psychophysical approach.[7] Equally, Helmholtz's "classical work" (Stumpf, 1883, p. v) served as a constant reference for his own enterprise of a *Psychology of Tone*. But if Helmholtz built his theory around a notion of a smallest element and from there proceeded in a synthetic manner toward more complex claims, Stumpf went in the opposite direction and tried to analyze the components present in given notions. As did Helmholtz, he extensively drew on music as a resource.

But if Helmholtz went from the rich descriptions available for the sounds of music to speculating about the physiology of hearing, Stumpf fastened on the rule-based nature of music as a system.

Music provided Stumpf with a set of notions that he could investigate from the standpoint of a philosopher. The comparison of the judgments by subjects who were familiar with these notions with those of subjects who failed to apply the code of Western music to sounds proved an invaluable methodological resource. Exposing both groups of subjects to phenomena that could be described in musical terms, such as the question of which of two tones was higher, he could assess the nature of the judgments and the degree of their variability. For this, no specific measurements on the stimuli were necessary, nor were any specific assumptions on brain anatomy or on the ear and its functioning. In his experiments, the seemingly "natural" distinctions, such as between consonance and dissonance, turned out to be "acquired, dependent on education, and malleable" (1883, p. 48). The comparison of judgments would thus produce a reevaluation of assumptions that had been taken for granted up to then.

The question regarding whether and how simple wave components merged into one tone in human cognition, which Stumpf had found unanswered in Helmholtz's work, took on a different aspect in Stumpf's analysis. Instead of asking how music listeners abstract from the components in order to produce a unitary tone, he discovered that the very distinction of single notes out of an assemblage of components was not evident to all subjects. Rather, the *amusoi* tended to hear only one sound, whereas the musically able detected several musical notes. Experimental subject "Dr. K," a former student in Stumpf's lectures in logics, took all notes played to him for one single sound, Stumpf reported. He had tried to make the multiplicity of notes easier to observe for Dr. K by playing them first separately before playing them together again; he increased their distance and number, or favored particularly dissonant combinations—"all in vain" (Stumpf, 1890, p. 363).

For Stumpf, subject Dr. K exhibited a zero level of the functions involved in musical cognition. What happened in the minds of the musically able subjects when listening to music happened on top of what happened in Dr. K's mind. This can be seen as different from today's research with subjects in which congenital amusia is assumed. According to ICD-10, R48.8 (World Health Organization, 2016), *congenital amusia* is the term for a deficit that concerns music in the broadest sense. At first glance, it implies a lack of a function. The notion of deficit seems to attribute normal functioning to musical hearing while at the same time pathologizing amusia. Stumpf, in contrast, held that any subject experiences at least what Dr. K experiences—and may be able to perform musical analysis.

On a second look, however, today's research in amusia is not that far from what has been described here as Stumpf's point of view. Not only do modern diagnostic tools such as the ICD avoid claims about etiology, but also the work with subjects who demonstrate comparable deficits to those encountered by Stumpf does not involve assumptions regarding pathology or other hypotheses about the origin of these deficits. Peretz (2006b, p. 8), for instance, has cautioned researchers that "the evidence for domain specificity, innateness and brain localization must be examined separately." When Peretz, Champod, and Hyde (2003) first published the *Montreal Battery of Amusia*, they rightly linked research in amusia to the history of localization research. Researchers such as the surgeon Jean-Baptiste Bouillaud, who used new evidence of brain lesions, began to explore localization of function in the brain, but they were far away from claiming fixed loci. As Peretz (2006b, p. 9) goes on to explain, "A domain may be as broad and general as auditory scene analysis and as narrow and specific as tonal encoding of pitch." Modular theories can model function without committing themselves to localization specifics or denying that domain-specificity can emerge from learning (Peretz, 2006b).

For Stumpf, Dr. K demonstrated that perceiving tones could not be taken for granted. Stumpf could now posit a limit in the ability to deal with musical notes, even though in other subjects, the degree of analytic abilities varied. Based on his experiments and tests, Stumpf assumed a separation of two levels in the processing of musical notes. He claimed that sound was sensed first without distinguishing anything like a single note in it. Based on this level of sensation (*Empfindung*), further analysis could then occur. In musically trained subjects, the analysis happened so quickly that they were not aware of sensing sound prior to analyzing it. This higher level of analysis (*Analyze*) did not occur in all subjects. The *amusoi* instead remained in the state of sensation. Stumpf did not assume, for the time being, that any of his subjects would be physically unable to eventually learn to analyze sound. He mentioned that three of his subjects, having learned to distinguish notes during the tests, were dismissed as "cured" (1890, p. 117) after the tests. Moreover, even the musically able would at times fail to recognize strongly consonant intervals, such as the octave, as being composed of two notes. In such cases, their ability to analyze failed, and they in turn fell back into the state of sensation.

The distribution of the judgments led Stumpf to propose his concept of fusion that featured prominently in the second volume of his *Tone Psychology* (1890). He declared fusion to be a mental process. Whether musically competent or incompetent, all subjects would in the first instance sense something genuinely mental when exposed to sound. Proceeding from the level of sensation, further analysis occurred according to the subject's ability to process musical features of the sound. In the case of intervals like the octave, subjects remained at the sensation stage with minimal ability to apply the code of music to the sound. In other cases, however, the musical subjects could easily come up with the corresponding judgment.

Fusion was thus not a property of musical notes or physical sounds, but a tendency in the subject's judgments to remain in the state of sensation. This explained for Stumpf why it had been difficult to distinguish sensation from analysis. The musically trained immediately processed their sensations in terms of tonality and musical harmony. The distinction of notes in sounds then seemed to occur naturally and was thus not detected as a distinct operation. To them, the result of this analysis appeared to be their immediate sensation.

They experienced two notes that were either consonant or dissonant. Whereas the latter feature was considered part of musical knowledge, the first seemed a naturally given sensation. The musically experienced thus ignored the fact that in order to apply the code of music, it was necessary to perceive the sound as consisting of two notes in the first place. For Stumpf the philosopher, the tension between expert judgments and the judgments of the amusoi revealed the mental process of perception and the role of analysis based on sensation. The comparison between the two groups yielded a clear result: sensation became apparent in the tendency to hear notes as fusing.

Preliminary Conclusion

The German language does not yield a noun that presents the opposite to *musicality*. As much as this language favors nouns and indefinite constructions of agency, the lack of musical talent has not made it into a noun of its own. Being *unmusikalisch* is always the property of something or someone. Looking back at Stumpf's writings with this in mind, one sees that he does not give prominence to musicality (*Musikalität*), either. The German word for *musicality* connotes excellence in musical talent and thus a level of music proficiency that is above average. This holds in particular for the nineteenth century, when listening to music depended on the presence of musicians making the music.

The earliest instances of discussing musicality in German occurred in the context of the rising interest in the development of the individual subject in idealist philosophy. Gembris (1998) has pointed to a treatise by Christian Friedrich Michaelis (1805), a student of Johann Gottlieb Fichte, who was among the first to work out suggestions for identifying talent in young children in a short treatise he published in a Berlin music journal. The interest of the growing bourgeois society in music as part of its educational ideals not only brought forth a greater quantity of musically trained individuals, but also more cases of failure. Robert Schumann addressed the education toward what he called musicality in an appendix to his *Album for the Youth* (1854). Musicality for him was the ability to understand music and to (re)produce it in a meaningful rather than merely mechanical way. Stumpf, however, did not study expert judgment for the sake of improving the quality of musical performance. Nor did he take the idealist concept of talent as a natural gift as his point of departure—as proposed in Kant's *Critique of Judgment* (§46; see Gembris, 1998). Stumpf was using music for understanding the mind.

Working with his amusoi, Stumpf's analytical method detected layers in cognition and thereby significantly shaped the notion of music cognition itself. No earlier investigator had wondered how music emerges in the mind. Far into the nineteenth century, music had been accepted as a phenomenon that is given in nature and in which humans with a musical talent participate in a privileged way. Stumpf pursued his work with amusical subjects as well as with musically experienced ones, not in the expectation that they would tell him anything about music but as a window on the workings of the human mind.

The Origins of Music

Background

Stumpf's next book, *The Origins of Music*, appeared in 1911. By this time, he held a position as professor of philosophy at Friedrich Wilhelm University in Berlin. His *Psychology of Tone* had earned him much acclaim in the scientific community and beyond. After positions in Würzburg, Prague, Halle, and Munich, Stumpf had accepted the position in Berlin because it was connected to the task of setting up a seminar for experimental psychology. He took up work there in 1894. The seminar quickly grew and was renamed the Institute of Psychology in 1900. It eventually became one of the best-equipped laboratories for psychology in Germany.[8] Stumpf still devoted much of his research to questions relating to music. His plan to write two more volumes of *Psychology of Tone* never materialized, but he did publish two more substantial articles on the topic that extended the argument of the first two parts to consonance in two simultaneous tones and in musical chords (1898, 1911b).

In several respects, *The Origins of Music* marks a turning point in Stumpf's work. First, it was to remain his last major publication on music, since after it, he turned to language as his main object of investigation. Second, the book carried forward two strands in his work that had emerged during the research for *Psychology of Tone*. One was his interest in theories of the origin of music itself. In 1885, he published an extensive review of the theories of Spencer, Darwin, and the two British psychologists John Sully and Edmund Gurney (Stumpf, 1885). Gurney especially attracted Stumpf's attention. This was not only because Gurney used Darwin's theory of sexual selection to counter Spencer's claim that music originated from language—a position with which Stumpf disagreed—but even more so because of Stumpf's problems with Gurney's "nativism" (1885, p. 432). In *The Power of Sound* (1880) Gurney posited a "musical faculty," which he defined as follows:

> I am assuming here, for the sake of clearness, and without further argument, that whatever explanations of musical effect turn out to be possible, the exercise of the musical faculty will present an ultimate and inexplicable element. (Gurney, 1880, p. 86)

According to Gurney, this particular faculty was what had enabled humans to go through the processes Darwin described. For Stumpf, such an assumption was problematic in two respects: Gurney claimed the "faculty of music" could not be analyzed, and in that he used it for explaining both the origins of music and its further development (Stumpf, 1885, p. 331). He was also not convinced by Gurney's claim that the music of Beethoven and Wagner and the earliest developments of music could be explained by one and the same unified trait.

When Gurney eventually did provide further explanation, he ran into contradictions, Stumpf noted. He disagreed in particular with a footnote in which he read that Gurney did not understand his musical faculty to refer to "some special sort of musical gift" but to the "ordinary power of perceiving and recognizing *tones*, of apprehending a melodic form as a whole by co-ordination of its parts" (Gurney, quoted in Stumpf 1885, pp. 333–334, my italics). For Stumpf, this only demonstrated that a rigid nativism could not apply here. The "powers" into which Gurney broke down his "musical faculty" were not specific for music, Stumpf argued. They belonged to sensation and perception rather than to a faculty that was meant exclusively for appreciating music. Therefore, these faculties could not account for a musical faculty in its own right.

Gurney's reference to Darwin also did not convince Stumpf either, who found in Darwin a particularly poor authority for a faculty of appreciating music's abstract qualities. "Resorting to history or Darwinism does not absolve us, as some seem to believe, from the obligation to define and individually derive a composed activity" (Stumpf, 1885, p. 334). In addition, Darwin had emphasized the sensory pleasure in the sound of music, not the pleasure in abstract musical form. If the difference might have escaped Darwin, who was known to Stumpf as being an amusos, it should not escape Gurney, the music expert.

In light of the notion of musicality proposed in this book, Stumpf's critique shows, first, that he pleaded for a separation between the phenomenal aspects of music in its historical development and a human faculty for music as a general trait and, second, treated music cognition as comprising a number of faculties not specific to music. His experimental subjects had shown to him that critical features of music were not recognized by every member of even so homogeneous a group as were his students, staff, and colleagues in terms of education and social standing. In fact, a distinction between those who could and those who could not apply these rules cut through Stumpf's own social surroundings, allowing him to gather sufficiently large groups of amusical subjects in less than a week (1883). Assuming that both of his groups—the musical and the amusoi—shared basic features of cognition, he did not venture to explain music through the mind but rather the mind through music. For Stumpf, if there was such a thing as a musical faculty, it had to be understood in terms of functions involved in sensing and perceiving more generally.

Parallel to the scrutiny of the English publications on music's origins, Stumpf came to realize that his experimental setup depended on a simplification with regard to his notion of music. All subjects, whether musical or amusical, responded to tones and rules that were part of the same musical tradition and that was also Stumpf's own. He became acutely aware of this when, in 1885, a group of Nuxalk singers from British Columbia visited Halle, where he then held a position. He attended their performance and was shocked not so much by their singing, which appeared as a mere howling to him, but by his own incompetence in following what they were doing.

Stumpf seized the opportunity to work with one of the singers in private sessions that was granted to him by the showmen who "exhibited" the "Bella Coola Indians": the adventurer and self-made ethnographer Johan Adrian Jacobsen and the zoo director Carl Hagenbeck. A member of the group by the name Nuskilusta, known to be the best singer among them, spent four consecutive days with Stumpf during which, supervised by the translator Jacobsen, Nuskilusta sang the melodies of the show for one or two hours while Stumpf tried to pencil them down. Stumpf's article, "Lieder der Bellakula-Indianer" (Songs of the Bella Coola Indians, 1886), gives a detailed account of the procedure and its asymmetry.

Until then, Stumpf had thought of himself as easily coping with the task of notating and analyzing musical performances on the spot. He possessed absolute pitch and had learned shorthand in his youth, and he used both extensively in his laboratory work. Neither of these skills worked properly in this case, however. The pitches that Nuskilusta consistently used did not conform to any Western European tonal key, nor did the rhythm, which did not conform to the metric structures familiar to Stumpf. Stumpf had Nuskilusta sing his songs over and over again while he took his notes and—rather casually—observed the ability of the singer to chunk, restart, reiterate, and transpose his melodies.

Although the procedure was a challenge for both participants, Stumpf nevertheless believed that it also bore advantages because it had made him aware of his mental presuppositions through learning about different ones. Being exposed to the unfamiliar Bella Coola repertoire, Stumpf had first found himself in the position of the amusos: he had been unable to apply the rules of the music. Yet when he visited a second performance after the training with Nuskilusta, he could follow the melodies, and he even noted some deviations among the singers, which made him remark that there were "probably a-musical individuals among the savage as well" (1886, p. 408).

This second strand in Stumpf's interest in music, that is to say his interest in non-European music, remained stable in his further work. His focus was not so much on the fascination with what today would be called "the Other," but concerned the comparison of mental presuppositions and their material and technical conditions. Although traditional notation had proven painstakingly unfit for conveying the songs of Nuskilusta, it had the advantage that it spoke to Stumpf's own scientific community, which for a long time to come would remain a community of readers. He made this explicit in his next article on non-European music, "Phonographirte Indianermelodien" (Phonographically recorded melodies of American Indians; Stumpf, 1892).

This article was a reaction to Benjamin Ives Gilman, who in turn commented on Native American melodies that J. Walter Fewkes had collected. Gilman, a former student of Charles Sanders Peirce, used European music notation, but in contrast to Stumpf, he wrote the melodies in such a way as to defamiliarize them for the European gaze. He achieved this by arbitrarily attributing new meanings to the available signs, including the accidentals. To Stumpf, the resulting melodies looked as if they stemmed from a modernist composing style with a preference for remote tonalities. Yet Gilman's notation did not refer to live performance but to recordings of them. The acoustic token of his notation was the recorded sound, and thus an object that was already part of a scientific endeavor. With Gilman's study, the investigation of non-European music shifted to a different operational mode. Phonographic recordings presented new objects of scholarly study, making the recorded music potentially accessible for circulation and repetition.[9]

Stumpf eventually opted for the phonograph as well. Together with one of his assistants, Otto Abraham, he made extensive recordings during the visit of a Siamese (now Thai) theater group to Berlin. He published a long article on these recordings under the title "Tonal System and Music of the Siamese" (Stumpf, 1901; cf. Schwörer-Kohl, 2011). Next to a score and analysis of the recorded pieces, this article presented an extensive discussion of recording technology. Stumpf not only commented on the advantages of microphony and magnetic recording, technologies that only dawned at this time, but he also connected his reflections on technology to the main psychological questions he had been dealing with up to then. More specifically, he referred back to the two functions he distinguished in music cognition in his earlier work on tone psychology. A researcher who used the phonograph did not have to rely on his own mental abilities to store his sensations for future analysis. The operations of perceiving and analyzing could now be separated and distributed among technical operations of recording, measuring, and notating. The analysis would then address these data, which in turn could be supported by reiterated listening. The decisive new aspect of phonographic recording for Stumpf was its ability to record and store music in an unanalyzed state. Whatever the appropriate code for notation might be, the phonographic recording was not subjected to it (discussed by Kittler, 1999).

By the time Stumpf published *The Origins of Music*, the interest in phonographic recording had taken on a new dimension at the Institute of Psychology. The recordings of the Siamese theater group eventually became the basis of an important collection of ethnological musical recordings from all over the world. In 1905, Erich Moritz von Hornbostel, one of Stumpf's former assistants and known today as one of the founding figures in ethnomusicology, was charged with organizing the collection. Being a chemist by training, he developed a technique for copying the fragile cylinders. Thus equipped, he worked out a strategy for how to gather music samples from all over the world as a collaborative enterprise involving musicologists, ethnologists, and experimental psychologists. Travelers were asked to make recordings, and they received phonographs and instructions for this purpose. Ethnologists from all over the world would send their cylinders to Berlin to have them copied, and one of the copies always remained in Berlin. When Stumpf (1908) publicly announced the Berliner Phonogramm-Archiv, the collection already held more than a thousand items, and by the time the phonograph was replaced by the tape recorder after World War II, the archive counted more than thirty thousand cylinders (for this collection, its methods, and its impact, see Ames, 2003; Klotz, 1998; Koch, 2013; Rehding, 2000, 2005; Ziegler, 2006).

The Argument of *The Origins of Music*

The Origins of Music has three parts. The main body of text was a lecture Stumpf had held at the Urania in Berlin, an establishment for the broader public interested in the latest findings of the sciences. The published version added an extensive critical apparatus to this text (Stumpf, 2012) and a discussion of music examples that Stumpf entitled "Songs of Primitive People." The tone of the book was set by the new technical context for research into music that had come about with phonographic recording, and the published version was replete with critical remarks about the naive trust in listening and writing as neutral tools in the comparative study of music from around the world.

Among others, Stumpf mentioned two examples of "primitive song" from Wundt's *Völkerpsychologie* (1908). One was reported there to stem from "among the man-eaters" in Australia and bore the rhythmic indication "tempo di valse"; the other was a "negro-melody" from the "heart of Africa" (Stumpf, 2012, p. 73), which Stumpf identified to be identical with the song "Hurra, Hurra, Huraleralera" that he found in a collection of songs for village schools around Berlin and in a soldiers' songbook. Whatever that said about Prussia, it promised little insight into African music.

Wundt's source, however, the book *In the Heart of Africa* (1874) by the renowned botanist and traveler Georg August Schweinfurth, seemed so trustworthy to Stumpf that he decided to approach the author and ask him how he had come to take down this melody. He promptly received an answer from Schweinfurth:

> I reproduced the melody as it struck my ear. Perhaps this stylized the melody unconsciously in a European way. At the time it made a great impression on me. I often hummed it to myself on my travels; it was always my habit to sing in half voice all the melodies I could get hold of while marching. I certainly believe, then, that I later reproduced it pretty accurately. A rhythmic song with a hundred voices must surely still have a melody that dominates, a diagonal cutting through, and I captured it in this way. ... The notation of the melody arose at that time (as I wrote my book) together with my brother Alexander, now deceased (whom Alexander Dorn declared an out-and-out musician). I performed the song several times for him; there was no other way to express it. (Stumpf, 2012, pp. 74–75)[10]

Stumpf (1911a, p. 75) commented, "I believe that everything necessary has been said." The episode was exemplary for the interference of mental presuppositions with the music heard in Schweinfurth's listening. Not only had Schweinfurth's musical ear perhaps betrayed him admittedly, the song was also likely to transform during the marching and perhaps even have mutated into the marching song for soldiers, but it had also undergone a last transformation when it was handed on to yet another European individual with even greater music skills than Schweinfurth's own: "What he sees as a guarantee for loyal transmission is in fact the opposite" (p. 75).

When *The Origins of Music* was published, Stumpf's efforts to take down the songs of Nuskilusta with paper and pencil seemed remote. This kind of research now could resort to phonographic recording. "There can hardly be better evidence than Wundt's two examples of how indispensable phonographic recordings are on location, and how little benefit there is—in terms of ethno-psychology—in getting advice from books alone," Stumpf now wrote (2012, pp. 75–76). Technically, at least, the phonograph solved the problem of the mental presuppositions. Given the abundant material of the phonogram archives in Berlin and elsewhere, a researcher now could compare these documents and try to make conclusions on the more general features of music.

The newly written second part of the book played an important role in this. While using a discourse that is hard to digest and rightly avoided in today's musicology, the idea behind this "journey around the earth" (p. 174) was to delineate a minimal concept of music. Stumpf aimed not so much at a hypothesis about music's development that used the "primitive" examples for showing what the West had left behind, but for demonstrating that all documented music displayed organization and variety and, first and foremost, all of the samples of music known by then used distinct pitches.

Stumpf started from the "simplest music" available: recordings of the Vedda from Ceylon (now Sri Lanka), who were known to use no musical instruments. Max Wertheimer, who would later acquire fame as one of the founders of Gestalt psychology, had transcribed the recordings held in the archive that had been collected by ethnographer Margarethe Selenka and published his findings in the journal of the International Society of Music (Wertheimer, 1909; about the collection see Ziegler, 2006, pp. 264–267, 376). These songs used clearly distinct pitch steps, as both Wertheimer and Stumpf noted. The songs sometimes alternated between only two pitches, but the pitches were clearly distinct and exhibited recurring patterns. For Stumpf, this indicated a threshold:

> These Vedda songs may present an example of that primordial or precursory stage of music that uses only small tonal steps. Neither consonance nor tonal relationships appear to have played a role in them. Nevertheless, they already have a certain structure, certain regularly recurring phrases with variations, and finally specific, distinctly shaped, closing formulae. (Stumpf, 1911a, p. 112; 2012, p. 154)

The songs of the Vedda exemplified the basic definition of music Stumpf applied in *The Origins of Music*: Whatever the function or cultural embedding of the recorded songs, they fulfilled the basic criterion of being organized in pitch steps. That music was characterized by the stepwise organization of pitch was not a new finding but an accepted opinion. Helmholtz had stated the same in *Sensations of Tone* using musicological writings such as those by François-Joseph Fétis as his source. Helmholtz (1863, p. 386) wrote, "In the music of all people—to the extent to which we know about it—melodies change stepwise in pitch rather than in continuous transition."[11] Helmholtz also pointed to the transposability of musical melodies, something he interpreted as a property of music as a conventional system rather than the reflection of a specific human faculty. He added a passage in the book's third edition of 1870 that stated a characteristic similarity between relations within the tonal space and geometrical space:

> It is an essential character of space that at every position within it like bodies can be placed, and like motions can occur. Everything that is possible to happen in one part of space is equally possible in every other part of space and is perceived by us in precisely the same way. This is the case

> also with the musical scale. Every melodic phrase, every chord, which can be executed at any pitch, can be also executed at any other pitch in such a way that we immediately perceive the characteristic marks of their similarity. (Helmholtz, 1875, p. 576)

Helmholtz put forward a notion of symbolic operations within geometrical space that also applied to music. The study of geometries with dimensions other than those of Euclidian space (Helmholtz, 1868) had convinced him that experience could cope with any given space, thereby revising Kant's notion of space as given a priori. For Helmholtz, spatial perception had to be acquired through experience. His notion of perception was distinct from the merely physiological sensation. Perception was part of higher-level cognitive functions, and these were acquired rather than innate (Kremer, 1993).

The assumption that hearing must be learned led Helmholtz to look at the conditions under which that happened. Because he could not prove the resonance theory of hearing he proposed in his work on physiological acoustics, he tried to test as much of the knowledge on hearing as possible against his hypothesis. Whereas the inner ear remained inaccessible, the knowledge of music in other times and other parts of the world was growing. Helmholtz thus discussed tonal scales from China and the Arab world, as well as Ancient Greek theory and the history of consonance and dissonance.

In this respect, Helmholtz's work could serve as a model for Stumpf. *The Origins of Music* certainly represents his boldest attempt to investigate music speculatively through using its history for insights into the human mind. As a psychologist, however, he was interested in the individual development of the human mind, and as a philosopher, his main requirement in this endeavor was consistency. He therefore used a rigid definition of music, asking how humans had come to use stepwise progression in pitch in their utterances. Like Helmholtz, he treated the history of music as a reservoir of knowledge about these questions, and he widened his scope to include prehistory.

To set the stage for his own remarks, Stumpf first surveyed the most prominent theories on the origin of music that had been put forth by then, starting with Darwin. Music seemed an anomaly against the background of the survival of the fittest, as it was unclear how music helped in the struggle of life. Darwin's answer from *The Descent of Man* was, in Stumpf's reformulation: "In the beginning was love" (2012, p. 34). Next he discussed Spencer's *The Origin and Function of Music* and the proposal to take "the word" as music's beginning, and the last proposal, "rhythm," came from the economist Karl Bücher, who in his book *Labor and Rhythm* (*Arbeit und Rhythmus*, 1896) included a chapter on the origin of music and poetry. All three theories could be easily refuted by reference to Stumpf's own definition of music. None of the theorists explained how the stepwise organization of music had come about. A theory that could not explain this feature was not a theory of music's origins in Stumpf's view.

As he had done in his earlier review, Stumpf again distinguished two levels in theories of origins of and faculties for music: the faculties that were a prerequisite for music but not specific for it and the properties of sound and auditory cognition that might have led to the emergence of pitch discretization. He wrote:

> If we understand music now to be the art whose material consists of fixed and transposable tonal steps and if we then try to grasp the origin of this art, we have to distinguish two questions: How did a faculty to recognize relations among sensations arise that is independent from the individual characteristics of these sensations? And: How had one come to the fixed intervals that we indeed find in the music of different peoples and times? (Stumpf, 1911a, p. 23; 2012, p. 44)

The answer to the first question was found in the faculties of abstraction and generalization that one had to assume in primordial humans if they were to arrive at music in Stumpf's sense. The second had to be split into two subquestions. One concerned the hypothetical opportunities to discover the advantages of a stepwise organization; another concerned the characteristic operations that made melodies appropriate for being transposed and recognized when starting on a different pitch.

Stumpf proposed that several ways to arrive at pitch discretization. The first was the need to communicate. Humans may have discovered that in calling each other, their voices gained in power when they remained on a stable pitch. Furthermore, they may have noticed that in calling together, they produce multiple sounds that get different qualities depending on the relation between the voices. In short, they might have encountered fusion. Finally, the curiosity one met not only in humans but also in primates might have taken them to find acoustic features of stable sounds in the use of tools. For music, Stumpf claims, one therefore had to assume origins rather than one single origin.

In sum, Stumpf's origins can be read as a consistent though speculative account on how it would have been possible to arrive at the features he took to be part of music's definition. This account was based on a separation between what might be attributed to human faculties and what might result from encounters, trial and error, and technological development. In all these developments, music had to be understood as a cultural phenomenon that only reconfigured faculties that had other uses as well.

Speaking and Singing

Background

In *The Origins of Music*, Stumpf discussed three phenomena in the music of "natural people" with regard to their connection to the possible origins of music: polyphony, rhythm, and what he called "intoned speech" (2012, p. 55). Whereas the first two were also characteristic for the music of the late nineteenth century, this was different for the last. Intoned speech was, for Stumpf, a way of using the voice as in speaking, but at the same time using definite pitch steps. The line Stumpf drew between speech and song coincided with that between tone and noises (*Geräusche*).[12] According to his own definition of both terms, tones could be transposed, whereas noises could not. Tones could thus be organized in steps, which also allowed stepwise transposition; noises would lose their character when transposed. Accordingly, speech would be nontransposable, whereas for music, transposability resulted from its definition: a stepwise organization of pitch included the feature of transposition.

Stumpf himself had noted that singers in other cultures were able to transpose their songs. Nuskilusta, for instance, the singer from British Columbia, experienced "no difficulty in starting his songs on a different pitch" (1885, p. 409), and the experience with phonographic recordings had shown that singers worldwide would start to sing on the pitch of the whistle the ethnographers used to gauge the recording speed—a tone on a standard pitch in the beginning of a recording had turned out to be the safest means to identify the rotation speed used for the recording (Kursell, 2012; Stumpf, 2012; Ziegler, 2006). Travelers who collected recordings for the Phonogramm-Archiv were therefore advised to blow the gauging pipe only after the singer had stopped in order to have him or her sing on a pitch of his or her own choice.

The faculty that made humans transpose their songs in the end drew the demarcation line between having music and not having it. The ability to abstract from absolute pitch enabled humans to not only react to music but recognize it, as Stumpf explained. Whereas animals might be seen to react to sounds repeatedly in the same way, he stated that in humans, "recognition also occurs in a truer sense, meaning not merely the same reaction but also recognition of the sameness or identity" (2012, p. 44). Although Stumpf did not exclude that some animals could be found that would be able to transpose a melody, he held this to be unlikely. Abraham, his assistant and colleague, as Stumpf reported tongue-in-cheek, had spent years trying in vain to teach his parrot to sing a melody starting on different pitches. Parrots lacked music, according to Stumpf's definition. The same held for humans who lacked this ability. One would be forced to "deny them a capacity for music in the narrower sense. Contrariwise, should talented animals just once verify or instill this ability, we would instantly consider them as our rightful brothers in Apollo" (2012, p. 37). His own experiments with the amusoi had shown to him that these subjects would need considerable support in catching up with Apollo and his siblings, and sometimes even failed to do so.

Abraham's futile attempts to teach his parrot true human singing may have had a different background, though. Brain physiologist Ludwig Edinger had been among the first to attend to the fact that sensory perception need not mean the same thing for different species in order to be subsumed under the same category. That fish do not have the same ears as humans do did not exclude that they perceived, though in their own way, what for humans was sound. When Edinger visited the Institute of Psychology, the Berlin psychologists convinced him to make recordings of his singing. The interest of this is unmistakable when one listens to the recordings today. Edinger was an amusos of the purest kind.

The phonogram archive preserves not only the ethnographical recordings but also some recordings labeled "experimental wax cylinders" (Ziegler, 2006, p. 84). A selection of them is available online.[13] Among these are some recordings of Edinger. One of them was labeled "*Wagner-Motive*" (Wagner motifs). The recording begin with an announcement that identified the singer.[14] After an announcement, he sings four different motifs, each performed twice. Edinger imitates instrumental parts on syllables such as "daa dadida dadida dadiii" and sings the opera's text for the vocal parts, such as "*Gold'ne Äpfel wachsen in Freya's Garten.*" He obviously sings from memory and sometimes changes the original lyrics. The singer's voice does not produce tones with a clear pitch, but resembles speaking with a strong rhythmical structure and an eccentrically overstated melodic contour, as opposed to mere speech. As the repetition demonstrates, these features did not occur by chance; they are characteristic of his performance.[15]

After this, the announcer enumerates the motives that have been heard, identifying them as being from *The Ring of the Nibelung*. The recording shows that Edinger is acquainted with Wagner's music—to the extent of knowing its leitmotifs. He utters them correctly in terms of rhythm, and he is even able to reproduce these well enough for the supervisor of the recording to recognize them.

What the Wagner motifs seem to convey is that the choice of the music was left to Edinger. He sang music that he knew well and felt comfortable reproducing. Yet Wagner's music is not meant for humming along. Edinger also attempted to sing motives that were written for instruments rather than voices and the range of which far exceeds the capacities of even a trained singer. In Wagner's music, the live experience of opera and its reproduction at home had drifted apart. Even before the phonograph, this music challenged the seemingly natural connection between the competences of performing and listening.

The fact that this singer could not reproduce intervals did not mean that he was unable to recognize them. The idea that recognizing melodies could be a distinct ability from producing the melodies' intervals had not occurred to Stumpf. In his review of the English literature on the origins of music, Stumpf even misread "tunes" in Gurney's remark to mean "tones." Whereas Gurney described the musical faculty to be based in modern terms on the recognition of melodic contour, Stumpf denied such a faculty on the basis that not all individuals recognized separate tones. In his own musical upbringing, a recognition of melodies that would have been independent of the recognition of intervals made little sense. Stumpf was a talented violin player and reported in his autobiography that he even had considered becoming a professional musician. Playing the violin without a notion of discrete pitches, however, was out of the question. The violin has no preset pitch steps in contrast to, for instance, the piano. An individual who possessed no mental representation of the steps to play would hardly get far on the violin.

To Stumpf, the fact that an individual reproduced a melody in a way that could hardly be distinguished from speech remained a bone of contention. After the publication of *The Origins of Music*, he would not return to music in his research and publications—with the exception of one text that dealt with the distinction between speaking and singing. There, he conceded a shortcoming in his earlier definition of music as the stepwise organization of sound. It was thinkable that a music could be invented that would build on the artful use of steadily raising and sinking pitch, although such a music seemed not the most desirable development to Stumpf as he remarked in a footnote: "I must admit that the polyphonic howl of wolfs and jackals that one sometimes hears during the day in the zoological gardens pleases to my ears that are calloused by futurist and atonal music" (Stumpf, 1924a, pp. 68–69). The counterpoint of their voices, enigmatic expressiveness, and the fact that they usually ceased after a short while promised that even such a thing was possible.

In hindsight, Stumpf realized he had grounded his search for the difference between speech and song in a concept of a smallest element of music. This concept had proven to be so rich that its limitations had not come to the fore. Especially the interaction between the mental presuppositions in musical listeners' minds and the various musical systems seemed only to gain in consistency the further he advanced. Already in his *Psychology of Tone*, Stumpf had described this feature in the listening of the musical subjects: he called this the "standpoint" of the musically able listeners. By this he meant that listeners with musical ability attributed any notes they heard to an implicit musical schema of tone relations. They found in tonal space, as he wrote, "clear and stable signaling poles for their acoustic geodesy" (1883, p. 150). The signaling poles guided them in locating notes, and they resorted to this aid even when the tones they were exposed to did not correspond to the European tonal system.

The singing of amusical subjects such as Edinger, however, indicated something different. Although the concept of the tonal geodesy applied to musical individuals, one had to assume that this geodesy somehow was founded on the recognition of melodic movement. In his article on speech and song, Stumpf therefore readjusted his theory of the signaling poles in two respects. On the one hand, one had to conceive of the distribution of these reference points within the octave as strictly historical (Stumpf, 1914, p. 323; 1924a, p. 72). On the other hand, one had to take into account that the mental presuppositions would be individual. A singer decided whether he or she sang or spoke, and the listener decided whether he or she heard singing or speaking. For both sides, this attitude (*Einstellung*) would shift the quality of the sounds in decisive ways.

To illustrate this case, Stumpf once more turned to Helmholtz, who had given the example of a spoken line in his book *On the Sensations of Tone*. Even for Helmholtz and all the more for his readers, Stumpf argued, to see the sentence in musical notation would turn them into music:

> The matter becomes essentially different, when the speaker shifts his attention during speaking to the intonation of his speech. Then the entire tonal material will change in character. … Who ever pronounces the two sentences "Did you go for a walk? I went for a walk." in the light of Helmholtz's notations of speech intonation or while remembering these, in order to test their correctness, is in a different state of mind, both with respect to the content and the functions. The notes linger before his mental eye and he has set himself a task, not that of hitting the notes, but that of comparing them. But can one avoid the influence of the notes? Hardly. Unwillingly, one will adjust one's

intonation more or less to the notes. Speech will thereby turn into singing like Daphne into a laurel. (Stumpf, 1924a, p. 60)

Stumpf concluded that speech and singing were phenomenally not specific. They differed only in degree, not in any basic property. Edinger's singing was singing because he meant it to be such. And Helmholtz's notated intonation was musical because everyone would join him in reading the notation in this way. Music was not only a culturally defined phenomenon, but one that is defined by the individual experiencing it.

Coda

"In the beginning was the deed"[16]—thus Stumpf speculated in his *Origins of Music* (1911a). The "deed" in the quote from Goethe's drama *Faust* stood for an origin comprising many origins. When the drama's hero tries to translate the New Testament, he stumbles over the first line. Could he accept that the word was in the beginning? The translation continues in the presence of the devil who, changed into a poodle, cringes more and more hearing Faust read the Bible. Was it not rather "sense," or "power" that stood at the beginning? When Faust decides that "deed" must be the word for that with which it all began, the poodle decides that it is time to interact with Faust. He turns into a traveling scholar, and the drama of what German media historian Friedrich Kittler (1990) has identified as the beginning of hermeneutics is launched.

Among middle-class *bildungsbürger* of Stumpf's times, it was a regular feature to know the core rhymed texts of the cultural canon by heart. Speaking about the origins of music to such an audience in Berlin's Urania, where his listeners expected popular lectures and scientific entertainment, Stumpf could allude to that desperate search for a beginning that lay beyond human understanding while at the same time casting this enterprise in slightly ironic light. The talk expressed his discontent with speculations that looked for a single reason that had made music emerge. Instead, the "deed" seemed a key to the question by opening more than one door. It offered to bring new concepts into the discussion of the origin of music. Rather than choosing between mutually exclusive causes, a term that centered on activities would allow him to embrace concepts of communication, social interaction, and the use of tools, all part of his own research agenda.

Stumpf himself had remained open to such a polymorphic concept of origins. During the almost fifty years of his research in music and auditory perception, he had tried to integrate new means of listening and readjust his conceptual framework to new findings that resulted not only from his own work but also from the generation that followed his. He remained open to the latest technological developments, thereby enabling research that pointed far beyond his own concepts and frames of reference. His methodology exhibits three characteristic features that made this possible. First, he worked with a principle that reminds modern readers of a knockout procedure. Testing the function of an unknown and unknowable element in a system could be done through comparing between systems that do or do not have what one assumes to be the key element for this function. For this, Stumpf chose the amusical subject. These subjects allowed him to bracket "music"—replacing music by a test that demonstrated that they did not share essential features of the musical subjects. This "experimental system" (Rheinberger, 1997) even enabled the amusos to speak back: Supported by new auditory technologies the amusical subjects even could tell Stumpf something about music. And he listened to them.

Acknowledgments

I thank the editor and especially Björn Merker for valuable comments on the draft of this chapter.

Notes

1. Ethnomusicologist Nettl (2000, p. 463), in his contribution to Wallin, Merker, and Brown (2000), translated the title of Stumpf's book as "The Beginnings of Music," which Markham (2014) also suggested as the more adequate translation.

2. The quote appears in volume 2 of *Principles of Psychology* and refers to theories of space. When Stumpf published his first book on this topic (1873), he also published a biographical sketch about James (1924b). On the relation between James and Stumpf more generally, see Fisette (2015).

3. All translations, where not indicated otherwise, are mine.

4. Lotze was not referring to the frequent examples of philosophers interested in music that Stumpf enumerates in the following (Herder, Herbarth, Schopenhauer), but to the predominance of Kant in nineteenth-century German epistemology, whose scarce and even ironic references to music may have contributed to the orientation on vision rather than hearing in epistemology. On this and Stumpf's relation to Kant and music, see Kaiser-el-Safti (2009).

5. Stumpf does not specify the number of subjects according to today's standards. From the numbers he gives, one can conclude that the groups in Würzburg, Prague, and Halle consisted of five to ten individuals.

6. For this, Stumpf is credited with having invented the first musicality test by von Maltzew, a Berlin-born Russian psychologist and Stumpf's former PhD student, in 1928 (von Maltzew & Serejsky, 1928). Today, it is used, for instance, as a subtest in the SDMA (Ullén, Mosing, Holm, Eriksson, & Madison, 2014).

7. For Stumpf's biography see his own account Stumpf (1924a, translated in Trippett, 2012a), and Sprung and Sprung (2006).

8. For a history of psychology in Germany, see Lück and Guski-Leinwand (2014); on the history of psychology in Berlin, see Sprung and Schönpflug (1992); on the Institute of Psychology at Friedrich Wilhelm University in the broader context of the history of twentieth-century psychology, see Ash (1995).

9. On the trade-off between using and preserving sound on wax rolls, see Fitzner (2014) and Trippett (2012b).

10. Alexander Dorn was a composer, piano teacher, and choir director who taught at the Berlin Conservatory.

11. Ellis's translation: "The first fact that we meet with in the music of all nations, so far as is yet known, is that alterations of pitch in melodies take place by intervals, and not by continuous transitions" (1875, p. 384).

12. The German *Geräusch* (concrete, specific, and mostly limited in duration) differs semantically from the English *noise*, which also encompasses *Rauschen* (continuous noise) and *Lärm* (disturbing noise). On this distinction in different languages and particularly for the early twentieth century, see Bijsterveld (2008) or Kursell (2003).

13. I had the opportunity to do this in the framework of the website http://vlp.mpiwg-berlin.mpg.de.

14. http://vlp.mpiwg-berlin.mpg.de/library/data/lit38590. Abraham (1923) comments on a comparable singer but makes no mention of the Edinger recordings.

15. http://vlp.mpiwg-berlin.mpg.de/library/data/lit38588.

16. Line 903, Harvard edition, http://www.bartleby.com/19/1/2.html (as of January 30, 2016).

References

Abraham, O. (1923). Tonometrische Untersuchungen an einem deutschen Volkslied [Tonometric investigations of a German folk song]. *Psychologische Forschung: Zeitschrift für Psychologie und ihre Grenzwissenschaften, 4*(1), 1–22. doi:10.1007/bf00410630.

Ames, E. (2003). The sound of evolution. *Modernism/Modernity, 10*(2), 297–325. doi:10.1353/mod.2003.0030

Ash, M. (1995). *Gestalt psychology in German culture, 1890–1967: Holism and the quest for objectivity*. New York: Cambridge University Press.

Bijsterveld, K. (2008). *Mechanical sound: Technology, culture, and public problems of noise in the twentieth century*. Cambridge, MA: MIT Press.

Boring, E. G. (1929). *A history of experimental psychology*. New York: Appleton-Century-Crofts.

Boring, E. G. (1942). *Sensation and perception in the history of experimental psychology*. New York: Appleton-Century-Crofts.

Brown, S., Merker, B., & Wallin, N. L. (2000). An introduction to evolutionary musicology. In N. L. Wallin, B. Merker, & S. Brown (Eds.), *The origins of music* (pp. 3–24). Cambridge, MA: MIT Press.

Bücher, K. (1899). *Arbeit und Rhythmus* [Labor and rhythm]. Leipzig: Teubner.

de la Motte-Haber, H. (2012). Carl Stumpf: Impulses towards a cognitive theory of musical evolution (D. Trippett, Trans.). In D. Trippett (Ed.), *Carl Stumpf: The origins of music* (pp. 3–16). Oxford: Oxford University Press.

Fechner, G. Th. (1860). *Elemente der Psychophysik*. Leipzig: Breitkopf und Härtel.

Fisette, D. (2015). Carl Stumpf. In E. N. Zalta (Ed.), *The Stanford Encyclopedia of Philosophy*. http://plato.stanford.edu/archives/spr2015/entries/stumpf/.

Fitzner, F. (2014). Die zergliederte Einheit: Aufschreibesystem und gestalt-theoretischer Anspruch bei Carl Stumpf und Erich M. von Hornbostel [The dissected unity: Discoursive networks and gestalt theoretical claims in Carl Stumpf and Erich M. von Hornbostel]. In E. Aschermann & M. Kaiser-el-Safti (Eds.), *Gestalt und Gestaltung in interdisziplinärer Perspektive* (pp. 191–203). Frankfurt: Lang.

Gembris, H. (1998). *Grundlagen musikalischer Begabung und Entwicklung*. Augsburg: Wißner.

Gurney, E. (1880). *The power of sound*. London: Smith, Elder & Co.

Helmholtz, H. (1863). *Die Lehre von den Tonempfindungen als physiologische Grundlage für die Theorie der Musik*. Braunschweig: Vieweg.

Helmholtz, H. (1867). *Handbuch der physiologischen Optik* [Handbook of physiological optics]. Leipzig: Voss.

Helmholtz, H. (1868). Ueber die Thatsachen, die der Geometrie zu Grunde liegen [On the facts on which geometry is based]. *Göttingen Nachrichten, 9*, 193–221.

Helmholtz, H. L. F. (1875). *On the sensations of tone as a physiological basis for the study of music* (A. J. Ellis, Trans.). London: Longmans, Green.

Hoffmann, C. (2001). Haut und Zirkel. Ein Entstehungsherd: Ernst Heinrich Webers Untersuchungen "Ueber den Tastsinn" [Skin and compass—a focus: Ernst Heinrich Weber's study "On the sense of touch"]. In M. Hagner (Ed.), *Ansichten der Wissenschaftsgeschichte* [Vistas on the history of science] (pp. 191–223). Frankfurt: Fischer.

Hoffmann, C. (2006). *Unter Beobachtung. Naturforschung in der Zeit der Sinnesapparat* [On a watch: The study of nature in the times of sensory apparatus]. Göttingen: Wallstein.

Honing, H., ten Cate, C., Peretz, I., & Trehub, S. E. (Eds.). (2015a). Biology, cognition and origins of musicality. *Philosophical Transactions of the Royal Society of London B: Biological Sciences, 370*(1664).

Honing, H., ten Cate, C., Peretz, I., & Trehub, S. E. (2015b). Without it no music: Cognition, biology, and evolution of musicality. *Philosophical Transactions of the Royal Society of London B: Biological Sciences, 370*(1664), 1–8. doi:10.1098/rstb.2014.0088

Hui, A. (2013). *The psychophysical ear: Musical experiments, experimental sounds, 1840*. Cambridge, MA: MIT Press.

Jackendoff, R., & Lerdahl, F. (2006). The capacity for music: What is it, and what's special about it? *Cognition, 100*(1), 33–72. doi:10.1016/j.cognition.2005.11.005

James, W. (1891). *The principles of psychology* (Vol. 2). London: Macmillan.

Kaiser-el-Safti, M. (2009). Der Witz (in) der Tonpsychologie Carl Stumpfs [The pun(ch line) of Carl Stumpf's tone psychology]. *Gestalt Theory, 31*(2), 143–174.

Kittler, F. A. (1999). *Gramophone, film, typewriter* (G. Winthrop-Young & M. Wutz, Trans.). Stanford, CA: Stanford University Press.

Kittler, F. A. (1990). *Discourse networks, 1800/1900* (Michael Metteer & Chris Cullens, Trans.). Stanford, CA: Stanford University Press.

Klotz, S. (1998). *Vom tönenden Wirbel menschlichen Tuns: Erich M. von Hornbostel als Gestaltpsychologe, Archivar und Musikwissenschaftler. Studien und Dokumente* [Of the sounding whirl of human action: Erich M. von Hornbostel as a Gestalt psychologist, archivist, and musicologist. Studies and documents]. Berlin: Schibri.

Klotz, S. (2008). Tonpsychologie und Musikforschung als Katalysatoren wissenschaftlich-experimenteller Praxis und der Methodenlehre im Kreis von Carl Stumpf [Tone psychology and music research as catalyzers of scientific experimental practices and of methodology in the circle of Carl Stumpf]. *Berichte zur Wissenschafts-geschichte, 31*, 195–210.

Koch, L.-C. (2013). Images of sound: Erich M. von Hornbostel and the Berlin Phonogram Archive. In P. V. Bohlman (Ed.), *The Cambridge history of world music* (pp. 475–497). Cambridge: Cambridge University Press.

Kremer, R. L. (1993). Innovation through synthesis: Helmholtz and color research. In D. Cahan (Ed.), *Hermann von Helmholtz and the foundations of nineteenth-century science* (pp. 205–258). Berkeley: University of California Press.

Kursell, J. (2003). *Schallkunst: Eine Literaturgeschichte der Musik in der frühen russischen Avantgarde* [Sound art: A literary history of music in early Russian avant-garde]. Munich: Kubon & Sagner.

Kursell, J. (2012). A gray box. The phonograph in laboratory experiments and field work, 1900–1920. In K. Bijsterveld & T. Pinch (Eds.), *The Oxford handbook of sound studies* (pp. 176–197). New York: Oxford University Press.

Kursell, J. (2013). Experiments on tone color in music and acoustics: Helmholtz, Schoenberg, and Klangfarbenmelodie. In A. Hui, J. Kursell, & M. W. Jackson (Eds.), *Music, sound, and the laboratory: From 1750–1980 (Osiris, 28)* (pp. 191–211). Chicago: University of Chicago Press.

Lück, H. E., & Guski-Leinwand, S. (2014). *Geschichte der Psychologie. Strömungen, Schulen, Entwicklungen.* Stuttgart: Kohlhammer.

Markham, E. J. (2014). The origins of music [review]. *Ethnomusicology Forum, 23*(3), 459–462. doi:10.1080/17411912.2014.958512

Michaelis, C. F. (1805). Ueber die Prüfung musikalischer Fähigkeiten. *Berlinische musikalische Zeitung, 55–58*, 222–230.

Nettl, B. (2000). An ethnomusicologist contemplates universals in musical sound and musical culture. In N. J. Wallin, B. Merker, & S. Brown (Eds.), *The origins of music* (pp. 463–472). Cambridge, MA: MIT Press.

Patel, A. D. (2008). *Music, language, and the brain*. Cambridge: Oxford University Press.

Peretz, I. (Ed.). (2006a). The nature of music [Special issue]. *Cognition, 100*(1).

Peretz, I. (2006b). The nature of music from a biological perspective. *Cognition, 100*(1), 1–32. doi:10.1016/j.cognition.2005.11.004.

Peretz, I., Champod, S., & Hyde, K. (2003). Varieties of musical disorders: The Montreal battery of evaluation of amusia. *Annals of the New York Academy of Sciences, 999*, 58–75.

Rameau, J.-P. (1750). *Démonstration du principe de l'harmonie, servant de base à tout l'art musical théorique et pratique* [Demonstration of the principle of harmony, serving as a basis for any theoretical and practical musical art]. Paris: Durand, Pissot.

Rehding, A. (2000). The quest for the origins of music in Germany circa 1900. *Journal of the American Musicological Society, 53*(2), 345–385. doi:10.2307/832011

Rehding, A. (2005). Wax cylinder revolutions. *Musical Quarterly, 88*(1), 123–160. doi:10.1093/musqtl/gdi004

Rheinberger, H.-J. (1997). *Toward a history of epistemic things: Synthesizing proteins in the test tube*. Stanford, CA: Stanford University Press.

Schumann, R. (1854). Musikalische Haus- und Lebensregeln. *Neue Zeitschrift für Musik, 32*(36) [Inset to Album for the Youth].

Schweinfurth, G. (1874). *Im Herzen von Afrika* [In the heart of Africa]. Leipzig: Brockhaus.

Schwörer-Kohl, G. (2011). Die Musikethnologie Carl Stumpfs am Beispiel der siamesischen Musik [Carl Stumpf's ethnomusicology, using the example of Siamese music]. In M. Ebeling & M. Kaiser-el-Safti (Eds.), *Die Sinne und die Erkenntnis.*[The senses and knowledge] (pp. 175–184). Frankfurt: Lang.

Sprung, L., & Schönpflug, W. (1992). *Zur Geschichte der Psychologie in Berlin.* Frankfurt: Lang.

Sprung, H., & Sprung, L. (2006). *Carl Stumpf: Eine Biographi* [Carl Stumpf: A biography]. Munich: Profil.

Stumpf, C. (1883). *Tonpsychologie* (Vol. 1). Leipzig: Barth.

Stumpf, C. (1885). Musikpsychologie in England: Betrachtungen über Herleitung der Musik aus der Sprache und aus dem thierischen Entwickelungsprocess, über Emprismus und Nativismus in der Musiktheorie [Music psychology in England: Reflections on the derivation of music from language and from animal evolution, on empiricism and nativism in music theory]. *Vierteljahrsschrift für Musikwissenschaft, 1*(3), 261–349.

Stumpf, C. (1886). Lieder der Bellakula-Indianer [Songs of the Bella Coola Indians]. *Vierteljahrsschrift für Musikwissenschaft, 2*, 405–426.

Stumpf, C. (1890). *Tonpsychologie* [Psychology of tone] (Vol. 2). Leipzig: Barth.

Stumpf, C. (1892). Phonographirte Indianermelodien [Phonographically recorded melodies of American Indians]. *Vierteljahrsschrift für Musikwissenschaft, 8*, 127–144.

Stumpf, C. (1898). Konsonanz und Dissonanz. *Beiträge zur Akustik und Musikwissenschaft, 1*, 1–108.

Stumpf, C. (1901). Tonsystem und Musik der Siamesen [Tonal system and music of the Siamese]. *Beiträge zur Akustik und Musikwissenschaft, 3*, 69–138.

Stumpf, C. (1908). Das Berliner Phonogramm-Archiv. *Internationale Wochenschrift für Wissenschaft: Kunst und Technik, 2*(22), 225–246.

Stumpf, C. (1911a). *Die Anfänge der Musik* [The origins of music]. Leipzig: Barth.

Stumpf, C. (1911b). Konsonanz und Konkordanz. Nebst Bemerkungen über Wohlklang und Wohlgefälligkeit musikalischer Zusammenklänge [Consonance and concordance. Along with notes on harmony and pleasantness of musical tone combinations]. *Beiträge zur Akustik und Musikwissenschaft, 6*, 116–150.

Stumpf, C. (1914). Ueber neuere Untersuchungen zur Tonlehre [On new investigations in the study of tone]. In F. Schumann (Ed.), *Bericht über den VI. Kongreß für Experimentelle Psychologie in Göttingen vom 15. bis 18. April 1914* (pp. 305–348). Leipzig: Barth.

Stumpf, C. (1924a). Singen und Sprechen [Singing and speaking]. *Beiträge zur Akustik und Musikwissenschaft, 9*, 38–74.

Stumpf, C. (1924b). Selbstdarstellung [Self-portrait]. In R. Schmidt (Ed.), *Philosophie der Gegenwart in Selbstdarstellungen* (pp. 205–265). Leipzig: Meiner.

Stumpf, C. (1928). *William James nach seinen Briefen* [William James according to his letters]. Berlin, Germany: Pan.

Stumpf, C. (2012). The origins of music (D. Trippett, Trans.). In D. Trippett (Ed.), *Carl Stumpf: The origins of music* (pp. 31–185). Oxford: Oxford University Press.

Trippett, D. (Ed.). (2012a). *Carl Stumpf: The origins of music* (D. Trippett, Trans.). Oxford: Oxford University Press.

Trippett, D. (2012b). The reluctant revolutionary (D. Trippett, Trans.). In D. Trippett (Ed.), *Carl Stumpf: The origins of music* (pp. 17–29). Oxford: Oxford University Press.

Ullén, F., Mosing, M. A., Holm, L., Eriksson, H., & Madison, G. (2014). Psychometric properties and heritability of a new online test for musicality, the Swedish Musical Discrimination Test. *Personality and Individual Differences, 63*, 87–93. doi:10.1016/j.paid.2014.01.057

Vitouch, O., & Ladinig, O. (Eds.). (2009–2010). Music and evolution. *Musicae Scientiae, 13*(Suppl. 2).

von Maltzew, C., & Serejsky, M. (1928). Prüfung der Musikalität nach der Testmethode [Examining musicality with the testing method]. *Psychotechnische Zeitschrift, 3*(4), 103–107.

Wallin, N. L., Merker, B., & Brown, S. (Eds.). (2000). *The origins of music*. Cambridge, MA: MIT Press.

Weber, E. H. (1835). Ueber den Tastsinn [On the sense of touch]. *Archiv für Anatomie, Physiologie und wissenschaftliche Medicin, 1*, 152–159.

Weber, E. H. (1846). Tastsinn und Gemeingefühl [Touch and sensorium commune]. In R. Wagner (Ed.), *Handwörterbuch der Physiologie mit Rücksicht auf physiologische Pathologie* [Encyclopedia of physiology with consideration of physiological pathology] (Vol. 3, Pt. 2, pp. 481–588). Braunschweig: Vieweg.

Wertheimer, M. (1909). Musik der Wedda [Music of the Vedda people]. *Sammelbände der Internationalen Musikgesellschaft, 11*, 300–309.

World Health Organization. (2016). *International statistical classification of diseases and related health problems* (10th ed.). http://apps.who.int/classifications/icd10/browse/2016/en#/R48.8.

Wundt, W. (1908). *Völkerpsychologie: Eine Untersuchung der Entwicklungsgesetze von Sprache, Mythus und Sitte* [Völkerpsychologie: An investigation of the developmental laws of language, myth and custom], Vol. 3: *Die Kunst* [Art] (2nd rev. ed.). Leipzig: Engelmann.

Zatorre, R. (2005). Music, the food of neuroscience? *Nature, 434*, 312–315. doi:10.1038/434312a

Ziegler, S. (2006). *Die Wachszylinder des Berliner Phonogramm-Archivs* [The wax cylinders of the Berlin Phonogram Archive]. Berlin: Ethnologisches Museum, Staatliche Museen zu Berlin—Preußischer Kulturbesitz.

Contributors

Jorge L. Armony, Department of Psychiatry, McGill University and Douglas Mental Health University Institute, Montreal, Quebec, Canada; International Laboratory for Brain, Music and Sound Research

Judith Becker, University of Michigan, Ann Arbor, Michigan, USA

Simon E. Fisher, Max Planck Institute for Psycholinguistics, and Donders Institute for Brain, Cognition and Behavior, Nijmegen, Netherlands

W. Tecumseh Fitch, Department of Cognitive Biology, University of Vienna, Austria

Bruno Gingras, Institute of Psychology, University of Innsbruck, Austria

Jessica Grahn, Brain and Mind Institute and Department of Psychology, University of Western Ontario, London, Ontario, Canada

Yuko Hattori, Primate Research Institute, Kyoto University, Japan

Marisa Hoeschele, Department of Cognitive Biology, University of Vienna, Austria

Henkjan Honing, Amsterdam Brain and Cognition and Institute of Logic, Language and Computation, University of Amsterdam, Netherlands

Dieuwke Hupkes, Institute of Logic, Language and Computation, University of Amsterdam, Netherlands

David Huron, School of Music and Center for Cognitive and Brain Sciences, Ohio State University, Columbus, USA

Yukiko Kikuchi, Institute of Neuroscience, Newcastle University Medical School, UK

Julia Kursell, Department of Musicology, University of Amsterdam, Netherlands

Marie-Élaine Lagrois, Department of Psychology, University of Montreal, Montreal, Quebec, Canada; International Laboratory for Brain, Music and Sound Research

Hugo Merchant, Department of Cognitive Neuroscience, Instituto de Neurobiología, Universidad Nacional Autonoma de México, Campus Juriquila, Querétaro, México

Björn Merker, Kristianstad, Sweden

Iain Morley, Institute of Social and Cultural Anthropology, University of Oxford, Oxford, UK

Aniruddh D. Patel, Department of Psychology, Tufts University, Medford, MA, USA, and Canadian Institute for Advanced Research (CIFAR), Azrieli Program in Brain, Mind, & Consciousness, Toronto, Canada

Isabelle Peretz, Department of Psychology, University of Montreal, Canada; International Laboratory for Brain, Music and Sound Research

Martin Rohrmeier, Institut für Kunst- und Musikwissenschaft, Technische Universität Dresden, Dresden, Germany

Constance Scharff, Institute of Biology, Freie Universität Berlin, Berlin, Germany

Carel ten Cate, Institute of Biology, Leiden University, Netherlands, and Leiden Institute for Brain and Cognition, Leiden University, Netherlands

Laurel J. Trainor, Department of Psychology, Neuroscience and Behavior, McMaster University, Hamilton, Ontario, Canada

Sandra E. Trehub, University of Toronto Mississauga, Mississauga, Ontario, Canada

Peter Tyack, Sea Mammal Research Unit, School of Biology, University of St. Andrews, St. Andrews, UK

Dominique T. Vuvan, Department of Psychology, University of Montreal, Montreal, Quebec, Canada; International Laboratory for Brain, Music and Sound Research

Geraint A. Wiggins, School of Electronic Engineering and Computer Science, Queen Mary University of London, London, UK

Willem Zuidema, Institute of Logic, Language and Computation, University of Amsterdam, Netherlands

Index